PROOF COMPLEXITY

Proof complexity is a rich subject drawing on methods from logic, combinatorics, algebra and computer science. This self-contained book presents the basic concepts, classical results, current state of the art and possible future directions in the field. It stresses a view of proof complexity as a whole entity rather than a collection of various topics held together loosely by a few notions, and it favors more generalizable statements.

Lower bounds for lengths of proofs, often regarded as the key issue in proof complexity, are of course covered in detail. However, upper bounds are not neglected: this book also explores the relations between bounded arithmetic theories and proof systems and how they can be used to prove upper bounds on lengths of proofs and simulations among proof systems. It goes on to discuss topics that transcend specific proof systems, allowing for deeper understanding of the fundamental problems of the subject.

Encyclopedia of Mathematics and Its Applications

This series is devoted to significant topics or themes that have wide application in mathematics or mathematical science and for which a detailed development of the abstract theory is less important than a thorough and concrete exploration of the implications and applications.

Books in the **Encyclopedia of Mathematics and Its Applications** cover their subjects comprehensively. Less important results may be summarized as exercises at the ends of chapters. For technicalities, readers can be referred to the bibliography, which is expected to be comprehensive. As a result, volumes are encyclopedic references or manageable guides to major subjects.

All the titles listed below can be obtained from good booksellers or from Cambridge University Press. For a complete series listing visit www.cambridge.org/mathematics.

122 S. Khrushchev *Orthogonal Polynomials and Continued Fractions*
123 H. Nagamochi and T. Ibaraki *Algorithmic Aspects of Graph Connectivity*
124 F. W. King *Hilbert Transforms I*
125 F. W. King *Hilbert Transforms II*
126 O. Calin and D.-C. Chang *Sub-Riemannian Geometry*
127 M. Grabisch *et al. Aggregation Functions*
128 L. W. Beineke and R. J. Wilson (eds.) with J. L. Gross and T. W. Tucker *Topics in Topological Graph Theory*
129 J. Berstel, D. Perrin and C. Reutenauer *Codes and Automata*
130 T. G. Faticoni *Modules over Endomorphism Rings*
131 H. Morimoto *Stochastic Control and Mathematical Modeling*
132 G. Schmidt *Relational Mathematics*
133 P. Kornerup and D. W. Matula *Finite Precision Number Systems and Arithmetic*
134 Y. Crama and P. L. Hammer (eds.) *Boolean Models and Methods in Mathematics, Computer Science, and Engineering*
135 V. Berthé and M. Rigo (eds.) *Combinatorics, Automata and Number Theory*
136 A. Kristály, V. D. Rădulescu and C. Varga *Variational Principles in Mathematical Physics, Geometry, and Economics*
137 J. Berstel and C. Reutenauer *Noncommutative Rational Series with Applications*
138 B. Courcelle and J. Engelfriet *Graph Structure and Monadic Second-Order Logic*
139 M. Fiedler *Matrices and Graphs in Geometry*
140 N. Vakil *Real Analysis through Modern Infinitesimals*
141 R. B. Paris *Hadamard Expansions and Hyperasymptotic Evaluation*
142 Y. Crama and P. L. Hammer *Boolean Functions*
143 A. Arapostathis, V. S. Borkar and M. K. Ghosh *Ergodic Control of Diffusion Processes*
144 N. Caspard, B. Leclerc and B. Monjardet *Finite Ordered Sets*
145 D. Z. Arov and H. Dym *Bitangential Direct and Inverse Problems for Systems of Integral and Differential Equations*
146 G. Dassios *Ellipsoidal Harmonics*
147 L. W. Beineke and R. J. Wilson (eds.) with O. R. Oellermann *Topics in Structural Graph Theory*
148 L. Berlyand, A. G. Kolpakov and A. Novikov *Introduction to the Network Approximation Method for Materials Modeling*
149 M. Baake and U. Grimm *Aperiodic Order I: A Mathematical Invitation*
150 J. Borwein *et al. Lattice Sums Then and Now*
151 R. Schneider *Convex Bodies: The BrunnMinkowski Theory (Second Edition)*
152 G. Da Prato and J. Zabczyk *Stochastic Equations in Infinite Dimensions (Second Edition)*
153 D. Hofmann, G. J. Seal and W. Tholen (eds.) *Monoidal Topology*
154 M. Cabrera García and Á. Rodríguez Palacios *Non-Associative Normed Algebras I: The Vidav–Palmer and Gelfand–Naimark Theorems*
155 C. F. Dunkl and Y. Xu *Orthogonal Polynomials of Several Variables (Second Edition)*
156 L. W. Beineke and R. J. Wilson (eds.) with B. Toft *Topics in Chromatic Graph Theory*
157 T. Mora *Solving Polynomial Equation Systems III: Algebraic Solving*
158 T. Mora *Solving Polynomial Equation Systems IV: Buchberger Theory and Beyond*
159 V. Berthé and M. Rigo (eds.) *Combinatorics, Words and Symbolic Dynamics*
160 B. Rubin *Introduction to Radon Transforms: With Elements of Fractional Calculus and Harmonic Analysis*
161 M. Ghergu and S. D. Taliaferro *Isolated Singularities in Partial Differential Inequalities*
162 G. Molica Bisci, V. D. Radulescu and R. Servadei *Variational Methods for Nonlocal Fractional Problems*
163 S. Wagon *The Banach–Tarski Paradox (Second Edition)*
164 K. Broughan *Equivalents of the Riemann Hypothesis I: Arithmetic Equivalents*
165 K. Broughan *Equivalents of the Riemann Hypothesis II: Analytic Equivalents*
166 M. Baake and U. Grimm (eds.) *Aperiodic Order II: Crystallography and Almost Periodicity*
167 M. Cabrera García and Á. Rodríguez Palacios *Non-Associative Normed Algebras II: Representation Theory and the Zel'manov Approach*
168 A. Yu. Khrennikov, S. V. Kozyrev and W. A. Zúñiga-Galindo *Ultrametric Pseudodifferential Equations and Applications*
169 S. R. Finch *Mathematical Constants II*
170 J. Krajíček *Proof Complexity*
171 D. Bulacu, S. Caenepeel, F. Panaite and F. Van Oystaeyen *Quasi-Hopf Algebras*

ENCYCLOPEDIA OF MATHEMATICS AND ITS APPLICATIONS

Proof Complexity

JAN KRAJÍČEK
Charles Univesity, Prague

CAMBRIDGE
UNIVERSITY PRESS

University Printing House, Cambridge CB2 8BS, United Kingdom

One Liberty Plaza, 20th Floor, New York, NY 10006, USA

477 Williamstown Road, Port Melbourne, VIC 3207, Australia

314–321, 3rd Floor, Plot 3, Splendor Forum, Jasola District Centre, New Delhi – 110025, India

79 Anson Road, #06–04/06, Singapore 079906

Cambridge University Press is part of the University of Cambridge.

It furthers the University's mission by disseminating knowledge in the pursuit of education, learning, and research at the highest international levels of excellence.

www.cambridge.org
Information on this title: www.cambridge.org/9781108416849
DOI: 10.1017/9781108242066

First published 2019

Printed and bound in Great Britain by Clays Ltd., Elcograf S.P.A.

A catalogue record for this publication is available from the British Library.

Library of Congress Cataloging-in-Publication Data
Names: Krajíček, Jan, author.
Title: Proof complexity / Jan Krajíček.
Description: Cambridge ; New York, NY : Cambridge University Press, 2019. |
Series: Encyclopedia of mathematics and its applications ; 170 | Includes bibliographical references and index.
Identifiers: LCCN 2018042527 | ISBN 9781108416849 (hardback : alk. paper)
Subjects: LCSH: Proof theory. | Computational complexity.
Classification: LCC QA9.54 .K72 2019 | DDC 511.3/6–dc23
LC record available at https://lccn.loc.gov/2018042527

ISBN 978-1-108-41684-9 Hardback

Let no one ignorant of logic enter.

Paraphrase of the sign on the gate to
the Academy at Athens,
c. 387 B.C.

Contents

Preface

The concept of propositional formulas and their logical validity developed over almost two and a half millennia and reached its current mathematical definition at the advent of mathematical logic in the second half of the nineteenth century. It is remarkably simple. We have propositional atoms and a few logical connectives to combine them into formulas. We identify truth assignments with functions assigning the values $\top$ (true) and $\bot$ (false) to atoms and use simple algebraic rules to quickly compute the truth value of any formula under a particular truth assignment. Logically valid formulas, tautologies, are then those formulas receiving the value $\top$ under any truth assignments. The definition of logical validity offers a mechanical way to decide whether a formula is a tautology: compute its truth value under all truth assignments to its atoms. The definition is amenable to countless variations, leading to formulations of various non-classical logics. In some of these alternative definitions a truth assignment may be a more complicated notion than in others but the idea of determining the truth value bottom up by a few rules is common to all.

The preoccupation of logic over the ages with logically sound modes of reasoning has led to various lists of logically valid inference schemes and eventually to the concept of a logical calculus. Aristotle and Frege are the main figures at the advent of logic and of mathematical logic, respectively. In such calculi one starts with initial tautologies, axioms, of a small number of transparent forms and uses a finite number of inference rules to derive increasingly more complicated tautologies. Typically one can simulate the mechanical process of going through all truth assignments by such a derivation.

A property of these specific derivations simulating computations is that they do not involve formulas more complicated than the one being proved. But one may prove a tautology entirely differently, deriving first intermediate formulas (lemmas), seemingly having nothing to do with the target formula, and eventually combining them into a valid derivation. It is a fundamental problem of logic to understand what a general derivation looks like and, in particular, when it can be significantly shorter than a trivial mechanical derivation and how can we find it.

When the truth values are represented by the bits 1 and 0, truth assignments can be identified with binary strings, truth functions with the properties of such strings and propositional formulas become algorithms for computing them. Thus, at once we find propositional logic at the heart of the theory of computing and the above questions about the efficiency of proofs become fundamental problems of computational complexity theory such as the P vs. NP problem or the NP vs. coNP problem.

The mathematical area studying the complexity of propositional proofs, *proof complexity*, provides at present one of the richest approaches to these problems. It is proof complexity which, together with its logical foundations, I emphasize in this book. Proof complexity has, in my view, at least as good a chance of achieving a significant breakthrough on the above-mentioned problems as any other approach contemplated at present. And, contrary to some other, even partial results, often lower bounds for specific proof systems, can be genuinely interpreted as saying something interesting and relevant about these problems.

We should also keep an open mind about how the eventual answers may turn out. Mathematics can be seen as a special kind of language developed for the world of virtual objects like points and lines, numbers and equations, sets, spaces, algebras and other occupants of this mathematical world and for our thoughts about them. This language is quite powerful and is akin in many respects to natural languages. And just as a talented poet can describe in few lines or even in few words a thought or a situation for which you or I may need volumes, miraculously short and eloquent proofs which nobody would have dreamt of may exist. I think that at present we have only a tentative understanding of the power of the language of mathematics.

Acknowledgements

I am grateful to Neil Thapen for a number of discussions about the book and about various topics covered in it, and for extensive comments on the draft. I am indebted to Pavel Hrubeš, Moritz Müller, Igor C. Oliveira, Ján Pich and Pavel Pudlák for offering comments or suggestions concerning significant parts of the draft.

I thank Albert Atserias, Olaf Beyersdorff, Sam Buss, Steve Cook, Gilda Ferreira, Michal Garlík, Azza Gaysin, Emil Jeřábek, Leszek Kolodziejczyk, Tomáš Krňák, Jakob Nordström, George Osipov, Theodoros Papamakarios, Zenon Sadowski and Iddo Tzameret for pointing out minor errors in the draft or for answering my questions and providing bibliographical information.

Introduction

The mathematical theory of computing grew out of the quest of mathematicians to clarify the foundations of their field. Turing machines and other models of computation are tied to formal logical systems and algorithmic unsolvability to unprovability. Problems whose solutions lead to this understanding have included Hilbert's Entscheidungsproblem (deciding the logical validity of a first-order formula) or his program of establishing by finitary means the consistency of mathematics. This development would not have been possible without the prior advent of mathematical logic in the late nineteenth century, guided by contributions from a number of people. The most important was, in retrospect, Frege's *Begriffsschrift* [188] realizing facets of an old dream of Leibniz about the *calculus ratiocinator*.

Eventually the problems received answers opposite to those that Hilbert and others had hoped for (the Entscheidungsproblem was found to be algorithmically undecidable by Church [143] and Turing [494], and the fact that a proof of the consistency of a theory formalizing a non-trivial part of mathematics requires axioms from outside the theory, Gödel [200]). But the leap in the development of logic that this study of the foundations of mathematics stimulated was enormous compared with the progress of the previous two millennia. All this is discussed and analyzed in a number of books; I have found most stimulating the volume [224], edited by J. Van Heijenoort, of translations of selected papers from the era accompanied by knowledgeable commentaries.

The next significant step from our perspective was the formalization of an informal notion of a feasible algorithm by the formal notion of a polynomial time algorithm in the late 1960s. This was accompanied by the introduction of complexity classes, reductions among problems and discoveries of natural complete problems, all quite analogous to the earlier developments in mathematical logic around Turing computability. A plethora of open questions about the new concepts were posed. The most famous of them, the P vs. NP problem, is now recognized as one of the fundamental problems of mathematics.

Some of these new problems were, in fact, foreshadowed by specific problems in logic which did not explicitly mention machines or time complexity. These include the spectrum problem in the model theory of Scholz [462] and Asser [25] (in the early 1950s), which, as we now know, asks for a characterization of NE sets (i.e. non-deterministic time $2^{O(n)}$ sets) and, in particular, whether the class NE is closed under complementation. They also include problems involving the rudimentary sets of Smullyan [476] and Bennett [65] (in the early 1960s). The first problem explicitly mentioning time complexity was perhaps the problem posed by Gödel to von Neumann in a 1956 letter [201]; there he asks about the time complexity of a problem we now know to be NP-complete (see the introduction to the volume of Clote and Krajíček [144] for the letter and its translation, and Buss [111] for the completeness result). Sipser [468] describes these developments clearly and succinctly.

In the light of this history it seems plausible that the P vs. NP problem has significant logical facets connected to the foundations of mathematics. It can even be seen as a miniaturization of the Entscheidungsproblem: replace first-order formulas by propositional formulas and ask for a feasible algorithm rather than just any algorithm. We do not know how hard the problem is. It may require just one trick (as some famous problems of the past did), one dramatically new idea (as was forcing for the independence of the continuum hypothesis) or a development of a whole part of mathematics (as was needed for Fermat's last theorem). In either case it seems sensible to try to develop an understanding of the logical side of the problem.

This thinking is, however, foreign to the canons of contemporary computational complexity theory. That field developed as a part of combinatorics and some early successes stimulated a lot of faith in that approach. However, although many interesting developments have taken place, progress on combinatorial insight into the original fundamental problems remains tentative. Despite this, complexity theorists sometimes go to great lengths to talk about logic without ever mentioning the word. Although it is remarkable that one can discuss provability or unprovability without using words like "a theory" or "a formula," it is hard to imagine that doing so is faithful to the topic at hand.

There are several research areas connecting mathematical logic and computational complexity theory; some of these connections are tighter than others. We shall concentrate on propositional proof complexity. Proof complexity (the adjective propositional is often left out, as it is tacitly understood) is now a fairly rich subject drawing its methods from many fields (logic, combinatorics, algebra, computer science, ...). The first results appeared in the 1960s, with a significant acceleration from the late 1980s to the early 2000s. For some years now, however, progress has lain more in sharpening and generalizing earlier advances and recasting them in a new formalism and in improving the technical tools at hand, but not in fact that much (at least not at the level of the progress in the 1990s) in developing completely new ideas. It thus seems a good time to write an up-to-date presentation of the field to enable newcomers to learn the subject and to allow active researchers to step back and contemplate a larger picture.

There are two approaches to proof complexity, which are complementary, in a sense, to each other. The first views problems, especially length-of-proof problems, as purely combinatorial questions about the existence and properties of various combinatorial configurations. Prominent early examples of this approach are Tseitin's [492] 1968 lower bound for regular resolution or Haken's [216] 1985 lower bound for general resolution proofs of the pigeonhole principle. This approach, when successful, typically gives qualitatively very detailed results. It works only for weak proof systems however.

The second approach views problems through the eyes of mathematical logic. Representative examples of this view are Cook's [149] 1975 and Paris and Wilkie's [389] 1985 translations of bounded arithmetic proofs into (sequences of) propositional proofs, Ajtai's [5] 1988 lower bound, obtained by forcing, for pigeonhole-principle proofs in constant-depth Frege systems or my 1991 idea of the feasible interpolation method formulated in [276] (1994, preprint circulating in 1991). This approach typically yields a general concept, and its applications in specific cases require combinatorics as well.

The two approaches are not disjoint and can be fruitfully combined. Nevertheless, there has been much emphasis in recent years on a purely combinatorial analysis of proof systems related to specific SAT-solving or optimization algorithms. Only a small number of researchers have involved themselves in mathematical logic or investigations aimed at explaining some general phenomena. Of course, the original fundamental – logic-motivated – problems are dutifully mentioned when the goals in this area are discussed in talks or papers, but in actual research programs these problems are rarely studied. It seems that not only are the original problems abandoned but some of the proof complexity theory developed around them is becoming exotic to many researchers. I suppose this is due in part to the demise of logic from a standard computer science curriculum. Being able to formulate a theorem or a proof does not provide sufficient education in logic, just as being able to draw a graph is not a sufficient education in combinatorics. Without at least an elementary logic background the only part of proof complexity that remains accessible is the algorithm analysis aspect. Of course, it is a perfectly sound research topic but it is not the whole of proof complexity; it is not even close to half of it.

I think that logic will play eventually a greater role also in proof complexity research motivated by applications. I am not competent to judge how significant a contribution current proof complexity makes to the analysis and practice of real-life algorithms. Such contributions often contain ingenious technical innovations and ought to shed light on the performance of the respective algorithms. But SAT solving (or the design of other algorithms) is mostly an engineering activity, with little theory, as we mathematicians understand it, involved. Even the basic issue of how to compare the efficiency of two algorithms is in reality approached in an ad hoc, heuristic, mathematically non-principled way. Therefore I think that proof complexity could contribute most by introducing new original theoretical concepts and ideas that could influence how the whole field of SAT solving is approached in future. A significant contribution may come even from technically simple ideas, for example, how to use stronger proof systems in proof search or how to combine the present-day optimization algorithms with a little logic. This could be, in my opinion, quite analogous to the fact that fairly general ideas of Turing and others influenced the thinking in computer science significantly. But from where else could genuinely new concepts and ideas, and not just technical innovations, come other than from considering the very foundational problems of the field?

In the book I stress my view of proof complexity as a whole entity rather than as a collection of various topics held together loosely by a few notions. The frame that supports it is logic. This is not a dogma but represents the current state of affairs. For this reason I choose as the motto at the start of the book a paraphrase of the famous sign on the gate to Plato's Academy, in order to underline the point, not to distress the reader.

Basic concepts, classical results, current work and possible future directions are presented. I have not attempted to compile a compendium of all known results, and have concentrated more on ideas than on the technically strongest statements. These ideas might be enlightening definitions and concepts, stimulating problems or original proof methods. I have given preference to statements that attempt some generality over statements that are strong in specific situations only.

The intended readers include researchers and doctoral students in mathematics (mathematical logic and discrete mathematics, in particular) and in theoretical computer science (in computational complexity theory, in particular). I do not necessarily present the material in the most elementary way. But I try to present it honestly, meaning that I attempt to expose the fundamental ideas underlying the topics and do not hide difficult or technically delicate things under the carpet. The book is meant to be self-contained, in a reasonable sense of the word, as a whole; it is not a collection of self-contained chapters and there is a lot of cross-referencing.

Organization of the Book

The book is divided into four parts. In Part I, "Basic Concepts," the first chapter recalls some preliminaries from mathematical logic and computational complexity and introduces the basic concepts and problems of proof complexity. The remaining six chapters in Part I give a number of examples of propositional proof systems and we prove there many of their fundamental properties.

The main topics of the second part, "Upper Bounds," are the relations between bounded arithmetic theories and proof systems and translations of arithmetic proofs into propositional logic, and how these can be utilized to prove length-of-proof upper bounds and simulations among proof systems.

The third part, "Lower Bounds," is devoted to known length-of-proof lower bounds. This is the royal subject of proof complexity, as the fundamental problems are formulated as questions about lengths of proofs. Some people reduce proof complexity to this topic alone.

The last part, "Beyond Bounds," discusses topics transcending the length-of-proof bounds for specific proof systems. I report on those developments attempting to address all proof systems. It is here where there is, I think, a chance to develop some deeper understanding of the fundamental problems. I present there topics for which some general theory has emerged in the past and discuss directions that seem to me to be promising for future research. In the last chapter we step back a little, look at what has been achieved and offer a few thoughts on the nature of proof complexity.

Each part is divided into chapters, which further divide into sections with, I hope, informative titles. This gives a structure to the material that should be apparent at first glance. There are various novel results, topics, constructions and proofs.

In the main narrative, typically I attribute to sources only the main results, concepts, problems and ideas discussed. The qualification *main* is subjective, of course, and does not necessarily refer only to the technically hardest topics but sometimes also to simple observations that have proved to be very useful: I rather agree with the classical dictum that *the really important ideas are also simple*. Each chapter ends with a section on bibliographical and other remarks, where I attempt to give full bibliographical information for all the results covered, point to other relevant literature and discuss at least some topics that are related but were omitted.

Occasionally I use the adjectives *interesting* and *important* and similar qualifiers. This should be understood as meaning *interesting* (or *important*) *for me*. In fact, it is a dangerous illusion, to which some people succumb, that these words have an objective meaning in mathematical literature.

Conventions and Notation

Throughout the book we use the standard symbols of propositional and first-order logic. Formulas are denoted either by capital Latin letters or by lower-case Greek letters. We also use the O-notation and the related o-, Ω- and ω-notations, which are very handy for expressing upper and lower bounds in complexity theory. In particular, for f, g: $\mathbf{N} \rightarrow \mathbf{N}$, $g = O(f)$ means that $g(n) \leq cf(n) + c$ for some constant $c \geq 1$ and all n, $g = o(f)$ means that $g(n)/f(n)$ goes to 0 as $n \rightarrow \infty$, $g = \Omega(f)$ means that $g(n) \geq \epsilon f(n)$ for some constant $\epsilon > 0$ and all n and finally $g = \omega(f)$ means that $g(n)/f(n)$ cannot be bounded by a constant for all n.

The symbol $[n]$ denotes the set $\{1, \ldots, n\}$. The symbols $\subseteq$ and $\subsetneq$ denote set-theoretic inclusion and proper inclusion, respectively. We use $n \gg 1$ as a shorthand for *n is large enough*. Most first-order objects we consider are finite words (ordered tuples) over an alphabet (numbers, binary words, formulas, truth assignments, proofs, etc.). When it is important that a variable ranges over tuples and we need to address their elements, I use the notation $\overline{b}, \overline{x}, \ldots$ and denote the elements at the ith position $b_i, x_i, \ldots$

Only a few notions and concepts that are exceptionally important in proof complexity have their numbered definitions; all other notions (the vast majority) are defined in the text and the defined notion appears in boldface. To refer to the latter definitions, we use the section number where they appeared (the sections are fairly short and it is easy to navigate them). I think that this treatment of definitions makes the structure of a long text more comprehensible. All lemmas, theorems and corollaries are syntactically distinguished and numbered in the conventional way.

All unfamiliar symbols are properly defined in the text and the special-symbol index near the end of the book lists for each such symbol the section where it appears first.

Remarks on the Literature

During the last ten years only two books about proof complexity have appeared: [155] by Cook and Nguyen and my text [304]. The Cook–Nguyen book develops various bounded arithmetic theories and their relations to weak computational complexity classes utilizing witnessing theorems, and defines associated propositional proof systems simulating the theories. The emphasis there is on the triple relation *theory/computational class/proof system*. However, no lower bounds are presented in that book, either old or more recent. My book [304] is a research monograph developing a particular construction of models of bounded arithmetic and how it can be applied to proof complexity lower bounds, but it reviews only snapshots of the proof complexity needed.

There is also a related book [421] by P. Pudlák about the foundations of mathematics. It contains one chapter (out of seven) about proof complexity (and one of its sections is on lower bounds), offering a few highlights. It is written in three levels of precision, with the top level giving details.

From older books the most used is perhaps my 1995 monograph [278], which presented the field in an up-to-date manner (relative to 1995). It may occur to the reader that it might have been more sensible to update that book rather than to write a new one. However, now I am putting a lot more emphasis on propositional proof systems while at least half the 1995 book was devoted to a development of bounded arithmetic theories, and only a smaller part of the latter is directly relevant to propositional proof complexity.

There is also an excellent book [145] from 2002 by P. Clote and E. Kranakis, which does contain one chapter (out of seven) about proof complexity lower bounds. It contains some results (and lower bounds, in particular) not covered in my 1995 book.

Handbook of Proof Theory (Elsevier, 1998) edited by S. R. Buss, contains a chapter on lengths of proofs by Pudlák [413]. The same volume includes introductory texts [115, 116] about first-order proof theory and theories of arithmetic, and we recommend those texts to all readers who need to learn that background. There are also a number of survey articles, usually addressing just a few selected topics; these include [497, 279, 281, 58, 117, 48, 297, 418, 120, 442].

Background

Although I review some elements of propositional and first-order logic in the first two sections of Chapter 1, this is mostly to set the scene and to introduce some formalism and notation. It is helpful if the reader has a basic knowledge of mathematical logic at the level of an introductory course in first-order logic and model theory. This should include the completeness and compactness theorems and Herbrand's theorem (although I do present a proof of this theorem). I do not assume any specific knowledge of bounded arithmetic or the model theory of arithmetic.

From computational complexity theory I assume a knowledge of the basic concepts of Turing machines and their time complexity and of the definitions of standard complexity classes (although some of this is briefly recalled in Section 1.3). I do not assume a particular knowledge of circuit complexity (it is explained when needed) but it could make some ideas and arguments more transparent.

Part I

Basic Concepts

Part I is devoted to the introduction of basic concepts and problems and to a number of examples of propositional proof systems. Chapter 1 also recalls some very basic notions of propositional and first-order logic (Sections 1.1 and 1.2) and of computational complexity theory (Sections 1.3 and 1.4). The former is aimed at those readers with a primarily computer science (CS) background without any logic education while the latter is aimed at readers who have avoided CS until now. These sections should give the reader enough intuition to follow the presentation in the later chapters but they cannot replace a proper education in these two subjects.

Section 1.5 introduces the key concepts of proof complexity and three fundamental problems. Chapters 2–5 present the main examples of logical proof systems and explain their basic properties, Chapter 6 introduces several examples of algebraic and geometric proof systems and Chapter 7 collects together a number of proof systems that do not fit into the previous categories.

1
Concepts and Problems

1.1 Propositional Logic

In this section we recall basic notions of propositional logic and some standard notation. We also formulate a few statements that will be used many times later.

The **DeMorgan language** of propositional logic consists of the following:

- An infinite set At of atoms (propositional variables) denoted by $p, q, \ldots, x, y, \ldots$;
- logical connectives, consisting of
  - the constants $\top$ (**true**) and $\bot$ (**false**),
  - a unary connective $\neg$ (**negation**),
  - binary connectives $\vee$ (**disjunction**) and $\wedge$ (**conjunction**);
- parentheses (,).

Propositional formulas are particular finite words over this alphabet, constructed by applying finitely many times in an arbitrary order the following rules:

- the constants $\top$ and $\bot$ are formulas and any atom is a formula;
- if α is a formula then so is $(\neg\alpha)$;
- if α, β are formulas then so are $(\alpha \vee \beta)$ and $(\alpha \wedge \beta)$.

We will also often use italic capital Latin letters $A, B, \ldots$ to denote formulas.

A **subformula** of a formula α is any sub-word of α that is also a formula. Equivalently (this uses Lemma 1.1.1 below), a subformula of α is any formula that occurs in the process of creating α via the three rules above.

The types of brackets used may be any, e.g. [,], to improve the readability. The reason for the abundant use of brackets is seen in the following lemma.

Lemma 1.1.1 (Unique readability) *If α is a propositional formula then exactly one of the following cases occurs:*

- *α is a constant or an atom;*
- *there is a formula β such that $\alpha = (\neg\beta)$;*
- *there are formulas β, γ such that $\alpha = (\beta \vee \gamma)$;*
- *there are formulas β, γ such that $\alpha = (\beta \wedge \gamma)$.*

An alternative way to write formulas is using a **prefix notation** that works without parentheses. For example, $(\beta \vee \gamma)$ can be written as $\vee\beta\gamma$. But this seems to be readable easily only by machines.

It is customary to employ some simplifications in the use of parentheses. These include the following.

- We do not use the very last (outside) pair of parentheses; for example, $(\alpha \vee \beta)$ is written just as $\alpha \vee \beta$.
- The negation $\neg$ has priority over $\vee, \wedge$; for example, $((\neg\alpha) \vee \beta)$ can be written as $\neg\alpha \vee \beta$.
- We do not use parentheses when the formula is built with repeated use of $\vee$ or of $\wedge$; for example, $((\alpha \vee \beta) \vee \gamma)$ thus becomes $\alpha \vee \beta \vee \gamma$.

A **literal** ℓ is an atom or its negation. The following notation is useful:

$$\ell^1 := \ell, \quad p^0 := \neg p \quad \text{and} \quad (\neg p)^0 := p. \tag{1.1.1}$$

A conjunction of literals is called a **logical term** (or just a term if there is no danger of confusion). A disjunction of literals is called a **clause**.

The expression $\alpha(p_1, \dots, p_n)$ means that all the atoms that *could* occur in α are included in $p_1, \dots, p_n$ but not all of them have to actually occur.

A **truth assignment** is any function

$$h\colon At \longrightarrow \{0, 1\}\ .$$

The values 1 and 0 are meant to represent the truth values **true** and **false**, respectively, and we often simply identify the bits 1, 0 with the **truth values**. An assignment h is extended to the map h^* giving the truth value to any formula by the following rules:

- $h^*(\top) = 1$ and $h^*(\bot) = 0$;
- $h^*(\neg\alpha) = 1 - h^*(\alpha)$;
- $h^*(\alpha \vee \beta) = \max(h^*(\alpha), h^*(\beta))$;
- $h^*(\alpha \wedge \beta) = \min(h^*(\alpha), h^*(\beta))$.

The extended map h^* will be often denoted by h as well. In accordance with the first item, the constants $\top$ and $\bot$ are written simply 1 and 0 if there is no danger of confusion. It is important to realize that the unique-readability lemma 1.1.1 is needed to make this definition correct; without it there could be several ways to evaluate a formula.

A formula that is valid under any truth assignment is **logically valid** and it is called **a tautology**, and such formulas form the set TAUT. A formula that is true under at least one truth assignment is called **satisfiable**. Satisfiable formulas form the set SAT and unsatisfiable formulas form the set UNSAT.

A function from $\{0, 1\}^n$ to $\{0, 1\}$ is called a **Boolean function**. The truth value that an assignment h gives to a formula $\alpha(p_1, \dots, p_n)$ depends only on the values that h gives to $p_1, \dots, p_n$. If $h(p_i) = b_i \in \{0, 1\}$ then we denote $h^*(\alpha)$ just by $\alpha(b_1, \dots, b_n)$.

Any formula $\alpha(p_1, \ldots, p_n)$ thus defines a Boolean function, the **truth table function**, defined by

$$\mathbf{tt}_\alpha\colon \overline{b} = (b_1, \ldots, b_n) \in \{0,1\}^n \longrightarrow \alpha(b_1, \ldots, b_n) \in \{0,1\}.$$

Now observe that for $\overline{b} = (b_1, \ldots, b_n) \in \{0,1\}^n$ the term

$$p_1^{b_1} \wedge \cdots \wedge p_n^{b_n}$$

is true only under the assignment $p_i := b_i$. Hence the truth table function defined by the formula

$$\bigvee_{\overline{b} \in \{0,1\}^n : f(\overline{b})=1} p_1^{b_1} \wedge \cdots \wedge p_n^{b_n}$$

equals f. Similarly, the truth table function of the formula

$$\bigwedge_{\overline{b} \in \{0,1\}^n : f(\overline{b})=0} p_1^{1-b_1} \vee \cdots \vee p_n^{1-b_n}$$

is also equal to f. A formula that is a disjunction of terms is said to be in a **disjunctive normal form** (DNF), and dually a conjunction of clauses is in a **conjunctive normal form** (CNF).

Hence we have:

Lemma 1.1.2 *Let f: $\{0,1\}^n \longrightarrow \{0,1\}$ be any Boolean function. Then there exists a* DNF-*formula $\alpha(p_1, \ldots, p_n)$ and a* CNF-*formula $\beta(p_1, \ldots, p_n)$ such that*

$$f = \mathbf{tt}_\alpha = \mathbf{tt}_\beta.$$

A function f is **monotone** if and only if $\overline{a} \leq \overline{b}$, meaning that $\bigwedge_i (a_i \leq b_i)$ implies $f(\overline{a}) \leq f(\overline{b})$. Observe that if f is monotone then a DNF-formula α representing f can be written without the negation. For example, if $f(0, a_2, \ldots) = 1$ then also $f(1, a_2, \ldots) = 1$ and the two terms

$$p_1^0 \wedge p_2^{a_2} \wedge \cdots \quad \text{and} \quad p_1^1 \wedge p_2^{a_2} \wedge \cdots$$

can be merged into one:

$$p_2^{a_2} \wedge \cdots .$$

Repeating this reduction process, first for all occurrences of p_1 as long as it is possible and then for p_2, etc., gets rid of all the literals p_i^0, i.e. of all negations.

If all terms in a DNF-formula contain at most k literals, the formula is said to be kDNF; analogously we have kCNF formulas. The set of kCNF-formulas in SAT is denoted by kSAT.

A formula $\alpha(p_1, \ldots, p_n)$ is **logically implied** by a set of formulas T, written as $T \models \alpha$, if and only if every truth assignment making true all formulas in T also makes α true. It follows that if T consists of just one formula $\beta(p_1, \ldots, p_n)$ then $T \models \alpha$, abbreviated to $\beta \models \alpha$, if and only if $\mathbf{tt}_\beta$ is majorized on $\{0,1\}^n$ by $\mathbf{tt}_\alpha$.

Further, this is equivalent to the condition that $\neg\beta \vee \alpha$ is a tautology. Two formulas α, β are **logically equivalent** if and only if both $\beta \models \alpha$ and $\alpha \models \beta$ hold.

The representability of all Boolean functions by formulas has an interesting immediate consequence.

Theorem 1.1.3 (Craig [163] and Lyndon [345]) *Let $\overline{p}, \overline{q}, \overline{r}$ be disjoint (possibly empty) tuples of atoms. Assume that*

$$\alpha(\overline{p}, \overline{q}) \models \beta(\overline{p}, \overline{r}).$$

In particular, the atoms occurring in both formulas are among $\overline{p}$.

- *Then there exists a formula $\gamma(\overline{p})$ such that*

$$\alpha \models \gamma \quad \text{and} \quad \gamma \models \beta\,.$$

- *If the atoms $\overline{p}$ occur only positively (i.e. in the scope of an even number of negations) in α or only negatively (in the scope of an odd number of negations) in β then there exists such a formula γ without $\neg$.*

The formula γ is called an **interpolant** *(or, in the case of the second bullet point, a* **monotone interpolant***) of α and β.*

Proof Assume that the $\overline{p}$ are n-tuples and define two sets $U, V \subseteq \{0,1\}^n$ by

$$U := \{\overline{a} \in \{0,1\}^n \mid \exists \overline{b}\ \mathbf{tt}_\alpha(\overline{a}, \overline{b}) = 1\}$$

and

$$V := \{\overline{a} \in \{0,1\}^n \mid \exists \overline{c}\ \mathbf{tt}_\beta(\overline{a}, \overline{c}) = 0\}.$$

The assumption $\alpha \models \beta$ is equivalent to $U \cap V = \emptyset$. Take any Boolean function f on $\{0,1\}^n$ that is identically 1 on U and 0 on V, and let $\gamma(\overline{p})$ be any formula defining it. It is clear that such a formula is an interpolant.

In the monotone case take for f either the function which equals 1 on U and 0 outside U, if the $\overline{p}$ appear positively in α, or the function that equals 1 on V and 0 outside V if the $\overline{p}$ appear only negatively in β. Such an f is monotone and the observation after Lemma 1.1.2 then yields a monotone interpolant. □

We postulated at the beginning of the section that the language of propositional logic is the DeMorgan language. That is a customary choice but it is often useful to enlarge the language by other connectives that allow one to shorten DeMorgan formulas. The **implication** $\alpha \to \beta$ and the **equivalence** $\alpha \equiv \beta$ are binary connectives whose truth values are computed by the formulas $\neg\alpha \vee \beta$ and $(\alpha \wedge \beta) \vee (\neg\alpha \wedge \neg\beta)$, respectively.

The equivalence connective is particularly useful for talking about logically equivalent formulas: α, β are logically equivalent if and only if $\alpha \equiv \beta$ is a tautology.

This allows us to express pairs of logically equivalent formulas as logical equations, meaning equivalences that are logically valid. These include the **DeMorgan laws**

$$\neg(\alpha \vee \beta) \equiv (\neg\alpha \wedge \neg\beta) \quad \text{and} \quad \neg(\alpha \wedge \beta) \equiv (\neg\alpha \vee \neg\beta),$$

properties of the negation such as

$$\neg\neg\alpha \equiv \alpha, \quad \neg\top \equiv \bot, \quad \neg\bot \equiv \top$$

and algebraic properties such as the commutativity and associativity of both $\vee$ and $\wedge$ and the distributivity of one of these connectives over the other.

The implication and the equivalence connectives are defined by the DeMorgan formulas and one can reverse the process: $\neg\alpha$ can be defined as $\alpha \to \bot$ and the DeMorgan laws allow us to define one of $\vee$, $\wedge$ from the other and from $\neg$, etc. There are even binary connectives which alone define all the DeMorgan connectives and hence, by Lemma 1.1.2, all Boolean functions. The most famous is the so-called **Sheffer's stroke** $\alpha|\beta$, denoted in computer science by NAND and defined as $\neg(\alpha \wedge \beta)$.

There are also connectives which are often used in circuit complexity and which are best defined as having an unbounded arity. The most important are the **parity** connective

$$\oplus\,(a_1, \ldots, a_n) = 1 \quad \text{if and only if} \quad \sum_i a_i \equiv 1 \pmod 2, \tag{1.1.2}$$

where the a_i are bits or their generalizations to the **counting-modulo-m** connectives

$$\mathrm{MOD}_{m,j}(a_1, \ldots, a_n) = 1 \quad \text{if and only if} \quad \sum_i a_i \equiv j \pmod m, \tag{1.1.3}$$

with $0 \le j < m$, or the **threshold connective**

$$\mathrm{TH}_{n,k}(a_1, \ldots, a_n) = 1 \quad \text{if and only if} \quad \sum_i a_i \ge k, \tag{1.1.4}$$

with $0 \le k \le n$.

There is one issue that arises when passing from one language to another: formulas may grow exponentially in size. But we need to clarify first what the size of a formula is. At the surface level this is obvious: any formula is, in particular, a finite string in some alphabet and its size is its length. But this is not an entirely satisfactory definition, however. We shall consider formulas (and other syntactic objects like proofs) also as inputs for algorithms, and hence they need to be represented by words over a *finite* alphabet. In propositional logic we use infinitely many names from the set At for atoms. Hence we need to represent At itself by a set of words over a finite language. For example, atoms could be words starting with the letter p followed by a string from $\{0, 1\}^*$. This does change the length: if a formula contains n atoms (we may assume without loss of generality that they are $p_1, \ldots, p_n$) then they will now be represented by strings of length up to $1 + \log n$ and the string representing the whole formula may grow by that multiplicative factor. This is admittedly a technicality, but we need to recognize the need to deal with it and then largely ignore it: almost all our questions ask about the existence of objects, often proofs, of polynomially (or exponentially) bounded size and the log factor is irrelevant. We shall continue to use the definition of size that refers to the original propositional language with the set At, but with the understanding that for algorithmic purposes the size is slightly larger.

In fact, there is an elegant way to picture formulas, as finite labeled trees, and this gives rise to the most natural size measure. For DeMorgan formulas the trees will be binary; for formulas in a language with connectives up to arity k the trees would be k-ary. Let us discuss the DeMorgan case in detail.

Given a DeMorgan formula α we construct a finite, binary, labeled tree S_α by induction on the size of α, as follows.

1. If α is an atom or a constant, S_α consists of one node, labeled by the atom or by the constant, respectively. It is also the root of the tree.
2. If $\alpha = (\neg\beta)$, S_α has one node a, the root, labeled by the connective $\neg$. The root has one child b, which is the root of the subtree S_β. There is an edge from b to a.
3. If $\alpha = (\beta \circ \gamma)$, S_α has one node a, the root, labeled by the connective $\circ = \vee, \wedge$. The root has two children: the left child b, which is the root of the subtree S_β, and the right child c, which is the root of the subtree S_γ. There are edges from b and c to a.
4. The tree is partially ordered by directed paths: node u is less than node v if and only if there is a (non-empty) directed path from u to v.

The definition of the **size of a formula** α that we shall use is the cardinality of its tree S_α, i.e. the number of vertices of the tree. Because each connective comes with two parentheses, the length of the string representing α and using names from At is

$$3i + (|S_\alpha| - i) \leq 3|S_\alpha|,$$

where i is the number of non-leaves in S_α, and this increases slightly to

$$3i + (|S_\alpha| - i)\ \log n \leq (3 + \log n)|S_\alpha| \leq (3 + \log|S_\alpha|)|S_\alpha|,$$

when α has n atoms represented by binary words.

This definition can be further simplified for formulas in **negation normal form**: the negations are applied only to atoms, and there are no constants (a DeMorgan formula may be transformed into such a form without an increase in size by the DeMorgan laws and by contracting subformulas with constants). For such a formula each inner node in S_α branches into exactly two subtrees and we may define the size of the formula to be the number of leaves in S_α. This definition has the nice property that the size of a disjunction or of a conjunction is exactly the sum of the sizes of the disjuncts or of the conjuncts, respectively, and it is proportional to the cardinality of S_α ($|S_\alpha|$ is twice the size of formula minus 1).

Now let us return to the issue of translating formulas from one language to another. Consider a formula using just the binary connective $\oplus$. Subformulas $\alpha \oplus \beta$ can be replaced by $(\alpha \wedge \neg\beta) \vee (\neg\alpha \wedge \beta)$, going from simpler to more complex ones. But both α and β appear in this defining formula twice and so, if we iterate the removal of $\oplus$ in a formula, its size may double in each step. For example, the formula $p_1 \oplus (p_2 \oplus (p_3 \oplus \cdots) \cdots)$ computing the parity of n atoms has size n but after the above translation it will be greater than 2^n. In particular, if the parity connectives are nested in ℓ levels then the size of the translation will be between 2^ℓ and $2^{\ell+1}$.

Next we need to define the notion of the logical depth of a formula. The **logical depth** of a formula α in language L, denoted by $\ell\text{dp}(\alpha)$, is defined as follows:

- the ℓdp of atoms and constants is 0:
- $\circ(\beta_1, \ldots, \beta_k) = 1 + \max_i \ell\text{dp}(\beta_i)$, for $\circ$ a k-ary connective in L.

There is a simple but very useful result, Lemma 1.1.5 below, about the logical depth allowing the balance of formulas. We first prove a lemma about finite k-ary trees, i.e. trees branching at each node into at most k subtrees. Think now of trees as ordered downwards from the root towards the leaves. Let us denote the ordering by a generic symbol $\leq$. For a k-tree T and a node a in T, denote by T_a the subtree of T consisting of nodes b such that $b \leq a$. Denote by T^a the tree $(T \setminus T_a) \cup \{a\}$, i.e. the tree consisting of nodes b such that $b \not\leq a$.

In the following lemma we shall denote by $|T|$ the *number of leaves* of the tree T. The lemma is usually formulated for binary trees only.

Lemma 1.1.4 (Spira's lemma [481]) *Let T be a finite k-ary tree and $|T| > 1$. Then there is a node $a \in T$ such that*

$$(1/(k+1))|T| \leq |T_a|, |T^a| \leq (k/(k+1))|T|.$$

Proof Walk on a path through T, starting at the root and always walking within the biggest subtree (if there are more options then choose one arbitrarily). The size s of a current subtree can decrease in one step only to $s' \geq (s-1)/k$.

Continue in this fashion until the first node a such that the subtree T_a has size $|T_a| \leq (k/(k+1))|T|$ is reached. Then we also have $(1/(k+1))|T| \leq |T_a|$, because the previous subtree can have size (by the bound to s' above) at most $s \leq k|T_a|$: if $|T_a| < (1/(k+1))|T|$ then the previous subtree would have had size $\leq (k/(k+1))|T|$ and the process would have stopped at that point.

As $|T^a| = |T| - |T_a|$, the inequalities $(1/(k+1))|T| \leq |T^a| \leq (k/(k+1))|T|$ hold too. □

We will need the notion of **substitution**. A substitution of formulas for atoms in a formula $\alpha(p_1, \ldots, p_n)$ is any map σ assigning to each p_i some formula β_i. The formula $\sigma(\alpha)$ arising from applying the substitution to α is also denoted by

$$\alpha(p_1/\beta_1, \ldots, p_n/\beta_n)$$

or just by

$$\alpha(\beta_1, \ldots, \beta_n),$$

and it is created by simultaneously replacing p_i by β_i, $i = 1, \ldots, n$. It is allowed that the atoms $\overline{p}$ occur in the formulas β_i.

Now we state the promised lemma about the logical depth.

Lemma 1.1.5 *Let α be a formula in a language consisting of at most k-ary connectives and assume its size is s. Then there is a logically equivalent DeMorgan formula β of logical depth $\ell\mathrm{dp}(\beta) \leq O(\log_{(k+1)/k}(s)) = O(\log s)$.*

Proof Assume that the atoms of α are $\overline{p}$ and let q be a new atom and $\gamma(\overline{p}, q)$ and $\delta(\overline{p})$ be two formulas such that after substituting δ for q in γ one obtains;

$$\alpha = \gamma(q/\delta).$$

Then α can be equivalently written also as

$$(\gamma(\overline{p}, 1) \wedge \delta) \vee (\gamma(\overline{p}, 0) \wedge \neg\delta).$$

The point of this is that the logical depth of the new formula is

$$2 + \max(\ell\mathrm{dp}(\gamma), \ell\mathrm{dp}(\delta)).$$

By Spira's lemma 1.1.4 we can choose γ such that both γ and δ have sizes at most a fraction $k/(k+1)$ of the size of α, and by induction we can assume that the statement holds for formulas of any size smaller than s. □

For a formula with $\oplus$ we can first balance it by Spira's lemma and then replace all occurrences of $\oplus$ by the DeMorgan formula above defining it.

Corollary 1.1.6 *Let α be any formula in the DeMorgan language augmented by the binary connective $\oplus$ and assume that the size of α is s. Then there is a DeMorgan formula logically equivalent to α and of size at most $2^{O(\log s)} \leq s^{O(1)}$.*

1.2 First-Order Logic and Finite Structures

In this section we shall review some material from first-order logic and recall some properties of finite structures.

Languages of first-order logic contain some common logical symbols and also a specific vocabulary. The **logical symbols** are:

- the DeMorgan propositional logical connectives and various brackets (but not propositional atoms or propositional constants);
- the universal $\forall$ and the existential $\exists$ quantifiers;
- a countably infinite set of variables $x, y, \ldots$;
- the binary relation symbol $=$ for equality.

A **vocabulary** of a language L contains three sets:

- C_L, the set of constants;

- F_L, the set-of-symbols function, each symbol $f \in F_L$ coming with a positive arity;
- R_L: the set of relation symbols, each symbol $R \in R_L$ comes with a positive arity.

Any or all of these sets may be empty. A language is **relational** if $C_L = F_L = \emptyset$ and **relational with constants** if $F_L = \emptyset$.

An L-**term** is a finite string of these symbols obtained by the following rules:

- variables and constants are terms;
- if $f \in F_L$ is a symbol of arity k and $t_1, \ldots, t_k$ are terms, so is $f(t_1, \ldots, t_k)$.

Note that even if the vocabulary is empty we still have some terms: the variables.

We do not have propositional atoms but we have **atomic formulas**. For any terms $t_1, \ldots, t_k$ these are:

- $t_1 = t_2$;
- $R(t_1, \ldots, t_k)$, if $R \in R_L$ is of arity k.

First-order formulas are built from atomic formulas by the propositional connectives and by quantification: if A is a formula then so are $(\exists x A)$ and $(\forall x A)$. It is not required that the variable x actually occurs in A. The parentheses are left out when there is no doubt about how to read the formula.

Variables have **free or bound occurrences** in a formula. This is defined inductively as follows. All occurrences of variables in an atomic formula are free; the combination of formulas by propositional connectives does not change the qualification for any variable occurrence in them. In $(\exists x A)$ and $(\forall x A)$ all occurrences of x are bounded and occurrences of other variables do not change the qualification. This notion is important when we want to substitute a term for a variable.

An L-**structure** $\mathbf{A}$ is a non-empty set A (sometimes denoted by $|\mathbf{A}|$) called the **universe** of $\mathbf{A}$ together with the **interpretation** of L on $\mathbf{A}$:

- there is an element $c^{\mathbf{A}}$ for each $c \in C_L$;
- there is a function $f^{\mathbf{A}}: A^k \longrightarrow A$ for each $f \in F_L$ of arity k;
- there is a relation $R^{\mathbf{A}} \subseteq A^k$ for each $R \in R_L$ of arity k.

This generalizes the truth assignment of propositional logic and determines the satisfaction relation, which we will define in a moment. We will use a convention analogous to that in propositional logic (see Section 1.1) that the notation $t(x_1, \ldots, x_k)$ means that all variables occurring in t are among $x_1, \ldots, x_n$, and $B(x_1, \ldots, x_k)$ means that all variables that have a free occurrence in B are among $x_1, \ldots, x_k$, but not all these variables need to actually occur.

First, note that any term $t(x_1, \ldots, x_n)$ defines naturally on any structure $\mathbf{A}$ an n-ary function $t^{\mathbf{A}}: A^n \longrightarrow A$. For an atomic formula $s = t$ with terms $s(x_1, \ldots, x_n)$ and $t(x_1, \ldots, x_n)$ and an n-tuple $\overline{a} = (a_1, \ldots, a_n) \in A^n$,

- $\mathbf{A}$ satisfies $s = t$ for the assignment $x_i := a_i$ if and only if $s^{\mathbf{A}}(\overline{a}) = t^{\mathbf{A}}(\overline{a})$. This is denoted by $\mathbf{A} \models s(\overline{a}) = t(\overline{a})$ if it holds, and by $\mathbf{A} \not\models s(\overline{a}) = t(\overline{a})$ if it does not.

This is then extended to all first-order formulas of L via **Tarski's** so-called **truth definition**, by induction on the logical depth of formulas, in first-order logic also called the logical complexity: the **logical complexity** is defined as for propositional formulas but with the additional clause that a quantifier increases the logical depth by 1. For the formulas $B(x_1, \ldots, x_n)$ and $C(x_1, \ldots, x_n)$ and $\overline{a} \in A^n$ define:

- $\mathbf{A} \models \neg B(\overline{a})$ if and only if $\mathbf{A} \not\models B(\overline{a})$;
- $\mathbf{A} \models B(\overline{a}) \wedge C(\overline{a})$ if and only if $\mathbf{A} \models B(\overline{a})$ and $\mathbf{A} \models C(\overline{a})$;
- $\mathbf{A} \models B(\overline{a}) \vee C(\overline{a})$ if and only if $\mathbf{A} \models B(\overline{a})$ or $\mathbf{A} \models C(\overline{a})$.

The inductive clauses for the quantifiers generalize those for $\vee$ and $\wedge$. For a formula $B(x_1, \ldots, x_n, y)$ and $\overline{a} \in A^n$ define:

- $\mathbf{A} \models \exists y B(\overline{a}, y)$ if and only if there is a $b \in A$ such that $\mathbf{A} \models B(\overline{a}, b)$;
- $\mathbf{A} \models \forall y B(\overline{a}, y)$ if and only if for all $b \in A$ it holds that $\mathbf{A} \models B(\overline{a}, b)$.

The following lemma is proved by induction on the logical depth of the formula.

Lemma 1.2.1 *Let L be finite and let $B(x_1, \ldots, x_n)$ be a first-order L-formula. Then there exists a deterministic log-space machine that upon receiving a finite L-structure $\mathbf{A}$ (i.e. its code) and an n-tuple $\overline{a}$ of its elements decides whether the satisfaction relation $\mathbf{A} \models B(a_1, \ldots, a_n)$ holds.*

The notions of log-space machines and of the NP machines used below are recalled in Section 1.3.

Second-order logic allows us to quantify also over functions or relations on the universe of a structure. This is a notion that appears natural at first but is quite foggy: properties of the class of all relations or all functions on an infinite set depend on the axioms of set theory and are not in any sensible way canonical. In particular, for a non-trivial vocabulary one can write second-order sentences equivalent to various set-theoretic statements that are undecidable in Zermelo–Fraenkel set theory with choice (ZFC).

When we will say second-order logic we will actually mean first-order logic with variables of several sorts, one of which is for the elements of the structure (as intended in first-order logic). Such a many-sorted structure has the universe of elements and also the universes of some k-ary functions and relations on the element universe. The quantification over k-ary functions then refers to ordinary first-order quantification over the corresponding sort. In particular, Tarski's definition of the truth of second-order quantifiers refers to elements of the function or relation universes and not to functions or relations on the universe of the structure. This may seem just a cosmetic change but it is not: one escapes set theory and maintains all the many wonderful properties of first-order logic (its completeness and compactness, in particular).

If an L-structure $\mathbf{A}$ is finite, we can represent any k-ary relation by a finite k-dimensional array of bits and any k-ary function by the array representing its graph. Hence, if L is finite, we can represent the whole structure by a finite string of bits and present it as an input to a machine. Note that the length of such a string will be polynomial in the size of the universe. But there is one subtler technical point. Any such representation implicitly defines an order on the universe of **A** but there may be no order relation in L. However, having a linear order on A determines the lexicographic order on the arrays and hence selects uniquely the final binary code of the structure. The solution adopted in finite model theory is that it is not required that an ordering is in L but it is demanded that the machine gives the same answer no matter which ordering is used in the encoding. This is admittedly an unpleasant technicality but it seems unavoidable.

Now we present two topics, Fagin's theorem and the so-called propositional translation of first-order formulas over finite structures. In later chapters we shall need generalizations (or rather, strengthenings) of both.

The class of second-order formulas may be stratified by the quantifier complexity. We shall define just one class now (and add others later). The class of **strict Σ^1_1-formulas**, denoted by $s\Sigma^1_1$, consists of all formulas of the form

$$\exists X_1 \exists X_2 \dots \exists X_k B(X_1, \dots, X_k),$$

where the X_i are second-order variables (i.e. variables over the relation sort or over the function sort) and B is first-order. We say that a class of structures is $s\Sigma^1_1$-definable if and only if it contains exactly the structures satisfying a fixed $s\Sigma^1_1$-sentence.

For finite structures we take a priori for the function or relation sorts the classes of all functions and all relations, respectively, of the appropriate arities on the universe.

Theorem 1.2.2 (Fagin's theorem [178]) *Let L be a finite language. A class of finite L-structures is $s\Sigma^1_1$-definable if and only if it is in the computational class NP.*

The "only-if" direction is simpler: the NP machine guesses witnesses for the second-order existential quantifiers (i.e. the tables of bits defining them) in the defining sentence and then checks in polynomial time (by Lemma 1.2.1) that they do indeed witness the formula. For the "if" direction one needs to encode into a finite number of relations an accepting computation of the machine (if it exists) and express in a first-order way that it is correct. This is tedious and quite analogous to the usual proofs of the NP-completeness of SAT.

Now we turn to propositional translation. Let L be a finite relational language. Recall that for $n \geq 1$ we denote by $[n]$ the set $\{1, \dots, n\}$.

Let $B(x_1, \dots, x_k)$ be an L-formula. We shall define by induction on the logical complexity of B the propositional formula $\langle B(i_1, \dots, i_k)\rangle_n$ for any choice of $i_1, \dots, i_k \in [n]$ as follows.

- If B is a k-ary relation symbol R then

$$\langle R(i_1, \dots, i_k)\rangle_n := r_{i_1,\dots,i_k},$$

where the $r_{i_1,\ldots,i_k}$ are n^k new propositional atoms associated with R and indexed by the k-tuples of elements of $[n]$.

- The translation $\langle\ldots\rangle_n$ commutes with the propositional connectives:
  - $\langle\neg B(i_1,\ldots,i_k)\rangle_n := \neg\langle B(i_1,\ldots,i_k)\rangle_n$,
  - $\langle B(i_1,\ldots,i_k)\circ C(i_1,\ldots,i_k)\rangle_n := \langle B(i_1,\ldots,i_k)\rangle_n \circ \langle C(i_1,\ldots,i_k)\rangle_n$ for $\circ = \vee, \wedge$.
- $\langle\exists y B(i_1,\ldots,i_k,y)\rangle_n := \bigvee_{j\in[n]}\langle B(i_1,\ldots,i_k,j)\rangle_n$.
- $\langle\forall y B(i_1,\ldots,i_k,y)\rangle_n := \bigwedge_{j\in[n]}\langle B(i_1,\ldots,i_k,j)\rangle_n$.

Now we slightly generalize this definition. Let $L_0 \subseteq L$ and let $\mathbf{A}$ be an L_0-structure with the universe $[n]$. For an L-formula B and a k-tuple $i_1,\ldots,i_k$ as above, define the translation

$$\langle B(i_1,\ldots,i_k)\rangle_{n,\mathbf{A}}$$

by the following substitution.

- For any k-ary relation symbol R from L_0 and any $i_1,\ldots,i_k \in [n]$, substitute in $\langle B(i_1,\ldots,i_k)\rangle_n$ for $r_{i_1,\ldots,i_k}$ the value $\top$ if $\mathbf{A} \models R(i_1,\ldots,i_k)$, and substitute $\bot$ otherwise.

The following lemma should be clear from the definitions. It uses the notion of the **expansion** of a first-order L-structure $\mathbf{A}$, which is an L'-structure for some $L' \supseteq L$ that has the same universe as $\mathbf{A}$ and interprets the symbols from L in the same way. That is, it is $\mathbf{A}$ together with an interpretation of some symbols not in the original language.

Lemma 1.2.3 *Let $L \supseteq L_0$ be finite relational languages and $B(x_1,\ldots,x_k)$ an L-formula. Let $n \geq 1$, $i_1,\ldots,i_k \in [n]$, and let $\mathbf{A}$ be an L_0-structure with the universe $[n]$.*

Then the following statements are equivalent:

- *$B(i_1,\ldots,i_k)$ holds in some L-expansion of $\mathbf{A}$;*
- *$\exists S_1 \ldots \exists S_t B(i_1,\ldots,i_k)$ holds in $\mathbf{A}$, where $\{S_1,\ldots,S_t\} = L \setminus L_0$;*
- *$\langle B(i_1,\ldots,i_k)\rangle_{n,\mathbf{A}}$ is satisfiable.*

This lemma, together with Fagin's theorem, implies the NP-completeness of SAT. Applying it to Skolemized formulas yields the NP-completeness of kSAT for some k. Of course, it can be seen that this is not surprising if one uses the idea of the proof of the NP-completeness of SAT in the proof of Fagin's theorem.

1.3 Complexity Classes

The machine model defined by Turing is not the only one considered in complexity theory, but it is the most intuitive and elementary and it is the model best suited for the purpose of defining basic complexity classes.

A **Turing machine** operates on a tape divided into infinitely many unit squares ordered identically as integers; the squares can be numbered by integers when discussing the machine but the machine has no a priori access to such a numbering. Initially all the squares are empty except possibly a finite interval of squares each holding one symbol from a non-empty **input alphabet** Σ. It is convenient to model the empty squares by saying that they contain the **blank symbol**, which is not a member of Σ. We will also assume that the two bits 0 and 1 are in Σ. Thus we imagine the initial tape configuration to be a bi-infinite string of blanks interrupted by a word, the **input**, from Σ^*.

The machine has a **head**, a read–write device, that initially scans the first non-blank symbol on the tape (if there is one), i.e. the first symbol of the input. If there are only blanks, we say that the input is the empty word Λ.

The set-up so far can be altered in a number of ways (and we will need to do that when defining space complexity below) but what will remain the same are:

- finite control;
- the locality of computations.

The phrase *finite control* means that the machine is always in one state from a finite set of states Q and the action taken at any given step depends only on:

- the symbol currently scanned by the head; and
- the state the machine is in.

The *locality of computations* means that:

- at every step the machine can replace only the symbol currently scanned by the head, the new symbol being from a **working alphabet** Γ that contains Σ as well as the blank symbol;
- the head can be moved only one square left or right;
- the machine state can be changed.

To define a machine M formally one needs to specify a 5-tuple

$$\Sigma, \Gamma, Q, q_0, I,$$

where

- Σ is a finite alphabet containing $0, 1$ and not containing the blank symbol,
- Γ is a finite alphabet extending Σ and containing the blank symbol,
- Q is a finite set of states and $q_0 \in Q$ is the **initial state**,
- $I \subseteq \Gamma \times Q \times \Gamma \times \{L, R\} \times Q$ is a set of **instructions**.

The machine M then operates as follows:

- if M's head scans symbol s when M is in a state q and a 5-tuple

$$(s, q, s', X, q')$$

is in I, then M

- replaces the symbol s in the current square by s',
- moves its head one square to the left if $X = L$ and to the right if $X = R$,
- and then changes its state to q'.

The machine is **deterministic** if for each pair $(s, q) \in \Gamma \times Q$ there is at most one instruction in I starting with this pair. Otherwise the machine is **non-deterministic**.

The computation of the machine stops when there is no instruction that could be used. When this happens we define the output of the machine to be the longest word from Σ^* written on the tape such that the head scans its first symbol. If the head scans a symbol not in Σ we define the output to be the empty word.

The number of steps taken by the machine in a particular computation that stopped is the **time** of the computation. If the computation did not stop, we may say for convenience that the time is ∞. For $w \in \Sigma^*$, define $\text{time}_M(w)$ to be the smallest time of some computation on w (if M is deterministic there is at most one such).

The **time complexity** of M is the function t_M: $\mathbf{N} \to \mathbf{N} \cup \{\infty\}$ defined by

$$t_M(n) \; := \; \max\{\text{time}_M(w) \mid w \in \Sigma^n\}.$$

A deterministic M **decides** a language $L \subseteq \{0, 1\}^*$ if and only if M computes the characteristic function

$$\chi_L \colon w \in \{0, 1\}^* \to \{0, 1\},$$

which is equal to 1 if and only if $w \in L$. A non-deterministic machine M **accepts** such an L if and only if, for all $w \in \{0, 1\}^*$,

$$w \in L \quad \text{if and only if} \quad \text{time}_M(w) \neq \infty.$$

That is, there is some finished computation on w.

Now we are in a position to define several basic computational complexity classes. For a function f: $\mathbf{N} \to \mathbf{N}$ define Time(f) to be the class of all languages L that can be decided by a machine with time complexity $O(f(n))$. Similarly, define NTime(f) to be the class of all languages L that can be accepted by a machine with time complexity $O(f(n))$.

The basic classes are:

- $\mathrm{P} := \bigcup_{c \geq 1} \mathrm{Time}(n^c)$ and $\mathrm{NP} := \bigcup_{c \geq 1} \mathrm{NTime}(n^c)$;
- $\mathrm{E} := \bigcup_{c \geq 1} \mathrm{Time}(2^{cn})$ and $\mathrm{NE} := \bigcup_{c \geq 1} \mathrm{NTime}(2^{cn})$;
- $\mathrm{EXP} := \bigcup_{c \geq 1} \mathrm{Time}(2^{n^c})$ and $\mathrm{NEXP} := \bigcup_{c \geq 1} \mathrm{NTime}(2^{n^c})$.

It is obvious that

$$\mathrm{P} \subseteq \mathrm{E} \subseteq \mathrm{EXP} \quad \text{and} \quad \mathrm{NP} \subseteq \mathrm{NE} \subseteq \mathrm{NEXP}$$

but, in fact, a suitable diagonalization, as in the well-known halting problem, establishes the following theorem. A time-constructible function is one which is the time complexity of some deterministic machine on the input $1^{(n)}$.

Theorem 1.3.1 (The time hierarchy theorem [221, 148]) *For $f, g\colon \mathbf{N} \to \mathbf{N}$ time-constructible such that $n \le f(n)$ and $f(n+1) = o(g(n))$ it holds that*

$$\mathrm{Time}(f) \subsetneq \mathrm{Time}(g \log g) \quad \textit{and} \quad \mathrm{NTime}(f) \subsetneq \mathrm{NTime}(g),$$

where $\subsetneq$ is the proper subset relation. In particular,

$$\mathrm{P} \subsetneq \mathrm{E} \subsetneq \mathrm{EXP} \quad \textit{and} \quad \mathrm{NP} \subsetneq \mathrm{NE} \subsetneq \mathrm{NEXP}\,.$$

Furthermore, an exhaustive search argument going through all possible computations of an NP machine shows that

$$\mathrm{NP} \subseteq \mathrm{EXP}.$$

Thus we know that

$$\mathrm{P} \subseteq \mathrm{NP} \subseteq \mathrm{EXP}$$

and that the last class is strictly larger than the first, but we do not know which of the two inclusions is or are strict. It is generally conjectured that both are.

Informally, the space of a computation is the number of tape squares that the machine used; however, we want to count only the squares used in the computation proper and not those on which the input is stored or the output is written. Because of this consideration, for the purpose of defining the space complexity, we consider a modification of the Turing machine model by adding:

- an **input tape** on which the input is written and which is scanned by a read-only head;
- an **output tape** on which the machine eventually writes the output and which is scanned by a write-only head;
- a **working tape** that is scanned by a read–write head operating as in the single-tape model.

The **space** of a computation is the number of squares on the working tape visited by the head during the computation. We shall denote by $\mathrm{space}_M(w)$ the minimum space of a computation of M on an input w. Then, analogously to the time complexity, define the **space complexity** of machine M to be the function $s_M\colon \mathbf{N} \to \mathbf{N} \cup \{\infty\}$ defined by

$$s_M(n) \;:=\; \max\{\mathrm{space}_M(w) \mid w \in \Sigma^n\}.$$

The classes $\mathrm{Space}(f)$ and $\mathrm{NSpace}(f)$ consist of languages that can be decided or accepted, respectively, by a machine with space complexity bounded by $O(f(n))$. The basic classes are:

- $\mathrm{L} := \mathrm{Space}(\log n)$ and $\mathrm{NL} := \mathrm{NSpace}(\log n)$;
- $\mathrm{PSPACE} := \bigcup_c \mathrm{Space}(n^c)$ and $\mathrm{NPSPACE} := \bigcup_c \mathrm{NSpace}(n^c)$.

For a class $\mathcal{C}$ denote by $co\mathcal{C}$ the class of complements of $\mathcal{C}$-languages:

$$co\mathcal{C} \;:=\; \{\{0,1\}^* \setminus L \mid L \in \mathcal{C}\}.$$

A function is space-constructible if and only if it is the space complexity of some machine.

Theorem 1.3.2 *Let f, g be space-constructible functions.*

(i) For $f = o(g)$, $\text{Space}(f) \subsetneq \text{Space}(g)$
(the **space hierarchy theorem** *[220]).*

(ii) For $f \geq \log n$, *NSpace(f)* $\subseteq \text{Space}(f^2)$ *and hence* PSPACE = NPSPACE
*(***Savitch's theorem** *[459]).*

(iii) For $f(n) \geq \log n$, $\text{NSpace}(f) = co\text{NSpace}(f)$
(the **Immermann–Szelepcsényi theorem**, *[485, 240]).*

We have also the simple inclusions

$$\text{L} \subseteq \text{NL} \subseteq \text{P} \subseteq \text{PSPACE},$$

the last class being strictly larger than the first but, as with the time classes earlier, we do not know which of the three inclusions is or are proper.

Finally, we recall in this section reductions and complete problems. A **polynomial-time reduction** (a **p-reduction** for short) of a language L_1 to a language L_2 is a p-time function $f: \{0,1\}^* \to \{0,1\}^*$ such that, for all $w \in L_1$,

$$w \in L_1 \quad \text{if and only if} \quad f(w) \in L_2.$$

The notation for this is $L_1 \leq_p L_2$.

For a class $\mathcal{C}$, a language L is $\mathcal{C}$**-hard** if and only if $L' \leq_p L$ holds for all $L' \in \mathcal{C}$, and it is $\mathcal{C}$**-complete** if and only if it is $\mathcal{C}$-hard and is itself in $\mathcal{C}$.

The existence of complete problems for various computational classes containing P is a simple consequence of the existence (and its properties) of universal Turing machines. What is deeper is that some *natural* languages having combinatorial or logical descriptions are complete for some of the classes defined earlier. We shall mention here just the most famous example.

Theorem 1.3.3 (Cook's theorem [147]) *SAT is* NP-*complete and hence TAUT is* coNP-*complete.*

The second part follows from the first part because, for a propositional formula α, it holds that

$$\alpha \in \text{TAUT} \quad \text{if and only if} \quad \neg\alpha \notin \text{SAT} .$$

The general argument for the existence of complete problems shadowing universal machines can be applied to the so-called *syntactic classes*. Informally, in such a class, some a priori (syntactic) restrictions on the machines defining languages in the class (e.g. a clock imposing a bound on time) guarantee that the language is indeed in the class. There are also the so-called *semantic classes*, to which such an argument does not seem to apply, and often it is open whether such a class has a complete problem.

An example is the class BPP, the **bounded probabilistic polynomial-time** class consisting of all languages L for which there is a Turing machine M with two inputs x, y and a constant $c \geq 1$ such that:

- on inputs $x := w \in \{0,1\}^n$ and $y := r \in \{0,1\}^*$ the machine runs in time at most n^c and outputs either 0 or 1;
- for all $w \in \{0,1\}^n$;
  - if $w \in L$ then $Prob_r[M(w,r) = 1] \geq 2/3$,
  - if $w \notin L$ then $Prob_r[M(w,r) = 0] \geq 2/3$,

  where r is uniformly distributed over $\{0,1\}^{n^c}$.

The two conditions involving the probability are not guaranteed to hold by any a priori condition on how M is programmed but instead refer to how M behaves (hence the qualifier "semantic") on the space of all r. It is not known whether BPP has a complete problem.

It is easy to see that

$$\text{P} \subseteq \text{BPP} \subseteq \text{EXP}$$

and, in fact, it is often conjectured that P = BPP.

1.4 Boolean Circuits and Computations

A formula whose truth table function is f: $\{0,1\}^n \to \{0,1\}$ can be used as an algorithm to compute f: given an input, evaluate the formula using its inductive definition. In this process the same subformula may appear several times (as in the example with the parity function in Section 1.1). But it is clearly redundant to compute its truth value again and again: the truth value of a subformula remains the same irrespective at which place in the whole formula we encounter it. This wastefulness is remedied by the notion of a **Boolean circuit**.

We shall define circuits over the DeMorgan language – that is the canonical choice. But the definition applies analogously to other propositional languages (and is even more general: there are arithmetic circuits, algebraic circuits, etc.). A circuit with inputs $x_1, \ldots, x_n$ is a sequence of instructions about how to compute Boolean values $y_1, \ldots, y_s$. Each instruction has one of the following forms:

- $y_i := \top$ or $y_i = \bot$;
- $y_i := x_j$, for any $1 \leq j \leq n$;
- $y_i := \neg y_j$, where $j < i$;
- $y_i := y_j \circ y_k$, where $j, k < i$ and $\circ = \vee, \wedge$.

The **size** of such a circuit is s and y_s is its output. The circuit defines a Boolean function on $\{0,1\}^n$: on an assignment $x_j := a_j \in \{0,1\}$ evaluate sequentially all the y_i using the instructions and output the value computed for y_s. If the circuit is called C the output value is denoted by $C(\overline{a})$.

It will often be useful to represent the instructions of a circuit C as above by a 3CNF formula, a set of 3-clauses; we shall denote the set by $\mathrm{Def}_C(\overline{x},\overline{y})$. For example, the instruction $y_i := x_j$ is represented by two clauses,

$$\{y_i, \neg x_j\}, \quad \{\neg y_i, x_j\}$$

and the instruction $y_i := y_u \wedge y_v$ by three clauses,

$$\{\neg y_i, y_u\}, \quad \{\neg y_i, y_v\}, \quad \{y_i, \neg y_u, \neg y_v\}.$$

The number of clauses in Def_C is at most $3s$ and for any $\overline{a} \in \{0,1\}^n$, the clauses of $\mathrm{Def}_C(a,\overline{y})$ are satisfied by a unique assignment to the variables y_i – the computation of C on the input $\overline{a}$.

Just as we represented the DeMorgan formulas by trees in Section 1.1, we can represent circuits by directed acyclic graphs (called often dags). A **dag** is defined analogously to a tree:

1. the nodes are $1,\ldots,s$ and they correspond to the instructions $y_1,\ldots,y_s$;
2. y_i is labeled by the type of the instruction y_i: the values $\bot$ or $\top$, the inputs x_j or one of the connectives $\neg$, $\vee$ or $\wedge$,
3. if the value y_j is used in the instruction defining y_i then there is an edge from j to i.

Formulas can thus be viewed as circuits that happen to be trees.

It is easy to see that the parity operator $\oplus_n$ has a circuit of size $O(n)$. In fact, a major problem in theoretical computer science is to produce an explicit example of a Boolean function that does not have a linear-size circuit. Often the informal qualification "explicit" means "to be in NP" but an example is unknown even for E. This is in contrast with the following fact.

Theorem 1.4.1 (Shannon's estimate [466]) *With probability going to* 1 *as* $n \to \infty$, *a random Boolean function on* $\{0,1\}^n$ *needs a circuit of size at least* $\Omega(2^n/n)$.

The importance of the task of establishing strong lower bounds for the size of circuits stems from the following theorem.

Theorem 1.4.2 (Savage [458]) *Let M be a deterministic Turing machine computing a predicate on* $\{0,1\}^*$ *and assume its time complexity is* $t_M(n)$. *Then for* $n = 1,2,\ldots$ *there are circuits* C_n *with inputs* $x_1,\ldots,x_n$ *which compute the predicate on* $\{0,1\}^n$ *and such that the size of* C_n *is bounded above by* $O(t_M^2(n))$.

In addition, the description (i.e. the list of instructions) of circuits C_n *can be computed by an algorithm from the input* $1^{(n)}$ *in a time polynomial in* $t_M(n)$.

It follows that if we could demonstrate a super-polynomial lower bound on the size of any circuits computing a problem in NP, the separation $\mathrm{P} \neq \mathrm{NP}$ would be established. At present we do not know such a problem even in the first-level NEXP of the exponential time hierarchy (recall that we even have $\mathrm{P} \subsetneq \mathrm{EXP}$ and $\mathrm{NP} \subsetneq \mathrm{NEXP}$ by Theorem 1.3.1). The counting argument behind Shannon's estimate 1.4.1 does not yield any example in NEXP.

The fact that time lower bounds can be reduced to size lower bounds for circuits leads to a combinatorial approach to the P vs. NP problem as well as to other separation problems. In this approach one disregards the reference to Turing machines (or to some other uniform algorithms) as solving a problem for all lengths. Instead one considers that a language is computed for each input length by a circuit specific for that length. The circuits are bound together only by a common form of an upper bound on their size. This leads to the definitions of several circuit classes. The most natural is the class $P/poly$. It consists of all languages $L \subseteq \{0,1\}^*$ such that there is a $c \geq 1$ and a sequence of circuits C_n, $n \geq 1$, such that:

- C_n has n inputs and computes the characteristic function χ_L on $\{0,1\}^n$;
- for all $n > 1$, $|C_n| \leq n^c$.

Theorem 1.4.2 implies that $P \subseteq P/poly$ but it is easy to see that $P \neq P/poly$, as the latter class contains languages that are not algorithmically decidable (for any $A \subseteq \mathbf{N}$ consider $L := \{w \in \{0,1\}^* \mid |w| \in A\}$: it is undecidable if A is undecidable but it has circuit complexity 1). The problem of demonstrating a super-polynomial lower bound for the size of circuits deciding some language in NP can thus be rephrased as the task of showing that $NP \not\subseteq P/poly$ (again, to show that $NP \neq P/poly$ is simple).

The problem of establishing such a lower bound appears to be very hard and researchers have considered, instead of general circuits, various restricted classes of circuits. Five prominent classes, important also for proof complexity, are as follows.

- AC^0, the languages computed on $\{0,1\}^n$ by circuits C_n in the DeMorgan language with $\bigvee$ and $\bigwedge$ of an unbounded arity, of logical depth bounded by a constant d common to all lengths $n \geq 2$ and with size bound $|C_n| \leq n^c$ as in $P/poly$.
- $AC^0[m]$, defined as AC^0 except that the circuits can also use the unbounded-arity connectives $MOD_{m,j}$ defined in (1.1.3); in particular, circuits defining languages in $AC^0[2]$ use the parity connective (1.1.2).
- TC^0, defined as AC^0 but the circuits can use also the unbounded-arity connectives $TH_{n,k}$ defined in (1.1.4).
- NC^1, languages computed on $\{0,1\}^n$ by *formulas* F_n in the DeMorgan language with binary $\vee$ and $\wedge$, having size $|F_n| \leq n^c$ for $n \geq 2$, for some constant $c \geq 1$.
- $mP/poly$ (m is for *monotone P/poly*), circuits in the definition of $P/poly$ that are not allowed to use the negation.

Strong lower bounds for some of these restricted classes of circuits are known. We state for an illustration the following theorem (see Section 1.6 for more references). For $n \geq k \geq 2$ the Boolean function $\text{Clique}_{n,k}$ is defined on $\{0,1\}^m$, where $m = \binom{n}{2}$ and where we identify a string $w \in \{0,1\}^m$ with an undirected simple graph on $[n]$; $\text{Clique}_{n,k}(w) = 1$ if and only if the graph determined by w contains a clique (i.e. a complete subgraph) of size at least k.

Theorem 1.4.3

(i) $\{\oplus_n\}_n \notin AC^0$ *(Ajtai [4], Furst, Saxe and Sipser [192]).*

(ii) $\{\mathrm{TH}_{n,n/2}\}_n \notin \mathrm{AC}^0[2]$ *(Razborov [431]).*
(iii) *If* $p \neq q$ *are primes then* $\{\mathrm{MOD}_{q,0}\}_n \notin \mathrm{AC}^0[p]$ *(Smolensky [474]).*
(iv) *For* $2 \leq k \leq n/4$, $\mathrm{Clique}_{n,k} \notin m\mathrm{P}/poly$ *(Razborov [430]).*

Monotone circuits play an important role in the feasible interpolation method that we shall treat in Chapters 17 and 18.

There is an important connection of the class AC^0 to first-order logic on finite structures, introduced in Section 1.2. Strings from $\{0,1\}^n$ can be identified with subsets of $[n]$; the subsets of $[n]$ are, in particular, all possible interpretations of a unary relation symbol on the universe $[n]$. For $\overline{w} \in \{0,1\}^n$ put $V_w := \{i \in [n] \mid w_i = 1\} \subseteq [n]$.

If L_0 is an arbitrary finite relational language, denote by $L_0(V)$ its extension by one unary relation symbol $V(x)$.

Lemma 1.4.4 *Let* $L \subseteq \{0,1\}^*$ *be a language (in the computational sense). Then the following statements are equivalent:*

(i) $L \in \mathrm{AC}^0$*;*
(ii) *there is a finite relational language* L_0*, an* $L_0(V)$*-sentence* B *and, for all* $n \geq 1$*, there are* L_0*-structures* $\mathbf{A}_n$ *with a universe* $[n]$ *such that*
  - *for all* $w \in \{0,1\}^n$*,* $w \in L$ *if and only if the* $L_0(V)$*-structure* $(\mathbf{A}_n, V_w) \models B$*, where* V_w *interprets the predicate* V*.*

Proof By Lemma 1.2.3, B holds in $(\mathbf{A}_n, V_w)$ if and only if $\langle B \rangle_{n,(\mathbf{A}_n,V_w)}$ is true. The "if" part of the lemma follows by noting that the $\langle \dots \rangle$ translation produces a sequence of constant-depth formulas of a size polynomial in n (each quantifier is translated by a disjunction or a conjunction of arity n).

For the opposite direction assume that C_n are circuits certifying that $L \in \mathrm{AC}^0$. Because the depth of C_n is bounded by a constant we can unwind the circuits into formulas, repeating the subformulas as many times as needed. Further, we may assume without loss of generality that:

- all negations in C_n are in front of atoms (the negation normal form from Section 1.1);
- C_n has constant depth d, the top connective being the conjunction and the connectives at each level $d-1, d-2, \dots$ being of the same type and alternating from one level to the next between disjunctions and conjunctions;
- the arity of each $\bigwedge, \bigvee$ in C_n is exactly n^c for some constant $c \geq 1$.

Wires entering a disjunction or a conjunction are in a one-to-one correspondence with the tuples from $[n]^c$ and hence the elements of $([n]^c)^d$ are in a one-to-one correspondence with the paths in C_n from the top connective $\bigwedge$ to a literal at the bottom level 0 (the variable x_j or $\neg x_j$).

We take for L_0 the language consisting of two $(cd+1)$-ary relations P and N and define an L_0-structure $\mathbf{A}_n$ by interpreting them on $[n]$ as follows. For $(i_1^1, \dots, i_c^1, \dots, i_1^d, \dots, i_c^d, j) \in ([n]^c)^d \times [n]$,

- $P(i_1^1, \ldots, i_c^1, \ldots, i_1^d, \ldots, i_c^d, j)$ holds if and only if the input determined by the path $(i_1^1, \ldots, i_c^1, \ldots, i_1^d, \ldots, i^d)$ is the variable x_j, and
- $N(i_1^1, \ldots, i_c^1, \ldots, i_1^d, \ldots, i_c^d, j)$ holds if and only if the input determined by the path $(i_1^1, \ldots, i_c^1, \ldots, i_1^d, \ldots, i^d)$ is the literal $\neg x_j$.

The input variables x_j of the circuit C_n are interpreted by $V(j)$ and hence we have, for $w \in \{0,1\}^n$, $w \in L$ if and only if $C_n(w) = 1$ if and only if the structure $(\mathbf{A}_n, V_w)$ satisfies the sentence

$$\forall \overline{y}^1 \exists \overline{y}^2 \forall \ldots \ [(P(\overline{y}^1, \ldots, \overline{y}^d, x) \to V(x)) \wedge (N(\overline{y}^1, \ldots, \overline{y}^d, x) \to \neg V(x))],$$

where the $\overline{y}^i$ are c-tuples of variables. □

We note that analogous characterizations can be given for the classes $\mathrm{AC}^0[m]$ and TC^0 by extending the first-order logic by quantifiers that count (either modulo m or exactly).

Before we conclude this section we shall look again at how one can extract an algorithm from a formula. This time we shall use the notion of decision trees. A **decision tree** T is a binary tree with ordered edges whose every node has out-degree 2 (the **inner nodes**) or 0 (the **leaves**). There is a unique node with in-degree 0 (the **root**) and all other nodes have in-degree 1. The inner nodes are labeled by variables x_i ($i \in [n]$) and the two edges leaving such a node are labeled by $x_i = 0$ and $x_i = 1$, respectively. The leaves are labeled by 0 or 1.

Any $\overline{a} \in \{0,1\}^n$ determines a unique path $P_T(\overline{a})$ in T, starting in the root and using the edges whose labeling is valid for $\overline{a}$. It thus determines a unique leaf and its label will be denoted by $T(\overline{a})$. In this sense T defines a Boolean function on $\{0,1\}^n$.

Here is a useful lemma. The **depth of a tree**, often called its **height**, is the maximum number of edges on a path through the tree.

Lemma 1.4.5 *Assume that a Boolean function $f : \{0,1\}^n \to \{0,1\}$ is defined by a kDNF-formula and also by an ℓCNF-formula. Then it can be computed by a decision tree of depth at most $k\ell$.*

However, any function computed by a decision tree of depth d can be defined by both dDNF- and dCNF-formulas.

Proof The hypothesis of the first part means that f can be defined by a disjunction of some terms α_i of size $\leq k$ and that $1 - f$ can be defined by a disjunction of some terms β_j of size $\leq \ell$. In particular, no two terms α_i, β_j can be simultaneously satisfied. Hence, letting a tree ask on its first $\leq k$ levels about all the variables in α_1, in each of β_j at least one variable is answered on these levels. Doing this ℓ times with $\alpha_2, \ldots$ either exhausts all the α_i or all the variables in all the terms β_j. In either case each path determines the value of f.

The second part is easier. Each path corresponds to a term of size $\leq d$: include x_i for an edge labeled $x_i = 1$ and $\neg x_i$ for an edge labeled $x_i = 0$. The function f is defined by the disjunctions of the terms corresponding to the paths ending in a leaf labeled 1, and $1 - f$ is defined by a disjunction of the terms coming from the paths ending in 0. □

1.5 Proof Systems and Fundamental Problems

What is the difference we experience between simple proofs and complex ones? Perhaps the most notable difference is the time we need to spend on them: a simple proof is apprehended very quickly while to understand and verify a complex proof may take ages.

We may model the informal requirement that a proof of a statement should be something that needs no further creative input and should provide complete evidence for the truth of the statement by stipulating that any verifying procedure, a proof system as we shall call it later, is algorithmic. Such a verifier is any algorithm that receives a statement and a finite string (the purported proof) and either accepts or rejects the string as a proof of the statement. It is natural to require that it is sound (you cannot prove an invalid statement), and complete (you can prove all valid statements) although weakenings of both requirements have been considered fruitfully in computational complexity. The time complexity of such a verifier is then a formal model of the informal notion of proof complexity: the minimum complexity for proofs of a particular formula is the minimum time needed to verify some proof (i.e. the simplest proof) of the formula.

I think that the following questions are then quite natural, irrespective of the connections to computational complexity discovered much later than logical proof systems:

1. *Is there a verifier with respect to which every valid statement has a simple proof, i.e. the verifier accepts quickly at least one proof of the statement?*
2. *Is there a verifier which is best among all verifiers, i.e. any statement having a simple proof for some verifier has also a simple proof for the best verifier?*
3. *What is the best way to find a proof of a statement that a given verifier is willing to accept?*

When *statements* are understood as first-order sentences and *verifiers* as logic calculi for first-order logic then the negative solution to the Entscheidungsproblem by Church [143] and Turing [494] implies strongly negative answers to all three questions. However, for propositional logic their arguments do not apply and the problems, when formalized as below, are all open.

The actual formalization of proof systems for propositional logic is done slightly differently, although equivalently. Namely, it is required that a proof system is not only an algorithm but also a feasible algorithm, meaning polynomial-time. A general verifier which is not polynomial-time is turned into a polynomial-time one by padding short but complex proofs with some dummy symbol, so that the verifier runs, on the original pair a formula and a proof, in time polynomial in the length of the padded pair. This maneuver transforms the time that the original verifier needs to accept a proof into the length of the padded proof.

We will now proceed formally. Recall from Section 1.1 that TAUT is the set of propositional tautologies in the DeMorgan language.

Definition 1.5.1 (The Cook–Reckhow definition [156]) *(function version)* A **functional propositional proof system** is any polynomial time function $P\colon \{0,1\}^* \to \{0,1\}^*$ whose range is exactly *TAUT*. Any string w such that $P(w) = \alpha$ is called a P-proof of α.

(relation version) A **relational propositional proof system** is a binary relation $Q \subseteq \{0,1\}^* \times \{0,1\}^*$ such that:

- Q is p-time decidable;
- for any w, α, if $Q(w, \alpha)$ holds then $\alpha \in$ TAUT;
- for any $\alpha \in$ TAUT, there is a $w \in \{0,1\}^*$ such that $Q(w, \alpha)$ holds.

The second condition gives the soundness of Q and the third its completeness.

We will often shorten the terminology *propositional proof system* to **proof system**. The two definitions are equivalent in the following sense. On the one hand, given a functional proof system P, the graph of P is a binary relation that is a relational proof system. On the other hand, for a relational proof system Q, define a function P by the following: if w encodes a pair (u, v) such that $Q(u, v)$ then put $P(w) := v$; otherwise, put $P(w) := \top$. Note that in these transformations the set of proofs of a given formula (except $\top$) remains the same. Hence from now on we shall talk about a proof system without specifying whether it is functional or relational (unless the distinction is needed for a particular construction).

Consider now the various elementary logical propositional calculi that one may learn (Chapters 2, 3 and 5 will give examples). They are sound and complete by their design. The provability relation is p-time decidable as they are typically defined by axiom schemas and schematic inference rules and so the verifying algorithm essentially needs only to decide repeatedly whether a string is a formula (a clause, a sequent, etc.) or whether one string is a substitution instance of another (an axiom or an inference rule). In fact, this can be taken one level up, to first-order logic. If T is a theory with a rich enough language and capable of formalizing the syntax of propositional logic (we shall see examples in Part II) we may define a proof system P_T by the relation

- *π is a T-proof of the (formalized) statement "α is a tautology".*

Simple conditions on T ensure the soundness and the completeness of P_T, and if T is axiomatized using a finite number of axiom schemes (as is e.g. the Peano arithmetic PA or the set theory ZFC) then the provability relation is p-time.

The following notation will come in handy. Let P be a proof system and α a propositional formula. Define

$$\mathbf{s}_P(\alpha) \ := \ \min\{|\pi| \mid \alpha = P(\pi)\}$$

if $\alpha \in$ TAUT and $\mathbf{s}_P(\alpha) := \infty$ otherwise. Cook and Reckhow[156] defined a proof system P to be **p-bounded** if and only if there exists $c \geq 1$ such that for all tautologies α; $\mathbf{s}_P(\alpha) \leq (|\alpha| + c)^c$. They pointed out the following direct consequence of this definition and of the coNP-completeness of TAUT. (The additive term c in the upper bound is there to cover the situation when $|\alpha| = 1$; we shall leave such trivial cases out in future.)

Theorem 1.5.2 (The Cook–Reckhow theorem [156]) *A p-bounded proof system exists if and only if* NP = coNP.

In the light of this theorem the first informal question asked above becomes the main fundamental problem of proof complexity:

Problem 1.5.3 (Main problem: NP **vs.** coNP) *Does there exists a p-bounded propositional proof system? That is, does* NP = coNP *hold?*

The notion of p-bounded proof systems and the problem of their existence cuts to the heart of the matter, but it is a bit coarse for understanding the whole realm of proof systems. We need a notion that will allow as to compare such systems.

Definition 1.5.4 (p-simulation, [156]) Let P and Q be two propositional proof systems. A p-time function f: $\{0,1\}^* \rightarrow \{0,1\}^*$ is a p-simulation of Q by P if and only if for all strings w, α,

$$Q(w, \alpha) \rightarrow P(f(w, \alpha), \alpha).$$

For the function version of proof systems this can be written more succinctly as

$$Q(w) \ = \ P(f(w)).$$

We write $P \geq_p Q$ if there exists such a p-simulation.

The proof systems P and Q are p-equivalent, written $P \equiv_p Q$, if and only if

$$P \geq_p Q \wedge Q \geq_p P.$$

It is easy to see that $\leq_p$ is a quasi-ordering that becomes a partial ordering after factoring by the equivalence relation $\equiv_p$.

Krajíček and Pudlák [317] defined a proof system P to be **p-optimal** if and only if P p-simulates all other proof systems Q, and to be **optimal** if and only if for all proof systems Q there exists $c \geq 1$ such that, for all tautologies α,

$$\mathbf{s}_Q(\alpha) \ \leq \ \mathbf{s}_P(\alpha)^c.$$

When P has at most a polynomial slow-down over Q, as in the latter definition, we often say that P **simulates** Q (leaving out the adjective *polynomially* and the prefix p-), and denote this by $P \geq Q$.

As p-time functions are p-bounded, a p-optimal proof system is necessarily also optimal. The following problem formalizes the second informal question discussed above.

Problem 1.5.5 (The optimality problem, Krajíček and Pudlák [317]) *Does there exist a p-optimal or, at least, an optimal proof system?*

This problem is related in a deep way to a surprising number of different topics, ranging from classical proof theory and Gödel's theorems to structural complexity theory or to finite model theory, to name just a few. We will discuss the problem and all these connections in Chapter 21.

The third informal question listed earlier is harder to formalize and, in my view, we do not at present know the right formalization. But for definiteness of discussion let us pose:

Problem 1.5.6 (The proof search problem – informal formulation) *Is there an optimal algorithm among all deterministic algorithms that, for every tautology τ, will find a proof of τ in a given proof system P? Is there such an algorithm for a fixed P that is optimal among all proof search algorithms for any proof system?*

A proof search algorithm does not need to find the shortest proof; in fact, finding that is often an NP-hard task. But if you have one proof, can you always find one closer in size to the shortest proof? And does the time lower bound for search hold only in the worst-case model or also for the average-case model under some natural distribution for tautologies? Is there a combinatorially transparent subset of TAUT such that searching for proofs is hard already for formulas from this set? Is searching for proofs in a stronger proof system easier or harder than in a weaker one or, on the other hand, is there a fixed proof system P such that searching for P-proofs is as efficient as searching for proofs in any proof system? There is a plethora of variants of these questions and we shall discuss this topic in Section 21.5, where we propose a technical formulation of the problem. The three fundamental problems we formulated also have various relations among themselves, and we mention some of these in Part IV.

When investigating length-of-proof problems it will be important to have examples of propositional tautologies that have a clear meaning but are plausible candidates for being hard formulas, formulas requiring long or hard to find proofs. It seems, in fact, fairly difficult to come up with such examples for strong proof systems, and we shall study this in Part IV. Now we shall mention only one example, arguably the most famous formula in proof complexity.

The **pigeonhole (principle) formula**, denoted by PHP_n, is formed from $(n+1)n$ atoms p_{ij}:

$$p_{ij}, \quad i \in [n+1] \text{ and } j \in [n].$$

We think of the atom p_{ij} as expressing that pigeon i is sitting in hole j (or simply that i is mapped to j) and in this interpretation the formula says that it cannot happen that every pigeon sits in exactly one hole and no two pigeons share a hole. Equivalently, there is no injective map from $[n+1]$ into $[n]$. The PHP_n-formula is thus

$$\neg [\bigwedge_i \bigvee_j p_{ij} \wedge \bigwedge_i \bigwedge_{j \neq j'} (\neg p_{ij} \vee \neg p_{ij'}) \wedge \bigwedge_{i \neq i'} \bigwedge_j (\neg p_{ij} \vee \neg p_{i'j})] . \quad (1.5.1)$$

It is known that this formula is not hard for strong proof systems but it will serve as a hard example for many weaker systems. There are many variants of (1.5.1). For example, the second conjunct $\bigwedge_i \bigwedge_{j \neq j'} (\neg p_{ij} \vee \neg p_{ij'})$ inside the brackets is not needed to make PHP_n a tautology (it restricts the principle to graphs of functions rather than allowing multi-functions); on the other hand, one can add a conjunct $\bigwedge_j \bigvee_i p_{ij}$, thus formalizing the principle only for surjective maps. Useful modifications are the so-called **weak** PHP principles: the number m of pigeons is much larger than just $n+1$; for example, we could have $m = 2n, n^2$ or even 2^n. Some of the candidate hard formulas for stronger proof systems formalize principles related to PHP, albeit quite differently.

One should be optimistic and aim at a full solution of one or all of these problems. The results obtained in proof complexity so far are either statements about properties of the class of all proof systems (e.g. results about the optimality problem) or about a specific proof system (e.g. various length-of-proof lower bounds). While the former are, by their design, of some relevance to the three problems, does a lower bound for a specific proof system P also say something relevant? After all, unless there is an optimal proof system you cannot hope to prove that $\mathrm{NP} \neq \mathrm{coNP}$ by gradually proving super-polynomial lower bounds for stronger and stronger proof systems as that would be an infinite process. It is one of the strong messages of proof complexity that lower bounds for specific proof systems P do indeed have significant consequences and that this is true even for very weak systems P. Let us list some now, although rather informally at this point.

An algorithm M deciding whether a formula is satisfiable (a SAT algorithm) can be interpreted as a special kind of proof system: the computation of M on α ending with a rejection can be taken for a proof that $\neg\alpha$ is a tautology. Thus:

- *A length-of-proofs lower bound for any one proof system P implies time lower bounds for the class of all SAT algorithms that can be simulated (when considered as a proof system) by P.*

We will see that quite weak proof systems can simulate, in this sense, the most popular SAT algorithms and that we have time lower bounds for them.

A proof in a first-order theory T that every finite string has some coNP-property (a special kind of Π^0_1-statement) can be transformed into a sequence of short P_T-proofs of tautologies expressing the universal statement for strings of individual lengths $n = 1, 2, \ldots$ This is analogous, to an extent, to how a sequence of circuits results from one algorithm in Theorem 1.4.2. The proof system P_T depends on T but not on the statement. Thus:

- *Length-of-proof lower bounds for P_T-proofs of specific tautologies imply the independence of specific true Π^0_1-statements from the theory T.*

Mathematical logic offers several methods for proving the independence of true statements about natural numbers (i.e. binary strings) in a theory, but none of them applies to Π^0_1-statements expressing natural combinatorial or number-theoretic facts, in particular, to statements fundamentally different from consistency statements. The restriction to Π^0_1 does not limit the expressive power of such statements to some elementary facts. For example, the Riemann hypothesis can be expressed as a Π^0_1-statement. The above relation to the lengths of propositional proofs is the only method that allows such independence proofs, at least for some theories. The length-of-proof lower bounds that we have at present give such independence results unconditionally for weak subtheories of Peano arithmetic (PA) (the so-called bounded arithmetic theories) but not for PA (or even the set theory ZFC) itself. This looks disappointing until we notice that:

- *any strong (more than polynomial or quasi-polynomial, depending on T) lower bounds for P_T-proofs of any formulas imply that* $P \neq NP$ *is consistent with T*; and
- *a large part of contemporary complexity theory can be formalized in a weak bounded arithmetic theory.*

These and other related issues will be studied in Part IV.

1.6 Bibliographical and Other Remarks

Secs. 1.1 and 1.2 Propositional and first-order logic as we know them originated with Frege [188], although some notation he used did not survive (in particular, his two-dimensional way of writing formulas) and we are indebted to later authors, Hilbert and Bernays [229, 230] especially, for cleaning up and streamlining the formalism. Textbooks covering basics include Shoenfield [470] and Enderton [174].

Craig [163] proved the interpolation theorem and Lyndon [345] its monotone version. The statements are trivial for propositional logic but they proved them for first-order logic. Spira's lemma 1.1.4 for $k = 2$ is from [481]; the result for $k > 2$ is folklore.

Sec. 1.3 Turing machines were introduced in Turing [494]. Sipser [469] and Papadimitriou [377] are excellent introductions to basic computational complexity theory; although neither text is recent they are in my view still the best textbooks on the subject. The definition of the class P was the result of research by a number of people, notably J. Edmonds [173]. The class NP was defined by Cook [147] who also proved the NP-completeness of SAT. Cook used the unnecessarily general notion of Turing reductions (which does not distinguish between NP and coNP, in particular) and Karp [266] pointed out that NP-completeness holds under more strict many–one reductions and, in addition, he produced a number of examples of combinatorial NP-complete problems. Versions of these ideas appeared in Russia (then behind the iron curtain) and Levin [340], in particular, formulated analogous notions of search problems. Tseitin recognized at that time the universality of the propositional Entscheidungsproblem (i.e. TAUT) but he did not have formal concepts of reducibility and completeness. Accounts of this history are to be found in Sipser[468] or in Trakhtenbrot [491].

The time hierarchy theorem 1.3.1 originated with Hartmanis and Stearns [221] and Hartmanis, Lewis and Stearns [220] and the eventual result was proved by Žák [511].

Sec. 1.4 Boolean circuits and other material from the section are excellently and succinctly presented in Boppana and Sipser [101]. There has been much slower progress on circuit lower bounds since the heroic 1980s, which yielded the results reported in Theorem 1.4.3; [101] is not too outdated. In particular, bounds from Theorem 1.4.3 were improved to exponential (Yao [510], Hastad [222], Andreev [20] and Alon and Boppana [18]). Shannon's estimate 1.4.1 was complemented by an $(1 + o(1))(2^n/n)$ upper bound by Lupanov [344]. Additional evidence that NP $\not\subseteq$ P/*poly* was given by Karp and Lipton [267], who proved that NP $\subseteq$ P/*poly* would imply the collapse of the polynomial-time hierarchy. This is considered unlikely by most experts, but one should keep an open mind. It may be that the uniformity of P and the non-uniformity of P/*poly* make the crucial difference. Wegener [504] and Jukna [263] are monographs on Boolean complexity. A more detailed history of the relevant ideas can be found there.

Sec. 1.5 The definition 1.5.1 of proof systems is from Cook and Reckhow [156] as is the definition 1.5.4 of p-simulations. The definition of a mere simulation in the sense of having at most a polynomial slow-down $\mathbf{s}_Q(\alpha) \leq \mathbf{s}_P(\alpha)^c$ is due to Krajíček and Pudlák [317], who also first formulated and studied the optimality problem. The result in [317] points to other possible notions of a simulation; one could utilize the links between the provability of the soundness of one proof system in another system and simulations from [317] (we shall study this in Part II). These include, for example, *effectively p-simulations*, defined by Pitassi and Santhanam [402] or p-reductions between disjoint NP-pairs representing the soundness of the respective proof systems (we shall study this in Chapter 21).

The PHP_n-tautology was defined by Cook and Reckhow [156], and its weak variants, with the number of pigeons much larger than the number of holes, by Woods [507] and Paris, Wilkie and Woods [392].

An example in which the Riemann hypothesis is expressed as a Π^0_1-statement is due to Davis, Matiyasevich and Robinson [167] and was simplified by Lagarias [333].

2
Frege Systems

In this chapter we introduce several logical calculi for propositional logic. The word *calculus* is used in this connection because all these proof systems are defined by selecting a finite number of inference rules of a particular form and a proof is created by applying them in a mechanical – though not deterministic – way. This is similar to algebraic calculations or, in perhaps a better analogy, to forming sentences and longer texts in a language.

A proof in these calculi consists of *proof steps* (also called *proof lines*), which have a form depending on the specific calculus. The steps may be formulas, sequences of formulas (called **cedents** or **sequents** in the sequent calculus) in Chapter 3 or formulas of a specific form (e.g. clauses, in the resolution proof system in Chapter 5). Proofs in many proof systems studied in later chapters also proceed step by step.

For all these proof systems it is natural to consider the **number of steps** in a proof as another measure of its complexity besides the size. In fact, we shall see later that it is equally as important as the size measure. Our generic notation for the number of steps in a proof π will be $\mathbf{k}(\pi)$ and the **minimum number of steps** in the P-proof of a formula α will be denoted by $\mathbf{k}_P(\alpha)$. We shall not attempt to give a general definition of a step, however, owing to the varying nature of these calculi.

2.1 Frege Rules and Proofs

Our foremost example of logical calculi is a class of proof systems operating with formulas in some complete propositional language via a finite set of inference rules. In the established terminology of Cook and Reckhow [156] they are rightly called *Frege systems* but in the mathematical logic literature they are often called Hilbert-style calculi, referring to Hilbert's work with Ackermann and Bernays [228, 229, 230] in founding proof theory.

Let L be a finite and complete language for propositional logic. That is, L consists of a finite number of connectives (including possibly the constants $\top$ and $\bot$) interpreted by specific Boolean functions of the appropriate arity and having the property that any Boolean function of any arity can be defined by an L-formula (this property is the **completeness** of L). The main example is the DeMorgan language introduced in Section 1.1 and its extensions discussed after Theorem 1.1.3.

Definition 2.1.1 Let L be a finite complete language for propositional logic and let $\ell \geq 0$. An **ℓ-ary Frege rule** (tacitly in the language L) is any $(\ell + 1)$-tuple of L-formulas $A_0, \ldots, A_\ell$, written as

$$\frac{A_1, \ \ldots, \ A_\ell}{A_0},$$

such that

$$A_1, \ \ldots, \ A_\ell \ \models \ A_0.$$

The formulas $A_1, \ldots, A_\ell$ are called the **hypotheses** of the rule and A_0 is its **consequence**. A rule with no hypotheses (i.e. $\ell = 0$) is called a **Frege axiom scheme**. An axiom scheme is written simply as A_0.

A well-known example of a Frege rule in a language containing the implication connective is the so-called **modus ponens**:

$$\frac{p, \ p \to q}{q}.$$

An example of an axiom scheme is **tertium non datur**:

$$p \ \vee \ \neg p.$$

Definition 2.1.2 Let F be a finite set of Frege rules in a finite complete language L. An F-proof of an L-formula C from the L-formulas $B_1, \ldots, B_t$ is any sequence of L-formulas $D_1, \ldots, D_k$ such that:

- $D_k = C$;
- for all $i = 1, \ldots, k$,
  - either there is an ℓ-ary rule

$$\frac{A_1, \ \ldots, \ A_\ell}{A_0}$$

    in F, numbers $j_1, \ldots, j_\ell < i$ and a substitution σ such that

$$\sigma(A_1) = D_{j_1} \quad , \ldots, \quad \sigma(A_\ell) = D_{j_\ell} \ \text{ and } \ \sigma(A_0) = D_i,$$

  - or $D_i \in \{B_1, \ldots, B_t\}$.

We shall denote by

$$\pi : B_1, \ldots, B_t \ \vdash_F \ C$$

the fact that $\pi = (D_1, \ldots, D_k)$ is an F-proof of C from $B_1, \ldots, B_t$ (and will drop the subscript F from the $\vdash$ sign if it is clear from the context), and by the notation

$$B_1, \ldots, B_t \ \vdash_F \ C$$

the fact that there exists an F-proof of C from $B_1, \ldots, B_t$.

Definition 2.1.3 Let L be a finite complete language for propositional logic. A **Frege proof system** F in the language L is a finite set of Frege rules that is sound and **implicationally complete**, meaning that, for any L-formulas $B_1, \ldots, B_t, C$,

$$B_1, \ldots, B_t \models C \quad \text{if and only if} \quad B_1, \ldots, B_t \vdash_F C.$$

In future we shall assume tacitly that the language of any Frege system is finite and complete.

A prominent (and the first) example of a Frege system was given by Frege [188]. The system, as subsequently simplified by Lukasiewicz [343], uses the language $\{\neg, \rightarrow\}$, has modus ponens as the only rule of arity larger than 0 and three axiom schemes:

- $p \rightarrow (q \rightarrow p)$;
- $[p \rightarrow (q \rightarrow r)] \rightarrow [(p \rightarrow q) \rightarrow (p \rightarrow r)]$;
- $(\neg p \rightarrow \neg q) \rightarrow [(\neg p \rightarrow q) \rightarrow p]$.

We will often use the following two lemmas. The first is a simple technical statement.

Lemma 2.1.4 *Let F be a Frege system in any language L and let σ be a substitution of L-formulas for atoms. Let $\pi = (D_1, \ldots, D_k)$ and assume that*

$$\pi : B_1, \ldots, B_t \vdash_F C.$$

Then for $\sigma(\pi) := (\sigma(D_1), \ldots, \sigma(D_k))$, it holds that

$$\sigma(\pi) : \sigma(B_1), \ldots, \sigma(B_t) \vdash_F \sigma(C).$$

Define the **width** of a Frege proof π, denoted by $\mathbf{w}(\pi)$, to be the maximum size of a formula in π.

Lemma 2.1.5 (The deduction lemma) *Let F be a Frege system. Assume that*

$$\pi : A, B_1, \ldots, B_t \vdash C$$

and let $k = \mathbf{k}(\pi)$ and $w = \mathbf{w}(\pi)$.

Then there is an F-proof

$$\pi' : B_1, \ldots, B_t \vdash A \rightarrow C$$

with $O(k)$ steps and of width $O(w)$ and size $|\pi'| \leq O(|\pi|^2)$. Here the implication $\rightarrow$ is either in the language of F or it denotes a definition in that language.

(We shall introduce in the next section the notion of *tree-like* proofs; the reader will have no problem in verifying that the lemma holds for tree-like proofs as well.)

Proof We first establish a useful claim.

Claim *There is a constant $c \geq 1$ depending just on F such that whenever*

$$\frac{E_1, \dots, E_\ell}{E_0}$$

is an F-rule and q is an atom not occurring in any of the E_j, there is an F-proof

$$\eta \colon q \to E_1, \dots, q \to E_\ell \vdash q \to E_0$$

with number of steps and width at most c.

The claim follows by the completeness of F, as

$$q \to E_1, \dots, q \to E_\ell \models q \to E_0.$$

Let $\pi = D_1, \dots, D_k$ and assume that A is indeed used in π. Prove by induction on $i \leq k$ that there is an F-proof

$$\rho_i \colon B_1, \dots, B_t \vdash A \to D_i$$

with $O(i)$ steps and width $O(w)$ (the size estimate $O(iw)$ then follows). We need to distinguish three cases concerning how D_i was inferred in π:

1. $D_i = A$;
2. D_i is one of the B_j;
3. D_i was inferred via an F-rule.

In the first case $A \to D_i$ becomes $A \to A$ and has an F-proof with a constant number of steps and width $O(|A|) = O(w)$ by Lemma 2.1.4. The second case is similar: $A \to D_i$ becomes $A \to B_j$, which can be derived from B_j and $B_j \to (A \to B_j)$ by a proof that is a substitution instance of a fixed derivation of $p \to q$ from q and $q \to (p \to q)$. However, the first formula, B_j, is a hypothesis that we can use and the second formula, $B_j \to (A \to B_j)$, is an instance of the tautology $q \to (p \to q)$ and hence has a proof that is an instance of a fixed proof.

In the third case, use the claim. □

Two main facts about Frege systems that we shall establish in this chapter concern the robustness of the definition of the complexity of Frege proofs: the proof format can be restricted to tree-like proofs (Section 2.2) without increasing the size or the number of steps too much; and all Frege systems are (essentially) p-equivalent (Section 2.3).

2.2 Tree-Like Proofs

A Frege proof is called **tree-like** if every step of the proof is part of the hypotheses of at most one inference in the proof. For F a Frege system, F^* will denote the proof system whose proofs are exactly tree-like F-proofs. If we want to stress that we are talking about general proofs, not necessarily tree-like proofs, we may describe them as **sequence-like** or equivalently **dag-like**. "Dag" is an abbreviation for a *directed acyclic graph* and refers to the underlying graph of a proof, where the nodes are proof steps and arrows lead from the hypotheses of an inference to its conclusion. Note that this **proof graph** is a tree if the proof is tree-like (and it is called a **proof tree** in that case).

A natural measure of the complexity of tree-like proofs is their height but the definition makes sense even for dag-like proofs: the **height** of an F-proof π, denoted by $\mathbf{h}(\pi)$, is the maximum number of edges on a directed path in the proof graph. In the case of tree-like proofs this is simply the height of the proof tree. Given a formula A, $\mathbf{h}_{F^*}(A)$ is the minimum height of some tree-like F-proof of A.

We shall consider first the case of the DeMorgan language.

Theorem 2.2.1 (Krajíček [276]) *Let F be any Frege system in the DeMorgan language. Assume that $\pi = C_1, \ldots, C_k$ is an F-proof of A from $B_1, \ldots, B_t$ having k steps and size s.*

Then there exists an F^*-proof π^* of A from $B_1, \ldots, B_t$ such that:

(i) the number of steps in π^* is $\mathbf{k}(\pi^*) = O(k \log k)$;
(ii) the height of π^* is $\mathbf{h}(\pi^*) = O(\log k)$;
(iii) the size of π^* is $|\pi^*| = O(sk \log k) = O(s^2 \log s)$.

The constants implicit in the O-notation depend only on F. The tree-like proof π^* can be constructed by a polynomial-time algorithm from π; in particular, $F \equiv_p F^*$.

Proof We may assume without loss of generality that $B_1, \ldots, B_t$ are the first t formulas in π. Let $c \geq 1$ be the maximum number of hypotheses in a rule of F. Define

$$D_i := C_1 \wedge \cdots \wedge C_i,$$

bracketed in a balanced binary tree fashion; in particular, the height of this tree of conjunctions is about $\log i$. (If we used left-to-right bracketing this would change the factor $k \log k$ to k^2 in items (ii) and (iii) above.)

Claim 1 For any $i \leq k$ and $j_1, \ldots, j_c < i$ the conjunct

$$(\cdots(D_i \wedge C_{j_1}) \wedge C_{j_2}) \wedge \cdots) \wedge C_{j_c})$$

has an F^*-proof ρ_i from D_i with $O(c \log i)$ steps and size $O(c|D_i| \log i)$.

This amounts to rearranging D_i and extracting copies of $C_{j_1}, C_{j_2}, \ldots, C_{j_c}$. The claim implies

Claim 2 For $i < k$, D_{i+1} has an F^*-proof ρ_i from D_i with $O(c \log i)$ steps and size $O(c|D_{i+1}| \log i)$.

Now construct π^* as follows.

1. Derive D_t from $B_1, \ldots, B_t$ by a tree-like proof of height $O(\log t)$ and size $O(t|D_t|) \leq O(ts)$.

2. Derive $D_t \longrightarrow D_k$ (with $\longrightarrow$ defined as in the DeMorgan language) by inductions arranged in a binary tree. That is, $D_t \longrightarrow D_k$ is derived from two implications $D_t \longrightarrow D_u$ and $D_u \longrightarrow D_k$, where $u := \lfloor (t+k)/2 \rfloor$, etc. The subproofs at the top of this proof tree are provided by Claim 2.
3. From D_k derive A.

It is easy to check that π^* has the required parameters. □

If the language of F is not DeMorgan then we can use the formulas β_k from the following lemma to represent the nested conjunctions.

Lemma 2.2.2 *Let L be any finite complete propositional language and $\alpha(p,q)$ an L-formula such that α defines the conjunction $\mathbf{tt}_\alpha = \mathbf{tt}_\wedge$. Let L-formulas $\beta_k(p_1,\ldots,p_k)$ be defined for $k \geq 1$ by the conditions*

- $\beta_1(p_1) := p_1$ *and* $\beta_2(p_1,p_2) := \alpha(p_1,p_2)$,
- $\beta_{2k}(p_1,\ldots,p_{2k}) := \alpha(\beta_k(p_1,\ldots,p_k), \beta_k(p_{k+1},\ldots,p_{2k}))$ *for* $k \geq 1$,
- $\beta_{2k+1}(p_1,\ldots,p_{2k+1}) := \alpha(\beta_k(p_1,\ldots,p_k), \beta_{k+1}(p_{k+1},\ldots,p_{2k+1}))$ *for* $k \geq 1$.

Then the following holds for any Frege system F in the language L. For all $k \geq i \geq 1$ there is an F^*-proof

$$\pi_{k,i}\colon \beta_k(p_1,\ldots,p_k) \vdash_F \beta_{k+1}(p_1,\ldots,p_k,p_i)$$

such that

- $\pi_{k,i}$ *has* $O(\log k)$ *steps,*
- $|\pi_{k,i}| \leq O(|\beta_k| \log k)$.

The constants implicit in the O-notation depend just on α and F.

We now give a game-theoretical interpretation of tree-like proofs. The game, often called the **Buss–Pudlák game**, is played by two players, *Prover* and *Liar*. The game is determined by formulas $B_1,\ldots,B_t,C$ and we shall denote it by $G(B_1,\ldots,B_t;C)$. Prover wants to show that $B_1,\ldots,B_t \models C$ while Liar maintains that he has an assignment satisfying all B_i but not C. The play proceeds in rounds. At each round Prover asks about the truth value of some formula A under the assignment Liar claims to have, and Liar answers 1 or 0. The rules are as follows.

1. The formulas $B_1,\ldots,B_t$ must take the value 1.
2. The formula C must take the value 0.
3. If formulas A, D, E are given the truth values a, d, e by Liar then
   - if $A = \neg D$ then $a = \neg d$,
   - if $A = D \circ E$ then $a = d \circ e$, for $\circ = \wedge, \vee$.

The game stops when Liar cannot answer without breaking a rule. If indeed $B_1,\ldots,B_t \models C$, Prover has a winning strategy. She can ask for the values of all atoms in the formulas and then subsequently for the values of larger and larger subformulas. By the third rule Liar must answer correctly and hence eventually he has to violate the first or the second rule. Note that this takes at most $|C| + \sum_i |B_i|$ rounds.

For a Prover strategy S, let $r(S)$ be the maximum number of rounds she needs to beat any Liar using S, and let $w(S)$ be the maximum size of a formula she needs to ask for in some play using S.

Lemma 2.2.3 *Assume that π is an F^*-proof of C from $B_1, \dots, B_t$ of height $\mathbf{h}(\pi) = h$. Then there is a winning strategy S for Prover such that*

$$r(S) = O(h) \quad \text{and} \quad w(S) = O(\mathbf{w}(\pi)).$$

Moreover, S asks only for the truth values of some formulas that appear as subformulas in π.

However, for any winning Prover strategy S there is an F^*-proof π of C from $B_1, \dots, B_t$ such that

$$\mathbf{h}(\pi) = O(r(S)) \quad \text{and} \quad \mathbf{w}(\pi) = O(w(S)).$$

Moreover, all formulas occurring in π are constructed from the formulas queried by S in some play using at most c connectives, where c is a constant depending on F only. In particular, the maximum logical depth of a formula in π is the maximum logical depth of any query that S can make plus a constant.

Proof Let π be an F^*-proof of C from $B_1, \dots, B_t$ of height h and width w. Use π to define the following strategy S for Prover. Prover asks for the value of C (the last formula in π) and Liar must say 0. Assume C was derived in π by an instance

$$\frac{D_1, \dots, D_\ell}{C}$$

of some F-rule

$$\frac{A_1, \dots, A_\ell}{A_0}$$

given by a substitution $p_i := \sigma(p_i)$ for the atoms $p_1, \dots, p_n$ in the rule. Hence $D_j = \sigma(A_j)$.

Prover asks for the values of all formulas $E(\sigma(p_1), \dots, \sigma(p_n))$, where $E(\overline{p})$ is a subformula of one of the A_j. Note that the number of these formulas is bounded by a constant c depending just on the rule, not on the substitution σ.

Because $A_1, \dots, A_\ell \models A_0$ and $\sigma(A_0) = C$ receive the value 0, Liar either violates the third rule or gives the value 0 to at least one of the hypotheses D_j, $1 \leq j \leq \ell$, of the inference. Say D_j received the value 0. Prover then uses the inference in π yielding this formula D_j and analogously forces Liar to give the value 0 to one of its hypotheses. In this fashion Prover moves up in the proof tree and so in $O(ch) = O(h)$ rounds arrives either at one of the B_i, forcing Liar to violate the first rule, or at an axiom of F, forcing him to violate the third rule.

Now assume that S is a winning strategy for Prover, allowing her always to win in at most $r = r(S)$ rounds and requiring formulas of size at most $w = w(S)$. Picture S

as a labeled binary tree: the nodes are labeled by Prover's questions (i.e. by formulas) and the two edges outgoing from a node by the two possible truth values Liar may answer with. Hence the root is labeled by Prover's first question.

Paths in the tree from the root to the leaves correspond to finished plays against some Liar. As S is a winning strategy, the path has to determine the truth values to formulas on the path in a way that violates one of the rules. Let

$$E_0, e_0, \ldots, E_v, e_v \tag{2.2.1}$$

be a possible partial path in the tree, with the formulas E_j receiving the values e_j. Note that $v \leq r$. We shall extend the notation from (1.1.1) to all formulas E: $E^1 := E$ and $E^0 := \neg E$.

Claim *For every partial path of the form* (2.2.1) *there is an* F^**-proof*

$$\rho\colon B_1, \ldots, B_t, E_0^{e_0}, \ldots, E_v^{e_v} \vdash C$$

such that the height of ρ *is* $O(r - v)$ *and its width is* $O(w)$.

We prove the claim by induction on $v = r, \ldots, 0$. Assume that the path is maximal (this must be the case if $v = r$, in particular). Liar's answers,

$$E_0^{e_0}, \ldots, E_v^{e_v},$$

must contain a violation of one of the rules. If it is the second rule then, for some j, $E_j^{e_j} = C^1 (= C)$ and ρ exists with a constant height and width $O(|C|) \leq O(w)$. An analogous situation holds if the first rule was violated. If the third rule is violated, say $E_u = E_a \wedge E_b$ but $e_u \neq e_a \wedge e_b$, then

$$E_u^{e_u}, E_a^{e_a}, E_b^{e_b} \tag{2.2.2}$$

is a substitution instance of one of

$$\neg(p \wedge q), p, q, \quad p \wedge q, \neg p, q, \quad p \wedge q, p, \neg q, \quad p \wedge q, \neg p, \neg q,$$

each of which, and hence also (2.2.2), can be brought to a contradiction by a fixed F^*-proof in a constant number of steps and width $O(|E_u| + |E_a| + |E_b|) \leq O(w)$, by Lemma 2.1.4.

For the induction step assume that there are proofs ρ_a for $a = 0, 1$,

$$\rho_a\colon B_1, \ldots, B_t, E_0^{e_0}, \ldots, E_{v-1}^{e_{v-1}}, E_v^a \vdash C,$$

with the required properties, which yield, by the deduction lemma 2.1.5, proofs

$$\rho_a'\colon B_1, \ldots, B_t, E_0^{e_0}, \ldots, E_{v-1}^{e_{v-1}} \vdash E_v^a \to C$$

with the required parameters. However, there is a fixed tree-like derivation of q from $p \to q$ and $\neg p \to q$ by Lemma 2.1.4 and hence ρ_0' and ρ_1' can be combined into the required proof of C from $B_1, \ldots, B_t, E_0^{e_0}, \ldots, E_{v-1}^{e_{v-1}}$.

The same argument with the role of E_v^0 and E_v^1 played by E_0^0 and E_0^1 derives the lemma from the claim. □

2.3 Reckhow's Theorem

Reckhow's theorem asserts that all Frege systems, irrespective of their language, p-simulate each other. There is, however, something a bit ad hoc about it. If the language L of F does not include the DeMorgan language we must represent the DeMorgan formulas by L-formulas. We shall see in Chapter 5 how this can be done in the resolution proof system via the so-called limited extension, but that presupposes having $\neg$ and $\vee$ in the language. Moreover, the limited extension does not produce an equivalent formula but only equivalence with respect to *satisfiability*.

The transformation of the DeMorgan formulas into L-formulas ought to be polynomial-time in order to stay within the Cook–Reckhow framework but otherwise there do not seem to be natural requirements that would single out a canonical transformation. This adds a certain arbitrariness to the formulation of the theorem.

In the simulation of F_2 by F_1 we take an F_2-proof and first rewrite it so that all formulas in it are balanced and thus have small logical depth. Then the balanced formulas are translated into the language of F_1 and the small logical depth will yield only a polynomial-size increase, as in Lemma 1.1.5, using Spira's lemma 1.1.4. Finally, the resulting sequence of formulas is filled in with more formulas to make it a valid F_1-proof. This is the technical heart of the matter.

To avoid the arbitrariness described above we shall prove the Reckhow theorem for the case when the languages of both F_1 and F_2 contain the DeMorgan language. We start with the simple case of when the two Frege systems have the same language.

Lemma 2.3.1 *Let F_1 and F_2 be two Frege systems with a common language L. Then $F_1 \equiv_p F_2$.*

In fact, for all L-formulas α,

$$\mathbf{s}_{F_1}(\alpha) = O(\mathbf{s}_{F_2}(\alpha)), \quad k_{F_1}(\alpha) = O(k_{F_2}(\alpha)) \quad \text{and} \quad \ell\text{dp}_{F_1}(\alpha) \leq \ell\text{dp}_{F_2}(\alpha) + O(1)\,.$$

Proof Let

$$\frac{A_1, \ldots, A_\ell}{A_0}$$

be an F_1-rule. In particular, we have

$$A_1, \ldots, A_\ell \models A_0$$

and, because F_2 is, by definition, implicationally complete there is an F_2-proof π of A_0 from $A_1, \ldots, A_\ell$. Hence, whenever the instance of this rule given by a substitution σ is used in an F_1-proof, we can simulate it in F_2, using Lemma 2.1.4, by $\sigma(\pi)$. The estimates follow as such an F_2-proof π is fixed for each F_1-rule. □

The general case will follow from the next two lemmas.

Lemma 2.3.2 *For any two Frege systems F_1 and F_2 in languages containing the DeMorgan language there exists a function f such that, for all DeMorgan formulas A, if ρ is an F_2-proof of A with $k = \mathbf{k}(\rho)$ steps and of logical $\ell = \ell\text{dp}(\rho)$ depth then*

$$f(\rho): \ \vdash_{F_1} \ A$$

and $f(\rho)$ is constructed in time $k2^{O(\ell)}$. In particular, $|f(\rho)| \le k2^{O(\ell)}$.

Proof Let t be a translation of the language L_2 of F_2 into the language L_1 of F_1. That is, t assigns to any L_2-connective $\circ(p_1,\ldots,p_r)$ an L_1-formula $t(\circ)(p_1,\ldots p_r)$ defining the same Boolean function as $\circ$ (cf. Section 1.1). We assume that t is the identity on the DeMorgan connectives.

Denote by t also the resulting translation of L_2-formulas into L_1-formulas. We have:

Claim 1 For any L_2-formula B, $t(B)$ is constructed in time $2^{O(\ell\mathrm{dp}(B))}$. In particular, $|t(B)| = 2^{O(\ell\mathrm{dp}(B))}$.

Let $\rho = D_1,\ldots,D_k$. We shall construct $f(\rho)$ by taking the sequence

$$t(D_1),\ldots,t(D_k) \tag{2.3.1}$$

(in time $k2^{O(\ell\mathrm{dp}(\rho))}$) and filling in some L_1-formulas to make (2.3.1) a valid F_1-proof. This is done for each of the k inferences separately, in each case adding a constant number of steps of size $2^{O(\ell\mathrm{dp}(\rho))}$.

Obviously, if

$$\frac{E_1(\overline{p}),\ldots,E_v(\overline{p})}{E_0(\overline{p})}, \quad \overline{p} = p_1,\ldots,p_n,$$

is an F_2-rule then there is an F_1-proof η of $t(E_0)$ from $t(E_1),\ldots,t(E_v)$. Hence each inference in ρ that is an instance of such a rule determined by a substitution $p_i := \sigma(p_i)$, $i \le n$, can be simulated using η and the same substitution. The correctness of this follows from

Claim 2 For any L_2-formula $E(p_1,\ldots,p_n)$ and any substitution $p_i := \sigma(p_i)$ of L_2-formulas for atoms it holds that

$$t(\sigma(E)) = t(E)(t(\sigma(p_1)),\ldots,t(\sigma(p_n))) .$$

This claim is verified by induction on the size of E. □

The heart of the matter is the next lemma.

Lemma 2.3.3 *Let F_2 be a Frege system in a language containing the DeMorgan language. Then there exists a p-time function g such that, for all DeMorgan formulas A, whenever*

$$\pi: \ \vdash_{F_2} \ A$$

then

$$g(\pi): \ \vdash_{F_2} \ A,$$

$$|g(\pi)| = |\pi|^{O(1)} \text{ and } \ell\mathrm{dp}(g(\pi)) = O(\log|\pi|) + \ell\mathrm{dp}(A).$$

Proof Let $a \geq 2$ be the maximum arity of a connective in L_2. For an L_2-formula $C(\overline{p})$ denote by $b[C]$ the equivalent balanced formula produced by the construction in Lemma 1.1.5 (based on Spira's lemma 1.1.4). In particular, the construction finds in a canonical way (by performing a walk in the tree of the formula) a formula $\beta(\overline{p}, q)$ with exactly one occurrence of q and a formula $\gamma(\overline{p})$ such that C is equal syntactically, i.e. as strings, to $\beta(q/\gamma)$, the size of the formulas satisfies $|C|/(a+1) \leq |\beta|, |\gamma| \leq \frac{a}{a+1}|C|$ and it holds that

$$b[C] \; = \; ((b[\beta](\overline{p}, 1) \wedge b[\gamma]) \vee (b[\beta](\overline{p}, 0) \wedge \neg b[\gamma])) \tag{2.3.2}$$

(we use $=$ instead of the logical $\equiv$ to stress the syntactic equality) and $\ell\mathrm{dp}(b[C]) = O(\log |C|)$.

Let $\pi = D_1, \ldots, D_k$ be an F_2-proof of A. We would like to take the sequence

$$b[D_1], \ldots, b[D_k]$$

and fill in some formulas to get another proof with a small logical depth. Indeed, this is how we proceed but the "filling-in" part is much more difficult than it was in the proof of Lemma 2.3.2 because the balancing procedure does not commute with the connectives.

Claim 1 The formula

$$b[\beta](\overline{p}, q/b[\gamma]) \equiv \; ((b[\beta](\overline{p}, 1) \wedge b[\gamma]) \vee (b[\beta](\overline{p}, 0) \wedge \neg b[\gamma]))$$

has an F_2-proof of size $O((|b[\beta]|+|b[\gamma]|)^2)$ and of logical depth at most $\ell\mathrm{dp}(b[\beta])+\ell\mathrm{dp}(b[\gamma]) + O(1)$.

The claim is established by induction on $\ell\mathrm{dp}(b[\beta])$. It is a special case of a propositional transcription of first-order equality axioms,

$$\bigwedge_i G_i \equiv H_i \; \rightarrow \; K(G_1, \ldots) \equiv K(H_1, \ldots).$$

Here and in future claims the statements claiming the existence of F_2-proofs of size $z^{O(1)}$ and logical depth $\ell+O(1)$ may be conveniently verified by describing a winning strategy S for Prover in the associated Buss–Pudlák game: this strategy queries only formulas of size $O(z)$ and logical depth at most ℓ and takes at most $O(\log z)$ rounds. The wanted proof is then obtained by the second part of Lemma 2.2.3.

The following is the key claim.

Claim 2 Let $\circ(q_1, \ldots, q_u) \in L_2$ and let $B_1, \ldots, B_u$ be L_2-formulas. Then

$$b[\circ(B_1, \ldots, B_u)] \equiv \circ(b[B_1], \ldots, b[B_u])$$

has an F_2-proof of size $(\sum_i |B_i|)^{O(1)}$ and of logical depth at most $\max_i \ell\mathrm{dp}(b[B_i]) + O(1)$. Moreover, the proof can be constructed in polynomial time.

To prove this claim we will prove by induction a more general statement (Claim 4) but let us first see whether the claim suffices to derive the lemma. The claim readily implies

Claim 3 Let $E(p_1,\ldots,p_n)$ and let $B_1,\ldots,B_n$ be L_2-formulas. Then there is an F_2-proof of

$$b[E(B_1,\ldots,B_n)] \equiv E(b[B_1],\ldots,b[B_n])$$

of size $|b[E(B_1,\ldots,B_n)]|^{O(1)}$ and of logical depth bounded by

$$\ell\mathrm{dp}(E) + \max_i \ell\mathrm{dp}(b(B_i)) + O(1).$$

Moreover, the proof can be constructed in polynomial time.

Now consider the inference

$$\frac{E_1(B_1,\ldots,B_n),\ldots,E_v(B_1,\ldots,B_n)}{E_0(B_1,\ldots,B_n)},$$

yielding, in the original proof π, step D_i; this inference is an instance of an F_2-rule

$$\frac{E_1(\overline{p}),\ldots,E_v(\overline{p})}{E_0(\overline{p})}, \quad \overline{p}=p_1,\ldots,p_n. \tag{2.3.3}$$

We assume that we have already constructed an F_2-proof containing all $b[D_j]$ for $j < i$, with size polynomial in $\sum_{j<i}|D_j|$ and of logical depth at most $\max_{j<i}\ell\mathrm{dp}(b[D_j]) + O(1)$. Using Claim 3 we can derive all the formulas

$$E_j(b[B_1],\ldots,b[B_n]), \qquad \text{for } j = 1,\ldots v$$

and, by the rule (2.3.3), also

$$E_0(b[B_1],\ldots,b[B_n]),$$

which implies, again via Claim 3, the formula $b[D_i]$. Finally, from $b[D_k](= b[A])$ we derive (using Claim 3) the formula A itself.

To prove Claim 2 we shall prove a more general statement (this is needed for the induction argument). Claim 2 then follows by taking $G(\overline{p},\overline{q}) = \circ(\overline{q})$.

Claim 4 Let $G(\overline{p},q_1,\ldots,q_u)$ and $H_1(\overline{p}),\ldots,H_u(\overline{p})$ be L_2-formulas such that each q_i has exactly one occurrence in G. Then the equivalence

$$b[G(\overline{p},q_1/H_1,\ldots,q_u/H_u)] \;\equiv\; b[G](\overline{p},q_1/b[H_1],\ldots,q_u/b[H_u])$$

has an F_2-proof of size $O(|G(\overline{p},q_1/H_1,\ldots,q_u/H_u)|^{O(1)})$ and of logical depth at most $\ell\mathrm{dp}(b[G(\overline{p},q_1/H_1,\ldots,q_u/H_u)]) + O(1)$.

To ease the notation let us omit the reference to the atoms $\overline{p}$ and assume (without loss of generality) that $u = 2$. Denote $K := G(q_1/H_1,q_2/H_2)$. Let t be the minimum number such that $|K| \leq (\frac{a+1}{a})^t$ (which gives the nesting of iterations of Lemma 1.1.5 in the construction of $b[K]$). We prove the claim by induction on t, distinguishing two cases:

1. either the walk in the tree of K using Spira's lemma 1.1.4 in the construction of $b[K]$ in Lemma 1.1.5 stops in the G-part of K; or
2. the walk stops inside H_1 or H_2.

We shall treat the first case in detail; the second case is analogous. Case 1 splits into three subcases: when the walk stops then: neither of H_1, H_2 is below the subformula; only one of them is; or both are. Let us treat the general case when H_1 is not below the end-point of the walk but H_2 is. That means that there are formulas $\beta(q_1, r)$, for r a new atom, and $\gamma(q_2)$ such that

$$G = \beta(q_1, \gamma(q_2))$$

and

$$K = \beta(q_1/H_1, \gamma(q_2/H_2)),$$

and their sizes satisfy

$$|\beta(H_1, r)|, \ |\gamma(H_2)| \leq \left(\frac{a+1}{a}\right)^{t-1} .$$

Then $b[K]$ is given by

$$(b[\beta(H_1, 1)] \wedge b[\gamma(H_2)]) \vee (b[\beta(H_1, 0)] \wedge \neg b[\gamma(H_2)]) .$$

By the induction hypothesis, this is provably equivalent (by an F_2-proof obeying the requirements on the size and the logical depth) to

$$(b[\beta](b[H_1], 1) \wedge b[\gamma](b[H_2])) \vee (b[\beta](b[H_1], 0) \wedge \neg b[\gamma](b[H_2]))$$

and, by Claim 1, also to

$$b[\beta](b[H_1], b[\gamma](b[H_2])) .$$

We now want to derive the equivalence

$$b[G](q_1, q_2) \equiv b[\beta](q_1, b[\gamma](q_2)), \tag{2.3.4}$$

which readily yields the wanted equivalence in Claim 4.

Unfortunately we cannot assume that the size of G is at most $(a + 1/a)^{t-1}$ and thus we cannot use the induction hypothesis. But we may use it for the formulas β and γ. We apply a balancing procedure to G and assume that the walk in the tree of G stops at subformula ψ and decomposes G as $\varphi(\psi)$, the sizes of both φ and ψ being at most $(a + 1/a)^{t-1}$. We use this decomposition to prove (2.3.4).

Three cases may occur:

1. ψ is a proper subformula of γ;
2. γ is a subformula of ψ;
3. neither (i) nor (ii) occurs.

As before we shall treat in detail only case 1; the other two cases are analogous.

In this case $\gamma = \rho(\psi)$ for some ρ, and $\varphi = \beta(\rho)$. The balanced formula $b[G]$ ($= b[\varphi(\psi)]$) is by definition

$$(b[\varphi](1) \wedge b[\psi]) \ \vee \ (b[\varphi](0) \wedge \neg b[\psi]),$$

which is, by the induction hypothesis applied to φ, which equals $\beta(\rho)$, provably equivalent to

$$(b[\beta](b[\rho](1)) \wedge b[\psi]) \vee (b[\beta](b[\rho](0)) \wedge \neg b[\psi])$$

and, by Claim 1, to

$$b[\beta](b[\rho](b[\psi])) .$$

Applying the induction hypothesis to γ, we see that $b[\rho](b[\psi])$ is equivalent to $b[\gamma]$ and thus establish (2.3.4). □

Reckhow's theorem now follows from Lemmas 2.3.2 and 2.3.3, as $f \circ g$ is a p-simulation of F_2 by F_1.

Theorem 2.3.4 (Reckhow's theorem [445]) *Any two Frege systems F_1, F_2 in any complete languages containing the DeMorgan language p-simulate each other.*

2.4 Extended and Substitution Frege Systems

In this section we consider two possible ways to make Frege systems stronger. I write *possible* as it has not been proved that these modified systems are indeed strictly stronger than Frege systems, in the sense of p-simulations or simulations.

A **substitution Frege** system extends a given Frege system F by the **substitution rule**

$$\frac{A(p_1, \ldots, p_n)}{A(B_1, \ldots, B_n)},$$

where the formulas B_i are arbitrary and are substituted for the atoms p_i simultaneously. This rule does not fall under the category of Frege rules from Definition 2.1.1. The new proof system is denoted *SF*. The use of the rule is very natural from a logical point of view, so much so that it often gets overlooked (even by Frege himself in [188]).

The substitution rule is sound in the sense that it preserves the logical validity but it does not necessarily preserve the truth under a particular assignment. For example, using the rule we may infer $\neg p$ from p. Thus we consider it as a proof system in the sense of Definition 1.5.1 but not as a proof system for proving one formula from a set of formulas. We will see in Section 2.5 that SF allows us sometimes to reduce the number of steps in a proof exponentially in comparison with F. But whether it also gives a super-polynomial speed-up for the size is open.

An **extended Frege** proof system is often said to extend a given Frege system F by the extension rule but that is incorrect; there is no such inference rule in the sense of a mechanical recipe for how to derive a formula from other formulas and could be applied locally at any step of the proof. The definition of *EF*, a system extending a given F, is slightly more roundabout.

First we define EF-**derivations**. A sequence of formulas

$$C_1, \ldots, C_k$$

is an EF-derivation from the set $B_1, \ldots, B_t$ if every C_i is one of the B_j, or is obtained from some earlier formulas C_u by one of the rules of F, or – and this is what is informally called the **extension rule** – has the form

$$q \equiv D, \tag{2.4.1}$$

where $\equiv$ abbreviates its definition in the language of the particular system F, and

- the atom q does not occur in any $B_1, \ldots, B_t$,
- q does not occur in $C_1, \ldots, C_{i-1}$, and
- q does not occur in D.

The formula (2.4.1) is called an **extension axiom** and the atom q is called the **extension atom** introduced in the axiom.

This way of introducing new formulas into a derivation is sound, in the same sense as Frege rules are sound: any assignment to the atoms occurring in the formulas $C_1, \ldots, C_{i-1}$ (and to further atoms in D, if there are any) can be extended to an assignment to q making the extension axiom true.

An EF-derivation from the set $B_1, \ldots, B_t$ ending with the formula A is called an EF-**proof** of A from $B_1, \ldots, B_t$ if, in addition,

- *no extension atom introduced in the derivation occurs in A.*

Intuitively, the extension atom q may be used as an abbreviation for D later in the derivation and this may, in principle, shorten proofs significantly. We demonstrate the use of extension variables in the proof of the next lemma.

Lemma 2.4.1 *The* PHP_n *tautologies* (1.5.1) *have size* $O(n^4)$ *EF-proofs.*

Proof Let p_{ij} be the atoms of PHP_n, $(i,j) \in [n+1] \times [n]$. For pairs (i,j) from $[n] \times [n-1]$, introduce by the extension rule new extension variables

$$q_{ij} := p_{ij} \vee (p_{n+1j} \wedge p_{in}) .$$

Then it is easy to see that $\mathrm{PHP}_{n-1}(\overline{q})$ has a size $O(n^3)$ F-proof from $\mathrm{PHP}_n(\overline{p})$. Iterate this reduction to $n-2, \ldots$ down to PHP_1, which has a constant-size F-proof. □

If we formalize the same proof in Frege systems without the extension rule then the definitions of the new atoms of the PHP principle for $n-1, n-2, \ldots$ will compose and their size would grow exponentially. However, as shown by Buss but by a different argument [108], the PHP_n do have polynomial-size F-proofs (we shall show this in Sections 11.2 and 11.3).

In fact, whether F p-simulates EF is open. But we can at least show that, with respect to the number of steps, EF has no significant speed-up over F.

Lemma 2.4.2 *For any tautology A, $\mathbf{k}_{EF}(A) \leq \mathbf{k}_F(A) = O(\mathbf{k}_{EF}(A))$.*

Proof The first inequality is trivial. For the second assume that $\pi: C_1, \ldots, C_k$ is an *EF*-proof of $A(= C_k)$ with the extension axioms $E_1, \ldots, E_r$ introduced in this order and with the E_i having the form $q_i \equiv D_i$.

First substitute for all occurrences of q_r in π the formula D_r, then do the same with q_{r-1} and D_{r-1} etc. Each substitution may increase the size of the sequence of formulas by the multiplicative factor $\Omega(|D_i|)$ and hence we cannot get a polynomial upper bound on the size of the eventual sequence. But the number of steps in it remains the same and, by Lemma 2.1.4, it is an F-proof of A (as no extension atoms occur in it) except that it also contains lines of the form $U \equiv U$. But each such formula has an F-proof with a constant number of steps (e.g. by Lemma 2.1.4). □

In the next lemma we use the substitution rule to make the previous construction feasible with respect to the size increase also. The particular order in which we eliminated the extension atoms in the proof of Lemma 2.4.2 did not matter, but here it will.

Lemma 2.4.3 $SF \geq_p EF$.

Proof Let π be an *EF*-proof of A in which extension atoms $q_1, \ldots, q_r$ were introduced in extension axioms E_i having the form $q_i \equiv G_i$ and in this order (i.e. q_j does not appear in the extension axioms $E_1, \ldots, E_{j-1}$).

Applying the deduction lemma 2.1.5 r times will transform π into an F-proof of the implication

$$E_r \to (E_{r-1} \to (\cdots \to (E_1 \to A) \cdots)$$

but the size will grow exponentially with r. We need to proceed more economically and use a generalized form of the deduction lemma stated in Claim 1.

Let B_j, $1 \leq j \leq t$, be formulas and σ a permutation of $[t]$. The notation $\bigwedge_\sigma B_j$ means the conjunction of formulas $B_{\sigma(1)}, \ldots, B_{\sigma(t)}$ bracketed for the definiteness to the right: $(B_{\sigma(1)} \wedge (B_{\sigma(2)} \wedge (\cdots \wedge B_{\sigma(t)}) \cdots)$.

Claim 1 Assume that ρ is an F-proof of C from $B_1, \ldots, B_t$ of size s. Let *id* be the identity permutation on $[t]$. Then

$$\bigwedge_{id} B_j \to C$$

has an F-proof of size $O(s^2)$.

Let $\rho = D_1, \ldots, D_k$. The proof of the claim shadows the proof of Lemma 2.1.5 but instead of deriving formulas $A \to D_i$ as in that proof we derive $\bigwedge_{id} B_j \to D_i$.

For simulating the inferences we need to reorder the big conjunctions. For example, if $D_i = B_u$ we take any permutation σ such that $\sigma(1) = u$ and derive $\bigwedge_\sigma B_j \to B_u$: this formula is an instance of

$$(q \wedge r) \to q$$

which has a fixed F-proof. But this may introduce different permutations of the $\bigwedge$ and for simulating the inferences in ρ we may need to reorder them. This is done via the following claim proved easily by induction on t.

Claim 2 Let σ be a permutation of $[t]$. Then

$$\bigwedge_{id} B_j \rightarrow \bigwedge_{\sigma} B_j$$

has an F-proof of size $O((\sum_j |B_j|)^2)$.

Claim 2 allows one to derive from each $\bigwedge_\sigma B_j \rightarrow D_i$ the required formula $\bigwedge_{id} B_j \rightarrow D_i$ in size $O(|D_i| + (\sum_j |B_j|)^2)$.

Let us return to the original EF-proof π of A. It can be seen as a Frege proof of A from $E_r, \ldots, E_1$ and Claim 1 gives us a Frege proof of

$$(E_r \wedge (E_{r-1} \wedge (\cdots \wedge E_1) \cdots) \rightarrow A \tag{2.4.2}$$

of size $|\pi|^{O(1)}$.

Now apply to this formula the substitution rule with $q_r := G_r$ and with the other atoms left unsubstituted, obtaining

$$((G_r \equiv G_r) \wedge (E_{r-1} \wedge (\cdots \wedge E_1) \cdots) \rightarrow A).$$

Any Frege system F can prove $G_r \equiv G_r$ and derive from this formula in a constant number of steps the formula

$$(E_{r-1} \wedge (\cdots \wedge E_1) \cdots) \rightarrow A\,. \tag{2.4.3}$$

The derivation of (2.4.3) from (2.4.2) has a constant number of steps and size $O(|\pi|^{O(1)})$. Thus repeating the same process for q_i, $i = r-1, \ldots, 1$, yields an SF-proof of A of size bounded above by $O(|\pi|^{O(1)})$. □

Somewhat harder is the opposite simulation.

Lemma 2.4.4 $EF \geq_p SF$.

Proof Let $\pi = D_1, \ldots, D_k$ be an SF-proof and let $\overline{p} = (p_1 \ldots, p_n)$ be all the atoms appearing in it. For each $i < k$ introduce a new n-tuple of mutually distinct atoms $\overline{q}^i$ and also denote $\overline{p}$ by $\overline{q}^k$ to unify the treatment below. We use these new tuples to create copies of the steps in π; for $i \leq k$ set:

$$E_i := D_i(\overline{q}^i)$$

(hence $E_k = D_k$).

Informally, the idea is the following. Because the atoms in the copies E_i of different steps D_i are disjoint, we may informally think about E_i as saying that D_i is true for all assignments. We will eventually show that each E_i has a short EF-proof and hence, in a sense, that all steps in π are logically valid. (The reader may compare this with the proof of the lemma via the bounded arithmetic simulation of the soundness proof of SF in S^1_2 in Section 11.5 and see that it is similar.)

For $j \leq k$ define an n-tuple of formulas $\overline{B}^j$ as follows.

(B1) If $D_j(\overline{p})$ is an axiom or has been inferred by a rule other than the substitution rule, put $\overline{B}^j := \overline{q}^j$.

(B2) If $D_j(\overline{p})$ has been inferred by the instance

$$\frac{D_i(\overline{p})}{D_i(\sigma(\overline{p}))}$$

of the substitution rule from D_i, $i < j$, then, by the substitution σ, put $\overline{B}^j := \sigma(\overline{q}^j)$, i.e. $B^j_u := \sigma(q^j_u)$.

Let $\tilde{E}_{ij}$ abbreviate the conjunction

$$E_i \wedge \cdots \wedge E_j, \quad 1 < i \leq j < k,$$

and also set $\tilde{E}_{i,i-1} := 1$.

Now we start the simulation of π in EF. Introduce the n-tuples of the extension variables $\overline{q}^{k-1}, \ldots, \overline{q}^1$ via the extension rule by stipulating that

$$q^i_\ell := (\tilde{E}_{i+1,i} \wedge \neg E_{i+1} \wedge B^{i+1}_\ell) \vee \cdots \vee (\tilde{E}_{i+1,k-1} \wedge \neg E_k \wedge B^k_\ell) .$$

Then the implication

$$\tilde{E}_{i+1,j-1} \wedge \neg E_j \ \rightarrow q^i_\ell \equiv B^j_\ell, \quad i < j,$$

is logically valid and has a short Frege proof. Lemma 2.1.4 then gives us short proofs of

$$\tilde{E}_{i+1,j-1} \wedge \neg E_j \ \rightarrow D_i(\overline{q}^i) \equiv D_i(\overline{B}^j), \quad i < j,$$

which are just

$$\tilde{E}_{i+1,j-1} \wedge \neg E_j \ \rightarrow E_i \equiv E_i(\overline{B}^j), \quad i < j . \tag{2.4.4}$$

The simulation of π proceeds by deriving the formulas E_i for $i = 1, \ldots, k$; this will conclude the construction as $E_k = D_k$. To derive E_j assume that $E_1, \ldots, E_{j-1}$ have been proved already and consider three cases:

1. d_j is an axiom;
2. D_j follows from D_u, D_v, $u, v \leq j-1$ by a rule other than the substitution rule;
3. D_j has been inferred by the substitution rule.

In case 1, E_j is also an axiom. In case 2, derive $\tilde{E}_{u_t+1,j-1}$, $t = 1, \ldots, \ell$, if D_{u_t} are the hypotheses of the rule yielding D_j in π. The implication (2.4.4) for all $i = u_t$, $t = 1, \ldots, \ell$, yields

$$\neg E_j \rightarrow \bigwedge_{t \leq \ell} D_{u_t}(\overline{B}^j) .$$

Applying the same rule to the ℓ formulas in the conclusion we get

$$\neg E_j \rightarrow D_j(\overline{B}^j),$$

which is just $\neg E_j \to E_j$, and thus E_j follows.

In the third case assume that D_j was inferred by the substitution σ from D_i. In a similar way to that above we get $\neg E_j \to D_i(\overline{B}^j)$. But, by (B2) above,

$$D_i(\overline{B}^j) = D_i(\sigma(\overline{q}^j)) = D_j(\overline{q}^j) = E_j .$$

This completes the derivation of E_j. □

Let us turn to the structure of *EF*-proofs. It does not make much sense to consider tree-like *EF* proofs, as the extension atoms and the corresponding axioms may be reused – that is their whole point.

A statement analogous to Reckhow's theorem is true and has a simple proof. The extension atoms can be used to define the value of any formula in any language using extension axioms of total size proportional to the size of the circuit computing the formula; this is analogous to the sets of clauses Def_C from Section 1.4. In particular, the total size of the extension axioms for a formula A is bounded by $O(|A|)$ irrespective of the language. This allows us to translate *EF*-proofs from one language to another with just a linear size increase.

Lemma 2.4.5 *Let EF_1 and EF_2 be two extended Frege systems in two languages containing the DeMorgan language. Then $EF_1 \equiv_p EF_2$ and there is a p-simulation changing the size only linearly.*

The most informative structural result about *EF*-proofs is the following. For a formula A let *Ext*(A) be the set of formulas (extension axioms) constructed as follows.

1. For each subformula E of A (including all atoms and A itself) introduce a new atom q_E.
2. For each such E put into *Ext*(A) axioms as follows:
   - if $E = \neg G$, include $q_E \equiv \neg q_G$;
   - if $E = G \circ H$, include $q_E \equiv (q_G \circ q_H)$, for $\circ = \vee, \wedge$;
   - include $p \equiv q_p$ for all atoms p.

Lemma 2.4.6 *Let EF be an extended Frege system in any language and assume that π is an EF-proof of a formula A with k steps.*

Then there is an EF-proof ρ of q_A from Ext(A) such that

$$\mathbf{k}(\rho) = O(k) \quad \textit{and} \quad \mathbf{w}(\rho) \le O(1).$$

Proof Let $\pi = D_1, \ldots, D_k$. Note that for all $i \le k$ there is a Frege proof of q_{D_i} from

$$\bigcup_{j \le i} Ext(D_j) \ \cup \ \{q_{D_1}, \ldots, q_{D_{i-1}}\}$$

of a constant size. For example, if D_i was derived from D_u and $D_v = D_u \to D_i$ then by modus ponens q_{D_i} follows from q_{D_u} and q_{D_v} and the extension axiom $q_{D_v} \equiv (q_{D_u} \to q_{D_i})$.

Hence we have a Frege proof of q_{D_k}, i.e. of q_A, from

$$\bigcup_{j \le k} Ext(D_j) \tag{2.4.5}$$

with $O(k)$ steps and of constant width. The expression (2.4.5) can be interpreted as giving the extension axioms in an EF-proof ρ of q_A. □

We can use the extension axioms from $Ext(A)$ to derive A from q_A. This derivation has $O(|A|)$ steps, size $O(|A|^2)$, width $O(|A|)$ and logical depth at most $\ell dp(A) + O(1)$.

Corollary 2.4.7 *Under the same hypotheses as in Lemma 2.4.6 there is an EF-proof π' of A with $O(k + |A|)$ steps, width $O(|A|)$, size $O(k + |A|^2)$ and logical depth $\le \ell dp(A) + O(1)$.*

In particular, for all formulas A,

- $\mathbf{s}_{EF}(A) = O(\mathbf{k}_{EF}(A) + |A|)$,
- $\mathbf{w}_{EF}(A) = O(|A|)$,
- $\ell dp_{EF}(A) \le \ell dp(A) + O(1)$.

2.5 Overview of Proof Complexity Measures

Let us start by reviewing complexity notions for formulas. The complexity of a formula A is measured by its size $|A|$ or by its logical depth $\ell dp(A)$. There is also the notion of the **depth** of A, denoted by $dp(A)$, which is defined as the maximum number d such that there is a chain of subformulas of A whose top connectives change their type d-times. For the DeMorgan formulas this is equivalent to the following inductive definition:

- atoms and the constants have depth 0;
- $dp(\neg B) = dp(B)$, if B starts with $\neg$ and $dp(\neg B) = 1 + dp(B)$ otherwise;
- if $B(q_1, \dots, q_t)$ is built from atoms and disjunctions only and none of the formulas $C_1, \dots, C_t$ starts with a disjunction then

$$dp(B(C_1, \dots, C_t)) = 1 + \max_i dp(C_i);$$

- a condition holds for conjunction analogous to the previous item.

For the DeMorgan formulas a version of this definition is often used in which we demand that the formula is in negation normal form and we do not count the negations. However, this is not that important, as the chief use of the depth measure is to distinguish between classes of formulas of bounded depth and classes of unbounded depth, as this distinction remains unchanged for the modified definition. This is then used to define *bounded-depth* (equivalently, *constant-depth*) subsystems of Frege systems.

We include now a result about bounded-depth subsystems of Frege systems but later we shall study bounded-depth subsystems of the sequent calculus instead, as that allows for a much more elegant treatment. In particular, if we formulated the DeMorgan language as having unbounded-arity disjunction $\bigvee$ and conjunction $\bigwedge$ we could simply define

$$\mathrm{dp}(\bigvee_i A_i) = 1 + \max_i \mathrm{dp}(A_i) \quad \text{and} \quad \mathrm{dp}(\bigwedge_i A_i) = 1 + \max_i \mathrm{dp}(A_i).$$

However, formulating Frege systems in such a language is cumbersome, while it is very simple and natural in the sequent calculus (Section 3.4). Moreover, the sequent calculus formalism also allows for more precise estimates of the depth of proofs.

For a Frege system F and $d \geq 0$, let F_d be the subsystem of F allowing in proofs only formulas of depth at most d. It is not complete (as tautologies do not have bounded depth). But at least we have

Lemma 2.5.1 *Given a Frege system F, for every d_0 there exists a $d_1 \geq d_0$ such that every tautology of depth $\leq d_0$ has an F_{d_1}-proof.*

Such a proof simulates the evaluation of a formula on all assignments.

Let us turn now to measures of the complexity of proofs. For Frege proofs π we have defined so far the following measures:

- the number of steps, $\mathbf{k}(\pi)$;
- the size, $|\pi|$;
- the maximum logical depth of a formula in it, $\ell\mathrm{dp}(\pi)$;
- the width, $\mathbf{w}(\pi)$;
- the height, $\mathbf{h}(\pi)$.

Further, for a Frege proof system F and a formula A we considered the minimum measure (one of the following five) that an F-proof of A must have:

$$\mathbf{k}_F(A), \ \ \mathbf{s}_F(A), \ \ \ell\mathrm{dp}_F(A), \ \ \mathbf{w}_F(A) \ \text{and} \ \mathbf{h}_F(A) \ .$$

The main relations among them established so far are:

- $\mathbf{k}_F(A)$ and $\mathbf{k}_{EF}(A)$ are proportional to each other and are polynomially related to $\mathbf{s}_{EF}(A)$ and $\mathbf{s}_{SF}(A)$ (Lemmas 2.4.2, 2.4.3, 2.4.4 and Corollary 2.4.7);
- $\mathbf{k}_{F^*}(A) = O(\mathbf{k}_F(A) \log \mathbf{k}_F(A))$ and $\mathbf{h}_{F^*}(A) = O(\log \mathbf{k}_F(A))$ (Theorem 2.2.1);
- $\ell\mathrm{dp}_{EF}(A) \leq \ell\mathrm{dp}(A) + O(1)$ (Corollary 2.4.7);
- $\mathbf{w}_{EF}(A) = O(|A|)$ (Corollary 2.4.7);
- $\ell\mathrm{dp}_F(A) \leq O(\log \mathbf{s}_F(A))$ (Lemma 2.3.3).

We add here one more relation between $\ell\mathrm{dp}_F(A)$ and $k_F(A)$.

The **term unification** is the following problem. Let $T = \{(s_i, t_i) \mid i \leq u\}$ be u pairs of terms in some first-order language L. A **unifier** for T is a substitution σ of L-terms for variables such that, for all $i \leq u$; $\sigma(s_i) = \sigma(t_i)$. A **most general unifier** is a unifier σ_0 such that for any unifier σ there is a substitution σ_1 such that $\sigma = \sigma_1 \circ \sigma_0$.

If a unifier for T exists then there is also a most general unifier. In fact, we can estimate its complexity as follows (cf. Section 2.6 for a reference).

Lemma 2.5.2 *Let T be a finite set of pairs of terms that have a unifier. Then any most general unifier σ_0 satisfies*

$$dp(\sigma_0(s_j)), dp(\sigma_0(t_j)) \leq \sum_{i \leq u}(|s_i| + |t_i|) .$$

Theorem 2.5.3 (Krajíček [272, 278]) *Let F be a Frege system in any finite language. Assume that π is an F-proof with k steps. Then there is an F-proof $\tilde{\pi}$ and a substitution σ such that:*

- $\sigma(\tilde{\pi}) = \pi$;
- $\ell\text{dp}(\tilde{\pi}) \leq O(k)$, *where the constant implicit in the O-notation depends only on F.*

Proof Let $\pi = D_1, \ldots, D_k$. Introduce, for each $i \leq k$,

- a new variable q_i,
- for the F-rule

$$\frac{A_1(\overline{p}), \ldots, A_\ell(\overline{p})}{A_0(\overline{p})}$$

  with $\overline{p} = p_1, \ldots, p_n$ whose instance was used in π to derive D_i from $D_{j_1}, \ldots, D_{j_\ell}$, introduce new variables $p_1^i, \ldots, p_n^i$,

and form a set T of pairs of the DeMorgan formulas as follows. For every $i \leq k$ add to T the pairs

- $(q_{j_t}, A_t(p_1^i, \ldots, p_n^i))$ for $t = 1, \ldots, \ell$, and
- $(q_i, A_0(p_1^i, \ldots, p_n^i))$.

Thinking of the DeMorgan formulas as of terms in the DeMorgan language we know that T has a unifier (determined by π) and hence there is a most general unifier.

The pairs of formulas in T enforce that substitution by a unifier is an F-proof and Lemma 2.5.2 gives the depth upper bound $\sum\{|t| \mid t \text{ occurs in } T\}$. However, each term in T has a constant size (depending on F only) and we have constantly many pairs in T for each step in π. Hence the upper bound is $O(k)$. □

Corollary 2.5.4 *Let F be a Frege system in any finite language. Assume π is an F-proof with k steps:*

$$\pi : B_1, \ldots, B_t \vdash_F C.$$

Let $\ell := \max(\{\ell\text{dp}(B_j) \mid j \leq t\} \cup \{\ell\text{dp}(C)\})$.

Then there is an F-proof

$$\pi' : B_1, \ldots, B_t \vdash_F C$$

such that $\ell\text{dp}(\pi') \leq O(k) + \ell$.

Corollary 2.5.5 *Let $A_n := (\neg)^{(2^n)}(1)$. Then $\mathbf{k}_F(A_n) = \Omega(2^n)$ while $\mathbf{k}_{SF}(A_n) = O(n)$.*

Proof That lower bound is $\mathbf{k}_F(A_n)$ follows immediately from Theorem 2.5.3: the last formula in $\tilde{\pi}$ must be A_n.

For the upper bound define $B_m := p \to (\neg)^{(2^m)}(p)$ and note that B_{m+1} can be derived in SF from B_m in a constant number of steps: substitute $p := (\neg)^{(2^m)}(p)$ and use modus ponens. The formula A_n is then derived from B_n via the substitution $p := 1$ (and modus ponens). □

There are interesting relations between the minimum sizes of proofs in TC^0-Frege systems, in unrestricted Frege systems and in AC^0-Frege systems (see Section 2.6 for the terminology). We shall state the result now but give its proof using bounded arithmetic in Section 11.6. Both statements utilize the theorem of Nepomnjascij (Section 8.7) and the first is a corollary of the second and of the results in Section 11.3 (but its stand-alone proof is simpler).

Theorem 2.5.6

(i) (essentially Paris and Wilkie [389])

For $d \geq 0$, let $F_d(\mathrm{TH})$ be a Frege system in the DeMorgan language augmented by all connectives $\mathrm{TH}_{n,k}$ defined in (1.1.4) and allowing only formulas of depth at most d.

Then, for any $d_0 \geq 0$ and $\epsilon > 0$, there is a $d \geq d_0$ such that for any DeMorgan formula A of depth $dp(A) \leq d_0$ it holds that

$$\mathbf{s}_{F_d}(A) \leq 2^{\mathbf{s}_{F_{d_0}(TH)}(A)^{\epsilon}}.$$

(ii) (Filmus, Pitassi and Santhanam [186])

Let F be a Frege system in the DeMorgan language and $d_0 \geq 0$. Then for every $\epsilon > 0$ there is a $d \geq d_0$ such that for any formula A of depth $\mathrm{dp}(A) \leq d_0$ it holds that

$$\mathbf{s}_{F_d}(A) \leq 2^{\mathbf{s}_F(A)^{\epsilon}}.$$

The last proof complexity measure that we mention here was defined in [278] and although this is rarely mentioned explicitly, many lower bounds are actually lower bounds for this measure. For a Frege system F and an F-proof π, let $\ell(\pi)$ denote the **number of distinct formulas** that occur as subformulas in π. For a formula A let $\ell_F(A)$ be the minimum of $\ell(\pi)$ over all F-proofs of π of A. Obviously

$$\mathbf{k}_F(A) \leq \ell_F(A) \leq \mathbf{s}_F(A)$$

and, in fact, we have

Lemma 2.5.7 *For all Frege systems F and all formulas A, $\ell_F(A) = O(\mathbf{k}_F(A) + |A|)$.*

2.6 Bibliographical and Other Remarks

Sec. 2.1 The Definitions from Section 2.1 are from Cook and Reckhow [156]. Hilbert (as simplified by Kleene [268]) extended the Frege–Lukasiewicz system to handle also $\vee$ and $\wedge$, by adding two triples of axioms;

$$p \to (p \vee q), \quad q \to (q \vee p), \quad (p \to r) \to [(q \to r) \to ((p \vee q) \to r)]$$

and

$$(p \wedge q) \to p, \quad (p \wedge q) \to q, \quad p \to (q \to (p \wedge q)) .$$

A number of finer variants of the Deduction lemma 2.1.5 were studied by Buss and Bonet [91, 92].

Sec. 2.2 The Buss–Pudlák game described in Section 2.2 is from [132], although here we formulated the constructions somewhat differently.

Sec. 2.3 The only published proof of Reckhow's theorem 2.3.4, aside of his unpublished Ph.D. thesis [445], is in [278]. The key lemma 2.3.3 appears in [445] only implicitly; it was stated explicitly and proved in [278, Lemma 4.4.14]. Our proof follows to a large extent the proof from [278, Lemma 4.4.14], but with some simplifications.

Sec. 2.4 The extension rule was first used by Tseitin [492] in the context of resolution proof systems (cf. Chapter 5) and the extended Frege systems were defined by Cook and Reckhow [156]. The name "substitution Frege" is also from [156], although the system goes back to Frege [188]. Even if only substitutions of atoms and constants are allowed in the substitution rule, it is still p-equivalent to *SF*, cf. Buss [112]. Lemmas 2.4.2 and 2.4.3 follow [278] although they are proved in some form in [156] too.

Lemma 2.4.4 was derived first by Dowd [172] using bounded arithmetic (see Section 11.5). The explicit p-simulation we give follows Krajíček and Pudlák [317]. Lemma 2.4.5 is from [156] and Lemma 2.4.6 is from [278].

Sec. 2.5 Theorem 2.5.3 was specialized in [278] to Frege systems from a much more general statement about predicate calculus in Krajíček [272]. Our treatment follows a later argument via the term unification from Krajíček and Pudlák [315]. Lemma 2.5.2 and the argument it depends on was proved in [315, Lemma 1.2] by analyzing the algorithm for term unification devised by Chang and Lee [137, p. 77]. Corollary 2.5.5 separating $\mathbf{k}_F(A)$ and $\mathbf{k}_{SF}(A)$ was proved originally by Tseitin and Choubarian [493]; we have followed [273].

The original proof of Theorem 2.5.6 by Filmus, Pitassi and Santhanam [186] was combinatorial. Lemma 2.5.7 was proved in [278] using a technique not developed here.

The bounded-depth Frege subsystems of a Frege system in the DeMorgan language are sometimes collectively called AC^0-Frege systems, referring to the circuit class AC^0; loosely speaking, AC^0-F operates with AC^0-formulas. What one means by that when making a claim that some tautologies τ_n have polynomial-size AC^0-Frege proofs is that there is one fixed $d \geq 2$ such that the τ_n have polynomial-size F_d-proofs. Analogously, there are classes of $AC^0[m]$- and TC^0-Frege systems operating with formulas from the respective circuit class: extensions of the DeMorgan language by the connectives $MOD_{m,i}$ from (1.1.3) or $TH_{n,k}$ from (1.1.4). Note that the NC^1-Frege systems are, owing to Lemma 2.3.3, p-equivalent to full Frege systems. We shall return to all these proof systems in some detail in Part II. Ghasemloo [199] studied a formal definition of these collections of proof systems.

Avigad [42] found a Frege system with a particularly economic definition. Its language uses the NAND connective of an arbitrary arity, $\text{NAND}(\overline{\varphi})$ being true if and only if not all the φ_i are true. His system has one axiom scheme,

$$\text{NAND}(\overline{\psi}, \overline{\varphi}, \text{NAND}(\overline{\varphi})),$$

and one inference rule,

$$\frac{\text{NAND}(\overline{\psi}, \overline{\varphi}), \text{NAND}(\overline{\psi}, \text{NAND}(\overline{\varphi}))}{\text{NAND}(\overline{\psi})}.$$

3
Sequent Calculus

The sequent calculus for first-order logic was introduced by Gentzen [198]. We will see that its propositional part is p-equivalent to Frege systems. Unlike some Frege systems it is very elegant, with simple, transparent and symmetric inference rules. This syntactic clarity allows for deep proof-theoretic analysis of derivations in the calculus. Gentzen himself used it for a penetrating study of predicate calculus and of theories of arithmetic.

Gentzen's proof system LK refers to predicate calculus but we shall use the same name also for its propositional part, defined below (some authors call it PK). First-order LK underlines some arguments about bounded arithmetic, and we shall discuss it then (see Section 8.2).

3.1 Sequent Calculus LK

The sequent calculus LK uses the DeMorgan language. A **cedent** is a finite, possibly empty, sequence of formulas. We denote cedents by capital Greek letters $\Gamma, \Delta, \ldots$ A line in an LK-proof is a **sequent**: it is an ordered pair of cedents Γ, Δ written as

$$\Gamma \longrightarrow \Delta.$$

Here Γ is the **antecedent**, and Δ is the **succedent**, of the sequent. A truth assignment satisfies the sequent if and only if either it makes some formula in the antecedent false or it makes some formula in the succedent true. In particular, an assignment satisfies

$$\longrightarrow A_1, \ldots, A_t$$

if and only if it satisfies the disjunction $\bigvee_{i \leq t} A_i$, and it falsifies

$$A_1, \ldots, A_t \longrightarrow$$

if and only if it satisfies the conjunction $\bigwedge_{i \leq t} A_i$. Note that the empty sequent $\emptyset \longrightarrow \emptyset$, usually written as just $\longrightarrow$, cannot be satisfied.

Definition 3.1.1 The inference rules of the sequent calculus LK are the following.

- **Initial sequents** are sequents of the following form:

$$p \longrightarrow p, \quad 0 \longrightarrow, \quad \longrightarrow 1$$

  with p an atom.

- **Structural rules:**
  - the weakening rules

$$\text{left } \frac{\Gamma \longrightarrow \Delta}{A, \Gamma \longrightarrow \Delta} \quad \text{and} \quad \text{right } \frac{\Gamma \longrightarrow \Delta}{\Gamma \longrightarrow \Delta, A},$$

  - the exchange rules

$$\text{left } \frac{\Gamma_1, A, B, \Gamma_2 \longrightarrow \Delta}{\Gamma_1, B, A, \Gamma_2 \longrightarrow \Delta} \quad \text{and} \quad \text{right } \frac{\Gamma \longrightarrow \Delta_1, A, B, \Delta_2}{\Gamma \longrightarrow \Delta_1, B, A, \Delta_2},$$

  - the contraction rules

$$\text{left } \frac{\Gamma_1, A, A, \Gamma_2 \longrightarrow \Delta}{\Gamma_1, A, \Gamma_2 \longrightarrow \Delta} \quad \text{and} \quad \text{right } \frac{\Gamma \longrightarrow \Delta_1, A, A, \Delta_2}{\Gamma \longrightarrow \Delta_1, A, \Delta_2}.$$

- **Logical rules:**
  - $\neg$:introduction rules

$$\text{left } \frac{\Gamma \longrightarrow \Delta, A}{\neg A, \Gamma \longrightarrow \Delta} \quad \text{and} \quad \text{right } \frac{A, \Gamma \longrightarrow \Delta}{\Gamma \longrightarrow \Delta, \neg A},$$

  - $\wedge$:introduction rules

$$\text{left } \frac{A, \Gamma \longrightarrow \Delta}{A \wedge B, \Gamma \longrightarrow \Delta} \quad \text{or} \quad \text{left } \frac{A, \Gamma \longrightarrow \Delta}{B \wedge A, \Gamma \longrightarrow \Delta}$$

$$\text{and right } \frac{\Gamma \longrightarrow \Delta, A \quad \Gamma \longrightarrow \Delta, B}{\Gamma \longrightarrow \Delta, A \wedge B},$$

  - $\vee$:introduction rules

$$\text{left } \frac{A, \Gamma \longrightarrow \Delta \quad B, \Gamma \longrightarrow \Delta}{A \vee B, \Gamma \longrightarrow \Delta},$$

$$\text{right } \frac{\Gamma \longrightarrow \Delta, A}{\Gamma \longrightarrow \Delta, A \vee B} \quad \text{or} \quad \text{right } \frac{\Gamma \longrightarrow \Delta, A}{\Gamma \longrightarrow \Delta, B \vee A}.$$

- **Cut rule:**

$$\frac{\Gamma \longrightarrow \Delta, A \quad A, \Gamma \longrightarrow \Delta}{\Gamma \longrightarrow \Delta}.$$

An LK-**proof** is a sequence of sequents starting with initial sequents and using the inference rules. An LK^--**proof** is an LK-proof that does not uses the cut rule; such a proof is called **cut-free**. An LK-proof is **tree-like** if and only if every sequent is used as a hypothesis for at most one inference. We shall denote the tree-like subsystem of LK by LK^*, as for Frege systems.

We will use the same notation for various complexity measures as that used for Frege systems: for an LK-proof π, $\mathbf{k}(\pi)$ is its number of steps and $|\pi|$ is its size, $\mathbf{w}(\pi)$ is its width (the maximum size of a sequent in π) and $\mathbf{h}(\pi)$ is the height of π (almost always tree-like for this measure).

The following terminology comes in handy: the formula introduced in an inference is called the **principal formula**, the formulas from which it is inferred are the **minor formulas** of the inference and all other formulas occurring in the inference are called **side formulas**. A formula in Δ or Γ has two occurrences in the inference, in the lower sequent and in the upper sequent: these are called the **immediate ancestor** or the **immediate successor**, respectively, of each other. The immediate ancestor(s) of a principal formula of an inference is or are the minor formulas.

For Frege systems we considered a situation when a formula C is provable from formulas $B_1, \ldots, B_t$. For LK this can be represented in two ways, either by the fact that the sequent

$$B_1, \ldots, B_t \longrightarrow C$$

is provable or by adding a new type of initial sequents, **axioms**, of the form $\longrightarrow B_j$, for $j \leq t$, and asking for a proof of $\longrightarrow C$.

A syntactically simplified version of LK using only sequents with empty antecedents is usually called the *Tait-style sequent calculus*. This calculus uses just those rules of LK in which both the minor and the principal formulas are in the succedents. In particular, there are no introduction rules for the negation $\neg$ but instead one adds a new type of an initial sequent,

$$\longrightarrow A, \neg A,$$

for any formula A.

Sometimes authors define cedents as *sets* of formulas and not *sequences*, and then the contraction and the exchange rules are redundant. But such cedents are written (or coded for an algorithm or a formal system) as a finite word in some alphabet, and this means that the two rules are used implicitly anyway.

3.2 The Strength of LK

First we need to verify that LK is indeed a proof system. We will show a bit more.

Theorem 3.2.1 *Both the proof systems* LK *and* LK$^-$ *are sound and complete. In fact, tree-like* LK$^-$ *is also complete.*

(Tree-like LK$^-$ could be denoted by (LK$^-$)* but that would be rather cumbersome.)

Proof The soundness means that only logically valid sequents, i.e. those true under all truth assignment, can be proved. This is easily verified by induction on the number of steps in an LK-proof.

It remains to show that LK^- is complete, meaning that any logically valid sequent $\Gamma \longrightarrow \Delta$ has a cut-free proof. We will build bottom up a tree-like LK^--proof of the sequent. The last sequent has to be $\Gamma \longrightarrow \Delta$. Having a tree with nodes labeled by sequents that everywhere branches to at most two subtrees, extend it as follows (for this construction – and other related proofs – think of trees as growing upwards). Pick a leaf and consider several cases according to the form of its label.

- If the leaf is labeled by

$$\Pi \longrightarrow \Sigma, \neg A,$$

where $\neg A$ is a formula with the maximum logical depth in the sequent, then the node has only one child (the predecessor in the future proof), labeled by

$$A, \Pi \longrightarrow \Sigma.$$

- If the leaf is labeled by

$$\Pi \longrightarrow \Sigma, A \wedge B,$$

where $A \wedge B$ has the maximum logical depth, then the node has exactly two predecessors, labeled by

$$\Pi \longrightarrow \Sigma, A \quad \text{and} \quad \Pi \longrightarrow \Sigma, B$$

respectively.
- If the leaf is labeled by

$$\Pi \longrightarrow \Sigma, A \vee B,$$

where $A \vee B$ has the maximum logical depth, then it has one predecessor, labeled by

$$\Pi \longrightarrow \Sigma, A \vee B, A \vee B$$

which itself has one predecessor labeled by

$$\Pi \longrightarrow \Sigma, A, A \vee B,$$

which has also one predecessor labeled by

$$\Pi \longrightarrow \Sigma, A, B.$$

- If there are more formulas in the succedent having the maximum logical depth, choose any.
- If a formula having the maximum logical depth occurs only in the antecedent then the tree is extended analogously, exchanging the right rules for the left rules and switching the roles of $\vee$ and $\wedge$.

It is easy to see that all sequents appearing as labels of nodes are logically valid, assuming that $\Gamma \longrightarrow \Delta$ is. Hence when the tree can no longer be extended, the leaves must be labeled by sequents consisting of atoms which appear both in the antecedent and in the succedent or containing 1 (resp. 0) in the succedent (resp. in the antecedent). All these sequents can be derived from the initial sequents by the weakening rules.

We have not used the cut rule and so the resulting labeled tree is a tree-like cut-free proof. □

The size of the cut-free proof constructed in the previous theorem can be estimated as follows.

Corollary 3.2.2 *Assume that* $\Gamma \longrightarrow \Delta$ *is a logically valid sequent of size s. Then the sequent has a tree-like* LK$^-$*-proof of size at most* $O(s2^s)$.

This is not a very interesting upper bound. Much more important is the following structural property of the constructed proofs.

Corollary 3.2.3 *Any formula that appears in an* LK$^-$*-proof of a sequent* $\Gamma \longrightarrow \Delta$ *is a subformula of some formula in* $\Gamma \cup \Delta$.

Proof Note that all formulas appearing in a hypothesis of any LK-rule except the cut rule appear also as subformulas in its conclusion and hence, in a cut-free proof, also in the end-sequent. □

This property is very useful and has its own name: the **subformula property**. Very often this is the only consequence of having the cut-free proof that one needs in an argument.

Theorem 3.2.4 *The* LK *and Frege systems are p-equivalent.*

Proof By Reckhow's theorem 2.3.4 it suffices to show that LK is p-equivalent to a Frege system in the DeMorgan language. If

$$\frac{A_1, \ldots, A_\ell}{A_0} \tag{3.2.1}$$

is a Frege rule then the sequent

$$A_1, \ldots, A_\ell \longrightarrow A_0 \tag{3.2.2}$$

is logically valid and has an LK$^-$-proof. Fix one such proof.

Claim *Assume the* $\pi = D_1, \ldots, D_k$ *is a proof in a Frege system in the DeMorgan language. Then for each* $i \leq k$ *there is an* LK*-proof of the sequent* $\longrightarrow D_i$ *with* $O(i)$ *steps and size bounded above by* $O(\sum_{j \leq i} |D_j|)$ *and the logical depth of all formulas in the proof bounded above by* $\max_{j \leq i} \ell\mathrm{dp}(D_j)$.

This is established by induction on i. Assume that D_i is deduced in π by an instance of rule (3.2.1) given by the substitution σ:

$$\sigma(A_0) = D_i \text{ and } \sigma(A_t) = D_{j_t},\ t = 1, \ldots, \ell \text{ and } j_t < i.$$

By the induction hypothesis all $\longrightarrow D_{j_t}$ have LK-proofs with the required properties. Joining them by ℓ cuts with a suitable substitution instance of the fixed LK$^-$-proof of (3.2.2) gives us the required proof of $\longrightarrow D_i$. The claim establishes that LK $\geq_p F$, any Frege system F.

For the opposite simulation it suffices to construct a specific Frege system F p-simulating LK. The language of F will be the DeMorgan language augmented by the implication connective $\longrightarrow$ and one of the F-rules will be the modus ponens. Further, for any LK-rule we add to F a rule tailored after it and allowing for a simple simulation of the LK-rule. For example, for $\wedge$:right include the axiom scheme

$$p \longrightarrow (q \longrightarrow (p \wedge q))$$

and for $\vee$:right include both axiom schemes

$$p \longrightarrow (p \vee q) \quad \text{and} \quad q \longrightarrow (p \vee q)\ .$$

We also include rules for simple manipulations with nested disjunctions, for example

$$\frac{p \vee (q \vee r)}{q \vee (p \vee r)}\ .$$

Represent a sequent

$$S\colon A_1, \ldots, A_u \longrightarrow B_1, \ldots, B_v$$

by the disjunction $\bigvee S$:

$$\bigvee S\colon \neg A_1 \vee \cdots \vee \neg A_u \vee B_1 \vee \cdots \vee B_v \tag{3.2.3}$$

(arbitrarily bracketed) and simulate the LK-proof step by step by an F-proof of the translations $\bigvee S$ of the sequents S in the proof. □

Note that we have used the cut rule (and modus ponens) in an essential way in Theorem 3.2.4. The next lemma shows that the cut rule is indeed necessary.

Lemma 3.2.5 LK$^-$ *does not p-simulate* LK. *In fact, there are sequents having a size-s* LK*-proof such that every* LK$^-$*-proof requires* $2^{s^{\Omega(1)}}$ *steps.*

The representation of a sequent S by one formula in (3.2.3) will be used repeatedly, and we now state a simple technical lemma.

Lemma 3.2.6 *Let* $S = \Gamma \longrightarrow \Delta$ *be any sequent and* $\bigvee S$ *its representation* (3.2.3). *Then:*

(i) $\longrightarrow \bigvee S$ *has a tree-like cut-free proof from S of size* $O(|S|^2)$*;*
(ii) both sequents

$$\bigvee S, \Gamma \longrightarrow \Delta \quad \textit{and} \quad \Gamma \longrightarrow \Delta, \bigvee S$$

have tree-like cut-free proofs of size $O(|S|^2)$*;*
(iii) S has a tree-like proof from $\longrightarrow \bigvee S$ *of size* $O(|S|^2)$.

(Note that the proof in (iii) cannot be cut-free, owing to the subformula property of such proofs (Lemma 3.2.3).)

Proof The constructions use that, for each formula A, the sequents

$$A \longrightarrow A,$$

$$A, \neg A \longrightarrow \quad \text{and} \quad \longrightarrow A, \neg A$$

have size-$O(|A|^2)$ tree-like cut-free proofs; the first is constructed by induction on the size of A, the other two by the $\neg$-rules.

The proof of (i) is constructed by applying first a number of $\neg$:right inferences (moving the antecedent with the negations added to the succedent) followed by a number of $\vee$:right inferences (replacing each formula by $\bigvee S$) and by contractions.

The proof of the first part of (ii) is constructed by combining the proofs of all sequents $A, \neg A \longrightarrow$ for $A \in \Gamma$ with the proofs of all sequents $B \longrightarrow B$ for $B \in \Delta$ by some $\vee$:left inferences. The proof for the other part of (ii) is constructed analogously.

The proof of (iii) is obtained by a cut from $\longrightarrow \bigvee S$ and the proof of $\bigvee S, \Gamma \longrightarrow \Delta$ (in item (ii)). □

The following statement extends the simulation of Frege proofs by tree-like Frege proofs (Theorem 2.2.1) to LK. We shall state the fact for the record now but will prove a finer version in Section 3.4 and in Chapter 14 concerning optimal bounds for the depth of formulas in a tree-like proof. Recall that we use the star notation LK* for a tree-like subsystem of LK.

Theorem 3.2.7 LK* *and* LK *are p-equivalent.*

3.3 Interpolation for Cut-Free Proofs

In this section it will be useful to restrict to formulas in negation normal form (Section 1.1). We will consider cut-free proofs of sequents consisting of such formulas only. In the absence of cuts the negation rules can be applied only to atoms or constants. In that case one can modify the calculus by removing the negation rules altogether and allowing the following new types of initial sequents:

$$p, \neg p \longrightarrow\,, \quad \longrightarrow p, \neg p\,, \quad \neg p \longrightarrow \neg p\,, \quad \neg 1 \longrightarrow\,, \quad \longrightarrow \neg 0\,. \tag{3.3.1}$$

Let us call this variant of LK the **negation normal form** of LK (resp. of LK$^-$). It is easy to see that this form is p-equivalent with the respective original version as far as the provability of sequents consisting of negation normal form formulas is concerned.

Every valid sequent has an LK$^-$-proof by Theorem 3.2.1. Hence the following theorem offers another proof of the interpolation theorem 1.1.3.

Theorem 3.3.1 *Let π be a negation normal form tree-like* LK$^-$*-proof of a sequent*

$$\Gamma(\overline{p}, \overline{q}) \longrightarrow \Delta(\overline{p}, \overline{r}),$$

where $\overline{p}$, $\overline{q}$ and $\overline{r}$ are disjoint tuples of atoms.

Then there exists a formula $I(\overline{p})$ such that:

- *both implications $\Gamma \longrightarrow I$ and $I \longrightarrow \Delta$ are logically valid and their negation normal form tree-like* LK$^-$*-proofs can be constructed from π by a p-time algorithm;*
- *the size of I is at most $\mathbf{k}(\pi)$ (the number of steps in π);*
- *the logical depth $\ell\mathrm{dp}(I)$ of I is at most $\mathbf{h}(\pi)$ (the height of π).*

Moreover, if the atoms $\overline{p}$ occur only positively (i.e. without negations) in either Γ or Δ then I is monotone (i.e. it does not use any negation).

Proof Note that no $\overline{q}$ atoms occur in π in any succedent and no $\overline{r}$ atoms occur in any antecedent. Let

$$S = \Pi(\overline{p},\overline{q}) \longrightarrow \Sigma(\overline{p},\overline{r})$$

be a sequent in π. We shall define a formula I interpolating this sequent by induction on the number of sequents in π that occur before S.

All possible initial sequents in π are interpolated by a constant except for $p \longrightarrow p$ and $\neg p \longrightarrow \neg p$, which are interpolated by p and $\neg p$, respectively. Note that $\neg p$ cannot occur as an interpolant of an initial sequent if $\overline{p}$ appears only positively in either Γ or Δ, as then $\neg p \longrightarrow \neg p$ does not occur in π.

If S is derived by a unary rule, do nothing. If S is derived from sequents U,V by $\wedge$:right which are interpolated by formulas J and K, respectively, put

$$I := J \wedge K$$

and if it is derived by $\vee$:left, put

$$I := J \vee K\,.$$

It is easy to see that the formula I constructed in this way has all the stated properties. □

Now we shall drop the hypothesis that π is tree-like. Recall from the start of Section 1.4 the definition of the clauses Def_C representing a circuit C.

Theorem 3.3.2 *Let π be a negation normal form* LK$^-$*-proof of a sequent*

$$\Gamma(\overline{p},\overline{q}) \longrightarrow \Delta(\overline{p},\overline{r}),$$

where $\overline{p}$, $\overline{q}$ and $\overline{r}$ are disjoint tuples of atoms.

Then there exists a Boolean circuit C of size s with inputs $\overline{p}$ and nodes $y_1,\ldots,y_s$ (y_s being the output node) such that:

- *both implications*

$$\Gamma, Def_C \longrightarrow y_s \quad \text{and} \quad Def_C, y_s \longrightarrow \Delta \tag{3.3.2}$$

  are logically valid and their negation normal form LK*-proofs (not necessarily cut-free) can be constructed from π by a p-time algorithm;*
- *the size of C is at most $\mathbf{k}(\pi)$ (the number of steps in π).*

Moreover, if the atoms $\overline{p}$ occur only positively (i.e. without negations) in either Γ or Δ then C is monotone.

(The circuit C is called an **interpolating circuit**.)

Proof The proof is analogous to the proof of Theorem 3.3.1 but instead of defining a formula we define a circuit. Assign to all sequents Z in π a variable y_Z: we will use them in instructions in a program defining the eventual circuit C.

For any initial Z, the instructions for y_Z define it as either a constant or p or $\neg p$ as before (and note again that the instruction $y_Y := \neg p$ cannot occur if p appears only positively in either Γ or Δ).

If Z is derived from the sequent W by a unary rule, include the instruction $y_Z := y_W$. If Z is derived from the sequents U, V by $\wedge$:right, put

$$y_Z := y_U \wedge y_V \tag{3.3.3}$$

and if it is derived by $\vee$:left, put

$$y_Z := y_U \vee y_V\ .$$

It is easy to see that the circuit C corresponding to the end sequent has the required properties.

The proofs of the sequents in (3.3.2) are not necessarily cut-free because, for example, in order to derive from y_U and y_V the variable y_Z (and vice versa) defined by instruction (3.3.3) one needs cuts. □

Later, the property of LK^- established in these two theorems will be called the **feasible interpolation property**. We shall study feasible interpolation extensively in Chapters 17 and 18.

Next we derive from the preceding theorem a feasible version of Beth's implicit–explicit definability theorem. Beth's theorem (its propositional form) states that if some property $A(\overline{p}, \overline{q})$ defines $\overline{q}$ implicitly, so that

$$[A(\overline{p}, \overline{q}) \wedge A(\overline{p}, \overline{r})] \longrightarrow \bigwedge_i q_i \equiv r_i,$$

then $\overline{q}$ can be defined explicitly using $\overline{p}$, that is,

$$A(\overline{p}, \overline{q}) \longrightarrow \bigwedge_i q_i \ := \ B_i(\overline{p})$$

for some formulas B_i. This follows by the same argument (that every Boolean function is definable by a formula) that we used in Section 1.1. However, we shall use Theorems 3.3.1 and 3.3.2 to derive an effective version of Beth's theorem.

Theorem 3.3.3 *Let π be a negation normal form* LK^-*-proof of a sequent*

$$\Gamma(\overline{p}, \overline{q}), q_1 \ \longrightarrow \ \neg\Gamma(\overline{p}, \overline{r}),\ r_1, \tag{3.3.4}$$

where $\neg\Gamma$ is the cedent of (the negation normal form of) the negations of formulas from Γ and where $\overline{p}$, $\overline{q}$ and $\overline{r}$ are disjoint tuples of atoms.

Then there exists a Boolean circuit B with inputs $\overline{p}$ and nodes $y_1, \ldots, y_s$ (y_s is the output) such that:

- *both implications*

$$\Gamma(\overline{p}, \overline{q}),\ \mathrm{Def}_B,\ y_s \longrightarrow q_1 \quad \textit{and} \quad \Gamma(\overline{p}, \overline{q}),\ \mathrm{Def}_B,\ q_1 \longrightarrow y_s$$

  are logically valid and their negation normal form LK*-proofs can be constructed from π by a p-time algorithm;*
- *the size of B is at most $\mathbf{k}(\pi)$.*

Moreover:

- *if π is tree-like then B is, in fact, a formula and $\ell\mathrm{dp}(B) \leq \mathbf{h}(\pi)$;*
- *if the atoms $\overline{p}$ occur only positively in Γ then B is monotone.*

Proof Apply Theorems 3.3.1 and 3.3.2 to the proof of the sequent (3.3.4): the interpolant has the required properties. □

3.4 Constant-Depth Subsystems of LK

Let $\bigwedge$ and $\bigvee$ be symbols for the unbounded-arity conjunction and disjunction, respectively. Formally, if $r \geq 1$ is arbitrary and $A_1, \ldots, A_r$ are formulas then so are the strings

$$\bigwedge(A_1, \ldots, A_r) \quad \text{and} \quad \bigvee(A_1, \ldots, A_r)\ .$$

We shall use the less formal notation

$$\bigwedge_{i \leq r} A_i \quad \text{and} \quad \bigvee_{i \leq r} A_i$$

instead. Sometimes it is convenient to use the conjunction or the disjunction of the empty set of formulas; the former stands for the constant 1, the latter for 0.

The reason why we only touched upon the topic of constant-depth Frege systems in Section 2.5 is that it is easier and more elegant to modify the definition of LK to a language with the unbounded arity connectives $\bigwedge$ and $\bigvee$. In particular, define

- *unbounded* $\bigwedge$:introduction *rules*

$$\text{left}\ \frac{A, \Gamma \longrightarrow \Delta}{\bigwedge_i A_i, \Gamma \longrightarrow \Delta}$$

$$\text{and right}\ \frac{\Gamma \longrightarrow \Delta, A_1 \quad \cdots \quad \Gamma \longrightarrow \Delta, A_r}{\Gamma \longrightarrow \Delta, \bigwedge_{i \leq r} A_i},$$

- *unbounded* $\bigvee$:introduction *rules*

$$\text{left}\ \frac{A_1, \Gamma \longrightarrow \Delta \quad \cdots \quad A_r, \Gamma \longrightarrow \Delta}{\bigvee_{i \leq r} A_i, \Gamma \longrightarrow \Delta}$$

$$\text{and right } \frac{\Gamma \longrightarrow \Delta, A}{\Gamma \longrightarrow \Delta, \bigvee_i A_i},$$

where in $\bigwedge$:left and $\bigvee$:right the formula A is one of the formulas A_i.

By the definitions of $\bigwedge_{i\leq r} A_i$ and $\bigvee_{i\leq r} A_i$ the formulas A_i are ordered in them. But it is easy to see that for any permutation h of $[r]$ the sequent

$$\bigwedge_{j\leq r} q_{h(j)} \longrightarrow \bigwedge_{i\leq r} q_i$$

has a tree-like cut-free proof with $2r + 1$ steps: from each initial $q_i \longrightarrow q_i$ derive by $\bigwedge$:left the sequent $\bigwedge_{j\leq r} q_{h(j)} \longrightarrow q_i$ and join them by one $\bigwedge$:right inference. This means that without loss of generality the ordering of formulas inside $\bigwedge$ and $\bigvee$ is irrelevant and we may think of them as (multi)sets.

Recall from Section 2.5 the definition of the depth of a DeMorgan formula with the unbounded $\bigwedge$ and $\bigvee$ literals and the constants have the depth 0, the negation does not change the depth and

$$\mathrm{dp}(\circ(A_1,\ldots,A_r)) := 1 + \max_i \mathrm{dp}(A_i), \qquad \text{for } \circ = \bigwedge, \bigvee .$$

With this definition of the depth in mind, define for $d \geq 0$ the system LK_d to be the system defined as LK, but with the binary $\wedge$ and $\vee$ and the corresponding rules replaced by the unbounded $\bigwedge$ and $\bigvee$ and their introduction rules, and using only proofs with all formulas of the depth at most d. The symbol LK_d^* will denote its tree-like subsystem.

The completeness of the cut-free LK^- system and the subformula property (Theorem 3.2.1 and Corollary 3.2.3) readily imply

Lemma 3.4.1 *Let S be a logically valid sequent consisting of formulas of depth at most d. Then S has an LK_d^*-proof.*

The following is a quantitative version of Theorem 2.2.1; we formulate it for the constant depth subsystems of LK rather than of a Frege system, as the structure of LK allows for a tighter estimate of the depth.

Lemma 3.4.2 *Let $d \geq 0$ and assume the π is an LK_d-proof of a sequent S from sequents $Z_1,\ldots,Z_t$, each consisting of formulas of depth $\leq d$.*

Then there is an LK_{d+1}^*-proof ρ of S from $Z_1,\ldots,Z_t$ of size $|\rho| = |\pi|^{O(1)}$. In fact, for each fixed $d \geq 0$, ρ can be constructed from π by a p-time algorithm.

Proof Let $\pi = T_1,\ldots,T_k$ be an LK_d-proof of the sequent S from the sequents $Z_1,\ldots,Z_t$. Let $w := \max_i |T_i|$ be the width of π.

As in (3.2.3), for a sequent $U = \Gamma \longrightarrow \Delta$ we denote by $\bigvee U$ the disjunction of the formulas from Δ together with the negations of the formulas from Γ. The constructions in Lemma 3.2.6 established

Claim 1 Let U be a sequent consisting of formulas of depth $\leq d$. Then there are LK_{d+1}^*-proofs of U from $\longrightarrow \bigvee U$ and of $\longrightarrow \bigvee U$ from U, both of size $O(|U|^2)$.

The next claim is proved by considering the type of the inference yielding T_i in π, utilizing Claim 1.

Claim 2 For all $i = 1, \ldots, k$ there are LK^*_{d+1}-proofs σ_i of

$$\bigvee T_1, \ldots, \bigvee T_{i-1} \longrightarrow \bigvee T_i$$

from $Z_1, \ldots, Z_t$ with $|\sigma_i| = O(iw^2) = O(|\pi|^3)$.

From Claim 2 we get, by induction on $j - i$,

Claim 3 For all $0 \le i < j \le k$ there are LK^*_{d+1}-proofs $\rho_{i,j}$ of

$$\bigvee T_1, \ldots, \bigvee T_{i-1} \longrightarrow \bigvee T_j$$

from $Z_1, \ldots, Z_t$ with $|\rho_{i,j}| = O(|\pi|^4)$.

The required proof is $\rho_{0,k}$ combined with an LK^*_{d+1}-derivation of S from $\longrightarrow \bigvee S$ guaranteed by Claim 1. □

The tree-like proof constructed in Lemma 3.4.2 is very unbalanced. But we can balance any tree-like proof with respect to the cut inferences at the expense of increasing the maximum depth of a formula in the proof by 1.

Lemma 3.4.3 *Let π be an LK^*_d-proof of a sequent S from sequents $Z_1, \ldots, Z_t$. Let $s := |\pi|$, $k := \mathbf{k}(\pi)$ and $w := \mathbf{w}(\pi)$ be its size, the number of steps in it and its width (the maximum size of a sequent in π), respectively.*

Then there is a set T of tautological sequents of depth at most d and an LK^*_{d+1}-proof ρ of S from $\{Z_1, \ldots, Z_t\} \cup T$ such that:

- $|\rho| = |\pi|^{O(1)}$;
- *all cut inferences in ρ are below all other non-unary inferences ($\bigvee$:left and $\bigwedge$:right);*
- *on any path in ρ there are at most $O(\log(kw))$ cut inferences.*

In fact, for each fixed $d \ge 0$, ρ can be constructed from π by a p-time algorithm.

Proof For each $\bigwedge$:right inference in π,

$$\frac{\Gamma \longrightarrow \Delta, A_1 \quad \ldots \quad \Gamma \longrightarrow \Delta, A_r}{\Gamma \longrightarrow \Delta, \bigwedge_{i \le r} A_i},$$

add to T all sequents (there are at most $O(r) \le O(w)$)

$$A_1, A_2, \Gamma \longrightarrow \Delta, A_1 \wedge A_2, \quad A_3, A_4, \Gamma \longrightarrow \Delta, A_3 \wedge A_4, \quad \ldots$$

$$A_1 \wedge A_2, A_3 \wedge A_4, \Gamma \longrightarrow \Delta, A_1 \wedge A_2 \wedge A_3 \wedge A_4\,, \quad \ldots$$

and simulate the inference by a balanced tree of cuts. This subderivation has $O(r) = O(w)$ cuts in a tree of height $O(\log w)$. Treat all $\bigvee$:left inferences analogously.

This procedure will transform π into a LK^*_d-proof σ of size $|\sigma| \leq O(kw^2) \leq O(s^3)$ with number of steps $\mathbf{k}(\sigma) \leq O(kw) \leq O(s^2)$ and width $O(w)$, in which the only non-unary inferences are cuts.

For a proof ξ denote by $c(\xi)$ the number of cut inferences in ξ. Now use Spira's lemma 1.1.4 to find a sequent $U = \Pi \longrightarrow \Sigma$ in σ such that the subproof σ_U ending with U satisfies

$$\frac{1}{3}c(\sigma) \leq c(\sigma_U) \leq \frac{2}{3}c(\sigma). \tag{3.4.1}$$

Let σ^U be the rest of σ when σ_U is taken out: it is an LK^*_d-proof of $S(= \Gamma \longrightarrow \Delta)$ from $\{Z_1, \ldots, Z_t\} \cup T \cup \{U\}$, and $c(\sigma^U)$ satisfies (3.4.1) as well.

Add $\bigvee U$ to the antecedent of U, prove this new sequent by a cut-free LK^*_{d+1}-proof of size $O(|U|^2)$ (Lemma 3.2.6), and carry the formula as a side formula to the end-sequent S, obtaining a proof $(\sigma^U)'$ of

$$\bigvee U, \Gamma \longrightarrow \Delta\,.$$

Take the proof σ_U of U and derive from it, by a cut-free tree-like proof of size $O(|U|^2)$ (Lemma 3.2.6 again), the sequent

$$\longrightarrow \bigvee U.$$

Finally, naming this proof $(\sigma_U)'$, join it by a cut to $(\sigma^U)'$ to get a new LK^*_{d+1}-proof of S.

Now repeat the same procedure with the two maximal subproofs $(\sigma_U)'$ and $(\sigma^U)'$; it is possible to use Spira's lemma because all the non-unary inferences other than cuts are above all the cuts, i.e. from the point of view of the cuts the proof tree is still binary. □

Now we complement Lemma 3.4.2 by a form of converse.

Lemma 3.4.4 *Let $d \geq 0$ and assume ρ is an LK^*_{d+1}-proof of a sequent S from sequents $Z_1, \ldots, Z_t$, all consisting of formulas of depth $\leq d$.*

Then there is an LK_d-proof π of S from $Z_1, \ldots, Z_t$ of size $|\pi| = O(|\rho|^4)$. Moreover, π can be constructed from ρ by a p-time algorithm.

Proof Assume without loss of generality that all depth $d+1$ formulas in ρ are disjunctions; hence a typical sequent U in ρ looks like

$$\ldots, \bigvee_j A^i_j, \ldots, \Gamma \longrightarrow \Delta, \ldots, \bigvee_v B^u_v, \ldots \tag{3.4.2}$$

where Γ, Δ contain only formulas of depth $\leq d$. Let ρ_U be the subproof of ρ ending with U.

Claim *The depth $\leq d$ sequent*

$$\Gamma \longrightarrow \Delta, \ldots, B^u_v, \ldots$$

has an LK_d*-proof* σ *from* $\overline{Z}$ *together with the sequents*

$$\longrightarrow \ldots, A^i_j, \ldots, \quad \textit{one for each } i,$$

such that $|\sigma| = O(|\rho_U|^4)$.

The claim is proved by induction on the number of steps in ρ_U, showing that the required proof σ exists and has $O(|\rho_U|^2)$ steps, each of size $O(|\rho_U|^2)$. The assumption that ρ (and hence ρ_U) is tree-like is used in the estimate of the number of steps.

The claim implies the lemma as there are no depth $d+1$ formulas in S. □

Under an additional condition we obtain a tree-like proof.

Lemma 3.4.5 *If the proof* ρ *in Lemma 3.4.4 has in each sequent at most* c *depth* $d+1$ *formulas, for some parameter* $c \geq 0$*, then* π *can be also tree-like (i.e. it is an* LK^*_d*-proof) and of size at most* $O(|\rho|^{c+4})$.

Proof Let ρ be an LK^*_{d+1}-proof of the sequent S from the sequents $Z_1, \ldots, Z_t$, all of depth $\leq d$. As in the proof of Lemma 3.4.4 we assume without loss of generality that all depth-$(d+1)$ formulas in ρ are disjunctions and that a typical sequent U in ρ looks like (3.4.2). Assume that i and u range over $[r]$ and $[s]$, respectively, i.e. $r + s \leq c$, and, without loss of generality, that j and v for all i and u range over $[t]$.

Claim *For each choice* $j_1, \ldots, j_r \in [t]$ *there is an* LK_d*-proof* $\pi[j_1, \ldots, j_r]$ *of*

$$A^1_{j_1}, \ldots, A^r_{j_r}, \Gamma \longrightarrow \Delta, B^1_1, \ldots, B^1_t, \ldots, B^s_1, \ldots, B^s_t$$

such that the total size of these proofs satisfies

$$\sum_{j_1,\ldots,j_r} |\pi[j_1, \ldots, j_r]| \leq O(t^r \mathbf{w}(\rho_U)^2 \mathbf{k}(\rho_U)^2) \leq O(|\rho_U|^{c+4}).$$

The claim is proved by induction on the number of inferences in ρ_U, distinguishing the type of the last inference yielding U. If the principal formula of the inference has depth $\leq d$, the same inference applies (this is part of the multiplicative term $O(\mathbf{k}(\rho_U)^2)$) to all proofs $\pi[j_1, \ldots, j_r]$, for all choices of $j_1, \ldots, j_r$ (this is the multiplicative term t^r).

If the inference was an $\bigvee$:left introduction of a depth $d+1$ formula then there is nothing to do. If it was an $\bigvee$:right introduction, the inference is simulated by one or more weakenings.

The only case requiring some non-trivial simulation occurs when the last inference is a cut with cut formula a depth $d+1$ disjunction $\bigvee_i C_i$. The induction assumption gives us proofs of the sequents

$$C_i, \Pi \longrightarrow \Omega, \quad \text{for } i = 1, \ldots, r, \tag{3.4.3}$$

and a proof of

$$\Pi \longrightarrow \Omega, C_1, \ldots, C_r \tag{3.4.4}$$

and we use (3.4.3) for $i = 1, \ldots, r$ to cut out all formulas C_i from (3.4.4). We do this for at most t^r choices of (at most) r-tuples $j_i \in [t]$, and each simulation of the cut needs size $O(\mathbf{w}(\rho_U)^2)$.

The theorem follows from the claim for the end-sequent. □

Now we can strengthen Lemma 3.4.1 a little.

Corollary 3.4.6 *Let $d \geq 0$ and let S and $Z_1, \ldots, Z_t$ be sequents consisting of formulas of depth $\leq d$. Assume that*

$$Z_1, \ldots, Z_t \models S.$$

Then there is an LK_d*-proof of S from $Z_1, \ldots, Z_t$.*

Proof As in (3.2.3), for a sequent

$$U: A_1, \ldots, A_u \longrightarrow B_1, \ldots, B_v,$$

$\bigvee U$ denotes the disjunction $\neg A_1 \vee \cdots \vee \neg A_u \vee B_1 \vee \cdots \vee B_v$. The hypothesis of the corollary implies that the depth $\leq d+1$ sequent

$$\bigvee Z_1, \ldots, \bigvee Z_t \longrightarrow \bigvee S \tag{3.4.5}$$

is logically valid and hence has an LK^*_{d+1}-proof (that is even cut-free), by Lemma 3.4.1.

There are short LK^*_{d+1}-proofs of $\bigvee Z_i$ from Z_i, $i \leq t$, and of S from $\bigvee S$, as proved in Claim 1 in the proof of Lemma 3.4.2. Hence Lemma 3.4.4 applies. Note that we do not use here the feasibility of the simulation, as the starting proof of (3.4.5) may be exponentially long. □

For connections with bounded arithmetic, discussed in Part II, it is useful to consider a modification of the depth measure, the **Σ-depth**, and the corresponding subsystems of LK. The reason is that then there is an exact correspondence between levels of the hierarchy of bounded formulas and the classes of bounded Σ-depth propositional formulas: the $\Sigma^b_d(\alpha)$ bounded formulas translate to Σ-depth-d propositional formulas, as in Section 1.2. It is a simple concept that is often defined incorrectly in the literature, however.

Informally, a Σ-depth d formula (tacitly in the DeMorgan language with the unbounded $\bigvee$ and $\bigwedge$ connectives) is a depth $d+1$ formula with depth 1 subformulas that are conjunctions or disjunctions of *small* arity t. For example, t can be a constant or some parameter growing very slowly with the formula size. The subtle point in defining is not the growth rate of t but the quantity in relation to which t is supposed to be small. Authors often define $t = O(\log n)$ where n is the number of atoms or, even, the *natural* parameter of the formula, as is the n in PHP_n. The latter is not well defined unless the formula is a translation of a first-order sentence over a finite structure, as in Section 1.2; in that case n can be the size of the structure. The former is well defined but does not work well: this definition breaks the link with bounded formulas mentioned earlier.

The only correct approach seems to be to relate t to the size S of the formula. A propositional formula A is $\Sigma_d^{S,t}$ if and only if:

1. A is in negation normal form (i.e. the negations are applied only to atoms);
2. the top connective is $\bigvee$;
3. $\mathrm{dp}(A) = d + 1$;
4. $|A| \leq S$;
5. the depth-1 connectives $\bigvee$ and $\bigwedge$ have arity $\leq t$.

We say that A is $\Pi_d^{S,t}$ if and only if $\neg A$ is $\Sigma_d^{S,t}$ (after applying the DeMorgan rules to put $\neg A$ into negation normal form).

The **Σ-depth-d subsystem** of LK, denoted by $\mathrm{LK}_{d+1/2}$, is simply a version of the subsystem LK_{d+1} in which the size of the proof is measured differently: the size of an $\mathrm{LK}_{d+1/2}$-proof π is S if and only if $|\pi| \leq S$ and all formulas in π are $\Sigma_d^{S,t}$ with $t = \log(S)$.

Choosing t to be the logarithm of S works well for the connections to bounded arithmetic. Other choices for the rate of t will be considered in the context of resolution in Section 5.7.

The constructions underlying the proofs of Lemmas 3.4.2, 3.4.4 and 3.4.5 do not alter the depth 1 subformulas (if the depth is at least 2 in the latter) and hence the statements hold for the Σ-depth as well. We state this for future reference.

Lemma 3.4.7 *Lemmas 3.4.2, 3.4.4 and 3.4.5 are also valid for the systems* $\mathrm{LK}_{d+1/2}$ *and* $\mathrm{LK}^*_{d+3/2}$, *for all* $d \geq 0$.

The same argument is easily modified to the case when the stronger, tree-like, system is $\mathrm{LK}^*_{1/2}$. We shall treat this in Section 5.7 using the resolution proof system (Lemma 5.7.2).

3.5 Bibliographical and Other Remarks

Sec. 3.1 LK stands for *Logischer Kalkül*. Takeuti [486] is a comprehensive treatment of classical proof theory based on the sequent calculus LK, and Buss [106, 116] gave his modification LKB of the calculus for bounded arithmetic. Some authors, for example Cook and Nguyen [155], use the name PK for the propositional part of LK. I see no danger in using LK.

Sec. 3.2 The p-equivalence of F and LK (Theorem 3.2.4) was treated in Reckhow [445]. Cook and Reckhow [156] proved instead the p-equivalence of Frege systems with a natural deduction system (see Prawitz [406]).

Lemma 3.2.5 was first proved by Statman [482, 483] (see also [145] for an exposition of his construction) but he used also the implication connective, which we have not included in the language of LK. Urquhart [497] is a good source on these weak systems.

Sec. 3.3 The section stems from [280]. Beth's theorem (whose feasible version is Theorem 3.3.3) was proved by Beth [74, 75].

Sec. 3.4 Lemma 3.4.3 is from [276] and [278, Lemma 4.3.10] but the balancing was done there somewhat differently. The introduction rules $\vee$:left and $\wedge$:right were replaced by new *special* unary rules and new axioms, and the balancing was done with respect to the cut rule (the only left non-unary rule). Lemmas 3.4.4 and 3.4.5 are from [276, 278].

4
Quantified Propositional Calculus

The class of all **quantified propositional formulas**, denoted by Σ_∞^q, extends the class of the DeMorgan propositional formulas defined in Section 1.1 by adding two new formation rules:

- If $A(p, \overline{q})$ is a quantified propositional formula then so are $\exists x A(x, \overline{q})$ and $\forall x A(x, \overline{q})$.

The truth functions of $\exists x A(x, \overline{q})$ and $\forall x A(x, \overline{q})$ are the same as those of the disjunction $A(0, \overline{q}) \vee A(1, \overline{q})$ and those of the conjunction $A(0, \overline{q}) \wedge A(1, \overline{q})$, respectively. Quantified propositional formulas thus do not define new Boolean functions (after all, ordinary propositional formulas already define all such functions, by Lemma 1.1.2) but may define them by a shorter formula. For example,

$$\bigvee_{\overline{b} \in \{0,1\}^n} B(\overline{b}, \overline{q})$$

has size at least $2^n|B|$ while the same truth table function can be defined by a size $O(n + |B|)$ quantified propositional formula

$$\exists x_1 \ldots \exists x_n \, B(\overline{x}, \overline{q}).$$

4.1 Sequent Calculus G

The class Σ_∞^q can be stratified into levels, analogously to the stratification of predicate formulas by the quantifier complexity.

Definition 4.1.1 The class Σ_0^q equals the class Π_0^q and consists of all DeMorgan quantifier-free propositional formulas. For $i \geq 0$ the classes Σ_{i+1}^q and Π_{i+1}^q are the smallest classes satisfying the following closure conditions:

- $\Sigma_i^q \cup \Pi_i^q \subseteq \Sigma_{i+1}^q \cap \Pi_{i+1}^q$;
- both Σ_{i+1}^q and Π_{i+1}^q are closed under $\vee$ and $\wedge$;
- if $A \in \Sigma_{i+1}^q$ then $\neg A \in \Pi_{i+1}^q$ and if $A \in \Pi_{i+1}^q$ then $\neg A \in \Sigma_{i+1}^q$;
- Σ_{i+1}^q is closed under the existential quantification and Π_{i+1}^q is closed under the universal quantification.

Note that $\Sigma_\infty^q = \bigcup_i \Sigma_i^q = \bigcup_i \Pi_i^q$.

To define the proof system G for quantified propositional formulas we extend the calculus LK by quantifier rules analogous to the first-order version of Gentzen's original LK. The notions of a free and a bounded occurrence of a variable (an atom) are the same as in first-order logic (cf. Section 1.2).

Definition 4.1.2 The quantified propositional calculus G extends LK by allowing any Σ_∞^q-formulas in sequents and by accepting also the following quantifier rules:

- $\forall$:introduction

$$\text{left } \frac{A(B), \Gamma \longrightarrow \Delta}{\forall x A(x), \Gamma \longrightarrow \Delta} \quad \text{and} \quad \text{right } \frac{\Gamma \longrightarrow \Delta, A(p)}{\Gamma \longrightarrow \Delta, \forall x A(x)},$$

- $\exists$:introduction

$$\text{left } \frac{A(p), \Gamma \longrightarrow \Delta}{\exists x A(x), \Gamma \longrightarrow \Delta} \quad \text{and} \quad \text{right } \frac{\Gamma \longrightarrow \Delta, A(B)}{\Gamma \longrightarrow \Delta, \exists x A(x)},$$

where B is any formula such that no variable occurrence that is free in B becomes bounded in $A(B)$, and with the restriction that the atom p does not occur in the lower sequents of $\forall$:right and $\exists$:left.

The system G_i is a subsystem of G which allows only Σ_i^q-formulas in sequents, and G^* and G_i^* are tree-like versions of G and G_i, respectively.

See Section 4.4 for variants of this definition considered in the literature. The systems G_i may look as though they are defined asymmetrically with respect to the classes Σ_i^q and Π_i^q, but it is easy to see that allowing Π_i^q-formulas in G_i-proofs would not make the system stronger (instead of a Π_i^q-formula in an antecedent or a succedent, its negation would be kept in the opposite cedent). This definition simplifies some later arguments.

We are interested in G and in its fragments primarily as proof systems for TAUT but we may also compare their strength as proof systems for quantified propositional formulas. The relations $\leq_p$ and $\equiv_p$ are not quite adequate for this. Denote by TAUT_i the set of tautological Σ_i^q-formulas and, for proof systems P, Q for TAUT_i, denote by $P \leq_p^i Q$ the existence of a p-simulation of P-proofs of formulas from TAUT_i by Q-proofs. Similarly, $P \equiv_p^i Q$ means $P \leq_p^i Q \wedge Q \leq_p^i P$.

Theorem 4.1.3 *$EF \equiv_p G_1^*$ and, for all $i \geq 1$, $G_i \equiv_p^i G_{i+1}^*$.*

Proof We shall treat in detail the key case $i = 1$; the case $i > 1$ is analogous.

Let us observe first that G_1^* can simulate efficiently an instance of the substitution rule

$$\frac{A(p_1, \ldots, p_n)}{A(B_1, \ldots, B_n)},$$

with $A, B_1, \ldots, B_n$ quantifier-free formulas. Assume that we have already derived $\longrightarrow A(p_1, \ldots, p_n)$ and apply $\forall$:right to it n times to get

$$\longrightarrow \forall x_1 \ldots \forall x_n\, A(x_1, \ldots, x_n).$$

Then derive separately the sequent

$$A(B_1,\ldots,B_n) \longrightarrow A(B_1,\ldots,B_n)$$

and apply $\forall$: left to it n times to derive

$$\forall x_1 \ldots \forall x_n A(x_1,\ldots,x_n) \longrightarrow A(B_1,\ldots,B_n).$$

Using the cut rule, infer as required:

$$\longrightarrow A(B_1,\ldots,B_n)\ .$$

This simulation of the substitution rule is clearly enough to show that G_1^* polynomially simulates *SF*. However, $SF \geq_p EF$ by Lemma 2.4.3.

Now we describe the simulation of G_1^* by *EF*. Let $s\Sigma_1^q$ be the class of **strict** Σ_1^q**-formulas** consisting of all quantifier-free formulas and all Σ_1^q-formulas that begin with a block of existential quantifiers followed by a quantifier-free kernel.

Consider first a G_1^*-proof in which all formulas are $s\Sigma_1^q$. In such a proof the sequents look like

$$A_1(\overline{p}),\ldots,\exists\overline{x}_1 A_1'(\overline{p},\overline{x}_1),\cdots \longrightarrow B_1(\overline{p}),\ldots,\exists\overline{y}_1 B_1'(\overline{p},\overline{y}_1),\ldots, \tag{4.1.1}$$

where $A_1,\ldots,A_1',\ldots,B_1,\ldots,B_1',\ldots$ are quantifier-free formulas.

We shall call a sequence of formulas an *EF*-**sequence** if it satisfies all the conditions on *EF*-derivations given in Section 2.4, except possibly the requirement that the extension atoms cannot appear in the last formula.

The following claim is established by induction on k.

Claim *Assume that the sequent* (4.1.1) *has a* G_1^**-proof* π *in which all formulas are either quantifier-free or strict* Σ_1^q *and with* $k := \mathbf{k}(\pi)$ *sequents and size* $|\pi|$*. Then there is an EF-sequence* ρ *from* $\{A_1(\overline{p}),\ldots,A_1'(\overline{p},\overline{q}_j),\ldots\}$ *with extension atoms* $\overline{r}_1,\ldots,$ *with last formula*

$$\bigvee_s B_s(\overline{p}) \vee \bigvee_t B_t'(\overline{p},\overline{r}_t)$$

and with $O(k)$ *steps and size* $O(|\pi|)$*.*

The claim implies the lemma for G_1^*-proofs containing only strict Σ_1^q-formulas. The general case when some formulas may be in $\Sigma_1^q \setminus s\Sigma_1^q$ is more tedious but can be treated analogously. □

For completeness of discussion we state a lemma that is proved by an argument analogous to that in Section 2.2.

Lemma 4.1.4 G *and* G* *are p-equivalent with respect to proofs of all quantified propositional sequents.*

We conclude this section by a lemma which underlines the importance of Σ_1^q-tautologies among all quantified propositional tautologies and provides another reason why considering the systems G_i and G_i^* as proof systems for TAUT and $TAUT_1$ is central. We will give its proof via bounded arithmetic in Section 12.5.

Lemma 4.1.5 *Assume that all Σ_1^q-tautologies have polynomial-size proofs in G_i, for some $i \geq 1$. Then all Σ_i^q-tautologies have polynomial-size proofs in G_i too. The same holds with G_i^* in place of G_i.*

4.2 Relativized Frege systems and Boolean Programs

Let F be a Frege system in the DeMorgan language. For a collection L of function symbols f, each with an assigned arity $k_f \geq 0$, let $F(L)$ be the proof system defined as follows:

1. the language of $F(L)$ is the DeMorgan language augmented by the symbols from L as new connectives;
2. $F(L)$ has the same Frege rules as F but its instances are obtained by substituting for atoms any formulas in the language of $F(L)$;
3. $F(L)$ has also the following schemes for the **equality axioms**:

$$\bigwedge_{i \leq k_f} p_i \equiv q_i \longrightarrow f(\overline{p}) \equiv f(\overline{q})$$

for all $f \in L$.

We call this system a **relativized Frege system** and (in this context) $F(L)$-formulas **relativized formulas**. The following is an easy observation. Its proof needs the equality axioms.

Lemma 4.2.1 *Any relativized Frege system $F(L)$ is complete with respect to all logically valid formulas in the expanded language (i.e. formulas true under any interpretation of the symbols from L and under any truth assignment to atoms).*

We are primarily interested in relativized Frege systems (and their extensions) as in proof systems for TAUT (i.e. tautologies in the DeMorgan language). One reason is that satisfiability for the relativized formulas is p-reducible to SAT. Namely, given a relativized formula C, form the following set $S(C)$ of DeMorgan formulas (it is analogous to the set $Ext(A)$ from Section 2.4):

1. introduce a new atom q_D for each subformula D of C;
2. if $D = \neg E$, include in $S(C)$ the formula $q_D \equiv \neg q_E$;
3. if $D = E \circ G$ for $\circ = \vee, \wedge$, include in $S(C)$ the formula $q_D \equiv (q_E \circ q_G)$;
4. if D is an atom p, replace q_D by p everywhere;
5. for all D, E starting with the symbol f and having subformulas $D_1, \ldots, D_{k_f}$ and $E_1, \ldots, E_{k_f}$ respectively, include in $S(C)$ the formula

$$\bigwedge_{j \leq k_f} q_{D_j} \equiv q_{E_j} \longrightarrow q_D \equiv q_E;$$

6. include in $S(C)$ the atom q_C.

Then we have:

Lemma 4.2.2 *For any relativized formula C:*

$$C \text{ is satisfiable if and only if } \bigwedge S(C) \in \text{SAT}.$$

Proof If C is satisfiable, each subformula is given a truth value and all local conditions are satisfied, as well as all instances of the equality axiom in item 5 above, because f is interpreted by a Boolean function.

However, having a satisfying assignment $\overline{a}$ for the atoms in $S(C)$, interpret the functions f: $\{0, 1\}^{k_f} \to \{0, 1\}$ for all $f \in L$ by setting

$$f(\overline{a}(q_{D_1}), \ldots, \overline{a}(q_{D_{k_f}})) := \overline{a}(q_D)$$

for each subformula D starting with f, and arbitrarily for all other assignments. The formulas in item 5 imply that there is no conflict in this definition of f, and q_C in item 6 means that the whole formula is given the value 1 when the functions f are interpreted as above and the atoms are given the same value as in $\overline{a}$. □

A consequence of this proof is the following lemma.

Lemma 4.2.3 *Relativized Frege systems are implicationally complete for relativized formulas also.*

Proof Let E_i be relativized formulas that imply a relativized formula G. Put

$$C := \bigwedge_i E_i \longrightarrow G.$$

Then $\neg C$ is not satisfiable and hence by Lemma 4.2.2, $S(\neg C)$ is not satisfiable either. Therefore there is an F-proof π of 0 from $S(\neg C)$ and, substituting in π for each atom q_D the formula D, yields an $F(L)$-refutation of $\neg C$ and hence a proof of C. □

The notion of a Boolean program introduced by Cook and Soltys [157] is quite natural when defining sequentially more and more complex Boolean functions. A **Boolean program** is a sequence of function symbols $f_1, \ldots, f_s$, each f_i having a prescribed arity $k_i \geq 0$, and a sequence of formulas $A_1, \ldots, A_s$ satisfying the following:

- A_i is built from k_i atoms p^i_j, $j \leq k_i$, using the DeMorgan language augmented by the symbols $f_1, \ldots, f_{i-1}$.

These formulas are understood as giving **instructions** how to compute f_i:

$$f_i(\overline{p}^i) := A_i(\overline{p}^i).$$

The **size** of the program is $\sum_{i \leq s} |A_i|$.

Boolean programs are analogous, to an extent, to circuits as defined by sequences of instructions in Section 1.4. In the same way as a sequence of p-size circuits defines a language in non-uniform P (i.e. P/*poly*), a sequence of p-size Boolean programs defines a language in non-uniform PSPACE (i.e. PSPACE/poly), cf. Cook and Soltys [157].

Let P be a Boolean program introducing functions $f_1, \ldots, f_s$; we shall denote the collection of such programs by L_P. The symbol $F(P)$ will denote the proof system that extends the relativized Frege system $F(L_P)$ by:

- the instructions of P as new axiom schemes.

In particular, a substitution of any formulas in the language of $F(P)$ for the atoms in any instruction of P is a valid axiom instance of $F(P)$.

We shall call these extensions of Frege systems collectively BP-**Frege systems**, using a generic notation BPF.

Lemma 4.2.4 *Let P be a Boolean program, assume that*

$$f(\overline{p}) \equiv A(\overline{p}, 1)$$

is one of its instructions and let $B(\overline{p})$ be a quantifier-free formula. Then the formula

$$A(\overline{p}, q) \rightarrow A(\overline{p}, f(\overline{p})) \tag{4.2.1}$$

has an $F(P)$-proof of size $|A|^{O(1)}$, and its substitution instance

$$A(\overline{p}, B) \rightarrow A(\overline{p}, f(\overline{p}))$$

has an $F(P)$-proof of size $|A(B)|^{O(1)}$.

Proof The second part of the statement follows from the first part on substituting B for q in the $F(P)$-proof of (4.2.1). To construct a size $|A|^{O(1)}$ $F(P)$-proof of this formula we proceed as follows.

The first claim below is established by induction on the logical complexity of A, using the equality axioms if A starts with one of the symbols f_i.

Claim 1 The formula $q \equiv r \rightarrow A(\overline{p}, q) \equiv A(\overline{p}, r)$ has an $F(P)$-proof of size $O(|A|^2)$.

From Claim 1 follows

Claim 2 Both formulas

$$A(\overline{p}, 1) \rightarrow A(\overline{p}, A(\overline{p}, 1)) \quad \text{and} \quad A(\overline{p}, 0) \rightarrow A(\overline{p}, A(\overline{p}, 1))$$

have $F(P)$-proofs of size $|A|^{O(1)}$.

By Claim 1,

$$1 \equiv A(\overline{p}, 1) \rightarrow A(\overline{p}, 1) \equiv A(\overline{p}, A(\overline{p}, 1))$$

has a p-size $F(P)$-proof from which the first statement follows. For the second part, split the proof into two cases: if $A(\overline{p}, 1)$ holds then the first part applies; if not then $0 \equiv A(\overline{p}, 1)$ and Claim 1 applies.

Claim 3 The formula $A(\overline{p}, q) \rightarrow (A(\overline{p}, 0) \vee A(\overline{p}, 1))$ has a p-size $F(P)$-proof.

Claims 2 and 3 imply that the formula

$$A(\overline{p}, q) \to A(\overline{p}, A(\overline{p}, 1))$$

has a p-size $F(P)$-proof, from which (4.2.1) follows using Claim 1 and the axiom of $F(P)$ about f. □

4.3 Skolemization of G

In this section we shall investigate the possibility of generalizing the simulation of G_1^* by EF from Theorem 4.1.3 to G. A natural approach is to introduce definable Skolem functions and remove from G-proofs all quantifiers; the Skolem functions for the existential quantifiers in a Σ_1^q-formula then witness the quantifiers. This witnessing is very much related to witnessing methods in bounded arithmetic; in particular, the links between the witnessing of bounded existential consequences of fragments of bounded arithmetic and total NP search problems offer a more detailed information for the systems G_i and G_i^* as well; see [278, 155].

A Skolem function for a formula $\exists x A(\overline{p}, x)$, with $\overline{p} = (p_1, \ldots, p_n)$, is any Boolean function f: $\{0,1\}^n \to \{0,1\}$ such that the formula (4.2.1) from Lemma 4.2.4,

$$A(\overline{p}, q) \to A(\overline{p}, f(\overline{p})),$$

is logically valid. Then the equivalence

$$\exists x A(\overline{p}, x) \equiv A(\overline{p}, f(\overline{p}))$$

is logically valid too but the advantage of (4.2.1) over this formula is that (4.2.1) is quantifier-free. Observe that the function f defined by

$$f(\overline{p}) := A(\overline{p}, 1)$$

satisfies (4.2.1). To remove quantifiers from a more complex Σ_∞^q-formula we iterate this process of introducing Skolem functions from smaller subformulas to larger ones, i.e. we remove the quantifiers from inside out. For example, for

$$\forall x \exists y B(\overline{p}, x, y), \tag{4.3.1}$$

with B open, first introduce $f(\overline{p}, q)$ by

$$f(\overline{p}, q) := B(\overline{p}, q, 1)$$

and then, using that (4.3.1) is equivalent to

$$\neg \exists x \neg B(\overline{p}, x, f(\overline{p}, x)),$$

introduce the function $g(\overline{p})$ by

$$g(\overline{p}) := \neg B(\overline{p}, 1, f(\overline{p}, 1)) .$$

Formula (4.3.1) is then equivalent to

$$B(\overline{p}, g(\overline{p}), f(\overline{p}, g(\overline{p}))) .$$

We see that the Skolemized formula becomes syntactically complex very fast as new functions introduced later are substituted into terms involving the functions introduced earlier. Moreover, the formula may grow exponentially in its size, as the functions may perhaps be substituted for atoms having multiple occurrences. We shall thus proceed slightly differently, introducing further new functions as abbreviations of formulas in which an existential quantifier is replaced by a Skolem function.

Given a Σ^q_∞-formula A we shall define

- a Boolean program $Sk[A]$, and
- a quantifier-free formula A_{Sk} in the language of $Sk[A]$,

and call this process the **Skolemization** of A. It is defined by induction on the complexity of A as follows.

1. For $B(\overline{p})$ quantifier-free, $Sk[B] := \emptyset$ and $B_{Sk} := B$.
2. For $B(\overline{p}) = C(\overline{p}) \circ D(\overline{p})$,

$$Sk[B] := Sk[C] \cup Sk[D] \quad \text{and} \quad B_{Sk} = C_{Sk}(\overline{p}) \circ D_{Sk}(\overline{p}) ,$$

   for $\circ = \vee, \wedge$.
3. For $B(\overline{p}) = \neg C(\overline{p})$, $Sk[B] := Sk[C]$ and $B_{Sk} = \neg C_{Sk}(\overline{p})$.
4. For $B(\overline{p}) = \exists x C(\overline{p}, x)$, $f_B(\overline{p})$ is a new function symbol, $Sk[B]$ is $Sk[C]$ together with the instruction

$$f_B(\overline{p}) := C(\overline{p}, C(\overline{p}, 1))$$

   and $B_{Sk} := f_B(\overline{p})$.
5. For $B(\overline{p}) = \forall x C(\overline{p}, x)$, $Sk[B]$ and B_{Sk} are defined as $Sk[\neg\exists x \neg C]$ and $(\neg\exists x \neg C)_{Sk}$, respectively.

In item 4 we have combined two elementary steps into one: instead of defining the Skolem function f as $C(\overline{p}, 1)$ and then introducing a new function $g(\overline{p})$ to abbreviate $C(\overline{p}, f(\overline{p}))$ we do it in one step. The last item of the definition implies that, for $B(\overline{p}) = \forall x C(\overline{p}, x)$, the Boolean program $Sk[B]$ is $Sk[C]$ together with the function $f_{\exists x \neg C}$ introduced by the instruction

$$f_{\exists x \neg C} := \neg C_{Sk}(\overline{p}, \neg C_{Sk}(\overline{p}, 1))$$

and $B_{Sk} = \neg f_{\exists x \neg C}$. Note also that both $Sk[A]$ and A_{Sk} have sizes polynomial in $|A|$.

For a set $\mathcal{A}$ of quantified formulas, let $Sk[\mathcal{A}]$ be the union of all $Sk[A]$ for all $A \in \mathcal{A}$, where it is understood that the same formula occurring as a subformula at different places in $\mathcal{A}$ is assigned the same Skolem function symbol.

For a sequent S: $\Gamma \longrightarrow \Delta$, $Sk[S]$ is $Sk[\Gamma \cup \Delta]$ and S_{Sk}: $\Gamma_{Sk} \longrightarrow \Delta_{Sk}$ is the sequent with each formula B in $\Gamma \cup \Delta$ replaced by B_{Sk}.

Theorem 4.3.1 *Assume that π is a* G*-proof of the sequent S:* $\Gamma \longrightarrow \Delta$*. Then there is a Boolean program P of size $|\pi|^{O(1)}$ and an F(P)-proof σ of S_{Sk} of size $|\pi|^{O(1)}$. Moreover, P and σ can be constructed by a p-time algorithm from π.*

Proof By Lemma 4.1.4 we may assume without loss of generality that π is tree-like. Let $\mathcal{A}$ be the set of all formulas occurring in π and let $P := Sk[\mathcal{A}]$.

For each sequent U from π we construct an $F(P)$-proof of U_{Sk} of size $O(t|\pi|^{O(1)})$, proceeding by induction on the number of steps t used in π to derive U. As Skolemization commutes with the propositional connectives we need to consider only cases where U was derived via the quantifier rules. We shall consider the right rules only, the left rules being analogous.

The rule $\exists$:right. The sequent U is derived from some sequent V with a minor formula $C(D)$ and principal formula $\exists x C(x)$. By the induction hypothesis, V_{Sk} has a short tree-like $F(P)$-proof σ_V and its succedent contains the formula $(C(D))_{Sk}$, which is $C_{Sk}(D_{Sk})$.

By Lemma 4.2.4 we have a short $F(P)$-proof of

$$C_{Sk}(D_{Sk}) \longrightarrow (\exists x C_{Sk}(x))_{Sk}$$

and joining it to σ_V by a cut yields σ_U.

The rule $\forall$:right. Here the minor formula in V is $C(\overline{p}, q)$ and the principal formula is $\forall x C(\overline{p}, x)$. The Skolemization of $\forall x C(\overline{p}, x)$ is

$$\neg f_{\exists x \neg C}(\overline{p}) \ . \tag{4.3.2}$$

Substitute $q := \neg C_{Sk}(\overline{p}, 1)$ everywhere in σ_V to get a short proof of V_{Sk} with C replaced by $C(\overline{p}, \neg C_{Sk}(\overline{p}, 1))$ (this uses the fact that σ_V is tree-like). Using the instruction in P for $f_{\exists x \neg C}$ this formula can be further replaced by (4.3.2), thus obtaining U_{Sk}. □

Now we are ready for two statements generalizing, in a sense, Theorem 4.1.3.

Corollary 4.3.2 *Assume that π is a* G*-proof of a quantifier-free formula A. Then there are a size $|\pi|^{O(1)}$ Boolean program P and a size $|\pi|^{O(1)}$ F(P)-proof of A, and both can be found by a p-time algorithm, given π.*

That is, BPF p-simulates G.

Corollary 4.3.3 *Assume that $B(\overline{p}, \overline{q})$ is a quantifier-free formula, $\overline{q} = q_1, \ldots, q_\ell$, and π is a* G*-proof of*

$$\longrightarrow \exists x_1, \ldots, x_\ell B(\overline{p}, \overline{x}).$$

Then there are a size-$|\pi|^{O(1)}$ Boolean program P containing the functions

$$f_1(\overline{p}), \ldots, f_\ell(\overline{p})$$

and a size-$|\pi|^{O(1)}$ F(P)-proof of

$$\longrightarrow B(\overline{p}, f_1(\overline{p}), \ldots, f_\ell(\overline{p})).$$

Moreover, both the program P and the $F(P)$-proof can be found by a p-time algorithm from π.

Note that the existence of a small P computing the witness for the existential quantifiers is obvious, as they can be computed in PSPACE and as polynomial-size Boolean programs define all functions in PSPACE; see Cook and Soltys [157]. The extra information provided is the existence of a short *BPF*-proof of the correctness of the program.

We state a lemma complementing Corollary 4.3.2. It is easy to prove using bounded arithmetic via the methods to be explained in Part II; in particular, the lemma follows from Theorem 12.5.4(ii) as the theory U_2^1 is easily seen to prove the soundness of *BPF*.

Lemma 4.3.4 G *p-simulates BPF with respect to TAUT.*

One can view *BPF* as a *second-order version* of *EF*, just as Boolean programs generalize Boolean circuits from defining mere values to defining functions. By Lemmas 2.4.3 and 2.4.4, $EF \equiv_p SF$ and it springs to mind that the relativized Frege system augmented by a suitable *second-order* substitution rule could be p-equivalent to *BPF* and thus, by Corollary 4.3.2 and Lemma 4.3.4, to G.

A **second-order substitution** σ into a relativized formula $A(\overline{p},\overline{f})$ is determined as follows:

- σ assigns to each atom p_i a relativized formula B_i;
- σ assigns to each f with arity k_f a relativized formula C_f and a k_f-tuple of atoms $q_1,\ldots,q_{k_f}$ (which may or may not occur in C_f and, further, C_f may have other atoms);
- the substitution instance $\sigma(A)$ is obtained by simultaneously replacing each atom p_i by the formula B_i and each occurrence of $f(x_1,\ldots,x_{k_f})$ by the formula C_f, with the atom q_j playing the role of the jth argument x_j of f.

The **second-order substitution rule**, abbreviated as the S^2-rule, allows us to infer from a relativized formula A any substitution instance $\sigma(A)$ given by any second-order substitution σ. It is easy to see that if A is logically valid, so is $\sigma(A)$. Thus we can define a sound extension of a relativized Frege system by adding the S^2-rule. Let us denote by S^2F **a Frege system with the S^2-rule**.

Unfortunately the next lemma shows that this system is weaker than expected and thus unlikely to be p-equivalent to G.

Lemma 4.3.5 *SF p-simulates S^2F.*

Although a direct p-simulation is not difficult, its definition is syntactically cumbersome. We shall give instead a more conceptual proof of the lemma, using bounded arithmetic, in Section 11.5. It comes down to the fact that the set of logically valid relativized formulas is p-reducible to TAUT.

The reason why a simulation of BPF by S^2F apparently cannot be constructed analogously to the simulation of EF by SF in Lemma 2.4.3 is that, during the elimination of the instructions of a program P, one substitutes for function symbols the formulas where these symbols occur, but for varying lists of arguments (this avoids the use of some circuit formalism for the substitution and so keeps its description small).

4.4 Bibliographical and Other Remarks

Sec. 4.1 The proof systems G_i and G_i^* were later redefined by Cook and Morioka [154] in the following way:

- they allowed arbitrary quantified propositional formulas in proofs and restricted only the cut formulas to the class Σ_i^q; and
- they restricted the formulas B in $\forall$:left and $\exists$:right to be quantifier-free.

The first modification is seen to be natural when one studies the propositional translations of bounded arithmetic proofs, as we shall do in Part II. The second modification has the effect that the only rule that can decrease the maximum quantifier complexity of a formula in a sequent is the cut rule. This is a form of subformula property analogous to that in the proof of Corollary 3.2.3 and, in fact, it can be formulated in a much more precise way.

However, when considering the systems G_i and G_i^* primarily as proof systems for TAUT or TAUT_1 (the latter in connection with NP search problems) this modification does not matter, as the two versions of the systems p-simulate proofs of tautologies of each other. In places we will refer to the original literature, where the definitions were given as above and because of this we will stick to the original definition of the systems.

Note that we could have similarly modified the definition of the systems LK_d in Section 3.4, restricting only cut formulas to depth $\leq d$. However, in those systems we are primarily interested in proofs (or refutations) of constant-depth sequents, and the modified definitions, in my view, only obfuscate the problems studied by allowing one to produce simple solutions involving LK_d-proofs of unbounded-depth sequents.

Morioka [356] proved that cut formulas in G_i^*-proofs (and also in the modified systems) can be restricted to prenex formulas, and this was extended to G_i-proofs by Jeřábek and Nguyen [261].

Generalizations of the witnessing-Σ_1^q consequences of G_1^*-proofs from Theorem 4.1.3 to higher fragments of G and to more complex formulas (in the style of the KPT theorem, see Section 12.2) were given by Perron [393, 394].

Sec. 4.2 A similar proof system to the relativized Frege system of Section 4.2 but in the sequent calculus formalism was considered by Cook [150] under the name *relativized propositional calculus* and attributed there to Ben-David and Gringauze [64].

Sec. 4.3 Cook and Soltys [157] showed how to witness G_1^*-proofs by Boolean programs. This was generalized to the whole of G by Skelley [471, 472], who also defined an extension *BPLK* of LK by Boolean programs that was similar but not equal to *BPF* (but he included the substitution rule).

One may speculate that going up in the hierarchy of Boolean circuits and Boolean programs for functions to Boolean programs for functionals, etc. will define extensions of *BPF* that are p-equivalent to the hierarchy of the implicit proof systems i^2EF, i^3EF, ... created from *EF* by the implicit proof system construction in [293] (presented here in Section 7.3).

It would make sense to consider also a *second-order Frege system* F^2, an extension of the relativized $F(L)$ that allows all substitution instances of F-rules obtained by a second-order substitution but not the S^2-rule itself. Beyersdorff, Bonacina and Chew [76] and Beyersdorff and Pich [81] studied other extensions of Frege systems relating to fragments of G, and they also defined a Skolemization of these fragments in which the Skolem functions are computed by Boolean circuits.

Lemma 4.3.5 was pointed out to me by E. Jeřábek.

5
Resolution

In this chapter we introduce a very simple and elegant proof system, the resolution proof system R. It is very popular owing to its prominence in SAT solving and automated theorem proving. A quick definition could be that it is essentially the subsystem of the sequent calculus LK using from the connectives, only $\neg$, not $\vee, \wedge$ (for convenience we sometimes use also the constants 0 and 1, however). But we shall develop R without any reference to LK.

The simplicity and elegance of R has to be paid for somewhere, and the bill comes when one wants to interpret R as a proof system for all tautologies; in its natural form it is a proof system for DNF-formulas only. The procedure for including all formulas is called limited extension and will be described below. We shall note that R is sound (that will be obvious) and give two proofs that it is also complete. After that we shall investigate several proof-theoretic properties of R, including the notion of its width and the so-called feasible interpolation property. Resolution and its variants will feature in several later chapters in all parts of the book.

Recall the following handy notation for a literal ℓ and $\epsilon \in \{0, 1\}$:

$$\ell^{\epsilon} := \begin{cases} \ell & \text{if } \epsilon = 1 \\ \neg\ell & \text{if } \epsilon = 0, \end{cases}$$

from Section 1.1.

5.1 The Resolution Rule and Its Completeness

The resolution proof system, denoted by R, is very simple. Its lines are just clauses C: $\ell_1 \vee \cdots \vee \ell_w$ written not as disjunctions but simply as sets of literals $\{\ell_1, \ldots, \ell_w\}$ that form the disjunction. The clause $C \cup \{\ell\}$ is often abbreviated by C, ℓ. Clauses are understood as not being ordered and the exchange rule of LK is used implicitly.

A clause may contain a pair of complementary literals. This is often excluded in SAT-solving algorithms, as such clauses are redundant, but in proof theory it is allowed; otherwise applications of the resolution rule would have to be restricted by some conditions preventing the creation of such a clause. The only inference rule, the **resolution rule** is

$$\frac{C, p \qquad D, \neg p}{C \cup D},$$

where p is an atom.

The resolution rule is obviously sound in the sense that any truth assignment satisfying both hypotheses of a resolution inference will satisfy the conclusion too. Given an initial set $\mathcal{C}$ of clauses, we can use the rule to expand $\mathcal{C}$ by repeatedly deriving new clauses. In particular, we may manage to derive the **empty clause**, which we shall denote by $\emptyset$. In that case it follows that the clauses in the original set $\mathcal{C}$ are not simultaneously satisfiable. This allows us to interpret R as a **refutation proof system**: instead of proving that a formula is a tautology it proves that a set of clauses is not satisfiable. If it could prove this for all unsatisfiable formulas (not just sets of clauses) it would be as good as having a proof system because A is a tautology if and only if $\neg A$ is not satisfiable.

Consider first a formula in DNF, say $A = \bigvee_i A_i$ where each A_i is a term, i.e. a conjunction of literals of the form $\bigwedge_j \ell_{i,j}$. Then A is a tautology if and only if $\neg A = \bigwedge_i \neg A_i$ is not satisfiable, i.e. if the set of clauses

$$\{\ell_{i,j}^0 \mid \ell_{i,j} \in A_i\}, \qquad \text{for all } i,$$

is not satisfiable. Hence, in R refuting this set of clauses amounts to proving that A is a tautology. If A is not in DNF then we could first transform it into DNF by the DeMorgan rules and then use R on its negation. However, a difficulty is that the DNF version of a formula may be exponentially longer, and hence a proof system that would employ such a transformation would a priori need exponentially long proofs. For example, any DNF version of $\bigwedge_{i \leq k}(p_i \vee q_i)$ must have at least 2^k clauses.

The way to avoid this issue is the so-called **limited extension** introduced by Tseitin [492]. Let A be any DeMorgan formula in atoms $p_1, \ldots, p_n$. Introduce for each subformula B of A that is not a constant or a literal a new variable q_B and for convenience of notation denote 0, 1 and p_i by q_0, q_1 and q_{p_i}, respectively. Form a set of clauses *LimExt(A)* as follows.

- If $B = \neg C$, include in *LimExt(A)* the clauses q_B, q_C and q_B^0, q_C^0.
- If $B = C \vee D$, include in *LimExt(A)* the clauses q_B^0, q_C, q_D and q_C^0, q_B and q_D^0, q_B.
- If $B = C \wedge D$, include in *LimExt(A)* the clauses q_B^0, q_C and q_B^0, q_D and $q_C^0 q_D^0, q_B$.

Lemma 5.1.1 *For any DeMorgan formula A, A is a tautology if and only if the set*

$$LimExt(\neg A), q_{\neg A}$$

is not satisfiable.

Definition 5.1.2 Let $\mathcal{C}$ be a set of clauses. An R-refutation of $\mathcal{C}$ is a sequence of clauses $D_1, \ldots, D_k$ such that:

- for each $i \leq k$, either $D_i \in \mathcal{C}$ or there are $u, v < i$ such that D_i follows from D_u, D_v by the resolution rule;
- $D_k = \emptyset$.

The number of steps in the refutation is k. A refutation of the set $LimExt(\neg A), q_{\neg A}$ is also called an R-proof of A.

We want to establish that R is complete, and we will show that by two proofs, the second of which is given in the next section. We continue to use the star notation for tree-like versions of proof systems: thus R* is a **tree-like resolution**.

Theorem 5.1.3 *Assume that $\mathcal{C}$ is an unsatisfiable set of clauses. Then there is an* R*-refutation of $\mathcal{C}$ and hence also an* R**-refutation.*

Proof Assume that $\mathcal{C}$ contains literals corresponding to n atoms and let p be one of them. Split $\mathcal{C}$ into four parts:

- $\mathcal{C}_{0,0} := \{C \in \mathcal{C} \mid C$ does not contain either p or $\neg p\}$;
- $\mathcal{C}_{1,0} := \{C \in \mathcal{C} \mid C$ contains p but not $\neg p\}$;
- $\mathcal{C}_{0,1} := \{C \in \mathcal{C} \mid C$ does not contain p but does contain $\neg p\}$;
- $\mathcal{C}_{1,1} := \{C \in \mathcal{C} \mid C$ contains both p and $\neg p\}$.

Form a set $\mathcal{D}$ consisting of all clauses that can be obtained by applying the resolution rule to a pair of clauses, one of which is from $\mathcal{C}_{1,0}$ and the other from $\mathcal{C}_{0,1}$.

Claim *$\mathcal{C}_{0,0} \cup \mathcal{D}$ is unsatisfiable.*

Assume that there is an assignment to the atoms of $\mathcal{C}_{00} \cup \mathcal{D}$ that makes all clauses true. It would follow that this assignment makes either all clauses in $\mathcal{C}_{1,0}$ or all clauses in $\mathcal{C}_{0,1}$ true, and thus we can set the value for p in a way that satisfies all of $\mathcal{C}$. That would be a contradiction.

Each clause in $\mathcal{C}_{0,0} \cup \mathcal{D}$ is derivable from $\mathcal{C}$ by one application of the rule and the set has fewer atoms. Proceed by induction to eliminate all atoms and to derive $\emptyset$.

The constructed refutation will not be tree-like because the clauses in sets $\mathcal{C}_{0,1}$ and $\mathcal{C}_{1,0}$ are used many times in a resolution inference. But we can unwind the proof – top to bottom – always repeating the subproof of a clause, for each occasion on which it is used as an inference. □

Note that the second proof of completeness in the next section will yield an R*-refutation directly.

5.2 R* and the DPLL Procedure

Let us start this section with another construction proving Theorem 5.1.3. Assume that we are given $\mathcal{C}$ and consider the following procedure, which we shall picture as a binary tree. At the root pick any atom p, label the root by p and consider two arrows leaving it labeled p^1 and p^0, respectively. Going along the arrow p^b, $b \in \{0,1\}$, substitute $p := b \in \{0,1\}$ in all clauses of $\mathcal{C}$. Then pick another variable q to label the node, split the subtree into two and label the arrows q^0 and q^1 as before, etc. Every node v in the tree determines a partial truth assignment α_v by looking at the labels of the arrows leading from the root to v. The process stops at the node v (i.e. there the tree no longer branches) if α_v falsifies a clause from $\mathcal{C}$.

Denote by T_v the subtree starting at the node v. In particular, if the process stopped at v, T_v consists of just the one node v.

Claim *For each v, the clause*

$$E_v := \{p^{1-b} \mid p \in dom(\alpha_v) \wedge \alpha_v(p) = b\}$$

has an R*-*derivation from* $\mathcal{C}$ *whose proof tree is* T_v.

This is proved by induction on the size of T_v. If $|T_v| = 1$ then, by the stopping condition, E_v contains a member of $\mathcal{C}$. If v is labeled by q and has two successors e,f then, by the induction assumption, E_e and E_f have R*-refutations with underlying trees T_e and T_f, respectively. Join them by resolving the clauses E_e, E_f with respect to the atom q.

The procedure described in this construction is called DPLL after Davis and Putnam [168] and Davis, Logemann and Loveland [166]. We now observe that the converse of the construction also holds. That is, on unsatisfiable sets of clauses R*-refutations are exactly DPLL computations.

The first part of the next lemma restates explicitly what was done in the construction above; the second part is obvious.

Lemma 5.2.1 *Let* $\mathcal{C}$ *be an unsatisfiable set of clauses and let* $\tilde{\mathcal{C}}$ *be the set of all clauses containing as a subset a clause from* $\mathcal{C}$. *Then the following correspondence holds.*

(i) If σ *is the labeled tree of a* DPLL *computation on* $\mathcal{C}$ *then an* R*-*refutation of* $\tilde{\mathcal{C}}$ *is obtained by the following relabeling:*
  - *label each node v by the clause* E_v *(defined in the claim above);*
  - *at each node v use the resolution rule to resolve the clauses labeling the child nodes of v with respect to the variable that labels v in* σ.

(ii) If π *is an* R*-*refutation of* $\mathcal{C}$ *then construct a* DPLL *computation as follows:*
  - *the tree of the computation will be the tree of* π, *the root being the end clause* $\emptyset$ *and the arrows leading from a clause to the two hypotheses of the inference yielding it in* π;
  - *the branching variable of* σ *at a node is the variable which was resolved at that node in* π.

Talking in item (i) about refutations of $\tilde{\mathcal{C}}$ instead of $\mathcal{C}$ allows us to leave the underlying tree unchanged; otherwise, it would have to be trimmed because the leaf labels E_v may be in $\tilde{\mathcal{C}} \setminus \mathcal{C}$ and, when replacing it by a subset that is from $\mathcal{C}$, it may not be possible to perform some inferences along the path to the root (as some literals to be resolved may have disappeared). In the next section we shall introduce the weakening rule, and with it $\tilde{\mathcal{C}}$ would not be needed.

One can look at the DPLL procedure also as solving on a decision tree the following **search problem**, Search($\mathcal{C}$), associated with an unsatisfiable set $\mathcal{C} = \{C_1, \ldots, C_k\}$ of clauses:

- given a truth assignment α to the variables of $\mathcal{C}$, find some C_i falsified by α.

Lemma 5.2.1 implies that the minimum size of a decision tree solving Search($\mathcal{C}$) equals the minimum size of an R*-refutation of $\mathcal{C}$.

5.3 Regular Resolution and Read-Once Branching Programs

We now extend the correspondence between R*-refutations and decision trees to a class of dag-like R-refutations satisfying the **regularity condition** defined by Tseitin [492]. An R-refutation of $\mathcal{C}$ is **regular** if and only if, on every path through the refutation from the end-clause to an initial clause, every atom is resolved at most once.

The corresponding computational device will generalize decision trees as well. A **branching program** σ for solving Search($\mathcal{C}$) is a dag with one root, several inner nodes and several leaves, such that:

- the root and the inner nodes are labeled by the variables of $\mathcal{C}$;
- the root and the inner nodes have out-degree 2 and the two arrows leaving the node are labeled by p and $\neg p$, respectively, where p is the variable labeling the node;
- the leaves of σ are labeled by the clauses of $\mathcal{C}$.

Every assignment α determines a path P_α through σ from the root to a leaf. We say that σ solves Search($\mathcal{C}$) if and only if the clause labeling the leaf on P_α is falsified by α, for all α.

A branching program is **read-once** if and only if on every path through the branching program every variable occurs at most once as a label of a node.

Theorem 5.3.1 *Let $\mathcal{C}$ be an unsatisfiable set of clauses. Denote by k the minimum number of clauses in a regular* R*-refutation of $\mathcal{C}$ and by s the minimum number of nodes in a read-once branching program solving* Search($\mathcal{C}$).

Then $k = s$.

Proof Take a regular R-refutation π of $\mathcal{C}$. It determines a branching program σ by a relabeling analogous to that in Lemma 5.2.1: the end-clause of π becomes the root of σ and the initial clauses in π become the leaves of σ, labeled by themselves; the other steps in π become the inner nodes of σ; the two edges directed from a clause to the two hypotheses of the inference yield the clause in π. The non-leaf nodes are labeled by the atom that was resolved at them in π and the two arrows are labeled by $p, \neg p$, using the literal complementary to that occurring in the clause.

Now assume that σ is a read-once branching program of *minimum size* – this assumption will indeed be used in the argument. Analogously to the transformation of DPLL computations to R*-refutations, we shall associate with every node v of σ a clause E_v having the property that every assignment determining a path going through v falsifies E_v. The assignment will be, however, less straightforward to define than it was before.

- For a leaf v set E_v to be the clause from $\mathcal{C}$ labeling v in σ.

For a non-leaf v, assume that the node is labeled by the atom p and that the edge labeled p goes to the node v_1 and the edge labeled p^0 to the node v_0.

Claim *The clause E_{v_1} does not contain p and E_{v_0} does not contain $\neg p$.*

Here we use that σ is read-once and is of minimum size: no path reaching v determines the value of p and at least one path does indeed reach v. If the statement in the claim were not true, we could extend a path that reached v by setting $p := 1$ if $p \in E_{v_1}$ or $p := 0$ if $\neg p \in E_{v_0}$. This longer path would satisfy E_{v_0} or E_{v_1}, respectively. That contradicts the property of the construction.

The claim implies that:

1. *either* one of the clauses E_{v_1}, E_{v_0} is disjoint from $\{p, \neg p\}$
2. *or* $p \in E_{v_0} \wedge \neg p \notin E_{v_0}$ and $p \notin E_{v_1} \wedge \neg p \in E_{v_1}$.

In case 1 set E_v to be the clause containing neither literal p non literal $\neg p$. In case 2 set C_v to be the resolution of the two clauses E_{v_1} and E_{v_0} with respect to the variable p. The construction implies that no path through v satisfies E_v.

All paths in σ start at the root; hence it has to be assigned the empty clause and we have constructed a regular R-refutation as required. □

We would like to extend the correspondence to all R-refutations and all branching programs but that would not hold. Note that $s \leq k$ does hold without the restriction to regular and read-once (as the proof shows).

5.4 Width and Size

The size $|D|$ of a clause D is also called the **width** of D, denoted by $\mathbf{w}(D)$, i.e. it is the number of literals in D. For a set $\mathcal{C}$ of clauses, define $\mathbf{w}(\mathcal{C}) := \max_{D \in \mathcal{C}} \mathbf{w}(D)$. In particular, the width of a proof π, $\mathbf{w}(\pi)$, is the maximum width of a clause in the proof. We shall denote by $\mathbf{w}_R(\mathcal{C})$ and $\mathbf{w}_{R^*}(\mathcal{C})$, respectively, the minimum width of an R- or R*-refutation of $\mathcal{C}$.

The number of steps in a width-w proof can be bounded by the total number of all clauses of width w, $\binom{2n+2}{w}$. Our aim is to prove a sort of opposite statement: a short R-proof can be transformed into a narrow proof (a proof of a small width). This will allow us later to prove lower bounds for the size of proofs by proving sufficiently strong lower bounds for their minimum width.

We shall use partial truth assignments often called in circuit and proof complexity theory simply **restrictions**. Restrictions turn some literals into 1 or 0 and for this reason it is useful to allow clauses to contain these constants.

For a literal ℓ, $\epsilon \in \{0, 1\}$ and C a clause define the **restriction** of C by $\ell = \epsilon$ to be the clause

$$C \upharpoonright \ell = \epsilon \ := \ \begin{cases} C & \text{if neither } \ell \text{ nor } \neg\ell \text{ occurs in } C, \\ \{1\} & \text{if } \ell^\epsilon \in C, \\ C \setminus \{\ell^{1-\epsilon}\} & \text{if } \ell^{1-\epsilon} \in C. \end{cases}$$

In particular, $\{p, \neg p\}$ is restricted by $p = \epsilon$ to $\{1\}$. Similarly, for a set of clauses $\mathcal{C}$ write

$$\mathcal{C} \upharpoonright \ell = \epsilon \ := \ \{C \upharpoonright \ell = \epsilon \mid C \in \mathcal{C}\} .$$

Consider now the effect that a restriction, say $p = \epsilon$, has on a resolution inference

$$\frac{X \cup \{q\} \qquad Y \cup \{\neg q\}}{X \cup Y}.$$

If $p = q$ then the inference transforms into

$$\frac{X \qquad \{1\}}{X \cup Y} \quad \text{or} \quad \frac{\{1\} \qquad Y}{X \cup Y}, \tag{5.4.1}$$

which are not resolution inferences. But it can be replaced by the **weakening rule**

$$\frac{Z_1}{Z_2}, \qquad \text{provided that} \quad Z_1 \subseteq Z_2,$$

which is obviously sound (when simulating (5.4.1) we use only one hypothesis, X or Y).

We shall also allow $\{1\}$ as a new initial clause (axiom). Then either a restriction of a weakening inference is again an instance of the weakening rule or its conclusion is $\{1\}$, the new axiom.

If $p \neq q$ and $p^\epsilon \in X \cup Y$ then the inference becomes

$$\frac{(X \upharpoonright p = \epsilon) \cup \{q\} \qquad (Y \upharpoonright p = \epsilon) \cup \{\neg q\}}{\{1\}}$$

and hence can be replaced by the new axiom. In the remaining case $p \neq q$ and $p^\epsilon \notin X \cup Y$, the restriction transforms the inference into another resolution inference.

Let R_w be the proof system extending R by

- allowing the weakening rule,
- adding the new initial axiom clause $\{1\}$.

The point is that a restriction of an R_w-proof is again an R_w-proof (after transforming the resolution inferences as described above). Clearly, lower bounds for R_w-proofs apply, in particular, to R-proofs too.

The last two pieces of useful notation are $\mathbf{w}_{R_w}(\mathcal{C} \vdash A)$, denoting the minimum width of an R_w-derivation of a clause A from $\mathcal{C}$, and $\mathcal{C} \vdash_k A$, which stands for $\mathbf{w}_{R_w}(\mathcal{C} \vdash A) \leq k$.

Lemma 5.4.1 *If $\mathcal{C} \upharpoonright p = 0 \vdash_k A$ then $\mathcal{C} \vdash_{k+1} A \cup \{p\}$. If $\mathcal{C} \upharpoonright p = 1 \vdash_k A$ then $\mathcal{C} \vdash_{k+1} A \cup \{\neg p\}$.*

Proof We will prove only the first part, as the proof of the second part is identical. Assume that $\pi = D_1, \ldots, D_t$ is an R_w-derivation of A from $\mathcal{C} \upharpoonright p = 0$ having width k. Put $E_i := D_i \cup \{p\}$, for all $i \leq t$. We claim that $\pi' = E_1, \ldots, E_t$ is essentially an R_w-derivation of $A \cup \{p\}$. The qualification "essentially" will become clear in a moment.

Assume first that $D_i \in \mathcal{C} \upharpoonright p = 0$, say $D_i = C \upharpoonright p = 0$ for some $C \in \mathcal{C}$. Consider three cases.

1. $\neg p \in C$ Then $D_i = \{1\}$ and so $E_i = \{1, p\}$ can be derived from the axiom $\{1\}$ by a weakening.
2. $p \in C \wedge \neg p \notin C$ Then $D_i = C \setminus \{p\}$ and hence $E_i = C$ is an initial clause from $\mathcal{C}$.
3. $C \cap \{p, \neg p\} = \emptyset$ Then $D_i = C$ and so $E_i = C \cup \{p\}$ can be derived from C by a weakening.

Note that the extra line in the derivations of the clauses E_i has width bounded above by the width of clauses already in π'.

The case when D_i is derived in π by a resolution rule has already been discussed, when we motivated the extension of R to R_w. The case when D_i is obtained by the weakening rule is trivial. □

Lemma 5.4.2 *For $\epsilon \in \{0, 1\}$, assume that*

$$\mathcal{C} \upharpoonright p = \epsilon \vdash_{k-1} \emptyset \quad \textit{and} \quad \mathcal{C} \upharpoonright p = 1 - \epsilon \vdash_k \emptyset.$$

Then

$$\mathbf{w}_{R_w}(\mathcal{C} \vdash \emptyset) \leq \max(k, \mathbf{w}(\mathcal{C})).$$

Proof By Lemma 5.4.1 the first part of the hypothesis implies $\mathcal{C} \vdash_k \{p^{1-\epsilon}\}$. Resolve $\{p^{1-\epsilon}\}$ with all $C \in \mathcal{C}$ containing $\{p^{\epsilon}\}$; the width of all these inferences is bounded above by $\mathbf{w}(\mathcal{C})$. Therefore each clause $D \in \mathcal{C} \upharpoonright p = 1 - \epsilon$ has an R_w-derivation from $\mathcal{C}$ of width at most $\max(k, \mathbf{w}(C))$.

This, together with the second part of the hypothesis of the lemma, concludes the proof. □

Theorem 5.4.3 (Ben-Sasson and Wigderson [73]) *Let $\mathcal{C}$ be an unsatisfiable set of clauses in the literals $p_i, \neg p_i$, for $i \leq n$. Assume that $\mathcal{C}$ has a tree-like* R_w*-refutation with $\leq 2^h$ clauses.*

Then

$$\mathbf{w}_{R_w^*}(\mathcal{C} \vdash \emptyset) \leq \mathbf{w}(\mathcal{C}) + h\,.$$

Proof We shall proceed by a double induction on n and h. If $n = 0$ or $h = 0$ then necessarily $\emptyset \in \mathcal{C}$ and there is nothing to prove. Assume that for $h_0 \geq 0$ the statement is true for all $h \leq h_0$ and for all $n \geq 0$. We shall prove that it is true also for $h_0 + 1$ by induction on n. By the above, we may assume that $n > 0$; hence there is a literal in $\mathcal{C}$.

Assume that the last inference in a refutation π (having $\leq 2^{h_0+1}$ clauses) is:

$$\frac{\{p\} \qquad \{\neg p\}}{\emptyset}.$$

Hence one of the subproofs has size $\leq 2^{h_0}$. Assume that it is the subproof π_0 ending with $\{p\}$. Restrict π_0 by $p = 0$; it then becomes an R_w-refutation of $\mathcal{C} \upharpoonright p = 0$. By the induction hypothesis for h_0,

$$\mathbf{w}_{R_w^*}(\mathcal{C} \upharpoonright p = 0 \vdash \emptyset) \ \leq \mathbf{w}(\mathcal{C} \upharpoonright p = 0) + h_0.$$

Similarly, the restriction of the subproof π_1 ending with $\{\neg p\}$ by $p = 1$ becomes a refutation of $\mathcal{C} \upharpoonright p = 1$. It may have up to $\leq 2^{h_0+1}$ clauses but it has $\leq n - 1$ atoms. So the induction hypothesis for $n - 1$ implies that

$$\mathbf{w}_{R_w^*}(\mathcal{C} \upharpoonright p = 1 \vdash \emptyset) \ \leq \mathbf{w}(\mathcal{C} \upharpoonright p = 1) + h_0 + 1.$$

Applying Lemma 5.4.2 concludes the proof. □

Note that this immediately yields a lower bound for the size in terms of a lower bound for the width.

Corollary 5.4.4 *Every tree-like R_w-refutation of any $\mathcal{C}$ must have size at least*

$$2^{\mathbf{w}_{R_w^*}(\mathcal{C} \vdash \emptyset) - \mathbf{w}(\mathcal{C})}.$$

Much more interesting is the following statement, which shows that one can derive a lower bound for the size from a lower bound for the width, even for general, not necessarily tree-like, R_w-proofs.

Theorem 5.4.5 (Ben-Sasson and Wigderson [73]) *Let $\mathcal{C}$ be an unsatisfiable set of clauses in literals $p_i, \neg p_i$, for $i \leq n$. Then every R_w-refutation must have size at least*

$$2^{\Omega((\mathbf{w}_{R_w}(\mathcal{C} \vdash \emptyset) - \mathbf{w}(\mathcal{C}))^2/n)}.$$

Proof Let k be the number of clauses in an R_w-refutation π of $\mathcal{C}$. Let $h \geq 1$ be a parameter. Later we shall specify that $h := \lceil \sqrt{2n \log(k)} \rceil$ but this actual value is not used in the argument; it is only used at the end to optimize the bound.

We shall prove that

$$\mathbf{w}_{R_w}(\mathcal{C} \vdash \emptyset) \ \leq \ \mathbf{w}(\mathcal{C}) + O(\sqrt{n \log(k)}).$$

If $n = 0$ then $\emptyset \in \mathcal{C}$ and there is nothing to prove.

Suppose $n > 0$. Call a clause C in π *wide* if $\mathbf{w}(C) > h$. Let $s := (1 - h/2n)^{-1}$. By a double induction on n and on t we can prove that if the number of wide clauses in π is $< s^t$ then

$$\mathbf{w}_{R_w}(\mathcal{C} \vdash \emptyset) \leq \mathbf{w}(\mathcal{C}) + h + t\,.$$

First assume $t = 0$. Then there is no wide clause, i.e. $\mathbf{w}(\pi) \leq h \leq \mathbf{w}(\mathcal{C}) + h$.

Now assume $t > 0$. One of the $2n$ literals, say ℓ, has to appear in at least $s^t h/2n$ wide clauses. Restrict π by $\ell = 1$. The clauses containing ℓ will transform to $\{1\}$. Hence, in $\pi \restriction \ell = 1$, a refutation of $\mathcal{C} \restriction \ell = 1$, there remain less than

$$s^t \; - \; \frac{s^t h}{2n} \; = \; s^{t-1}$$

wide clauses. By the induction hypothesis for $t - 1$,

$$\mathbf{w}_{R_w}(\mathcal{C} \restriction \ell = 1 \vdash \emptyset) \leq \mathbf{w}(\mathcal{C} \restriction \ell = 1) + h + t - 1\,.$$

Now apply to π the dual restriction $\ell = 0$. This produces a refutation $\pi \restriction \ell = 0$ of $\mathcal{C} \restriction \ell = 0$ in which the number of wide clauses is still $< s^t$ but the number of atoms is $n - 1$. Hence, by the induction hypothesis for $n - 1$,

$$\mathbf{w}_{R_w}(\mathcal{C} \restriction \ell = 0 \vdash \emptyset) \leq \mathbf{w}(\mathcal{C} \restriction \ell = 0) + h + t\,.$$

By applying Lemma 5.4.2 we get

$$\mathbf{w}_{R_w}(\mathcal{C} \vdash \emptyset) \leq \mathbf{w}(\mathcal{C}) + h + t\,.$$

The particular value of the parameter h yields the wanted upper bound (using the trivial estimate $t \leq \log_s(k)$). □

In order to be able to prove some lower bounds on the size of resolution proofs via this theorem, we will need unsatisfiable sets of clauses of small width (perhaps even constant), which require wide R_w-proofs. We shall construct such sets of clauses later, in Section 13.3.

5.5 Height and Space

The height of an R-refutation π, denoted by $\mathbf{h}(\pi)$, is defined in an identical way to the height of Frege proofs in Section 2.2: it is the maximum length of a path through the proof-graph of π. Similarly, $\mathbf{h}_R(\mathcal{C})$ is the minimum height of an R-refutation of $\mathcal{C}$, and clearly $\mathbf{h}_R(\mathcal{C}) = \mathbf{h}_{R^*}(\mathcal{C})$ for R*. Its apparent significance is in estimating the longest run when using an R- or R*-refutation for solving the search problem Search($\mathcal{C}$) from Section 5.2: the maximum length of a path through the associated decision tree or branching program is the height. This, perhaps coupled with a Spira-type argument, can be used for simple lower bound proofs (see Chapter 13).

The height also relates to the space complexity of an R-proof. This is a measure that has a natural intuitive appeal for all proof systems that are based on manipulating lines via rules. For resolution and, in general, for R-like proof systems the height is claimed to have an algorithmic significance. There are several versions of this measure (see also Section 5.9) but we shall consider here only the *total space* and the *clause space*.

Let π be an R-refutation of $\mathcal{C}$. The space that π uses is, intuitively, the minimum number of symbols other than clauses from $\mathcal{C}$ that it suffices for us to remember at any given time in order to verify that π is indeed a valid refutation. Formally, consider a sequence of CNFs, which we picture as sets of clauses

$$\mathcal{D}_0, \ldots, \mathcal{D}_s,$$

called **configurations**, such that:

1. $\mathcal{D}_0 = \emptyset$;
2. for each $t = 1, \ldots, s$, $\mathcal{D}_t$ is obtained from $\mathcal{D}_{t-1}$ by one of the rules
   (a) $\mathcal{D}_t = \mathcal{D}_{t-1} \cup \{C\}$ for some $C \in \mathcal{C}$ (an axiom download),
   (b) $\mathcal{D}_t \subseteq \mathcal{D}_{t-1}$ (a clause erasure),
   (c) $\mathcal{D}_t = \mathcal{D}_{t-1} \cup \{G\}$, where G is a resolvent of two clauses $E, F \in \mathcal{D}_{t-1}$ and the inference occurs in π (a resolution inference);
3. $\emptyset \in \mathcal{D}_s$.

The **total space of** π, denoted by $\mathbf{Tsp}(\pi)$, is

$$\mathbf{Tsp}(\pi) \;:=\; \min(\max_t \sum_{D \in \mathcal{D}_t} |D|) \tag{5.5.1}$$

and the **clause space of** π, denoted by $\mathbf{Csp}(\pi)$, is

$$\mathbf{Csp}(\pi) \;:=\; \min(\max_t |\mathcal{D}_t|), \tag{5.5.2}$$

where the minima are taken over all sequences $\mathcal{D}_0, \ldots, \mathcal{D}_s$ satisfying the above conditions. The minimum total space and minimum clause space of any R-refutation of $\mathcal{C}$ will be denoted by $\mathbf{Tsp}_R(\mathcal{C})$ and $\mathbf{Csp}_R(\mathcal{C})$, respectively, and analogously for R*.

Lemma 5.5.1 *For any set of clauses* $\mathcal{C}$:

(i) $\mathbf{Tsp}_R(\mathcal{C})$ *and* $\mathbf{Csp}_R(\mathcal{C})$ *are defined by* (5.5.1) *and* (5.5.2), *where the minima are taken over all sequences* $\mathcal{D}_0, \ldots, \mathcal{D}_s$ *of configurations satisfying the conditions 1, 2 and 3 above with the condition 2(c) that the inference has to occur in* π *dropped;*
(ii) $\mathbf{Csp}_R(\mathcal{C}) \leq 1 + \mathbf{h}_R(\mathcal{C})$;
(iii) $\mathbf{Csp}_R(\mathcal{C}) \leq n + O(1)$, *where n is the number of variables in* $\mathcal{C}$.

(The sequences $\mathcal{D}_0, \ldots, \mathcal{D}_s$ from statement (i) are called **configurational refutations** of $\mathcal{C}$. The meaning of the statement is that when discussing the space complexity it is sufficient to consider configurational refutations. We say that a sequence $\mathcal{D}_0, \ldots, \mathcal{D}_s$ is a **configurational proof** of clause A from clauses $\mathcal{C}$ if and only if the condition 3 that $\emptyset \in \mathcal{D}_s$ is changed to $A \in \mathcal{D}_s$.)

Proof The first statement is obvious. For the second, assume that $h = \mathbf{h}_R(\mathcal{C})$ and take without loss of generality an R*-refutation π of $\mathcal{C}$ of height h. Say that a clause A in π is at level ℓ if and only if ℓ is the maximum length of a path from A to some initial clause. By induction on $\ell = 0, \ldots, h$ prove the following

Claim *Every clause in π at level ℓ has a configurational proof from $\mathcal{C}$ with clause space level at most $1 + \ell$.*

If A is an initial clause ($\ell = 0$) this is obvious. If A is at level $\ell + 1$ and was derived from E, F (necessarily at levels at most ℓ), first find a configurational proof of $\{E\}$ (of space $\leq \ell$) and use the erasure rule to keep just E in the memory, and then take a space $\leq \ell$ configurational proof of $\{F\}$, adding E to each configuration clause. This yields a configurational proof of $\{E, F\}$ of space $\leq 1 + \ell$, and in one step we derive $\{G\}$ itself.

The third statement follows from the second, as clearly $\mathbf{h}_R(\mathcal{C}) \leq n$. □

The definition of the space measure using configurations relates a priori to general, dag-like, refutations. For tree-like refutation we could use the same definition but with the condition 2(c) changed to

- $\mathcal{D}_t = (\mathcal{D}_{t-1} \setminus \{E, F\}) \cup \{G\}$, where G is a resolvent of two clauses $E, F \in \mathcal{D}_{t-1}$ and the inference occurs in π (resolution inference).

In other words, whenever we use a resolution inference to get the next configuration we have to erase the hypotheses of the inference.

An alternative and a quite elegant formulation uses a **pebbling** of the proof graph of π; the same definition applies then to both the dag-like and the tree-like refutations. Think of the nodes in a proof graph (i.e. clauses in π) as being in two possible states: **pebbled** or **not pebbled**. At the beginning no nodes are pebbled and we can pebble or un-pebble a node according to the following rules:

- a pebble can be placed on an initial clause at any time (the clause is then said to be pebbled);
- a pebble can be removed at any time;
- if both hypotheses of a resolution inference are pebbled then a pebble can be put on the consequence of the inference.

The pebbles can be reused. The aim is eventually to pebble the end-clause. The following lemma is straightforward (and can be also modified for **Tsp**).

Lemma 5.5.2 *For an* R*-refutation π, the minimum number of pebbles needed to legally pebble the end-clause of π equals* $\mathbf{Csp}(\pi)$. *The same holds for* R**-refutations if the configurational definition is altered as described above.*

To complement the second statement note that there are R*-refutations with number of steps and width bounded by 2^h and h respectively, for height h. Hence we get

Corollary 5.5.3 $\mathbf{Tsp}_R(\mathcal{C}) \leq \mathbf{h}_R(\mathcal{C})(1 + \mathbf{h}_R(\mathcal{C}))$.

We would like the form of an opposite estimate and it is clear that this ought to be possible. The configuration proofs carry in each configuration all information needed to progress in the proof. This is quite analogous to the construction of tree-like proofs from general Frege proofs in Theorem 2.2.1 (the formulas D_i in its proof are essentially configurations). Just as tree-like proofs can be further put into a balanced tree form, we can balance configurational proofs as well, getting an estimate on their height. This view of configurational proofs yields the following lemma. In its proof we shall use the relation between R* and DPLL as this is simpler than directly manipulating proofs.

Lemma 5.5.4 *It holds that*

$$\mathbf{h}_R(\mathcal{C}) \leq O(\mathbf{Tsp}_R^2(\mathcal{C}) \log n) ,$$

where n is the number of atoms in $\mathcal{C}$.

Proof Put $s = \mathbf{Tsp}_R(\mathcal{C})$ and let Θ: $\mathcal{D}_0, \ldots, \mathcal{D}_r$ be a configurational proof witnessing the space bound. We shall use Θ to devise a strategy for the DPLL procedure solving Search($\mathcal{C}$) (see Section 5.2).

Let $a = \lfloor r/2 \rfloor$ and ask for the values of all (at most s) variables in $\mathcal{D}_a$. If they make all clauses in the configuration true, concentrate on the second half, $\mathcal{D}_{a+1} \ldots, \mathcal{D}_r$; if they make some clause false, concentrate on the first half, $\mathcal{D}_0, \ldots, \mathcal{D}_a$. After at most $\log r$ such rounds (i.e. at most $s(\log r)$ variable queries) we must find a pair $\mathcal{D}_t, \mathcal{D}_{t+1}$ in which all clauses in the first configuration are true and in the second configuration there is some clause that is false. The only way in which this could happen is that the configuration was obtained by the axiom download rule, i.e. we have found a falsified clause in $\mathcal{C}$.

Lemma 5.2.1 finishes the proof, on noting that $r \leq (2n + 2)^s$. □

We conclude this section by a theorem linking width and clause space.

Theorem 5.5.5 (Atserias and Dalmau [31]) *It holds that:*

$$\mathbf{w}_R(\mathcal{C}) \leq \mathbf{Csp}_R(\mathcal{C}) + \mathbf{w}(\mathcal{C}).$$

Proof Assume ρ: $\mathcal{D}_0, \ldots, \mathcal{D}_r$ is a sequence of CNFs (configurations) witnessing the value $s = \mathbf{Csp}_R(\mathcal{C})$. We shall use ρ to construct an R-refutation having width bounded above by $s + \mathbf{w}(\mathcal{C})$ (in fact, by $s + \mathbf{w}(\mathcal{C}) - 1$).

Each $\mathcal{D}_i$ is of the form

$$\bigwedge_{u \leq s_i} E_u^i, \quad \text{where } E_u^i = \bigvee_v \ell_{u,v}^i \text{ and } \ell_{u,v}^i \text{ are literals} ,$$

with $s_i \leq s$; $\mathcal{D}_{i+1}$ follows simply from $\mathcal{D}_i$. The qualification *simply* means that there is a short tree-like depth-2 LK-proof of the sequent

$$\longrightarrow \mathcal{D}_{i+1}$$

from the sequents

$$\longrightarrow \mathcal{D}_i \text{ and } \longrightarrow C, \qquad \text{for all } C \in \mathcal{C} .$$

These proofs can be arranged into a balanced tree and, using the constructions of Lemmas 3.4.4 and 3.4.5, which can be further transformed into a resolution refutation of $\mathcal{C}$. To liberate the reader from the need to refresh that material (or to actually read it), we will present the construction in a self-contained manner. However, the constructions remains essentially the same.

Think of the derivation of

$$\mathcal{D}_i \longrightarrow \mathcal{D}_{i+1}$$

from $\mathcal{C}$ as saying that if we have witnesses (some literals $\ell^i_{u,v}$) for each disjunction E^i_u, for all $u \leq s_i$, we can prove that some s_{i+1}-tuple of literals witnesses all the E^{i+1}-formulas in $\mathcal{D}_{i+1}$. In resolution this is better phrased counter-positively.

For $i \leq r$, denote by $\mathcal{F}_i$ the clauses

$$\{\neg \ell^i_{u,v_u} \mid u = 1, \dots, s_i\}, \qquad \text{for all choices } v_1, \dots, v_{s_{i+1}} .$$

Note that their width is bounded above by $s_i \leq s$.

Claim *For each $i < r$, all the clauses of $\mathcal{F}_i$ can be derived in R from $\mathcal{F}_{i+1} \cup \mathcal{C}$ by a derivation of width at most $s + \mathbf{w}(\mathcal{C}) - 1$.*

This is fairly easy to check for the rules stating how one can derive $\mathcal{D}_{i+1}$ from $\mathcal{D}_i$. Let us verify the case in which the width increases by $\mathbf{w}(\mathcal{C}) - 1$, the download of an axiom $C = \{\ell_1, \dots, \ell_w\} \in \mathcal{C}$. This means that the clauses in $\mathcal{F}_{i+1}$ have the form $F, \neg \ell$ for all choices of $F \in \mathcal{F}_i$ and $\ell \in C$.

Consider ℓ_1 and resolve all the clauses $F, \neg \ell_1$ against C, obtaining all the clauses $F, \ell_2, \dots, \ell_w$. Then resolve all these clauses against all the clauses F, ℓ_2, obtaining all the clause $F, \ell_3, \dots, \ell_w$, etc.

The claim implies the theorem because the last configuration $\mathcal{D}_r$ contains $\emptyset$ thus we are starting with $\{1\}$, and the first configuration is empty; hence we get a refutation. □

5.6 Interpolation

The following notation will be handy; for a given clause D write:

- $\bigvee D$, the disjunction of all literals in D;
- $\neg D$, $|D|$ unit-size clauses each consisting of the negation of a literal from D;
- $\bigwedge \neg D$, the conjunction of the negations of all literals in D.

Let $A_1, \dots, A_m, B_1, \dots, B_\ell$ be clauses satisfying (for all $i \leq m$ and $j \leq \ell$)

$$A_i \subseteq \{p_1, \neg p_1, \dots, p_n, \neg p_n, q_1, \neg q_1, \dots, q_s, \neg q_s\} \tag{5.6.1}$$

and

$$B_j \subseteq \{p_1, \neg p_1, \ldots, p_n, \neg p_n, r_1, \neg r_1, \ldots, r_t, \neg r_t\} . \tag{5.6.2}$$

In particular, only the literals corresponding to $\overline{p}$ may occur in both A_i and B_j.

The set $\{A_1, \ldots, A_m, B_1, \ldots, B_\ell\}$ is unsatisfiable if and only if the implication

$$\bigwedge_{i \leq m} (\bigvee A_i) \longrightarrow \neg \bigwedge_{j \leq \ell} (\bigvee B_j) \tag{5.6.3}$$

is a tautology. By the Craig interpolation Theorem 1.1.3 it has an interpolant $I(\overline{p})$. The size of $I(\overline{p})$ constructed in the proof of Theorem 1.1.3 may be exponential in n, the number of atoms $\overline{p}$.

We proved in Theorems 3.3.1 and 3.3.2 that it is possible to estimate the complexity of the interpolant by the size of a cut-free proof of the implication. The following theorem proves an analogous result for R. That is rather surprising, as R is in a sense the opposite of cut-free LK: there are only cuts. We shall give two more proofs of this theorem in Chapter 17.

Theorem 5.6.1 (The feasible interpolation theorem, Krajíček [280]) *Assume that the set of clauses*

$$\{A_1, \ldots, A_m, B_1, \ldots, B_\ell\}$$

satisfies (5.6.1) and (5.6.2) and has a resolution refutation with k clauses.

Then the implication (5.6.3) has an interpolating circuit $I(\overline{p})$ whose size is $O(kn)$. If the refutation is tree-like, I is a formula.

Moreover, if all atoms $\overline{p}$ occur only positively in all A_i then there is a monotone interpolating circuit (or a formula in the tree-like case) whose size is $O(kn)$.

Proof Let $\sigma = D_1, \ldots, D_k$ be an R-refutation of clauses $A_1, \ldots, A_m, B_1, \ldots, B_\ell$. For a clause D, denote by D^p, D^q and D^r the subsets of D consisting of the literals corresponding to the atoms $\overline{p}, \overline{q}$ and $\overline{r}$, respectively. Let $\mathcal{S}$ be the set of the following sequents:

$$\bigvee A_u, \neg A_u^q \longrightarrow A_u^p, \quad 1 \leq u \leq m, \tag{5.6.4}$$

$$\longrightarrow B_v^p, B_v^r, \bigwedge \neg B_v, \quad 1 \leq v \leq \ell. \tag{5.6.5}$$

Claim *For $i \leq k$, there exist cedents Γ_i, Δ_i and negation normal form* LK$^-$*-proofs π_i of sequents S_i of the form*

$$S_i \colon \Gamma_i, \neg D_i^q \longrightarrow D_i^p, D_i^r, \Delta_i$$

from $\mathcal{S}$ such that it holds that:

(i) all initial sequents in π_i are in $\mathcal{S}$;

(ii) the atoms $\overline{r}$ do not occur in Γ_i and all formulas in Γ_i are either one of $\bigvee A_1, \ldots, \bigvee A_m$ or are logically valid;

(iii) the atoms $\overline{q}$ do not occur in Δ_i and all formulas in Δ_i are either one of $\bigwedge \neg B_1, \ldots, \bigwedge \neg B_\ell$ or are unsatisfiable;
(iv) π_i has at most i non-unary inferences and if σ is tree-like, so is π_i.

The claim is proved by induction on i. If D_i is an initial clause A_u then S_i (with $\Gamma_i = \{\bigvee A_u\}$ and $\Delta_i := \emptyset$) is in $\mathcal{S}$. An analogous statement holds when D_i is an initial clause B_v.

If D_i was inferred from D_u, D_v by resolving a $\overline{p}$ or an $\overline{r}$ atom x, say $x \in D_u$, $\neg x \in D_v$ and

$$D_i = (D_u \setminus \{x\}) \cup (D_v \setminus \{\neg x\}),$$

then put $\Delta_i := \Delta_u \cup \Delta_v \cup \{x \wedge \neg x\}$, and $\Gamma_i := \Gamma_u \cup \Gamma_v$ and from

$$S_u \colon \Gamma_u, \neg D_u^q \longrightarrow D_u^p, D_u^r, \Delta_u$$

and

$$S_v \colon \Gamma_v, \neg D_v^q \longrightarrow D_v^p, D_v^r, \Delta_v$$

derive

$$S_i \colon \Gamma_i, \neg D_i^q \longrightarrow D_i^p, D_i^r, \Delta_i$$

by some weakenings and one $\wedge$:right. If the resolved variable was a $\overline{q}$ variable then put $\Delta_i := \Delta_u \cup \Delta_v$, $\Gamma_i := \Gamma_u \cup \Gamma_v \cup \{x \vee \neg x\}$ and use some weakenings and one $\vee$:left.

The initial sequents in π_k are all in $\mathcal{S}$. If an initial sequent is in the part (5.6.4) of $\mathcal{S}$, its interpolant is $\bigvee A_u^p$. Note that it has size $O(n)$ and it is monotone if no negated $\overline{p}$ variable occurs in A_u. If an initial sequent is in (5.6.5) then its interpolant is the constant 1.

The required interpolant is then constructed from π_k as in the proofs of Theorems 3.3.1 and 3.3.2. It has the general form $K(s_1, \ldots, s_t)$, where the s_i are interpolants of the initial sequents and K is a monotone circuit (or a formula in the tree-like case) whose size is bounded by the number of non-unary inferences, i.e. by k. The total size is then $O(kn)$. □

5.7 DNF-Resolution

The DNF-**resolution** system (denoted by DNF-R), a proof system operating with DNF formulas, was defined in [288] (and called there R^+) in order to describe a proof system (to be denoted later by R(log)) corresponding precisely to the bounded arithmetic theory $T_2^2(\alpha)$ (Section 10.5). But this proof system and its fragments $R(f)$ have now a life of their own that is independent of bounded arithmetic.

In the context of DNF-R we may talk about DNF-formulas as clauses of logical terms (i.e. conjunctions of literals, cf. Section 1.1), to stress that DNF-R extends R by allowing in clauses not only literals but their conjunctions as well. The DNF-R system has the following rules:

$$\frac{C \cup \{\bigwedge_j \ell_j\} \quad D \cup \{\neg \ell'_1, \ldots, \neg \ell'_t\}}{C \cup D},$$

if $t \geq 1$ and all ℓ'_i occur among the ℓ_j, and

$$\frac{C \cup \{\bigwedge_{j \leq s} \ell_j\} \quad D \cup \{\bigwedge_{s < j \leq t} \ell_j\}}{C \cup D \cup \{\bigwedge_{i \leq s+t} \ell_i\}}.$$

The proof of the following statement is straightforward and is omitted.

Lemma 5.7.1 *Let $\mathcal{A}$ be a set of* DNF*-formulas. Then $\mathcal{A}$ is unsatisfiable if and only if it has a* DNF-R *refutation.*

*Moreover, if $\mathcal{A}$ consists of k-*DNF *formulas then there is such a refutation using k-*DNF *formulas only.*

For a constant $k \geq 1$, denote by R(k) the subsystem of DNF-R operating with k-DNF formulas. For the link with bounded arithmetic it is important to define systems in which the size of the terms in the formulas allowed in refutations may grow with the size of the initial clauses and the size of the refutation itself.

Let f: $\mathbf{N}^+ \longrightarrow \mathbf{N}^+$ be a non-decreasing function. Define the **R(f)-size** of a DNF-R refutation π to be the minimum s such that:

- π has at most s steps; and
- every logical term occurring in π has size at most $f(s)$.

For example, a size-s R(log)-refutation may contain terms of size up to $\log s$.

The bottom case of Lemma 3.4.7 about the relation of the tree-like and dag-like LK-proofs of a fixed Σ-depth is the mutual simulation of $LK_{1/2}$ (i.e. R(log)) and $LK^*_{3/2}$. Later, we shall need to go one step lower, as formulated in the next lemma.

Lemma 5.7.2 *The proof system* R *p-simulates* R*(log) *with respect to refutations of sets of clauses. In fact, if $\mathcal{C}$ is a set of clauses in n variables and of width $w_0 := \mathbf{w}(\mathcal{C})$ and ρ is its* R*(log)*-refutation with number of steps $k := \mathbf{k}(\rho)$, height $h := \mathbf{h}(\rho)$ and total size $s := |\rho|$, then there is an* R*-refutation π of $\mathcal{C}$ such that:*

- $\mathbf{k}(\pi) = O(kw_0)$;
- $\mathbf{w}(\pi) = \max(w_0, h \log(s))$.

Moreover, π can be constructed from ρ by a p-time algorithm.

Proof The construction of π follows the strategy employed in the proof of Lemma 3.4.4. Let D be a DNF-clause in ρ of the form

$$\bigwedge_{j \in J_1} \ell^1_j, \ldots, \bigwedge_{j \in J_r} \ell^r_j$$

where the ℓ_j^i are literals. Note that $|J_i| \leq \log s$. Let ρ_D be the subproof of ρ ending with D and assume that $k_D := \mathbf{k}(\rho_D)$ and $h_D := \mathbf{h}(\rho_D)$.

Claim *There is an* R*-refutation π_D of $\mathcal{C}$ together with the clauses*

$$\bigvee_{j \in J_1} \neg \ell_j^1, \ldots, \bigvee_{j \in J_r} \neg \ell_j^r \tag{5.7.1}$$

such that:

- $\mathbf{k}(\pi_D) = O(k_D) + O(|\mathcal{C}|)$*;*
- $\mathbf{w}(\pi_D) = \max(w_0, h_D \log s)$.

In particular, $|\pi_D| = O(s^3 \log s)$.

Let $\mathcal{C}'$ be $\mathcal{C}$ together with the clauses (5.7.1). To simulate the key R(log) inference of the form

$$\frac{D, \bigwedge_{j \in J} \ell_j \qquad D \cup \{\neg \ell_j \mid j \in J'\}}{D}, \qquad \text{for some } \emptyset \neq J' \subseteq J,$$

we assume that we have an R-refutation π_1 of $\mathcal{C}' \supseteq \mathcal{C}$ together with

$$\bigvee_{j \in J'} \neg \ell_j$$

and an R-refutation π_2 of $\mathcal{C}'$ and with $|J'|$ singleton clauses

$$\{\ell_j\}, \qquad \text{one for each } j \in J' .$$

Modify π_2 as follows: whenever any clause $\{\ell_j\}$ is used as one hypothesis of an inference, do nothing and keep the complementary literal $\neg \ell_j$ in the other hypothesis. This turns π_2 into an R-derivation π_2' of $\bigvee_{j \in J'} \neg \ell_j$ from $\mathcal{C}'$, with $\mathbf{k}(\pi_2)$ steps and width at most

$$\mathbf{w}(\pi_2) + |J'| \leq \mathbf{w}(\pi_2) + \log s \leq (\mathbf{h}(\pi_2) + 1) \log s .$$

Extending π_2' by π_1 yields π_D.

The additive factor $O(|\mathcal{C}|)$ in the estimate of $\mathbf{k}(\pi_D)$ comes from the estimate of the total number of steps needed to repeat, for each $C \in \mathcal{C}$, the refutation of

$$\mathcal{C} \cup \{\{\neg \ell\} \mid \ell \in C\}. \qquad \square$$

By organizing the construction in Spira-type fashion we can replace the multiplicative factor $\mathbf{h}(\rho)$ in the estimate for $\mathbf{w}(\pi)$ by $O(\log \mathbf{k}(\rho))$. This yields

Lemma 5.7.3 *Under the same assumptions as in Lemma 5.7.2 an* R*-refutation can be found with*

$$\mathbf{w}(\pi) = \max(w_0, (\log \mathbf{k}(\rho) \log s).$$

In particular, if $w_0 \leq \log s$ then $\mathbf{w}(\pi) \leq O(\log^2 s)$.

5.8 Extended Resolution

The second strengthening of R is Tseitin's [492] **extended resolution**, denoted by *ER*. It is defined analogously to *EF* but it dates, in fact, from some ten years earlier.

The system *ER* allows us to enlarge the set of initial clauses as follows.

1. Pick two literals ℓ_1 and ℓ_2 and a new atom q and add the following three clauses to the initial clauses:

$$\{\neg\ell_1, q\}, \quad \{\neg\ell_2, q\}, \quad \{\ell_1, \ell_2, \neg q\}. \tag{5.8.1}$$

2. Repeat step 1 an arbitrary number of times, for any literals ℓ_1, ℓ_2 but always choosing a new atom in the place of q.

The variables q introduced in step 1 are called **extension variables** and the three clauses are the **extension axioms** corresponding to them. Note that if the extension axioms are satisfied then necessarily $q \equiv (\ell_1 \vee \ell_2)$. Hence, by repeating the process one can reduce the size of clauses in a refutation.

Lemma 5.8.1

(i) ER is sound: no ER-refutable set of clauses is satisfiable.
(ii) Let π be an ER-refutation of a set $\mathcal{A}$ of initial clauses in n variables and assume that each clause in $\mathcal{A}$ has size at most w. Assume also that π has k steps.

Then there is an ER-refutation σ of $\mathcal{A}$ such that:

- $\mathbf{k}(\sigma) = O(k)$;
- *the size of any clause in σ is bounded by* $\max(3, w)$.

Proof The first statement is obvious: if b_1, b_2 are truth values of ℓ_1, ℓ_2 then q can be assigned the value $b_1 \vee b_2$ to satisfy all extension axioms. The second part is analogous to Corollary 2.4.7 regarding *EF*. □

We may interpret the definition of *ER* in terms of circuits. Consider the circuits $C(\overline{p})$ with inputs $\overline{p} = (p_1, \ldots, p_n)$ and inner nodes $\overline{q} = (q_1, \ldots, q_s)$, and the set of clauses Def_C introduced in Section 1.4.

Lemma 5.8.2 *Let $\mathcal{C}$ be a set of clauses in the variables $\overline{p} = (p_1, \ldots, p_n)$. Assume that π is an ER-refutation of $\mathcal{C}$. Then on the one hand there is a circuit $D(\overline{p})$ of size $s \leq |\pi|$ and a size-$O(|\pi|)$ R-refutation of $\mathcal{C} \cup Def_D$.*

On the other hand, for any circuit D and an R-refutation σ of $\mathcal{C} \cup Def_D$, there is an ER-refutation of $\mathcal{C}$ of size $O(|\sigma|)$.

Proof The extension axiom defines the extension variable q as a disjunction of literals (corresponding to initial atoms as well as to other extension variables introduced earlier), i.e. by a constant-size circuit from the variables of the literals. Now rewrite D using only the gates $\neg$ and $\wedge$; the instructions of D have the form of extension axioms. □

Next, we will gauge the power of *ER*.

Lemma 5.8.3 $ER \equiv_p EF$.

Proof That $ER \leq_p EF$ is obvious. The opposite simulation follows from the construction underlying Lemma 2.4.6 and Corollary 2.4.7. □

Finally, we state a useful technical lemma that is completely analogous to Lemma 2.4.6.

Lemma 5.8.4 *Assume that π is an ER-refutation of a set of clauses $\mathcal{C}$ in the variables $\overline{p} = (p_1, \ldots, p_n)$. Let $w_0 = \mathbf{w}(\mathcal{C})$, $k = \mathbf{k}(\pi)$ and $w = \mathbf{w}(\pi)$.*

Then there is an ER-refutation ρ of $\mathcal{C}$ such that

$$\mathbf{k}(\rho) = O(kw) \quad \text{and} \quad \mathbf{w}(\rho) \leq \max(3, w_0) .$$

In particular, $|\rho| \leq O(w_0|\pi|^2)$.

5.9 Bibliographical and Other Remarks

The prominence of resolution in SAT solving and automated theorem proving started with papers by Davis and Putnam [168] and Davis, Logemann and Loveland [166]. Contrary to the popular belief, R was not defined by these authors (or by Robinson [453] to whom it is sometimes attributed) but by Blake [86] some 30 years earlier.

Sec. 5.1 Resolution is also denoted *Res* by some authors. The system R*, i.e. tree-like R, p-simulates the analytic tableaux method (see Arai, Pitassi and Urquhart [22]) but it is stronger, as shown by Urquhart [497]. Cut-free LK$^-$ p-simulates regular resolution according to Arai [21]. Another subsystem considered in connection with SAT solving is *linear R*: each new inference has to use the last derived clause.

Secs. 5.2 and 5.3 Lemma 5.2.1 seems to be folklore. Theorem 5.3.1 is from [278, Theorem 4.2.3]; it was noted by several authors and first stated explicitly by perhaps Lovász *et al.* [341]. Tseitin's [492] lower bound for regular resolution was one of the first non-trivial length-of-proofs lower bounds.

Sec. 5.4 The width of R refutations was first considered by Krishnamurthy and Moll [329], who proved a width lower bound for refutations of a tautology related to Ramsey's theorem (see Section 13.2). Kullmann introduced the notion of *assymetric width* (see [80] for an exposition), which allows us to talk about narrow refutations of wide initial clauses and can be characterized in terms of families of partial assignments. Ben-Sasson and Wigderson [73] built on the work of Clegg, Edmonds and Impagliazzo [142], who showed a relation between PC degree (see Section 6.2) and resolution size, and on Beame and Pitassi [57]. The trivial estimate $\binom{2n+2}{w}$ of the number of steps in a width-w proof was shown to be essentially tight by Atserias, Lauria and Nordström [36]. The system R_w from Section 5.4 is denoted RWS in [145, 5.4] and the weakening rule is called the subsumption rule there (and in SAT solving).

Sec. 5.5 A *clause space* measure of the complexity of R-proofs was defined by Esteban and Toran [176]. Alekhnovich *et al.* [11] considered the space complexity of algebraic proof systems and of Frege systems, and defined the *total space* and also the *variable space*, which we did not discuss: you count only the number of distinct variables occurring in configurations. Clearly the minimum necessary variable space (needed to refute a set of clauses) is bounded by the minimum total space and it cannot be bounded by the minimum clause space: by Nordström [369] the number $\mathbf{Csp}_R$ may be constant while the minimum variable space may be $\Omega(n/\log n)$. It is open at present whether one can meaningfully estimate $\mathbf{Csp}_R$ by the minimum variable space. Urquhart [498] proved that the variable space can be bounded by the minimum height of a proof. Bonacina [88] bounded $\mathbf{w}_R$ by $O(\mathbf{Tsp}_R^{1/2}) + 2\mathbf{w}(\mathcal{C})$. The third item in Lemma 5.5.1 was pointed out by Esteban and Toran [176]. Lemma 5.5.4 was noted by Razborov [441] and put by him into the wider context of relations among various proof complexity measures for resolution. The term *configuration* is due – in the wider context of first-order logic – to Smullyan [477]. Theorem 5.5.5 was proved by Atserias and Dalmau [31] by a method close to the model-theoretic Ehrenfeucht–Fraissé games. Filmus *et al.* [184] gave a syntactic proof of the Atserias–Dalmau theorem analogous to the proof given here.

In connection with this, Nordström [369] proved that $\mathbf{Csp}_R$ is not $O(\mathbf{w}_R)$ (the latter can be constant while at the same time the former can be $\omega(n/\log n)$).

Esteban and Toran [177] gave a characterization of the space for *tree-like* refutations of a formula using the Prover–Delayer game of Impagliazzo and Pudlák [245] (see Section 13.7) played with the formula; the characterization does not use the notion of refutations and it is a purely combinatorial property of the formula. They also demonstrated an $\Omega(n)$ gap in the space complexity between dag-like and tree-like refutations.

Galesi and Thapen [193] defined a number of variants of the Prover–Delayer game and used it to characterize various subsystems of R and to formulate some lower bound proofs. They also defined an elegant proof system called **narrow resolution**, which uses the initial clauses $C := \{\ell_1, \ldots, \ell_e\}$ indirectly; you may derive a clause D once you proved all the clauses $D \cup \{\neg\ell_i\}$, $i \leq e$. The advantage is that proofs in this system may be narrow even if the initial clauses are not.

Among researchers concentrating on resolution, the height of an R-proof is often called the *depth*; however, we want to maintain a terminological unity for all proof systems, and the term *depth* is already prominently used with a different meaning in Frege and other stronger systems.

Sec. 5.6 This follows [280]; the proof of the feasible interpolation theorem 5.6.1 has been streamlined a little and gives a better bound.

Sec. 5.7 The proof system DNF-R and the notion of the R(*f*)-size were defined in [288]. In the subsequent literature many authors interpret R(log), and general R(*f*), by simply restricting the size of terms to $\log n$, where n is a "natural" parameter associated with the initial clauses (this is often not the number of variables). This is strictly speaking not correct, for the following three reasons at least.

1. There is no canonical definition, in terms of the set of initial clauses, of what n should be and, as some authors prove results about systems such as $R(\epsilon \log n)$, even a polynomial change to n makes a difference.
2. The meaning of lower bounds changes. For example, a size-2^{n^δ} lower bound for $R(\log n)$ means in a formal definition that there is such a lower bound for refutations in which terms of size up to n^δ may occur, while in an informal definition only terms of size $\log n$ are allowed.
3. The link to bounded arithmetic is lost.

I shall stick to the formal definition in this book but the reader should be aware that other authors may mean something a bit different.

Lauria [336] has unpublished notes presenting a proof of Lemma 5.7.3. Beyersdorff and Kullmann [80] offered a unified treatment of various proof complexity measures of resolution refutations, using a game-theoretic formulation.

Sec. 5.8 This follows Cook and Reckhow [156] and [278].

6

Algebraic and Geometric Proof Systems

The algebraic and geometric proof systems that we shall introduce in this chapter are very natural. They were studied first in fields other than proof complexity: operational research, optimization, robotics and universal algebra, to name only a few. With the exception of the cutting planes proof system these fields became the subject of studies in proof complexity at first only as a technical vehicle assisting investigations of the mutual relations of the counting principles over AC^0-Frege systems (see Section 15.5). This is true especially of the Nullstellensatz proof system. But once Pandora's box was open, many variants of the proof systems poured out and are now studied for their own sake. In fact, they are studied at least as extensively as any other logical proof system.

The general set-up for representing propositional logic in algebra or in geometry is the following. Let $C := \{\ell_1, \ldots, \ell_t\}$ be a clause. If we think of the truth values as the 0, 1 elements of a ring then $\neg\ell$ is simply $1 - \ell$. We can then express that C is true either by the polynomial equation

$$(1 - \ell_1) \cdot \cdots \cdot (1 - \ell_t) = 0 \tag{6.0.1}$$

over a field or by the integer linear inequality

$$\ell_1 + \cdots + \ell_t \geq 1 , \tag{6.0.2}$$

thinking in both cases just of $0/1$ solutions. The former representation leads to algebraic proof systems, the latter to geometric, although this division is somewhat arbitrary and blurred and semi-algebraic proof systems (Section 6.4) combine both.

It is easy to prove the completeness of all proof systems discussed in this chapter by simulating resolution proofs in them. However, in all cases completeness in the propositional sense is a special case of a more general completeness in some algebro-geometric sense: Hilbert's Nullstellensatz in the case of algebraic proof systems, the Positivstellensatz of Krivine [330] and Stengle [484] in the case of the semi-algebraic systems, the theorems of Chvátal [140] and Gomory [205] in the case of the cutting plane proof system and a theorem of Lovász and Schrijver [342] in the case of an intermediate system discussed in Section 6.5. These more general results are used in operational research, optimization algorithms and SAT solving and connections to these topics form important motivations for some researchers to study algebro-geometric propositional proof systems.

We shall start with the equational calculus as it is a natural link to the logical proof systems introduced in earlier chapters.

6.1 Equational Calculus

A prominent example of a complete propositional language different from the DeMorgan one is the language that, instead of the disjunction $\vee$ has the parity connective $\oplus$, also called **logical addition**. The reason for this terminology is that $\oplus$ defines addition over the two-element field $\mathbf{F}_2$ with elements $0, 1$. Conjunction and negation define multiplication and the function $1 - x$, respectively. It is thus natural to use the symbols of the **language of rings**:

$$0, 1, +, -, \cdot\,,$$

where $-x$ is the additive inverse symbol. The natural next step is then to consider proof systems in this language not just over $\mathbf{F}_2$ but over any field (or an integral domain, more generally). The role of formulas in logical systems is played here by terms in the language. This view of propositional logic was brought forward by Boole [100] and although it did not lead to the foundations of logic as he originally envisioned, it has played an important role in mathematical logic ever since.

The **equational calculus**, denoted by EC, is a specific proof system in the language of rings. We shall consider EC over any underlying field $\mathbf{F}$ and not just over $\mathbf{F}_2$. When we want to stress which field is being used, we will denote the system by EC/$\mathbf{F}$. If $\mathbf{F}_2 \neq \mathbf{F}$ then it is strictly speaking not a Frege system because it transcends the Boolean domain and also uses constants for the field elements.

An **F-term** is a term (see Section 1.2) of the ring language augmented by constants for all elements of $\mathbf{F}$. In the algebraic tradition we use letters $x, y, z, \ldots$ for variables in terms. The lines in the system EC/$\mathbf{F}$ are equations

$$t = s$$

between $\mathbf{F}$-terms. The axioms and the inference rules of EC/$\mathbf{F}$ are:

1. the equality axiom and the rules

$$\frac{}{t = t}, \quad \frac{t = s}{s = t}, \quad \frac{t = s \quad s = r}{t = r};$$

2. the additive group axioms

$$t + 0 = t, \quad t + (s + r) = (t + s) + r, \quad t + s = s + t, \quad t + (-t) = 0;$$

3. the multiplicative commutative semigroup axioms

$$t \cdot 1 = t, \quad t \cdot (s \cdot r) = (t \cdot s) \cdot r, \quad t \cdot s = s \cdot t;$$

4. the distributivity axiom

$$t \cdot (s + r) = (t \cdot s) + (t \cdot r);$$

5. the atomic diagram of **F**, given by

$$a + b = c \quad \text{and} \quad d \cdot e = f,$$

where a, b, c, d, e, f are elements of **F** that are constants and the respective equalities are valid in **F**;

6. the two rules

$$\frac{t = s \quad u = v}{t + u = s + v} \quad \text{and} \quad \frac{t = s \quad u = v}{t \cdot u = s \cdot v};$$

7. the Boolean axioms

$$x \cdot (x - 1) = 0$$

for all variables x.

Let $\mathcal{E}$ be a set of equations between terms. An **EC/F-proof** of an equation $u = v$ from $\mathcal{E}$ is a sequence $\pi = E_1, \ldots, E_k$ of equations

$$E_i\colon t_i = s_i$$

such that each step E_i is from $\mathcal{E}$ or is an axiom of EC/**F** or was derived from some earlier steps by one of the rules and E_k is the target equation $u = v$. An **EC/F-refutation** of $\mathcal{E}$ is an EC/**F**-proof of $1 = 0$ from $\mathcal{E}$.

If we think of a term t as a formula in the language of EC then the assertion that it is true is represented by the equation $t = 1$. However, an equation $r = s$ is true for a given assignment if and only if $(r - s) + 1 = 1$ is true, i.e. if and only if the formula $(r - s) + 1$ is true. Hence we may think of EC as of a collection of Frege rules.

Lemma 6.1.1 *The system* EC *is implicationally complete.*

The EC is therefore a Frege system and the following statement is then a corollary of Reckhow's theorem.

Lemma 6.1.2 *The system* EC/$\mathbf{F}_2$ *is p-equivalent to any Frege system.*

6.2 Algebraic Geometry and the Nullstellensatz

An EC-proof of an equation $t = 0$ from equations $s_i = 0$, $i = 1, \ldots, m$, can be interpreted as proving that the polynomial defined by t is in the ideal $\langle s_1, \ldots, s_m, x_1^2 - x_1, \ldots, x_n^2 - x_n \rangle$ of the polynomials generated by the polynomials s_i and $x_j^2 - x_j$ (where the x_j range over all the variables involved). If we disregard the term structure and think of the proof steps just as polynomials, the resulting system is the **polynomial calculus**, denoted by PC (or PC/**F** when stressing the underlying field **F**). The axioms of EC in the items 1–5 become irrelevant and what remains are

- the **Boolean axioms**, $x_j^2 - x_j$,

and two rules,

- the **addition rule**,

$$\frac{f \quad g}{f+g}$$

- and the **multiplication rule**

$$\frac{f}{f \cdot h}$$

where f, g, h are arbitrary polynomials from the ring $\mathbf{F}[\overline{x}]$. Note that the two rules correspond to the closure properties defining ideals of polynomials.

A PC/**F-proof** of a polynomial g from polynomials $\mathcal{F} = \{f_i \mid i \leq m\}$ is a sequence of polynomials

$$\pi = h_1, \ldots, h_k \tag{6.2.1}$$

such that either each h_i is from $\mathcal{F}$ or one of the Boolean axioms or it is derived from some earlier steps by one of the two rules, and $h_k = g$. A PC-**refutation** of $\mathcal{F}$ is a PC-proof of 1 from $\mathcal{F}$, demonstrating that the ideal generated by $\{f \mid f \in \mathcal{F}\} \cup \{x_1^2 - x_1, \ldots, x_n^2 - x_n\}$ is trivial.

There are two key issues we need to address.

1. How do we represent proofs so that it is feasible to recognize those that are valid?
2. Is PC complete?

The first issue is resolved by representing polynomials in a canonical way as an **F**-linear combination of monomials formed from the variables x_j occurring in $\mathcal{F}$ or in g. When **F** is finite or countable the linear coefficients can be represented by finite strings of bits and the whole proof can be thus represented. It is then a polynomial-time solvable task to check, for polynomials f, g, h whether $f + g = h$ or $f \cdot g = h$ (in the case of a countable **F** we require that its atomic diagram is polynomial-time decidable, as is the case for the rationals **Q**). But we shall occasionally consider also an uncountable **F**, such as the reals **R** or the complex numbers **C**, and in these cases the system is not a Cook–Reckhow proof system.

The second issue is resolved by Hilbert's Nullstellensatz (Theorem 6.2.2 below), although the next lemma shows that it is much simpler in propositional logic.

Lemma 6.2.1 *Let* **F** *be a field. Then there is a mapping assigning to an* R*-refutation* π *of some clauses* $C_1, \ldots, C_m$ *a* PC/**F***-refutation* σ *of polynomials* $f_{C_1}, \ldots, f_{C_m}$ *translating the clauses as in* (6.0.1). *In particular,* PC/$\mathbf{F}_2$ *is a complete propositional proof system.*

In fact, the mapping of π *to* σ *satisfies:*

- $\mathbf{k}(\sigma) = O(\mathbf{k}(\pi))$;

- *the maximum degree of a polynomial in σ is at most $2\mathbf{w}(\pi)$ (that quantity will be later defined as the degree $\mathbf{deg}(\sigma)$ of σ);*
- *if π is tree-like then so is σ, and $\mathbf{h}(\sigma) \leq 2\mathbf{h}(\pi)$.*

Proof The resolution inference

$$\frac{C,p \quad D,\neg p}{C \cup D}$$

corresponds to the inference

$$\frac{g(1-x) \quad hx}{gh},$$

which is simulated in PC by deriving from the hypotheses using the multiplication rule for two polynomials

$$gh(1-x) \quad \text{and} \quad ghx$$

and summing them together. □

The reason why the mapping from the proof of Lemma 6.2.1 is not a p-simulation is that the translation of a clause C into an equivalent polynomial f_C in (6.0.1) may blow up the size exponentially: t negated variables yield 2^t monomials. This is remedied by the system PCR (to be discussed in Section 7.1), which introduces new variables standing for the negations of the original variables.

The completeness of PC holds in a much more general sense, though.

Theorem 6.2.2 (Hilbert's Nullstellensatz) *Let $\mathbf{F}$ be any field and let $\mathcal{F} \subseteq \mathbf{F}[\overline{x}]$. Then the polynomial system*

$$f = 0, \quad \textit{for } f \in \mathcal{F},$$

has no solution in the algebraic closure $\mathbf{F}^{acl}$ of $\mathbf{F}$ if and only if the ideal $\langle f \mid f \in \mathcal{F}\rangle$ is trivial.

In particular, the system has no $0/1$ solution if and only if the ideal generated by $\mathcal{F} \cup \{x_j^2 - x_j \mid j \leq n\}$ is trivial.

We use this to show that the simulation of R respecting structure of proofs in Lemma 6.2.1 holds for a much stronger proof system.

Lemma 6.2.3 *Let $C_1, \ldots, C_m$ be clauses in n atoms and assume that π is their refutation in a Frege system in the DeMorgan language. Then there is a* PC*-refutation of $f_{C_1}, \ldots, f_{C_m}$ with $O(\mathbf{k}(\pi)n)$ steps.*

Proof Extend the translation (6.0.1) to all Frege formulas by:

- $f_\perp := 1$ and $f_\top := 0$;
- $f_{\neg E} := 1 - f_E$;
- $f_{D \vee E} := f_D f_E$;
- $f_{D \wedge E} := 1 - f_{\neg D} f_{\neg E}$.

Then, by Theorem 6.2.2, a formula E_0 follows from the formulas $E_1, \ldots, E_\ell$ if and only if the equation $f_{E_0} = 0$ follows from the equations

$$f_{E_1} = 0, \quad \ldots, \quad f_{E_\ell} = 0, \quad x_1^2 - x_1 = 0, \quad \ldots, \quad x_n^2 - x_n = 0$$

if and only if the polynomial f_{E_0} is an $\mathbf{F}[\overline{x}]$-linear combination of the polynomials f_{E_i} and $x_j^2 - x_j$, $i, j \geq 1$.

In particular, every Frege inference with ℓ hypotheses can be simulated in PC by $2(\ell + n)$ inferences. □

Now we consider another formalization of ideal membership proofs. A proof of $g \in \langle f \mid f \in \mathcal{F} \rangle$ in the **Nullstellensatz proof system**, denoted by NS or NS/$\mathbf{F}$ when stressing the field, is an $\mathcal{F}$-tuple π of polynomials $h_f, f \in \mathcal{F}$, written as an $\mathbf{F}$-linear combinations of monomials and such that

$$g = \sum_{f \in \mathcal{F}} h_f f \tag{6.2.2}$$

holds in the ring of polynomials. In propositional logic we add, as before, the Boolean axioms, the polynomials $x_j^2 - x_j$ for all variables x_j, that a proof is a tuple of polynomials h_f, as before, and polynomials r_j such that

$$g = \sum_{f \in \mathcal{F}} h_f f + \sum_j r_j (x_j^2 - x_j). \tag{6.2.3}$$

A polynomial can be expressed as an $\mathbf{F}[\overline{x}]$-linear combination of polynomials from $\mathcal{F}$ and $x_j^2 - x_j$ if and only if it is in the ideal generated by $\mathcal{F} \cup \{x_j^2 - x_j \mid j \leq n\}$. Hence NS/$\mathbf{F}$ is sound and complete.

In proof complexity the main measure of the complexity of PC and NS proofs is their size, as is the case for all other proof systems. But the most *natural measure* in the algebraic context is the **degree of the proof**. For a PC-proof π as in (6.2.1), define

$$\mathbf{deg}(\pi) := \max_{i \leq k} deg(h_i)$$

and for a NS-proof π as in (6.2.2), put

$$\mathbf{deg}(\pi) := \max_{f \in \mathcal{F}} deg(h_f f)$$

or

$$\mathbf{deg}(\pi) := \max(\max_{f \in \mathcal{F}} deg(h_f f), \max_{j \leq n} deg(r_j(x_j^2 - x_j)))$$

in the Boolean case (6.2.3). A polynomial of degree d in n variables has at most $\binom{n+d}{d}$ monomials and so an upper bound d to the degree of an NS-proof implies, for $d \leq n$, an upper bound $O((|\mathbf{F}| + n)dn^d)$ to its size for finite $\mathbf{F}$ (for any $\mathbf{F}$ as well if we count the size of the coefficients from $\mathbf{F}$ as 1). This upper bound is also a lower bound when the representation of polynomials is what is called **dense**: it lists the coefficients of all monomials of degree at most d, even if the degree is 0.

The number of steps in a PC-proof is well defined (there are no steps in NS-proofs) but it makes little sense to study its value on its own. Every refutable system of m polynomials in n variables has a PC-refutation with $2(m+n)-1$ steps; take the NS-proof (6.2.3) and derive first in $n+m$ steps all $h_f f$ and $r_j(x_j^2 - x_j)$ and then sum them up in $n+m-1$ steps. However, when a priori we restrict the degree, the number of steps in a PC-proof becomes a non-trivial measure.

The following lemma gives a simple link between the number of steps and the degree. We shall use the star notation PC* to denote the tree-like version of PC, as we did for other proof systems earlier.

Lemma 6.2.4 *Let $\mathcal{F}$ be a set of polynomials from $\mathbf{F}[\overline{x}]$ and let $g \in \mathbf{F}[\overline{x}]$. Then:*

(i) if there is a degree-d NS-*proof of g from $\mathcal{F}$ then there is also a degree d* PC*-*proof of g from $\mathcal{F}$;*
(ii) if there is a PC*-*proof π of g from $\mathcal{F}$ in which each instance of the multiplication rule is multiplied by a polynomial of degree at most d then there is an* NS-*proof of g from $\mathcal{F}$ of degree $(d \cdot h) + d_{\mathcal{F}}$, where $h = \mathbf{h}(\pi)$ is the height of the proof tree of π and $d_{\mathcal{F}} := \max(deg(f) \mid f \in \mathcal{F})$.*

Proof The first part is a trivial consequence of the construction of a PC-proof from an NS-proof, given earlier. For the second part, express all polynomials p in π (by induction on the number of steps in the subproof π' yielding p) as linear combinations of the elements of $\mathcal{F}$ where the coefficients are some polynomials of degree at most $(dh') + d_{\mathcal{F}}$, where $h' = \mathbf{h}(\pi')$ is the height of π'. □

6.3 Integer Linear Programming and the Cutting Plane Method

A particular case of **integer linear programming** (ILP) is the task of finding an integer solution of a system of linear inequalities over rationals $\mathbf{Q}$ or, without loss of generality, of inequalities over integers $\mathbf{Z}$:

$$\mathbf{A} \cdot \overline{x} \geq \overline{b}, \tag{6.3.1}$$

where A is an $m \times n$ integer matrix, $\overline{x}$ is an n-tuple of variables and $\overline{b}$ is an m-tuple of integers considered as a vector. The ordering $\geq$ between vectors means that the inequality holds for every coordinate.

The decision problem, i.e. whether the system has an integer solution, is NP-hard. This is true even when the problem is restricted to Boolean solutions, in which case it is NP-complete (Karp [266]). This is readily seen by translating CNFs into systems like (6.3.1), as described at the beginning of this chapter.

Utilizing a result of Gomory [205], Chvátal [140] introduced the **cutting plane proof method** of how to solve the decision ILP problems, and proved its completeness. The propositional proof system **cutting planes** (CP) uses his method but considers only 0/1 solutions. Showing the completeness of CP as a propositional proof system is much easier (Lemma 6.3.1 below).

The lines in a CP-**proof** are integer inequalities of the form

$$\sum_i a_i x_i \geq b,$$

where the a_i and b are integers. This is the normal form but we may also write some absolute terms on the left-hand side, when convenient. The **axioms** are

$$x \geq 0 \quad \text{and} \quad -x \geq -1$$

for all variables x, forcing $0 \leq x \leq 1$. The CP system has three inference rules:

- the **addition rule,**

$$\frac{\sum_i a_i x_i \geq b \quad \sum_i d_i x_i \geq d}{\sum_i (a_i + c_i) x_i \geq (b + d)};$$

- the **multiplication rule,**

$$\frac{\sum_i a_i x_i \geq b}{\sum_i (a_i \cdot c) x_i \geq b \cdot c},$$

  if $c \geq 0$; and
- the **division rule,**

$$\frac{\sum_i a_i x_i \geq b}{\sum_i (a_i / c) x_i \geq \lceil b/c \rceil},$$

  if $c > 0$ divides all a_i.

The last rule is called the Gomory–Chvátal cut.

Cutting planes is a refutation system: it **refutes** the solvability of the initial set of linear inequalities in 0, 1 by deriving $0 \geq 1$.

Lemma 6.3.1 CP $\geq_p$ R.

Proof Let $C_1, \ldots, C_m$ be initial clauses and assume the $\pi: D_1, \ldots, D_k (= \emptyset)$ is their R-refutation. For E, a clause, let L_E be the integer linear inequality that represents E. By induction on i we can prove the following

Claim *The inequality L_{D_i} has a* CP*-derivation from $L_{C_1}, \ldots, L_{C_m}$ with $O(i)$ steps and such that all coefficients occurring in the derivation have absolute value at most* 2.

Assume that

$$\frac{(A =) U, x \quad (B =) V, \neg x}{U \cup V}$$

is an inference in π. For $U = \{\ell_1, \ldots, \ell_t\}$ write L_U as $\ell_1 + \cdots + \ell_t \geq 1$, and similarly for $A, B, V = \{\ell'_1, \ldots \ell'_s\}$. Hence we want to derive $L_{U \cup V}$ from

$$L_A: \ell_1 + \cdots + \ell_t + x \geq 1 \quad \text{and} \quad L_B: \ell'_1 + \cdots + \ell'_s + (1 - x) \geq 1 .$$

Sum both inequalities and subtract 1 from both sides: this will yield an inequality where the right-hand side is 1 and the left-hand side is a sum of literals from $U \cup V$, some having coefficients 1 and some possibly 2 (if they appear in both U and V). Add to this inequality the inequalities $\ell \geq 0$ for all ℓ having coefficient 1, and then divide the result by 2 to get $L_{U \cup V}$. (Note that crucially this uses rounding up in the division rule: $1/2$ gets rounded up to 1.) □

Let us give an example of a short proof in CP. In Section 1.5 we defined the propositional pigeonhole principle PHP_n in (1.5.1). It is translated into integer linear inequalities with variables x_{ij}, $i = 1, \dots, n+1$ and $j = 1, \dots, n$ as

$$\sum_j x_{ij} \geq 1 \quad \text{and} \quad -x_{i_1 j} - x_{i_2 j} \geq -1 \tag{6.3.2}$$

for all $i \in [n+1]$, $i_1 \neq i_2 \in [n+1]$ and $j \in [n]$. The third conjunct in $\neg\mathrm{PHP}_n$ becomes

$$-x_{i,j_1} - x_{i,j_2} \geq -1$$

for $i \in [n+1]$ and $j_1 \neq j_2 \in [n]$, but it is not needed in the proof below.

Lemma 6.3.2 *The system* CP *refutes* (6.3.2) *in* $O(n^2)$ *lines in which all coefficients are bounded in absolute value by* n *and hence the total size is at most* $O(n^4 \log n)$.

Proof By summing up the left-hand sides of the inequalities in (6.3.2) over all i we get

$$\sum_i \sum_j x_{ij} \geq n+1 \,. \tag{6.3.3}$$

This needs n inferences and all coefficients are 0 or 1.

Now we shall deduce also

$$\sum_i \sum_j x_{ij} \leq n \,. \tag{6.3.4}$$

In the format of CP-lines we write $-\sum_i \sum_j x_{ij} \geq -n$, and summing this inequality with (6.3.3) yields $0 \geq 1$, i.e. the refutation of the PHP-inequalities (6.3.2).

Let us rewrite for convenience the right-hand side inequalities from (6.3.2) as

$$x_{i_1 j} + x_{i_2 j} \leq 1 \,.$$

We shall derive for each j

$$\sum_i x_{ij} \leq 1 \tag{6.3.5}$$

and summing these inequalities (in $n-1$ steps) over all j yields (6.3.4) as required.

To prove (6.3.5), show by induction on $s = 1, \dots, n+1$ that

$$\sum_{i \leq s} x_{ij} \leq 1 \,. \tag{6.3.6}$$

For $s = 1$, this is one of the Boolean axioms.

For the induction step, assume that we know (6.3.6). Multiplying this by $s-1$ and adding to it the axioms

$$x_{ij} + x_{(s+1)j} \leq 1$$

for all $i \leq s$ yields

$$\sum_{i \leq s+1} s x_{ij} \leq 2s - 1,$$

from which (6.3.5) follows by rounded division by s (note that we have the relation $\leq$ and hence $2 - 1/s$ rounds *down* to 1).

The induction step needed $s+2$ steps, so in total the derivation of (6.3.5) used $O(n^2)$ steps, in which all coefficients were bounded in absolute value by n. □

In principle, a CP-refutation can apply the division rule with any $c > 0$ dividing all a_i and there is also no bound to the size of the integer coefficients that may occur. The next statement offers a useful reduction.

Theorem 6.3.3 (Buss and Clote [121])

(i) *For $c > 1$, the subsystem* CP_c *of* CP *allowing division only by c is p-equivalent to* CP.

(ii) *Assume that $A \cdot \overline{x} \geq \overline{b}$ is a system of m integer linear inequalities and that every coefficient in the system (i.e. in A as well as in $\overline{b}$) is bounded in absolute value by B.*

If there is a CP*-refutation of the system with k steps then there is another* CP*-refutation σ such that:*

- *σ has $O(k^3 \log(mB))$ steps;*
- *every integer occurring in σ has absolute value bounded above by $O(k2^k mB)$.*

6.4 Semi-Algebraic Geometry and the Positivstellensatz

Reasoning over the real closed ordered field $\mathbf{R}(0, 1, +, -, \cdot, \leq)$ combines in a sense, algebraic reasoning as formalized by PC (we are in a field) and (linear) geometric reasoning as formalized by CP (we have the ordering in the language). In the structure $\mathbf{R}(0, 1, +, -, \cdot, \leq)$ one cannot define the ring of integers but the Boolean axioms still single out the values 0 and 1. In the presence of these axioms we can represent a clause either by the equation (6.0.1) or by the linear inequality (6.0.2).

In **semi-algebraic calculus**, to be denoted by SAC, the lines in proofs are polynomials from $\mathbf{R}[\overline{x}]$, the line f representing the inequality $f \geq 0$ (we will use the explicit notation with the inequality sign when necessary). Semi-algebraic calculus has the **non-negativity of squares** axiom;

$$f^2 \geq 0,$$

the axiom

$$1 \geq 0,$$

and the following rules:

- the **addition rule**

$$\frac{f \geq 0 \quad g \geq 0}{af + bg \geq 0},$$

  for all $a, b \in \mathbf{R}^+$, and
- the **multiplication rule**

$$\frac{f \geq 0 \quad g \geq 0}{f \cdot g \geq 0}.$$

For propositional logic we also include the Boolean axioms, as earlier, written now as two inequalities:

$$x^2 - x \geq 0 \quad \text{and} \quad x - x^2 \geq 0 \tag{6.4.1}$$

for all variables x.

To justify the claim that reasoning over the reals incorporates the algebraic reasoning of PC, we prove the following simple lemma.

Lemma 6.4.1 SAC $\geq_p$ PC/**R**. *In particular,* SAC *is a complete propositional proof system.*

Proof Assume that $\pi = h_1, \ldots, h_k$ is a PC-proof of g from $f_1, \ldots, f_m$, all polynomials over **R**. Show by induction on $i \leq k$ that both

$$h_i \geq 0 \quad \text{and} \quad -h_i \geq 0$$

have SAC-proofs σ_i^+, σ_i^-, respectively, from

$$f_j \geq 0 \quad \text{and} \quad -f_j \geq 0, \quad \text{for all } j \leq m,$$

with $O(i)$ steps and degree at most $\mathbf{deg}(\pi)$. This is non-trivial only when h_i is derived from $h_j, j < i$, by multiplying it by p:

$$h_i = h_j \cdot p\,.$$

Write $p = p^+ - p^-$, where both p^+ and p^- have only non-negative coefficients.

Claim 1 Both the inequalities $p^+ \geq 0$ and $p^- \geq 0$ have SAC-proofs with $O(|p|)$ lines and degree $\leq deg(p)$.

First note that $x \geq 0$ follows by summing up the Boolean axiom $x - x^2 \geq 0$ and the squares axiom $x^2 \geq 0$, for each variable x. Use the multiplication rule, then derive $x_{j_1} \cdots x_{j_t} \geq 0$ for all monomials $x_{j_1} \cdots x_{j_t}$ in p and finally use the addition rule to get the claim.

Now, to construct the proof σ_i^+, use Claim 1 and by the multiplication rule derive from $h_j \geq 0$ and $-h_j \geq 0$ both

$$h_j \cdot p^+ \geq 0 \quad \text{and} \quad -h_j \cdot p^- \geq 0,$$

from which $h_i (= h_j p)$ follows by the addition rule. □

Lemma 6.2.1 implies

Corollary 6.4.2 SAC $\geq_p$ R.

We would like to justify the earlier claim that SAC subsumes linear geometric reasoning by showing that SAC p-simulates cutting planes, CP. But that is an open problem. To prove such a simulation it would clearly suffice to simulate the rounded division rule of CP and show that SAC is closed (with efficient proofs) under the following (sound) rule:

$$\frac{f \geq 1/k}{f \geq 1},$$

where $f \in \mathbf{Z}[\overline{x}]$ is a linear polynomial and $k > 1$ is an integer. In fact, if we want to simulate only CP-refutations (rather than CP derivations of general inequalities) this may not be needed.

Polynomial calculus, PC, is complete in a stronger sense than just as a propositional proof system, as guaranteed by the Nullstellensatz (Theorem 6.2.2). Similarly, we can consider SAC more generally as a proof system starting with an initial combined system of polynomial equations and inequalities

$$f = 0 \text{ for } f \in \mathcal{F} \quad \text{and} \quad g \geq 0 \text{ for } g \in \mathcal{G}, \tag{6.4.2}$$

where $\mathcal{F}, \mathcal{G} \subseteq \mathbf{R}[\overline{x}]$ and $\overline{x} = x_1, \dots, x_n$.

In this context the notion analogous to an ideal is a **cone**. A cone $c(g_1, \dots, g_\ell)$ generated by polynomials $g_i \in \mathbf{R}[\overline{x}]$ is the smallest subset of $\mathbf{R}[\overline{x}]$ that contains all g_i and all squares h^2 from $\mathbf{R}[\overline{x}]$ and that is closed under addition and multiplication. If we denote by SOS the subset of $\mathbf{R}[\overline{x}]$ consisting of the **sums of squares**, the elements of $c(g_1, \dots, g_\ell)$ are exactly those polynomials that can be expressed as

$$\sum_J h_J \cdot (\prod_{j \in J} g_j)$$

where J ranges over all subsets of $\{1, \dots, \ell\}$ (including the empty set) and all h_J are from SOS.

The following theorem is a Nullstellensatz-type statement in this context. From the mathematical logic point of view, both the Nullstellensatz and this statement can be seen as consequences of the model-completeness of algebraically closed fields and of the real closed ordered field, respectively.

Theorem 6.4.3 (Positivstellensatz) *The system of equations and inequalities* (6.4.2) *has no solution in* $\mathbf{R}$ *if and only if there are polynomials*

$$f \in \langle \mathcal{F} \rangle \quad \textit{and} \quad g \in c(\mathcal{G})$$

such that $f + g = -1$.

Let us consider two special cases of this theorem: $\mathcal{G} = \emptyset$ and $\mathcal{F} = \emptyset$. In the former case the cone $c(\emptyset)$ is just the set SOS and hence we get the following corollary.

Corollary 6.4.4 *The system of equations*

$$f = 0 \quad \text{for } f \in \mathcal{F} \tag{6.4.3}$$

has no solution in $\mathbf{R}$ *if and only if there are polynomials*

$$f \in \langle \mathcal{F} \rangle \quad \text{and} \quad g \in SOS$$

such that $f + g = -1$.

For $\mathcal{F} = \{f_1, \ldots, f_m\}$, this means that there are polynomials h_i, $i \leq m$, and a polynomial g from SOS such that

$$g + \sum_{i \leq m} h_i f_i = -1. \tag{6.4.4}$$

The $(m+1)$-tuple $(g, h_1, \ldots, h_m)$ is called a **sum-of-squares proof**, an SOS-proof, of the unsolvability of the system (6.4.3). It is a special case of more general Positivstellensatz proofs (see below), where we allow $\mathcal{G} \neq \emptyset$.

Consider now the latter case when $\mathcal{F} = \emptyset$ and assume $\mathcal{G} = \{g_1, \ldots, g_\ell\}$.

Corollary 6.4.5 *The system of inequalities*

$$g_i \geq 0, \quad \text{for } i \leq \ell\,, \tag{6.4.5}$$

has no solution in $\mathbf{R}$ *if and only if there are polynomials* $h_J \in$ SOS *for all* $J \subseteq \{1, \ldots, \ell\}$ *such that*

$$-1 = \sum_J h_J \prod_{j \in J} g_j\,.$$

In this situation the SOS-proof of the unsolvability of (6.4.5) is the 2^ℓ-tuple of polynomials h_J.

Note that we can always arrange that either $\mathcal{F}$ or $\mathcal{G}$ is empty: on the one hand the equation $f = 0$ can be replaced by the two inequalities $f \geq 0$ and $-f \geq 0$ and, on the other hand, the inequality $g \geq 0$ can be replaced by the equation $g + y^2 = 0$, where y is a new variable.

In general, a **Positivstellensatz proof** of the unsolvability of the combined system (6.4.2) is the $(m + 2^\ell)$-tuple

$$(h_i)_{i \leq m}, (h_J)_{J \subseteq [\ell]} \tag{6.4.6}$$

with the polynomials h_J from SOS, for $|\mathcal{F}| = m$ and $|\mathcal{G}| = \ell$, certifying the unsolvability of the combined system in the sense of Theorem 6.4.3. The *Positivstellensatz proof system* is thus the NS version of the SAC calculus.

As for PC and NS, besides the size of a proof, an important complexity measure is the degree. The **degree** of an SAC-proof is the maximum degree of a polynomial occurring in it and the degree of a *Positivstellensazt* refutation (6.4.6) is

$$\max(\max_i deg(h_i f_i), \max_j deg(h_j \prod_{j \in J} g_j)) .$$

6.5 Between the Discrete and the Continuum

From the duality of linear programming over $\mathbf{R}$ it follows that if *linear* inequalities $g_i \geq 0$, $i \leq \ell$, imply a linear inequality $g \geq 0$ then g is a positive linear combination of g_i and 1. In addition, a suitable linear combination can be found by any feasible algorithm for linear programming.

Of course, this does not hold over the Boolean domain. We can force Boolean values by the Boolean axioms written as the two inequalities (6.4.1), which are, however, not linear but quadratic. Hence, in order to use them we must use some multiplication inferences. There are several hierarchies of such proof systems that attempt to treat the discrete and the Boolean cases as being close to the continuous case, all stemming from various specific algorithms for the ILP problem. We shall discuss this in Section 6.6. In this section we present the most studied of these proof systems.

The **Lovász–Schrijver proof system**, to be denoted by LS, is derived from the Lovász–Schrijver method for solving the ILP. It is a subsystem of a degree-2 subsystem of SAC. The proof lines in LS are polynomials from $\mathbf{R}[\overline{x}]$ of degree at most 2; a line f is understood to stand for the inequality $f \geq 0$. The LS system incorporates Boolean axioms by two inequalities,

- the **integrality conditions** $x^2 - x \geq 0$ and $x - x^2 \geq 0$,

and hence also has the axioms

- $x \geq 0$ and $1 - x \geq 0$

for all variables x. The rules of LS are

- **positive linear combinations**

$$\frac{f \quad g}{af + bg},$$

  where $a, b \in \mathbf{R}^+$, and
- the **lifting rules**

$$\frac{f}{xf} \quad \text{and} \quad \frac{f}{(1-x)f},$$

  where x is a variable and f is linear.

The lifting rules are special cases of the multiplication rule of SAC. In the multiplication rule, allowing just multiplication by x or $1 - x$ rather than by any previously derived g (hence $g \geq 0$) is stipulated in order to link the rules better to the LS method, as discussed in Section 6.6.

The proof system LS_+ augments LS by the non-negativity of squares axioms

- $f^2 \geq 0$.

An LS-**proof** (respectively an LS_+-**proof**) of g from $f_1, \ldots, f_t$ is a sequence of polynomials of degree at most 2 and such that each is one of f_i on one of the axioms of the system or was derived from some earlier steps by one of the rules. An LS-**refutation** (respectively an LS_+-**refutation**) is a proof ending with -1 (i.e. deriving the false inequality $-1 \geq 0$).

For a clause $C = \{\ell_1, \ldots, \ell_t\}$, denote the linear inequality (6.0.2) representing it by $f_C \geq 1$, i.e.

$$f_C := \ell_1 + \cdots + \ell_t$$

(we are using non-trivial right-hand sides of equations for better transparency). An LS-refutation of $C_1, \ldots, C_m$ is an LS-refutation of $f_{C_1} \geq 1, \ldots, f_{C_m} \geq 1$; similarly for LS_+.

Lemma 6.5.1 *Both* LS *and* LS_+ *are sound and complete as propositional proof systems. In fact,* SAC $\geq_p$ LS_+ $\geq_p$ LS $\geq_p$ R.

Proof The soundness follows from SAC $\geq_p$ LS_+, which is obvious. The completeness follows from LS $\geq_p$ R, which we demonstrate now.

By induction on i show that if $D_1, \ldots, D_k$ is an R-refutation of $C_1, \ldots, C_m$ then each integer linear inequality $f_{D_i} \geq 1$ has an LS-proof from $f_{C_j} \geq 1, j \leq m$, with $O(i)$ lines that can be written as linear combinations of literals or products of two literals with all coefficients of absolute value at most 2. As $D_k = \emptyset, f_{D_k} \geq 1$ becomes $0 \geq 1$ as is required. Further, the LS-refutation can be constructed by a p-time algorithm.

Clearly it is enough to show how to simulate one resolution inference,

$$\frac{C, x \quad D, \neg x}{C \cup D}.$$

Assume that $f_C + x \geq 1$ and $f_D + (1 - x) \geq 1$ have been derived and that we want to infer from them $f_{C \cup D} \geq 1$. Summing the two polynomials yields $f_C + f_D \geq 1$. This differs from $f_{C \cup D} \geq 1$ because literals from $C \cap D$ (if any) appear with the coefficient 2 rather than 1. The following claim solves that.

Claim *Let g be a linear polynomial. Then, for any literal ℓ, $g + \ell \geq 1$ can be derived in LS from $g + 2\ell \geq 1$ in a constant number of steps and in a way such that all coefficients of all literals and products of two literals have absolute value at most 2.*

Consider the case $\ell = x$. From $g + 2x \geq 1$ derive $g + x \geq 1 - x$ and, by the lifting rule, $g(1-x) + x(1-x) \geq (1-x)^2$. Using the Boolean axioms derive $g \geq g(1-x)$, $x(1-x) = 0$ and $x^2 = x$ and use them to obtain $g \geq 1 - x$, i.e. $g + x \geq 1$. □

6.6 Beyond Propositional Logic

The cutting plane CP and Lovász–Schrijver LS_+ proof systems reformulate the original algorithms for ILP as logical calculi. The original algorithms are given a polytope P in $\mathbf{R}$ or $\mathbf{Q}$ defined by the linear inequalities

$$\overline{a}_j^\top \cdot \overline{x} \geq b_j, \quad j \leq \ell \tag{6.6.1}$$

in n variables and aim at finding the convex closure of $P \cap \mathbf{Z}^n$, the *integer hull* P_I of P:

$$P_I := conv(P \cap \mathbf{Z}^n) .$$

If we are interested in Boolean values only then we include the inequalities $x \geq 0$ and $-x \geq -1$ in (6.6.1).

A procedure using the Gomory–Chvátal cuts produces from P a new polytope $C(P)$ defined by the inequalities obtained from (6.6.1) as follows:

1. first, take all positive linear combinations of inequalities in (6.6.1),
2. then round any inequality $\overline{a}^\top \cdot \overline{x} \geq b$ to $\overline{a}^\top \cdot \overline{x} \geq \lceil b \rceil$.

The polytope $C(P)$ is defined by the resulting inequalities. Chvátal [140] proved that this process eventually terminates with P_I.

When $P_I = \emptyset$ the process can be pictured as a CP*-refutation of the system (6.6.1). An important measure of such a refutation is its **rank**, defined similarly to the height of the proof tree but counting only rounded division inferences.

Lemma 6.6.1 *Assume that (6.6.1) is unsatisfiable in* $\mathbf{Z}$. *Then:*

(i) the minimum number of inferences in a CP*-refutation of (6.6.1) equals the minimum number of linear combinations and roundings needed to derive the empty set via the Gomory–Chvátal cuts;*

(ii) the minimum number t such that $C^{(t)}(P) = \emptyset$ equals the minimum rank of a CP**-refutation of (6.6.1).*

The system LS_+ corresponds analogously to the lift-and-project method of Lovász and Schrijver [342]. In this method the inequalities are first *lifted* by products introducing some degree-2 monomials x_i^2 or $x_i x_j$ and then linearized and *projected* by essentially forgetting these. In more detail, given a polytope P defined by (6.6.1) do the following:

1. first, lift all inequalities by multiplying them by all x_i and by all $1 - x_i$, and add all Boolean axiom inequalities (6.4.1);

2. take positive linear combinations;
3. delete all non-linear inequalities.

The resulting linear inequalities define a new polytope $N(P)$. Lovász and Schrijver [342] proved that this process eventually terminates and that $N^{(t)}(P) = P_I$ for some $t \leq n$.

Define the **rank** of a tree-like LS 2- or LS_+-refutation π, denoted by $\mathbf{rk}(\pi)$, to be the maximum number of lifting inferences on a path through the proof tree (the former is sometimes called the **Lovász–Schrijver rank**, the latter the **semidefinite rank**). As in the case of CP we have a simple lemma.

Lemma 6.6.2 *Assume that* (6.6.1) *is unsatisfiable in* $\mathbf{Z}$. *Then we have the following:*

(i) The minimum number of inferences in an $(LS_+)^*$*-refutation (i.e. a tree-like* LS_+*-refutation) of* (6.6.1) *equals the minimum number of linear combinations and liftings needed to derive the empty set via the Lovász–Schrijver method.*
(ii) The minimum number t such that $N^{(t)}(P) = \emptyset$ *equals the minimum rank of an* $(LS_+)^*$*-refutation of* (6.6.1).

Other algorithms were proposed by Sherali and Adams [467] and Lasserre [334], extending LS and LS_+ respectively, by allowing liftings up to degree-d polynomials but stipulating that all these inferences must come before all linear combination inferences. These are essentially the degree-d variants of tree-like LS and LS_+, respectively (Section 6.7). There is a number of variants of these basic algorithms and they correspond to a similar number of subsystems of the SAC or the Positivstellensatz proof systems. Grigoriev, Hirsch and Pasechnik [209] surveyed and studied some of them.

6.7 Bibliographical and Other Remarks

Sec. 6.1 Equational calculi with terms were considered in many parts of mathematics before proof complexity, including universal algebra or mathematical logic. A reference for algebraic geometry with an eye for algorithms is Cox, Little and Shea [162].

Sec. 6.2 The Nullstellensatz proof system NS was first considered explicitly as a propositional proof system by Beame *et al.* [52] and the polynomial calculus PC by Clegg, Edmonds and Impagliazzo [142] (under the name *Gröbner basis calculus*). They restricted the multiplication rule h to constants or variables, as that allows for an easier analysis of how the degree of polynomials in a refutation evolves.

A subsystem of PC called *binomial calculus* was studied in connections with degree lower bounds (see Chapter 16) in Buss *et al.* [122] and Ben-Sasson and Impagliazzo [69]. Buss *et al.* [124] contains an exposition of the relations of PC and

NS to the $AC^0[m]$-Frege systems and also introduces **extended** NS (ENS), which we shall discuss in Section 7.5.

Sec. 6.3 The proof system CP was defined first explicitly by Cook, Coullard and Turán [161]. They proved Lemma 6.3.1 and also noted a short proof of PHP_n in CP (Lemma 6.3.2). The specific formulation of part (ii) of Theorem 6.3.3 (Buss and Clote [121]) is from Clote and Kranakis [145].

Sec. 6.4 A reference for model theory of the real closed ordered field mentioned in the text is van den Dries [500] and, for semi-algebraic geometry, Bochnak, Coste and Roy [87]. Theorem 6.4.3 originated with Hilbert's 17th problem asking whether all semidefinite real multivariate polynomials are sums of squares of rational functions. This was solved by Artin [24] and given a simple model-theoretic proof by A. Robinson [452]. The Positivstellensatz (Theorem 6.4.3) was proved first by Krivine [330] using model theory and ten years later by Stengle [484] using semi-algebraic geometry.

The semi-algebraic calculus SAC is denoted sometimes just by S or SA. A subsystem of SAC called *Positivstellensatz calculus* and denoted by $PC_>$ and the Positivstellensatz proof system (in particular, the sum-of-squares system SOS) were first studied in proof complexity by Grigoriev and Vorobjov [210, 207]. The Positivstellensatz calculus allows one to derive the polynomials f and g from the Positivstellensatz Theorem 6.4.3 *sequentially* from the equations in $\mathcal{F}$ and the inequalities in $\mathcal{G}$, respectively. Unlike SAC, it does not allow a sequential derivation of the equality $f+g = -1$. For refutations of sets of equations (i.e. $\mathcal{G} = \emptyset$) this is like PC augmented by SOS, so the degree of a $PC_>$ refutation may be, in principle, smaller than the degree of any Positivstellensatz (SOS) refutation.

Sec. 6.5 The last part of Lemma 6.5.1 (LS $\geq_p$ R) is stated in Pudlák [414]. In the same paper he also constructed short LS-refutations of $\neg PHP_n$ (as formalized by (6.3.2)).

A number of authors have considered subsystems of SAC extending LS or CP in a natural way but allowing a higher degree of the polynomials forming lines in a proof. In particular, Grigoriev, Hirsch and Pasechnik [209] introduced many variants of LS and LS_+: LS^d and LS^d_+ (allowing polynomials up to degree d), CP^2 (allowing quadratic polynomials in CP), LS_* (allowing the multiplication rule of SAC), LS_{split} (allowing strict inequalities $f > 0$), many combinations like $LS^d_{+,*}$ and others.

Sec. 6.6 The Sherali–Adams method and the Lasserre method correspond to $(LS^d)^*$ (tree-like systems LS^d) and $(LS^d_+)^*$ (tree-like LS^d_+), respectively. The minimum degree d in these systems such that there exists such a refutation of a system of linear inequalities is called the **Sherali–Adams degree** and the **Lasserre degree**, respectively. These minimum degrees coincide with the minimum degree of a Positivstellensatz refutation which does not or does, respectively, allow initial inequalities of the form $Q^2 \geq 0$. Laurent [335] offered a comparison of the methods.

Barak and Steurer [47] surveyed the links of the SOS proof system to various related optimization algorithms.

The Sherali–Adams method was studied from a model-theoretic point of view by Atserias and Maneva [38]. Among other things they showed that the hierarchy of the Sherali–Adams systems defined by degree interleaves with the levels of indistinguishability of logic with counting quantifiers and a bounded number of variables, studied in finite model theory. They used this connection to derive a number of striking algorithmic consequences, including a lower bound on the Sherali–Adams degree.

7
Further Proof Systems

We shall discuss in this chapter a number of proof systems that have been considered by some authors but do not belong to one of the general classes of proof systems studied in earlier chapters. Nevertheless they are of some interest, I think. Combined proof systems (Section 7.1) show how even a little logic reasoning added to algebro-geometric systems strengthens them and, in particular, allows them to escape the direct combinatorial analysis prevalent in the current research (Part III will present some). Circuit proof systems (Section 7.2) precisely formalize reasoning with circuits rather than with formulas, a topic discussed often informally (and incorrectly), and implicit proof systems (Section 7.3) offer a combinatorial description of a large class of proof systems stronger than any other proof system defined without a reference to first-order theories. Sections 7.4 and 7.5 give examples of various stand-alone proof systems of either a logic or an algebro-geometric nature. Examples of proof systems that are outside the realm of the Cook–Reckhow definition but are nevertheless linked with those within it are presented in Section 7.6. Section 7.7 mentions some leftover examples not fitting into any previous section.

7.1 Combined Proof Systems

If we have two proof systems P and Q we may combine them in various ways. A straightforward way is to define a new system $P \vee Q$ whose class of proofs is the union of the classes of P-proofs and of Q-proofs. This system is obviously the least common upper bound for P, Q in the quasi-ordering of proof systems by p-simulations. But it does not seem to be of much other interest.

A more intrinsic way to combine two proof systems, both of which create proofs line by line via some inference rules, is to replace atoms in the lines used by one proof system by the lines used by the other one, and to introduce suitable inference rules allowing the manipulation of these composed lines. We shall not attempt to define this in a general way but will instead consider several examples illustrating this construction.

7.1.1 Resolution over Cutting Planes: R(CP)

Lines in this proof system are clauses of the form

$$\neg E_1, \ldots, \neg E_u, F_1, \ldots, F_v,$$

where each E_a and F_b is a CP-**inequality**, an integer linear inequality of the form

$$\sum_i a_i \cdot x_i \geq b \tag{7.1.1}$$

(i.e. a_i and b are integers). We shall abbreviate $\sum_i a_i \cdot x_i$ by $\overline{a} \cdot \overline{x}$. The inference rules of R(CP) are the following (the letters Γ and Δ stand for clauses of CP-inequalities):

- the resolution rule

$$\frac{\Gamma, E \quad \Delta, \neg E}{\Gamma \cup \Delta},$$

  where E is a CP-inequality;
- the rules CP
  - the addition rule

$$\frac{\Gamma, \overline{a} \cdot \overline{x} \geq b \quad \Delta, \overline{a}' \cdot \overline{x} \geq b'}{\Gamma \cup \Delta, (\overline{a} + \overline{a}') \cdot \overline{x} \geq b + b'},$$

  - the multiplication rule

$$\frac{\Gamma, \overline{a} \cdot \overline{x} \geq b}{\Gamma, (c\overline{a}) \cdot \overline{x} \geq cb},$$

    where $c \geq 0$,
  - the rounded division rule

$$\frac{\Gamma, \overline{a} \cdot \overline{x} \geq b}{\Gamma, (1/c)\overline{a}) \cdot \overline{x} \geq \lceil b/c \rceil},$$

    where $c > 0$ divides all a_i,
  - and the Boolean axioms

$$\frac{}{x_i \geq 0} \quad \text{and} \quad \frac{}{-x_i \geq -1};$$

- the negation rules (new)

$$\frac{}{\overline{a} \cdot \overline{x} \geq b\,,\ -\overline{a} \cdot \overline{x} \geq -b + 1}$$

  and

$$\frac{}{\overline{a} \cdot \overline{x} \not\geq b\,,\ -\overline{a} \cdot \overline{x} \not\geq -b + 1}$$

  for all integer $\overline{a}$ and b.

7.1.2 First-order logic over CP: LK(CP)

Consider a first-order language where the atomic formulas are the CP-inequalities (7.1.1): the variables are x_i and these can be quantified but the parameters a_i and b are never quantified. We shall call such first-order formulas CP-**formulas**.

The proof system LK(CP) operates with sequents formed by arbitrary CP-formulas (let us stress that first-order quantifiers are allowed in them) and has in addition to the rules of the propositional LK from Chapter 3 the following first-order quantifier rules:

- $\forall$:introduction

$$\text{left } \frac{A(t), \Gamma \longrightarrow \Delta}{\forall x A(x), \Gamma \longrightarrow \Delta} \quad \text{and} \quad \text{right } \frac{\Gamma \longrightarrow \Delta, A(x)}{\Gamma \longrightarrow \Delta, \forall x A(x)},$$

- $\exists$:introduction

$$\text{left } \frac{A(x), \Gamma \longrightarrow \Delta}{\exists x A(x), \Gamma \longrightarrow \Delta} \quad \text{and} \quad \text{right } \frac{\Gamma \longrightarrow \Delta, A(t)}{\Gamma \longrightarrow \Delta, \exists x A(x)},$$

where t is any term such that no variable occurrence free in t becomes bounded in $A(t)$, and with the restriction that the variable x does not occur in the lower sequents of $\forall$:right and $\exists$:left.

The LK(CP) system has, in addition, the following rules inherited from CP and written in the sequential formalism:

- the addition rule

$$\frac{\Gamma \longrightarrow \Delta, \overline{a} \cdot \overline{x} \geq b \quad \Gamma \longrightarrow \Delta, \overline{a}' \cdot \overline{x} \geq b'}{\Gamma \longrightarrow \Delta, (\overline{a} + \overline{a}') \cdot \overline{x} \geq b + b'},$$

- the multiplication rule

$$\frac{\Gamma \longrightarrow \Delta, \overline{a} \cdot \overline{x} \geq b}{\Gamma \longrightarrow \Delta, (c\overline{a}) \cdot \overline{x} \geq cb},$$

  where $c \geq 0$;
- the rounded division rule

$$\frac{\Gamma \longrightarrow \Delta, \overline{a} \cdot \overline{x} \geq b}{\Gamma \longrightarrow \Delta, (1/c)\overline{a} \cdot \overline{x} \geq \lceil b/c \rceil},$$

  where $c > 0$ divides all a_i;
- the Boolean axioms

$$\frac{}{\longrightarrow x_i \geq 0} \quad \text{and} \quad \frac{}{\longrightarrow -x_i \geq -1};$$

- the negation rules

$$\frac{}{\longrightarrow \overline{a} \cdot \overline{x} \geq b, -\overline{a} \cdot \overline{x} \geq -b + 1}$$

and

$$\overline{\overline{a} \cdot \overline{x} \geq b, -\overline{a} \cdot \overline{x} \geq -b+1 \longrightarrow},$$

for all integer $\overline{a}$ and b.

7.1.3 Resolution over Linear and Polynomial Calculi: R(LIN) and R(PC)

The proof system R(LIN) operates with clauses of linear equations over $\mathbf{F}_2$. A line C in an R(LIN)-proof thus looks like a clause

$$f_1, \ldots, f_k,$$

where the $f_i \in \mathbf{F}_2[x_1, \ldots, x_n]$ are linear polynomials. The clause C is true for $\overline{a} \in \{0,1\}^n$ if and only if the Boolean formula

$$\bigvee_{i \leq k} f_i$$

in the language with $\bigvee, \oplus, 0, 1$ is true.

The inference rules combine the rules of resolution and of the **linear equational calculus** (LEC), i.e. the linear algebra rules:

- the Boolean axiom

$$\frac{}{h, h+1};$$

- the weakening rule

$$\frac{C}{C, f};$$

- the contraction rule

$$\frac{C, 0}{C};$$

- the addition rule

$$\frac{C, g \quad D, h}{C, D, g+h+1}.$$

Allowing general polynomials (or polynomials of degree at most d) and adding to R(LIN) one more inference rule,

- the multiplication rule

$$\frac{C, g}{C, gh+h+1},$$

defines **resolution over** PC, R(PC), or its degree d subsystems R(PC_d).

7.1.4 Polynomial Calculus with Resolution: PCR

When translating a clause C into a polynomial f_C as in (6.0.1) the size may blow up: $\neg x$ is translated as $1 - x$ and a product of such terms may have many monomials. An elegant way to avoid this issue is to introduce new variables that will translate $\neg x$.

The system PCR has for each variable x a **complementary variable** x' and it extends PC by the axiom

$$1 - x + x' = 0;$$

it also has the Boolean axioms $(x')^2 - x' = 0$ for all new variables. Note that it follows that $(x')' = x$ by a constant-size degree-1 PC-proof.

For the sake of completeness we now state a lemma; its first assertion is obvious and the second will be demonstrated in Part II.

Lemma 7.1.1 *All the systems* R(CP), LK(CP), R(LIN), R(PC) *and* PCR *p-simulate* R. *Further,* R(CP) *p-simulates* CP *and* R(PC) *and* PCR *p-simulate* PC.

Furthermore, all proof systems introduced in this section are p-simulated by Frege systems, in fact, by their TC^0*-Frege subsystems.*

7.2 Circuit and WPHP Frege Systems

According to folklore, extended Frege systems can be described as Frege systems operating with circuits or even as P/*poly*-Frege systems. In Section 2.6 we mentioned classes of proof systems suggestively named AC^0-Frege systems or $\mathrm{AC}^0[m]$-Frege systems, and the like. For constant-depth circuit classes, a formal definition is not so important, as one can without loss of generality restrict to *formulas* in the respective class without affecting the size more than polynomially. For P/*poly*, however, this is not known and it is conjectured to be in general impossible (at least nobody knows how to transform circuits into formulas without an exponential blow-up).

We shall need two technical notions about circuits, using the definition of circuits (tacitly in the DeMorgan language) from Section 1.4 and the notation $\mathrm{Def}_C(\overline{x}, \overline{y})$. Let D be a circuit with inputs $\overline{z} = (z_1, \ldots, z_m)$ and instructions $\overline{w}$ and let C_i, $i \leq m$, be circuits having all inputs $\overline{x}$ and let C_i have instructions $\overline{y^i}$ with the output $y^i_{s_i}$.

The **substitution** of C_i for z_i in D, denoted by $D(C_1, \ldots, C_m)$, is the circuit obtained by identifying the output nodes $y^i_{s_i}$ with z_i and taking the union of all instructions of all circuits $D, C_1, \ldots, C_m$.

For a circuit C, we can unwind it in a *unique* way to a possibly much larger formula. Using induction on $i \leq s = |C|$ define for each instruction y_i of C the formula φ_i, by the following procedure:

- if y_i is defined as a constant $b \in \{0, 1\}$ or an input variable x_j, put $\varphi_i := b$ or $\varphi_i := x_j$, respectively;

- if $y_i := \neg y_j$, put $\varphi_i := \neg \varphi_j$;
- if $y_i := y_j \circ y_k$ and $j < k$, put $\varphi_i := \varphi_j \circ \varphi_k$.

We say that two circuits C, D are **similar** if and only if $\varphi_C = \varphi_D$ (i.e. they are *identical* formulas, not just equivalent ones). If two circuits are not similar we can guess a partial path through them, going from the output back towards the inputs via the reverse edges, where on at least one instruction they do not agree. Hence we have

Lemma 7.2.1 *The similarity of two circuits C, D is in* coNLOG $\subseteq$ P.

Let F be a Frege system in the DeMorgan language. A proof of a circuit A from the circuits $B_1, \dots, B_t$ in the **circuit Frege** proof system CF is a sequence of circuits $\pi = C_1, \dots, C_k$ such that:

- each C_i,
  - is one of B_j,
  - *or* is derived from some $C_{j_1}, \dots C_{j_\ell}, j_1, \dots, j_\ell < i$ by an F-rule

$$\frac{E_1, \dots, E_\ell}{E_0}, \tag{7.2.1}$$

  meaning that there is a substitution σ of some *circuits* into the formulas E_u such that $\sigma(E_u) = C_u$ for $u = 0, \dots \ell$,
  - *or* is similar to C_j for a $j < i$.
- $C_k = A$.

Lemma 7.2.1 implies that CF is a Cook–Reckhow proof system. The next lemma makes precise the informal claims about EF mentioned earlier.

Lemma 7.2.2 *CF and EF are p-equivalent.*

Proof We shall show first that $CF \geq_p EF$. Let π be an EF-proof of A and assume for simplicity that there are no hypotheses B_j. Let $q_1, \dots, q_t$ be the extension variables in π, each q_s being introduced by the extension axiom

$$q_s \equiv C_s.$$

Given $0 \leq r < s \leq t$, define circuits $D_s^r(q_1, \dots, q_r)$ by the conditions

- $D_r^{r-1}(q_1, \dots, q_{r-1}) := C_r(q_1, \dots, q_{r-1})$,
- $D_s^{r-1}(q_1, \dots, q_{r-1}) := \tilde{D}_s^r(q_1, \dots, q_{r-1}, q_r / C_r(q_1, \dots, q_{r-1}))$,

where $\tilde{D}_s^r$ is D_s^r with all occurrences of the input q_r identified with one node (and hence $|\tilde{D}_s^r| \leq |D_s^r| + |C_r|$). Finally, for $1 \leq s \leq t$, set

$$E_s := D_s^0 .$$

Note that E_s may have as inputs the variables of A but no extension variables.

Claim 1 E_s is similar to $C_s(E_1, \ldots, E_{s-1})$ for all $1 \leq s \leq t$.

Now define a substitution σ: $q_s := E_s$ and let $\sigma(\pi)$ be the substitution instance of π.

Claim 2 $\sigma(\pi)$ is a *CF*-proof of *A* from the formulas $E_s \equiv E_s$, $1 \leq s \leq t$.

To see this, note that no substitution violates the Frege inference rules of F and that the σ-substitution instance of the extension axiom $q_s \equiv C_s(q_1, \ldots, q_{s-1})$ is the circuit

$$E_s \equiv C_s(E_1, \ldots, E_{s-1}),$$

which is similar to $E_s \equiv E_s$. As each such equivalence has a size $O(|E_s|)$ F-proof we get a *CF*-proof of *A*.

For the opposite p-simulation, $EF \geq_p CF$, assume that $\rho = C_1, \ldots, C_k$ is a *CF*-proof of *A*, with *A* a formula. Assign to every subcircuit *D* occurring anywhere in ρ an extension variable q_D and define it from the other extension variables (for smaller circuits) by the same instruction (i.e. by an extension axiom) as that by which *D* was defined from its subcircuits. In addition, identify the variables q_D for all similar circuits. Let $\mathcal{E}$ be all the extension axioms introduced in this way.

Claim 3 Each q_{C_i} has a constant-size F-proof from $q_{C_1}, \ldots, q_{C_{i-1}}$.

Assume that C_i was inferred using the rule (7.2.1) by a substitution σ: $C_{j_t} = \sigma(E_t)$ for $1 \leq t \leq \ell$ and $C_i = \sigma(E_0)$, and write $\sigma(E_t)$ simply as $E_t(G_1, \ldots, G_n)$. From each q_{C_t}, $1 \leq t \leq \ell$, derive by a constant-size F-proof the formula

$$E_t(q_{G_1}, \ldots, q_{G_n}),$$

using the extension axioms $\mathcal{E}$ as hypotheses, and by the same Frege rule derive also

$$E_0(q_{G_1}, \ldots, q_{G_n}),$$

from which q_{C_i} follows (again using $\mathcal{E}$).

If C_i was derived from C_j, $j < i$, by the similarity condition then $q_{C_i} = q_{C_j}$ and there is nothing to do.

It follows that there is a size-$O(|\rho|)$ F-proof of q_A from $\mathcal{E}$, from which we can derive, using $\mathcal{E}$, the formula *A* itself by a size-$O(|A|^2)$ F-proof. □

Now we turn to an extension of *CF*, the **weak** PHP **Frege system**, denoted by *WF*. It corresponds – in the sense to be presented in Part II – to an important bounded arithmetic theory playing a significant role in formalizations of randomized constructions.

The qualification "weak" in weak PHP refers to the fact that this principle, denoted by WPHP, talks about maps F: $U \rightarrow V$, where *U* is much larger than *V*. In ordinary PHP we have $|U| = 1 + |V|$. The definition of *WF* actually uses its *dual* version, dWPHP:

- *if $G\colon U \to V$, where $|U| < |V|$, then G is not surjective.*

In the original PHP we have $U = [n+1]$ and $V = [n]$ and we formalize it as PHP_n (see (1.5.1)) using the atoms p_{ij} to represent the atomic statements $F(i) = j$. In the formalization of dWPHP below we consider $U = \{0,1\}^n$ and $V = \{0,1\}^m$ for $m > n \geq 1$ and thus we cannot use the atomic statements $G(i) = j$ as there are exponentially many of them. Instead we consider G to be computed by Boolean circuits,

$$G\colon \overline{x} \in \{0,1\}^n \to (C_1(\overline{x}), \ldots, C_m(\overline{x})) \in \{0,1\}^m,$$

and we formalize the statement of dWPHP as

$$\exists \overline{y} \in \{0,1\}^m \forall \overline{x} \in \{0,1\}^n, \quad \bigvee_{i \leq m} C_i(\overline{x}) \neq y_i. \tag{7.2.2}$$

A *WF*-**proof** of A from $B_1, \ldots, B_t$ in the proof system *WF* (the acronym stands for *WPHP* Frege) is a *CF*-proof that can also use the following rule.

- *For any $1 \leq n < m$ and any collection $\mathcal{C}$ of m circuits $C_i(\overline{x})$ with n inputs $\overline{x}$, we can introduce a new m-tuple of atoms $\overline{r} = (r_1, \ldots, r_m)$ attached to $\mathcal{C}$ and not occurring in any of the formulas $A, B_1, \ldots B_t, \mathrm{Def}_{C_1}, \ldots, \mathrm{Def}_{C_m}$, and for any circuits $D_1, \ldots, D_n$ (which may contain $\overline{r}$), we may use the axiom*

$$\bigvee_{i \leq m} C_i(D_1, \ldots, D_n) \neq r_i. \tag{7.2.3}$$

The m-tuple $\overline{r}$ serves as a canonical witness for the existential quantifier in (7.2.2). As (7.2.2) is valid, we have

Lemma 7.2.3 *WF is a sound propositional proof system.*

It is unknown whether *CF* p-simulates *WF* but we have at least some upper bound on its strength.

Lemma 7.2.4 $G_2 \geq_p WF$.

The lemma will be proved in Section 12.6 using bounded arithmetic. It involves a formalization of a proof of dual WPHP for P/*poly* maps G in a particular bounded arithmetic theory.

7.3 Implicit Proof Systems

We will learn in Part II how to translate a certain type of universal first-order statement $\forall x \Phi(x)$ about natural numbers into a sequence of tautologies expressing the statement for numbers of bit length $n = 1, 2, \ldots$: $\forall x(|x| = n)\Phi(x)$. For a fairly general class of first-order theories T (Peano arithmetic and set theory ZFC among them), we will be able to translate a T-proof of such a universal statement into a sequence of short propositional proofs of the associated tautologies. The resulting proofs will be proofs in propositional proof systems attached to the theories. These proof systems are very strong even for relatively weak theories. The proofs all share a remarkable property: they are very uniform. The qualification *very* will be quantified in Part II. This can be perceived as a certain measure of the easiness of the proofs despite the fact that the proof systems are strong. One of the motivations beyond implicit proof systems is to understand what happens if we allow proofs to be very large (i.e. to be hard in the sense of proof complexity) but easy in the sense that they are uniform and the ambient proof system is simple.

Let P, Q be two Cook–Reckhow proof systems. A proof π in a proof system Q of a tautology τ is a particular string. If it is very long, say it has $\ell = 2^k$ bits, we can try to represent it implicitly by a circuit β with k inputs: β computes from $i \in [\ell] = \{0, 1\}^k$ the ith bit π_i. The defining circuit β may be, in principle, exponentially smaller than π. But the circuit on its own does not suffice to constitute a proof in a Cook–Reckhow proof system. In order to obtain that we supplement the circuit β with a P-proof α of the fact that β indeed describes a valid Q-proof of the formula τ. The resulting pair (α, β) is a proof of τ in a proof system to be denoted by $[P, Q]$.

Now we formalize this definition. Consider the function version of the proof system in Definition 1.5.1: Q is a polynomial-time map from $\{0, 1\}^*$ onto TAUT. Assume that Q is computed by a deterministic Turing machine (also denoted by Q) as defined in Section 1.3 and running in time n^c. The computation of Q on an input $w \in \{0, 1\}^n$ will be coded by a tableaux W of all (at most $t \leq n^c$) **instantaneous descriptions** of the computation ending after t steps. The ith row ($i \leq t$) W_i is the ith instantaneous description: it lists the content of the tape visited by the machine head during the computation up to that time, the current state of the machine and the position of its head. The length of W_i is bounded by $O(t)$. Thus we can view W as a $0/1$ $t \times O(t)$ matrix. Take a suitable $k := \log(t) + O(1) = O(\log n)$ and assume without loss of generality that both the rows i and the columns j are indexed by elements of $\{0, 1\}^k$.

The property that W encodes a valid computation of Q on some input w can be expressed by quantifying universally over the coordinates i,j and stating that for each such pair i,j the local conditions of a valid computation are obeyed. Hence if $\beta(i,j)$, $i = (i_1, \ldots, i_k)$ and $i = (j_1, \ldots, j_k)$, is a circuit with $2k$ inputs that describes a potential computation W, we can express by a propositional formula $Correct^Q_\beta$ the property that:

- *the tableaux $W_{i,j} := \beta(i,j)$ is a valid computation of Q on some input (encoded in the first row W_1).*

Note that although the size of W is $O(t^2)$, the size of $Correct^Q_\beta$ is $O(|\beta|)$ and this can be much smaller (even $O(\log t)$).

Now we can define **implicit proof systems** formally. Let P, Q be two proof systems and assume that P contains the resolution system R. Then, a proof of a tautology τ in the **proof system** $[P, Q]$ is a pair (α, β) such that:

- *β is a Boolean circuit with $2k$ inputs $(i_1, \ldots, i_k, j_1, \ldots, j_k)$, for some $k \geq 1$;*
- *β defines a valid computation W of Q on some input w whose output is τ;*
- *α is a P-proof of the tautology $Correct^Q_\beta$.*

Note that the fact that the output encoded in W is τ needs no P-proof, as that is a true Boolean sentence. For a P containing R, we define **implicit** P, denoted by iP, by

$$iP := [P, P] .$$

The technical condition that P contains R is included because R, in a short proof (i.e. in a proof with a number of clauses that is proportional to the circuit size), proves that a computation of a circuit is unique (see Lemma 12.3.1).

The definition of iP looks rather too formal; we would like β to output the steps of a P-proof and not just bits. However, if a proof is long, so also can be the formulas forming its steps. For example, the steps in a Frege proof may be almost as large as the proof itself. But some proof systems escape this unboundedness of the width. In resolution R we know a priori that each clause contains at most $2|\tau| + 2$ literals and constants and in Extended resolution, ER, we know by Lemma 5.8.1(item (ii)) that the width can be bounded by $O(|\tau|)$. Consider as a case study ER. Define an **f-implicit** ER-refutation of τ to be a pair (α, β) where the following hold.

- *β is a Boolean circuit with inputs $i_1, \ldots, i_k$ outputting a string (i.e. it has several output gates).*
- *The sequence $\beta(\overline{0}), \ldots, \beta(i), \ldots, \beta(\overline{1})$, where $\{0,1\}^k$ is ordered lexicographically, describes an ER-refutation of τ with $\beta(i)$ containing also information about the earlier two steps (and the resolved atom) that were used in the derivation of the ith clause, or about the form of the extension axiom, if that is what it was.*
- *α is an ER-proof of a tautology $Correct_\beta(x_1, \ldots, x_k)$ expressing that β describes a valid ER-refutation.*

In general, an exponentially long (in k) ER-refutation may introduce exponentially many extension atoms and hence could contain exponentially long clauses that could not be described by a β of size polynomial in k. However, by Lemma 5.8.4 we may restrict to ER-refutations that have width at most that of τ, and such refutations can be described by β.

With this in mind the next lemma ought to be clear.

Lemma 7.3.1 *Implicit and f-implicit ER p-simulate each other.*

The next lemma summarizes two simple properties of the bracket operation $[P, Q]$.

Lemma 7.3.2 *(i) For all P, we have $P \leq_p [P, R^*]$.*
(ii) For all $P \geq_p R$, we have $iP \equiv_p [iP, P]$.

Implicit ER is very strong; we shall derive the following in Section 12.7 using bounded arithmetic.

Lemma 7.3.3 $iER \geq_p$ G.

It is open whether the opposite simulation also holds but the links with bounded arithmetic suggest that perhaps it does not. The bracket operation can be iterated. By Lemma 7.3.2(ii) we know that it makes no sense to iterate at the first argument. Hence we put

$$i_1 ER := iER \quad \text{and} \quad i_{k+1} ER := i(i_k ER) ,$$

where i_k means that the implicit construction is iterated k times. The topic of iterations is, I think, rather impossible to discuss without bounded arithmetic and we shall leave it. But we may look at ER itself and how it can itself be obtained from a weaker proof system by the bracket operation.

Lemma 7.3.4 $ER \equiv_p [R, R^*]$.

Proof We shall prove $ER \leq_p$ [R, R*] here and leave the opposite p-simulation to Part II (Lemma 11.5.3). Let $\mathcal{C}$ be a set of clauses and let π be its ER-refutation. Assume that $q_1, \ldots, q_t$ are extension atoms introduced in this order via extension axioms $E_1, \ldots, E_t$, as in (5.8.1):

$$E_i\colon \{\neg\ell_1, q_i\},\ \{\neg\ell_2, q_i\},\ \{\ell_1, \ell_2, \neg q_i\}, \tag{7.3.1}$$

where the literals ℓ_1 and ℓ_2 do not contain $q_i, \ldots, q_t$. Let π^* be a tree-like R-refutation of $\mathcal{C} \cup \{E_1, \ldots, E_t\}$; it is not an ER-refutation of $\mathcal{C}$ as the extension axioms appear in the proof several times. The refutation π^* is obtained by unwinding the proof-graph (a dag) of π into a tree: each partial path through the dag yields one node in the proof tree of π^* and the paths can be used as the addresses of the nodes. Clearly there is a p-time algorithm A that given π and the address of a node in π^* (i.e. a path in π), computes the clause labeling that node: just travel in π along the path. The same algorithm can also output the address of the two hypotheses of the resolution inference (and the name of the resolved atom) used at the node or the form of the extension axiom.

We shall construct a sequence of R*-refutations σ of $\mathcal{C} \cup \{E_1, \ldots, E_i\}$ for $i = t, t-1, \ldots, 1$ and, to stress the specific set of initial clauses, we shall denote such a σ by $\sigma[q_1, \ldots, q_i]$. Hence we denote π^* by $\pi^*[q_1, \ldots, q_t]$.

Having $\sigma[q_1, \ldots, q_i]$ we denote by $\sigma[q_1, \ldots, q_{i-1}, 1]$ the refutation of

$$\mathcal{C} \cup \{E_1, \ldots, E_{i-1}\} \cup \{\ell_1, \ell_2\},$$

where ℓ_1, ℓ_2 are from E_i in (7.3.1), obtained by substituting $q_i := 1$, and we denote by $\sigma[q_1, \ldots, q_{i-1}, 0]$ the refutation of

$$\mathcal{C} \cup \{E_1, \ldots, E_{i-1}\} \cup \{\neg\ell_1\} \cup \{\neg\ell_2\}$$

obtained by the dual substitution $q_i := 0$. In fact, these refutations are found in the system R^*_w from Section 5.4 but we shall ignore this technical detail.

Define now specific R*-refutations $\sigma[q_1, \ldots, q_i]$ by induction on $i = t, \ldots, 1$ by taking

$$\sigma[q_1, \ldots, q_t] := \pi^*[q_1, \ldots, q_t]$$

and, for $i \leq t$, defining $\sigma[q_1, \ldots, q_{i-1}]$ by the following process.

1. Modify $\sigma[q_1, \ldots, q_{i-1}, 0]$ into an R*-proof ξ_i of $\{\ell_1, \ell_2\}$ from $\mathcal{C} \cup \{E_1, \ldots, E_{i-1}\}$ by adding the literal ℓ_1 (resp. ℓ_2) to the initial clause $\{\neg \ell_1\}$ (resp. $\{\neg \ell_2\}$) and carrying the two extra literals along the proof to the empty end-clause.
2. Modify $\sigma[q_1, \ldots, q_{i-1}, 1]$ into $\sigma[q_1, \ldots, q_{i-1}]$ by attaching, above each occurrence of the initial clause $\{\ell_1, \ell_2\}$, the proof ξ_i.

In the same way, let ρ be the R*-refutation of $\mathcal{C}$ obtained at the end of the induction process from $\sigma[q_1]$. The nodes in the proof tree of ρ are naturally addressed by

- a string from $\{0, 1\}^t$, which says for each i whether the node is in the $\sigma[q_1, \ldots, q_{i-1}, 1]$-part or whether it is in one of the ξ_i-parts of $\sigma[q_1, \ldots, q_{i-1}]$, in which case the address contains the path in $\sigma[q_1, \ldots, q_{i-1}, 1]$ to identify the particular copy of ξ_i,
- followed by a path in $\sigma[q_1, \ldots, q_{t-1}, 1]$ or $\sigma[q_1, \ldots, q_{t-1}, 0]$ attached at the end.

Using the algorithm A describing π^*, an algorithm describing ρ can be defined.

Claim 1 There is a polynomial-time algorithm β that takes π as an "advice" and computes the label of any node in the proof tree of ρ (together with the information about the inference).

Claim 2 There is an R-proof α of size $|\beta|^{O(1)}$ of the fact that the circuit β describes an R*-refutation on $\mathcal{C}$.

The statement that α is supposed to prove has variables for the address of a node and for the computation of β and to assert that the form of each inference is valid and that the initial clauses are only from $\mathcal{C}$. To construct α we use the extra information computed by β about each node, i.e. the hypotheses of the inference, or the resolved variable or the form of an initial clause. It comes down, via an induction argument, to the fact that π is a valid ER-refutation and hence also that π^* and all $\sigma[q_1, \ldots, q_i]$ are R*-proofs. We leave the details to the reader. □

7.4 Auxiliary Logic Proof Systems

In this and the next section we collect a few examples of proof systems that have a natural definition (or, at least, appear in a natural context) but – with the exception of P_{PA} – do not fit into a wider class of proof systems and, in particular, do not fit into any class discussed so far. This section presents three such logical examples; the next section will consider algebraic examples, and more examples are given in Section 7.7.

7.4.1 Monotonic LK

This proof system, denoted by MLK, is a subsystem of LK from Chapter 3 that forbids the negation. In particular, the rules of MLK are the same as those of LK but there are no $\neg$:introduction rules.

There are only trivial negation-free tautologies (essentially just the constant 1 and its various disguises as $p \vee 1$, etc.) but there are non-trivial tautological *sequents* that use no negation. For example, the PHP principle (1.5.1) can be written as such:

$$\bigwedge_i \bigvee_j p_{ij} \longrightarrow \bigvee_i \bigvee_{j \neq j'} (p_{ij} \wedge p_{ij'}), \bigvee_{i \neq i'} \bigvee_j (p_{ij} \wedge p_{i'j}), \tag{7.4.1}$$

where i, j range over the same domains as in (1.5.1). Also, some other tautologies prominent in connection with the feasible interpolation method (see Chapter 17) can be written without the negation.

The proof of Theorem 3.2.1 immediately yields the completeness of MLK with respect to all tautological negation-free sequents. The key fact about MLK is that it is essentially as strong as unrestricted LK and hence, by Theorem 3.2.4, as Frege systems.

Theorem 7.4.1 ((i) and (ii), Atserias *et al.* [34]; (iii), Buss *et al.* [126])

(i) MLK (tree-like MLK) quasi-polynomially simulates LK with respect to the proofs of all negation-free sequents. In particular, if a negation-free sequent with n atoms has an LK-proof of size s then it also has an MLK*-proof of size* $s^{O(1)} n^{O(\log n)} \leq s^{O(\log s)}$ *and with* $s^{O(1)}$ *steps.*

(ii) LK is p-bounded if and only if MLK is p-bounded (with respect to tautological negation-free sequents).

(iii) MLK p-simulates LK with respect to the proofs of all negation-free sequents.

A corollary of this theorem is that the PHP sequent (7.4.1) has a quasi-polynomial MLK*-proof (see the remark after Lemma 2.4.1) and a polynomial MLK proof.

Lemma 7.4.2 (7.4.1) *has a polynomial-size MLK-proof.*

7.4.2 OBDD Proof System

An OBDD (the acronym stands for *ordered binary decision diagram*) is a special kind of a branching program (see Section 5.3) computing a Boolean function: on each path through the program, each variable is queried at most once and the queries are ordered consistently with one global (but arbitrary) linear ordering of all variables. This notion was defined by Bryant [103, 104] and it has the interesting property that (a suitably defined) reduced form of an OBDD is unique and, moreover, can be computed by a polynomial-time algorithm. Various properties of OBDDs can be decided in polynomial time using this reduced form (see also Wegener [505]), for example, whether the Boolean f_P defined by the OBDD P majorizes the function f_Q defined by Q. They also compute from P, Q the reduced OBDD R computing $f_P \wedge f_Q$

and so on. These properties make OBDDs useful for representing Boolean functions and they have found many applications across computer science.

Atserias, Kolaitis and Vardi [35] defined a proof system for handling instances of the constraint satisfaction problem that uses OBDDs. The OBDD **proof system** is a refutation proof system. Let $\mathcal{C}$ be a set of clauses. An OBDD-refutation of $\mathcal{C}$ is a sequence of OBDDs

$$P_1, \ldots, P_k$$

such that:

- *for every* $i \leq k$,
  - *either* P_i *computes the Boolean function defined by a clause in* $\mathcal{C}$,
  - *or there are* $u, v < i$ *such that* $f_{P_u} \wedge f_{P_v} \leq f_{P_i}$;
- f_{P_k} *is the identically zero function.*

To supply some information about the strength of the system, we state without a proof the following lemma.

Lemma 7.4.3 *The OBDD-proof system p-simulates resolution but not* LK_d, *for large enough* $d \geq 1$, *and no* LK_d *p-simulates it. Frege systems do, however, p-simulate OBDD.*

We shall consider OBDD proof systems again in Part III (Section 18.5).

7.4.3 Peano Arithmetic as a Proof System

Peano arithmetic, denoted by PA, serves here as the canonical example of a first-order theory that is able to develop finite mathematics smoothly. This informal statement can be substantiated by the fact that PA is mutually interpretable with the theory of finite sets, the set theory ZFC with the axiom of infinity replaced by its negation. The advantage of PA over set theory is that it has terms and hence a simple way to represent individual finite binary strings.

The language of PA, L_{PA}, consists of the constants 0 and 1, the binary function symbols $+$ and $\cdot$, and the binary ordering relation symbol $\leq$; the equality $=$ is understood to belong to the ambient first-order logic. The **standard model** of PA is the set of natural numbers $\mathbf{N}$ with the *standard* interpretation of the language. The axioms of PA consist of a finite list forming **Robinson's theory** Q:

1. $x + 1 \neq 0$,
2. $x + 1 = y + 1 \rightarrow x = y$,
3. $x + 0 = x$,
4. $x + (y + 1) = (x + y) + 1$,
5. $x \cdot 0 = 0$ and $x \cdot 1 = x$,
6. $x \cdot (y + 1) = (x \cdot y) + x$,
7. the axioms of discrete linear orders for $\leq$ with $x + 1$ the successor of x,
8. $x \leq y \equiv (\exists z, x + z = y)$,

accompanied by all (infinitely many) instances of the **induction scheme** IND:

$$[A(0) \wedge \forall x(A(x) \rightarrow A(x+1))] \rightarrow \forall x A(x), \tag{7.4.2}$$

for all L_{PA}-formulas A (it may contain free variables other than x).

There are canonical closed terms, the **numerals** s_n, whose values range over all $\mathbf{N}$:

$$s_0 := 0, \quad s_1 := 1 \quad \text{and} \quad s_{n+1} := s_n + 1.$$

This is simple but rather inefficient; it is essentially the unary notation. Better are the **dyadic numerals** $\underline{n}$:

$$\underline{0} := 0, \quad \underline{1} := 1, \quad \text{and } \underline{2n} := (1+1) \cdot \underline{n} \quad \text{and} \quad \underline{2n+1} := (\underline{2n} + 1).$$

Note that the length of the term $\underline{n}$ is $O(\log n)$.

The class of **bounded formulas**, denoted by Δ_0, is the smallest class of L_{PA}-formulas that contain all quantifier-free formulas and are closed under the DeMorgan connectives and under **bounded quantification**: if $A(\overline{x}, y)$ is a Δ_0-formula and $t(\overline{x})$ is an L_{PA}-term, both the formulas

$$\exists y \leq t(\overline{x})\, A(\overline{x}, y) \quad \text{and} \quad \forall y \leq t(\overline{x})\, A(\overline{x}, y)$$

are Δ_0 too. Here $\exists y \leq t(\overline{x})\, A(\overline{x}, y)$ abbreviates $\exists y,\ y \leq t(\overline{x}) \wedge A(\overline{x}, y)$ and $\forall y \leq t(\overline{x})\, A(\overline{x}, y)$ abbreviates $\forall y,\ y \leq t(\overline{x}) \rightarrow A(\overline{x}, y)$. A Σ_1^0-formula has the form

$$\exists z B(\overline{x}, z),$$

where $B \in \Delta_0$. We shall use two classical facts of mathematical logic.

1. A relation $R \subseteq \mathbf{N}^k$ is recursively enumerable if and only if there is a Σ_1^0-formula $A(\overline{x})$, $\overline{x} = x_1, \ldots, x_k$, such that for all k-tuples $\overline{m} \in \mathbf{N}^k$,

$$\overline{m} \in R \quad \text{if and only if} \quad \mathbf{N} \models A(\overline{m})\ .$$

2. For all Σ_1^0-formulas $A(\overline{x})$ and for $\overline{m} \in \mathbf{N}^k$,

$$\mathbf{N} \models A(\overline{m}) \quad \text{if and only if} \quad \text{PA} \vdash A(\underline{m_1}, \ldots, \underline{m_k}).$$

Note that R is recursive if and only if both R and $\mathbf{N}^k \setminus R$ are Σ_1^0-definable.

For a string $w \in \{0,1\}^*$, define the closed term $\ulcorner w \urcorner$ as follows:

$$\ulcorner \Lambda \urcorner := 0,$$

where Λ is the empty word, and, for $w = w_{t-1} \ldots w_0 \in \{0,1\}^t$, $t \geq 1$, put

$$\ulcorner w \urcorner := \underline{n}, \qquad \text{for } n = 2^t + \sum_{j<t} w_j 2^j.$$

By fact 1 above there is a Σ_1^0-formula $Taut(x)$ such that for all $w \in \{0,1\}^*$,

$$w \in \text{TAUT} \quad \text{if and only if} \quad \mathbf{N} \models Taut(\ulcorner w \urcorner)$$

and, by fact 2,

$$w \in \text{TAUT} \quad \text{if and only if} \quad \text{PA} \vdash Taut(\ulcorner w \urcorner)\ .$$

Note that for any formula τ: $|\ulcorner \tau \urcorner| = O(|\tau|)$.

This allows us to define the propositional proof system P_{PA} by:

$$P_{PA}(\pi, \tau) \quad \text{if and only if} \quad \pi \text{ is a PA-proof of Taut}(\ulcorner \tau \urcorner).$$

Lemma 7.4.4 *P_{PA} is a Cook–Reckhow proof system.*

Proof Peano Arithmetic PA holds true in the standard model, and hence the existence of a proof π means that $Taut(\ulcorner \tau \urcorner)$ is true in $\mathbf{N}$ and hence, by the definition of the formula *Taut*, $\tau \in$ TAUT. The soundness and the completeness follow by fact 2.

It remains to verify that the provability predicate for P_{PA} is a polynomial-time recognizable relation. This boils down to the fact that it is a polynomial-time question whether an L_{PA}-formula is an axiom of PA: one needs to check it against a finite list of axioms of Q or decide whether it is a substitution instance of the axiom scheme IND. □

The proof system P_{PA} is incredibly strong. It will follow from the theory exposed in Part II that it p-simulates all the proof systems introduced in Part I.

7.5 Auxiliary Algebraic Proof Systems

7.5.1 Extended Nullstellensatz

Consider an algebraic proof system over a field $\mathbf{F}$; for simplicity we take here $\mathbf{F} := \mathbf{F}_2$ (a more general case of finite prime fields can be treated analogously). If $f_1, \ldots, f_t$ are polynomials over $\mathbf{F}_2$, the Boolean function defined as

$$\bigvee_i f_i$$

is equivalent to the polynomial

$$1 - \prod_i (1 - f_i).$$

However, this polynomial has degree $\sum_i deg(f_i)$ and as we saw in Section 6.2, keeping the degree small is a key issue. Extended Nullstellensatz represents the disjunction without a large increase in degree: it introduces *new* variables $r_1, \ldots, r_t$ and represents the disjunction as

$$r_1 f_1 + \cdots + r_t f_t.$$

Clearly $\sum_i r_i f_i = 1$ implies that at least one of the f_i is 1, and to enforce the opposite implication we include new axioms

$$f_i(1 - (r_1 f_1 + \cdots + r_t f_t)) = 0, \qquad \text{for all } i \leq t. \tag{7.5.1}$$

Using such axioms we derive from $r_1 f_1 + \cdots + r_t f_t = 0$ that $f_i = 0$ for all i. The degree of (7.5.1) and of this simple derivation is at most $2d + 1$, where $d := \max_i deg(f_i)$.

The **extended** NS proof system, denoted by ENS, is a proof system that proves the unsolvability of $g_1 = 0, \ldots, g_k = 0$ for all $g_i \in \mathbf{F}_2[\overline{x}]$ and $\overline{x} = x_1, \ldots, x_n$, as follows. For a set of variables *Var*, $\mathbf{F}_2[Var]$ is the ring of polynomials in those variables. We first enlarge the set of initial polynomials as follows.

1. Put $Var := \{x_1, \ldots, x_n\}$ and $Init := \{g_1, \ldots, g_k, x_1^2 - x_1, \ldots, x_n^2 - x_n\}$.
2. Select any $\mathbf{F}_2[Var]$-polynomials $f_1, \ldots, f_t$, choose *new* **extension variables** $r_1, \ldots, r_t$ (i.e. they are not in *Var*), add to the initial polynomials *Init* all **extension polynomials** (7.5.1) and then add to *Var* all $r_1, \ldots, r_t$.
3. Repeat process 2 arbitrarily many times, always choosing new extension variables. Let Var' and $Init'$ be the final sets of variables and initial polynomials.

An ENS-proof of the unsolvability of $\{g_i = 0\}_{i \leq k}$ is a sequence of polynomials $h_f \in \mathbf{F}_2[Var']$, for $f \in Init'$; such that

$$\sum_{f \in Init'} h_f f = 1.$$

Lemma 7.5.1 *The proof system ENS is sound and complete.*

Proof The completeness follows as ENS contains NS. The soundness follows as any assignment $\overline{x} := \overline{a} \in \mathbf{F}_2$ can be extended to an assignment to all variables in Var' making all extension axioms in $Init'$ true. In particular, for (7.5.1), define

$$r_i := \begin{cases} 1 & \text{if } i \text{ is the minimum index such that } f_i = 1, \\ 0 & \text{if } i \text{ is not the above or all } f_j = 0. \end{cases}$$

□

In Section 15.6, the particular proof system defined above will be called leveled ENS of accuracy 1, because the extension axioms (and the associated extension variables) can be partitioned into levels, axioms at level $\ell + 1$ correspond to polynomials using only variables at levels $\leq \ell$. The term "accuracy" refers to the form of (7.5.1): it is a linear combination of the f_i.

Let us now formalize the idea of how to deal with disjunctions differently. Given $f_1, \ldots, f_t$, let $r_i^j, j = 1, \ldots, h$ and $i = 1, \ldots, t$, be mutually different variables that do not occur in any of $f_1, \ldots, f_t$. The **unleveled extension axioms of accuracy** h are

$$f_i \prod_{j \leq h} (1 - (r_1^j f_1 + \cdots + r_t^j f_t)) = 0, \quad \text{for all} \quad i \leq t. \tag{7.5.2}$$

A proof of the unsolvability of $\{g_i = 0\}_{i \leq k}$ in **unstructured ENS**, denoted by UENS, is a set $\mathcal{A}$ of any polynomials (7.5.2) such that $2^h > |\mathcal{A}|$ and a sequence of polynomials (in the variables of all the polynomials involved) h_f for $f \in uInit :=$

$\{g_1, \ldots, g_k, x_1^2 - x_1, \ldots, x_n^2 - x_n\} \cup \mathcal{A}$ such that

$$\sum_{f \in uInit} h_f f = 1.$$

The UENS system can be defined in a more elegant (and even less structured but equivalent) way, but we postpone this to Section 15.7.

Lemma 7.5.2 *The UENS system is sound and complete.*

Proof The completeness is simple as before. For the soundness, let $\overline{x} := \overline{a} \in \mathbf{F}_2$ be any assignment to the x-variables and choose an assignment $\overline{b}$ to all the remaining variables uniformly at random. Then, for any $p \in \mathcal{A}$,

$$Prob_{\overline{b}}[p(\overline{a}, \overline{b}) \neq 0] \leq 2^{-h}.$$

Using the cardinality condition imposed on $\mathcal{A}$ it follows by averaging that there is some $\overline{b}$ such that $p(\overline{a}, \overline{b}) = 0$ for all $p \in \mathcal{A}$. Hence $\overline{a}$ cannot solve all the equations $g_i(\overline{x}) = 0$. □

The following lemma summarizes what is known about the strength of these systems, and we shall derive it (in fact, in a much more precise form) in Section 15.6. When comparing logical and algebraic proof systems, we consider that the same set of initial clauses $\mathcal{C}$ is represented by (canonical) polynomial equations, as in (6.0.1). In fact, there is a more general correspondence, which we will present in Section 15.6.

Lemma 7.5.3 *Let $\mathcal{C}$ be a set of constant-width clauses. A depth ℓ, size s $F(\oplus)$-refutation of $\mathcal{C}$ can be simulated by ENS over $\mathbf{F}_2$ with extension axioms leveled into $\ell + O(1)$ levels and degree bounded by $(\log s)^{O(1)}$.*

An EF-refutation of size s can be simulated by a degree-$(\log s)^{O(1)}$ UENS-proof and UENS can be p-simulated by WF.

It is open whether *EF* itself p-simulates UENS.

7.5.2 Ideal Proof System

Let $\mathbf{F}$ be a field and $\mathcal{F} := \{f_1, \ldots, f_m\} \subseteq \mathbf{F}[\overline{x}]$, $\overline{x} = x_1, \ldots, x_n$, and assume that the polynomials $x_i^2 - x_i$ are among $\mathcal{F}$. Consider now the variant (often adopted) of PC in which

- *only multiplication by constants or single variables is allowed.*

By Lemma 6.2.4, a PC*-refutation π of $\mathcal{F}$ of height h then yields an NS-refutation of $\mathcal{F}$

$$g_1 f_1, + \cdots + g_m f_m = 1, \tag{7.5.3}$$

of degree at most $h + \max deg(\mathcal{F})$. If the proof tree of π is balanced, i.e. $h \approx \log(k)$, $k = \mathbf{k}(\pi)$, then this upper bound may be useful. If π is unbalanced then h can be larger than n, for example. However, even in this case we have some information about the complexity of the coefficient polynomials g_i. Namely, each is computed by an algebraic formula (i.e. **F**-term, see Section 6.1) A_i whose underlying tree is the proof tree of π: the leaves holding f_i are labeled by 1 and the other leaves by 0 and all other nodes perform the same operation as in π. In fact, even if π is not tree-like, the same construction defines an **algebraic circuit** B_i computing g_i, and its size $|B_i|$ is at most k ($= \mathbf{k}(\pi)$). Hence we can write the left-hand side of (7.5.3) as one circuit

$$C(\overline{x}, f_1, \ldots, f_m),$$

where

$$C(\overline{x}, y_1, \ldots, y_m) \ := \ C_1(\overline{x})y_1 + \cdots + C_m(\overline{x})y_m \, ,$$

is a particular algebraic circuit of size at most $mk \leq k^2$. The qualification *particular* refers to the fact that, having C, we can reconstruct the original PC-refutation π.

The property whereby (7.5.3) witnesses the fact that $\mathcal{F}$ generates the trivial ideal can be expressed by two properties of C in $\mathbf{F}[\overline{x}]$:

1. $C(\overline{x}, 0, \ldots, 0) = 0$;
2. $C(\overline{x}, f_1, \ldots, f_m) = 1$.

The idea behind the **ideal proof system**, denoted by IPS, due to Grochow and Pitassi [211], is to accept *any* algebraic circuit $C(\overline{x}, y_1, \ldots, y_m)$ satisfying properties 1 and 2 as a proof that $\mathcal{F}$ is unsolvable, i.e. not confined to the circuits obtained from PC-refutations.

The drawback of this definition is that we do not know how to recognize such circuits in p-time and hence this may not be a Cook–Reckhow proof system. However, if **F** is large enough, say $|\mathbf{F}| \geq 2n$ (or better still, **F** is infinite), there is a probabilistic polynomial-time algorithm (see Section 7.7 for more discussion). We conclude by a simple statement gauging the strength of IPS.

Lemma 7.5.4 *IPS p-simulates EF.*

Proof Consider an equational calculus EC/**F** refutation π of $\mathcal{F}$ (where we take $\mathcal{F}$ to represent a CNF in order that the discussion makes sense for *EF* too). The same circuit construction as that above for PC yields in this situation also an algebraic circuit C (i.e. the terms in π are algebraic formulas) forming an IPS proof of the unsolvability of $\mathcal{F}$. Its size is bounded by the number of *different* formulas (terms) in π.

By Lemma 6.1.2 (and the underlying construction), EC p-simulates Frege systems, and the number of different terms in the resulting EC-refutation is proportional to the number of different formulas in the starting Frege refutation. By Lemma 2.5.7 the minimum number of different formulas in a Frege refutation is bounded by the minimum number of steps plus the size of $\mathcal{F}$, and that is proportional to the minimum size of an EF-refutation of $\mathcal{F}$. See Section 2.5 for a summary of these relations among the measures. □

Let us remark that the degree of the resulting IPS-proof may be exponential but when simulating only F it can be bounded by a polynomial.

7.5.3 Dehn Calculus

Let $G = \langle A, R\rangle$ be a finitely presented group. That is, R is a finite set of words (called **relators**) over the alphabet $A \cup A^{-}$, $A^{-1} := \{a^{-1} \mid a \in A\}$, and G is the free group generated by A and factored by all the relations $r = 1$, $r \in R$. Let $W(G)$ be the set of words w from $(A \cup A^{-})^*$ such that $w = 1$ in G (the unit is represented by the empty word); this set is called the **word problem set** of G. It is a fundamental task of combinatorial group theory, going back to Dehn, to characterize these sets. As defined the task depends on the presentation $\langle A, R\rangle$ and not just on G, but for different presentations the sets are p-time reducible to each other. It is well known that $W(G)$ can be very complex: Novikov [374] and Boone [98, 99] constructed finitely presented groups with an undecidable word problem. The fundamental theorem about the word problem is Higman's embedding theorem [227], by which a group is recursively presented if and only if it can be embedded into a finitely presented group.

A way to prove that $w \in W(G)$ is to sequentially insert or delete in w conjugates of the relators from R until the empty word is reached. Let us call this non-deterministic algorithm the **Dehn calculus**. It is analogous to logical calculi, and there are a number of other similarities between finitely presented groups and logical theories, some at the surface, some deeper. An example of the latter is a theorem by Birget *et al.* [85], which can be seen as a feasible version of Higman's theorem. For $w \in W(G)$ denote by $D_0(w)$ the minimum number of steps in the Dehn calculus needed to reduce it to 1 and define the **Dehn function**

$$D_G(n) := \max_{w \in W(G), |w|=n} D_0(w).$$

Birget *et al.* [85] proved that, for any non-deterministic algorithm $\mathcal{A}$ accepting $W(G)$ and running in time $t(n)$, we can embed the group G into another (perhaps much larger) finitely presented group H such that the Dehn function $D_H(n)$ is polynomially bounded by $t(n)$. That is, up to an embedding, the Dehn calculus is optimal. This is also analogous to propositional logic; for example, the Frege system p-simulates any other proof system when one adds a simple set of new axioms. The qualification *simple* will indeed mean "simple" and we shall study this in Part IV.

The appeal of the Dehn function comes from links to geometry via the so called **van Kampen diagrams**. For technical reasons assume that R is closed under cyclic permutations of words and under inverses and does not contain any word aa^{-1} or $a^{-1}a$. A van Kampen diagram is a planar, connected, simply connected, labeled, two-dimensional complex. The (directed) edges are labeled by $A \cup A^{-1}$ and it is required that, for any cell and any point on its border, reading the labels in one or the other direction yields a word from R. For a van Kampen diagram we define a word as follows: pick a point on the diagram's boundary and read the word, say w, in one direction going around and back to the starting point. The cells of the diagram can be taken out from the diagram one by one by until one point is reached, and this corresponds to a Dehn calculus reduction of w to 1. The van Kampen lemma states that the minimum area (i.e. the number of cells) in a van Kampen diagram for w is $D_0(w)$ (see Section 7.7 for more references).

The drawback of using the Dehn calculus in propositional logic is that there are no known natural examples (i.e. examples not encoding any machines) of finitely presented groups with a coNP-complete word problem. Hence one needs to use reductions into natural finitely presented groups with a coNP-hard word problem, and that spoils the elegance of the original set-up.

7.6 Outside the Box

This section presents three examples that are not Cook–Reckhow proof systems but are, I think, nevertheless relevant to them. The first example fails Definition 1.5.1 because the target language is not TAUT but another coNP-complete language. That is admittedly a technicality but the proof system does not fit well into earlier sections and hence it is given here. The remaining two examples fail the condition that one should be able to recognize proofs in p-time.

7.6.1 Hajós Calculus

TAUT is coNP-complete and any other coNP-complete set can, in principle, replace TAUT as the target of proof systems. However, not many coNP-complete sets from outside logic or algebra are associated with a calculus allowing one to construct all its members (and nothing else) by a few rules. A notable example is the *Hajós calculus* for constructions of non-3-colorable graphs.

The Hajós calculus starts with K_4, the complete graph on four vertices, and can use any of the following three operations.

1. Add a new edge or a new vertex.
2. Identify any two non-adjacent vertices.
3. The Hajós construction: having two graphs G and H and an edge in each, (a, b) in G and (u, v) in H, form a new graph as follows:
   (a) identify vertices a and u;
   (b) delete the edges $(a, b), (u, v)$;
   (c) add a new edge (b, v).

Theorem 7.6.1 (Hajós's theorem [215]) *A graph (tacitly undirected and simple) is not 3-colorable if and only if it can be constructed by the Hajós calculus.*

The following theorem is based on the same construction but is for vertex-labeled graphs and uses a p-reduction between TAUT and the class of non-3-colorable graphs from NP-completeness theory.

Theorem 7.6.2 (Pitassi and Urquhart [404]) *EF is p-bounded if and only if the Hajós calculus is p-bounded.*

7.6.2 Random Resolution

One motivation for introducing this system is the link of propositional logic to bounded arithmetic (the topic of Part II). As an informal example, consider the dual WPHP from Section 7.2 where we defined the system *WF*. Assume that $f : \{0,1\}^n \rightarrow \{0,1\}^m$ is p-time and that $m > n \geq 1$. The rule of *WF* allows us to introduce new variables for a witness to the existential quantifier in

$$\exists \overline{y}\, \overline{y} \in \{0,1\}^m \setminus rng(f).$$

An alternative way is to choose $\overline{b} \in \{0,1\}^m$ at random and use in the proof a propositional formula expressing

$$\overline{b} \neq f(\overline{x}). \tag{7.6.1}$$

Such a $\overline{b}$ serves as a witness, *assuming* that it actually has this property, i.e. that (7.6.1) is a tautology. This happens with probability at least $1 - 2^{n-m} \geq \frac{1}{2}$. The system we shall define now allows us to make such an informal construction precise.

Random resolution, denoted by RR, refutes a set of clauses $\mathcal{C}$ by adding first to it another, randomly chosen, set of clauses and then operates as ordinary R. The formal definition is as follows. For $\delta > 0$, a δ**-random resolution refutation distribution** of $\mathcal{C}$ is a random distribution

$$\Pi = (\pi_{\overline{r}}, \mathcal{D}_{\overline{r}})_{\overline{r}}$$

such that the $\overline{r}$ are random strings, $\pi_{\overline{r}}$ is a resolution refutation of $\mathcal{C} \cup \mathcal{D}_{\overline{r}}$ and it is required that

- for a truth assignment $\overline{a} \in \{0,1\}^n$ to all atoms, it holds that

$$Prob_{\overline{r}}[\overline{a} \text{ satisfies } \mathcal{D}_{\overline{r}}] \geq 1 - \delta.$$

The number of clauses in Π is defined to be the maximum number of clauses among all $\pi_{\overline{r}}$.

By the condition above, any truth assignment satisfying $\mathcal{C}$ would satisfy also some $\mathcal{D}_{\overline{r}}$ and hence $\pi_{\overline{r}}$ would be an R-refutation of a satisfiable set of clauses, which is impossible. Thus we have

Lemma 7.6.3 *The RR proof system is sound: any RR-refutable set is unsatisfiable.*

The following theorem shows that RR is in one sense strong but in another weak.

Theorem 7.6.4 (Pudlák and Thapen [424])

(i) *With probability tending to* 1 *a random 3CNF with n variables and* 64*n clauses admits a constant-size* 1/2*-RR refutation (i.e.* $\delta = 1/2$*).*
(ii) *If* P $\neq$ NP *then* 1/2*-RR cannot be p-simulated by any Cook–Reckhow proof system.*
(iii) *RR does not p-simulate* R(2) *(see the discussion near the start of Section 5.7).*

7.6.3 Proof Systems with Advice

Assume we have short proofs in *EF* of propositional formulas φ_n representing the universal statement

$$\varphi_n := A(f(\overline{x}), \overline{y}) \vee A(g(\overline{x}), \overline{z}) \tag{7.6.2}$$

where the three tuples have all length $n = 1, 2, \ldots$, A is a p-time relation and f and g are p-time functions. The formula has size $n^{O(1)}$ and it is a tautology if, given $n \geq 1$, either $f(\overline{a})$ or $g(\overline{a})$ witnesses the existential quantifier in

$$\exists u(|u| = n) \forall v(|v| = n)\, A(u, v) \tag{7.6.3}$$

for any $\overline{a} \in \{0,1\}^n$. If *EF* could get a *little advice* (one bit), telling it which of the two disjuncts is a tautology in each φ_n, it could prove either

$$A(f(\overline{x}), \overline{y})$$

or

$$A(g(\overline{x}), \overline{z}).$$

Somewhat more advice would be needed if instead of (7.6.2) we had

$$A(f(\overline{x}), \overline{y}) \vee A(g(\overline{x}, \overline{y}), \overline{z}).$$

This means that either $f(\overline{a})$ is a valid witness in (7.6.3) or if not and $\overline{b} \in \{0,1\}^n$ witnesses the failure

$$\neg A(f(\overline{a}), \overline{b})$$

then $g(\overline{a}, \overline{b})$ is a valid witness. Here the potential advice for *EF*, for a given n, would have to indicate not only which of the two cases occurs but, if it is the latter, also yield some $\overline{b}$. This and further considerations in bounded arithmetic lead to a definition of proof systems with advice.

We have a relational version and a functional version of the Cook–Reckhow definition 1.5.1, and they are equivalent. Similarly there are relational and functional versions of proof systems with advice, but they are not necessarily equivalent. Below we shall use the functional version.

A (functional) **proof system with** $k(n)$ **bits of advice** is a system P: $\{0,1\}^* \to \{0,1\}^*$ whose range is exactly TAUT and such that P is computable in polynomial time using $k(n)$ bits of advice on inputs of length n.

The following statements are the two main facts about these proof systems that can be stated now without a reference to bounded arithmetic.

Theorem 7.6.5 (Cook and Krajíček [153])

(i) TAUT $\in$ NP$/k(n)$ *if and only if there exists a p-bounded proof system with advice* $k(n)$.

(ii) There exists a proof system with advice 1 *(i.e. one bit) that simulates all proof systems with advice* $k(n) = O(\log n)$ *and, in particular, simulates all Cook–Reckhow proof systems. The simulation is computed in p-time with* $k(n)$ *bits of advice.*

We shall return to proof systems with advice in Chapter 21.

7.7 Bibliographical and Other Remarks

Sec. 7.1 The proof systems R(CP) and LK(CP) were defined in [284] and contain a *CP with deduction*, considered by Bonet, Pitassi and Raz [95, 96]. The proof system LK(CP) is related to model theory and via that to the LLL algorithm from lattice theory, see [284] for references.

The system R(LIN) was defined somewhat differently but equivalently by Itsykson and Sokolov [252] (denoted by Res-lin there); our definition, and the definition of R(PC) and R(PC_d), are from [312]. It is easy to generalize the definition to any field. Linear equational calculus (denoted by LEC in [280]) is sometimes dubbed *Gaussian calculus* (and denoted by GC), see [145]. Raz and Tzameret [427] earlier considered resolution over linear functions with integer coefficients and also studied a multi-linear variant of polynomial calculus and relations between these two systems.

Let us note that the systems R(LIN), R(CP) and other similarly constructed systems are complete not only as propositional proof systems but also in the implicational sense: each R(LIN) (resp. R(CP) etc.) clause that semantically follows from a set of initial clauses can be derived from it.

The system PCR was defined by Alekhnovich *et al.* [12]. They also extended to it the space measure defined for R by Esteban and Toran [176] and showed that the relation between the width and the size in R-proofs translates into a relation between the degree and the size for PCR-proofs.

Pudlák [414] considered the system LS + CP by adding to LS the rounding rule. Krajíček [282] defined a system $F_d^c(\mathrm{MOD}_p)$ operating like PC with (some) polynomials of degree at most c built from the depth d DeMorgan formulas. Garlík and Kolodziejczyk [197] further generalized and investigated these hybrid systems.

Sec. 7.2 The definition of *CF* and *WF* and all statements about them in Section 7.2 are due to Jeřábek [256]. The proof system *WF* is also related to the theory of proof complexity generators (cf. Section 19.4) and this connection suggests that it is unlikely that *CF* p-simulates *WF* (cf. [291]). The dual WPHP principle dWPHP is often called also *surjective* WPHP.

There are a number of ways to incorporate into a proof system a true Σ_2-statement that are different from how *WF* does it. For example, let h be a polynomial-time map that is a permutation of each $\{0,1\}^n$. Then a rule allowing one to infer $A(\overline{x})$ from $A(h(\overline{x}))$ is sound and in a sense replaces the Σ_2-statement that h is surjective.

Sec. 7.3 Implicit proof systems were defined in [293]. In Section 7.3 we considered the systems $[P, Q]$ as proof systems for tautologies in the ordinary sense, but we could have also considered tautologies defined only implicitly by a circuit, thus being possibly exponentially large. This was put to use in [292, 284, 310]. Lemma 7.3.1 is [293, Lemma 3.2] and Lemma 7.3.2 combines [293, Lemmas 4.1 and 4.2]. More facts are pointed out about modified f-implicit systems at the end of [293]. The implicit construction can also be iterated internally, adding to a proof (α, β) a parameter $1^{(m)}$ describing the length of the iteration. In fact, there does not seem to be a canonical way to do this: you can describe the iteration length by the length of the truth table of a circuit (or of a circuit described itself implicitly by a circuit), etc. This ambiguity is reminiscent of the ambiguity in classical proof theory of how to define *natural* ordinal notations. Lemma 7.3.4 is due to Wang [503]; the proof there is different. Šanda [457] constructed implicit R-proofs of PHP_n.

Sec. 7.4 The study of MLK was proposed by Pudlák [414] in the hope of extending the feasible interpolation method to it. The first two items in Theorem 7.4.1 are from Atserias, Galesi and Pudlák [34]; the third item is from Buss, Kabanets, Kolokolova and Koucký [126]. The improvement from a quasi-polynomial to a polynomial upper bound was possible because [126] formalized, in a theory corresponding to Frege systems, a construction of expanders; this could then be combined with an earlier formalization of a sorting-network construction by Jeřábek [260]. It is not known whether there is also a p-simulation by *tree-like* MLK.

Lemma 7.4.2 was originally proved – with a quasi-polynomial upper bound – by Atserias, Galesi and Gavalda [33], who gave a short proof in MLK (by quasi-polynomial-size proofs) of the natural counting properties of suitable monotone formulas defining the threshold connectives $\mathrm{TH}_{n,k}$ from (1.1.4). This was subsequently improved in [34].

Another natural modification of LK from the logical point of view is the **intuitionistic propositional calculus**, denoted by LJ. Following Takeuti [486] we define LJ as follows:

- add the implication connective $\rightarrow$ and two introduction rules for it,

$$\text{left}\ \frac{\Gamma \longrightarrow \Delta, A \quad B, \Gamma \longrightarrow \Delta}{A \rightarrow B, \Gamma \longrightarrow \Delta} \ \text{and right}\ \frac{A, \Gamma \longrightarrow \Delta, B}{\Gamma \longrightarrow \Delta, A \rightarrow B};$$

- allow only proofs in which the succedent of each sequent contains *at most one formula*.

The second condition makes the right versions of the exchange and the contraction rules redundant.

The class of intuitionistically valid formulas is defined using one of a number of semantics, the most popular being perhaps the Kripke semantics. We shall not define this semantics here but will mention that LJ is sound and complete with respect to it. We see that all LJ-proofs are also LK-proofs, and so the class of intuitionistically valid formulas is a proper subclass of classical tautologies in language having the implication (a famous classical tautology not intuitionistically valid is $p \vee \neg p$). It is, presumably, much more complex as it is PSPACE-complete. There is, however, a simple interpretation of classical logic in the intuitionistic case, the **double negation translation**. For each formula A in the DeMorgan language with implication define a new formula A^* as follows:

- $p^* := \neg\neg p$, for any atom p;
- * commutes with $\neg, \wedge, \rightarrow$;
- $(A \vee B)^* := \neg(\neg A^* \wedge \neg B^*)$.

It holds that LK proves a sequent $A_1, \ldots, A_u \longrightarrow B_1, \ldots, B_v$ if and only if LJ proves $A_1^*, \ldots, A_u^*, \neg B_1^*, \ldots, \neg B_v^* \longrightarrow$; see Takeuti [486]. In particular, A is a classical tautology if and only if $\neg\neg A^*$ is an intuitionistic tautology.

The OBDD proof system was defined and investigated by Atserias, Kolaitis and Vardi [35] and Lemma 7.4.3 is from there. Their definition was different from the more general one used here (which comes from [299]): they defined the initial OBDDs and three inference rules stating how to manipulate proof lines. Itsykson *et al.* [250] considered additional rules; [299] defined a combined proof system R(OBDD) and Mikle-Barat [354] defined and studied a DPLL-like procedure for R(OBDD).

The language and the axiomatizations of Robinson's arithmetic differ a little in literature but the important thing is that this arithmetic is finite. For this, PA theory and the general facts stated about it, see Shoenfield [470] or Hájek and Pudlák [214]. In Part II we shall consider proof systems attached to any *reasonable* theory.

One can interpret the first-order predicate calculus as a propositional proof system: atomic formulas serve as atoms. Henzl [225] proved, in particular, that in this sense predicate calculus is p-equivalent to Frege systems while the theory stating that the universe has at least two elements p-simulates the quantified propositional logic G. In fact, one can interpret individual first-order structures as proof systems; the relevant notion is the *combinatorics of a structure*, *Comb*(**A**), introduced in [294].

Sec. 7.5 The proof systems ENS and UENS were defined in Buss *et al.* [124]; the soundness and the completeness and the first two parts of Lemma 7.5.3 are from there. The last part of Lemma 7.5.3 (UENS $\leq_p$ *WF*) is from Jeřábek [256].

The ideal proof system IPS was defined by Grochow and Pitassi [211]. The testing of whether two polynomials over some field (e.g. a large finite field) are identical, the **polynomial identity problem** (PIT), can be done by a probabilistic-polynomial time algorithm using the so-called Schwartz–Zippel lemma. A deterministic algorithm is unknown but many experts believe that $P = BPP$, in which case such an algorithm would exist. Grochow and Pitassi [211] discuss the PIT-axioms and state that short provability of these in *EF* (or in any sufficiently strong proof system) implies that *EF* p-simulates IPS. The axioms simply state that a certain algorithm solves the PIT problem in a sound way and thus assert the soundness of IPS; the p-simulation is then an instance of a general relation between the soundness statements and p-simulations that we will study in Parts II and IV. Also, the PIT axioms do not state some property of all algorithms but state it only for some particular but *unknown* algorithm.

Dehn calculus (under the perhaps long-winded name "group-based proof systems") was studied in connection with proof complexity in [290]. A number of correspondences between propositional proof systems and finitely presented groups are discussed there. Basic texts about combinatorial and geometric group theory are Bridson [102] or Lyndon and Schupp [346]. Polynomial-time reductions of languages $L \subseteq \{0, 1\}^*$ to the word problem sets of finitely presented groups were defined by Birget [84].

Tzameret [495] considered proof systems operating with *non-commutative* formulas and compared them with some classical calculi. The non-commutativity allows one to encode any syntax into formulas and thus fairly direct simulations of (some) logical calculi.

Cavagnetto [134] interpreted resolutions (both R and R*) as rewriting systems, attached to them planar diagrams analogous to van Kampen diagrams and used them to interpret geometrically the size, the width and the space in R-proofs.

Sec. 7.6 Pitassi and Urquhart [404] used reductions from Garey, Johnson and Stockmeyer [194] in their investigation of the Hajós calculus.

Random resolution was suggested by Dantchev (as reported in [128]) and formally defined by Buss, Kolodziejczyk and Thapen [128]. Pudlák and Thapen [424] considered several variants of the definition. Theorem 7.6.4 combines Propositions 3.2 and 3.3 and Theorem 6.5 of [424].

Proof systems with advice were defined by Cook and Krajíček [153] in connections with bounded arithmetic and the consistency of NP $\not\subseteq$ P/*poly*. Theorem 7.6.5 is from that paper (Thms.6.5 and 6.6. there). We shall study these topics in Part IV.

Pudlák [419] defined quantum proof systems and quantum Frege systems, in particular. He showed that quantum Frege systems do not have super-polynomial speed-up over classical Frege systems but, unless factoring is p-time computable, are not p-simulated by them. A number of open problems and some conjectures are formulated in [419]; in particular, it is open whether there is a quantum Frege system simulating other quantum Frege systems or whether these can be simulated by classical Frege systems when the simulation is allowed to be computed in quantum p-time.

Müller and Tzameret [360] formalized in the so called TC^0-Frege system (cf. Sections 2.6 and 10.2) an incomplete proof system invented by Feige, Kim and Ofek [179]. This proof system proves in polynomial size a random 3CNF of the density (the number of clauses) $\Omega(n^{1.4})$ meaning that such a short proof exists with a high probability but not always.

Other incomplete (and unsound) proof systems are the *pseudo proof systems* from [304] (they are just an auxiliary notion there). Such a pseudo proof system is a polynomial-time function P whose range is formulas (but not necessarily just tautologies) and which is such that an arbitrary polynomial-time function f finds from a string w a falsifying assignment $f(w)$ for the formula $P(w)$ with a negligible probability over any $\{0, 1\}^n$, for $n \gg 0$. Malý and Müller [349, 350] related certain properties of these systems to the existence of hard sequences for SAT algorithms.

Part II

Upper Bounds

In this part we shall present general methods for constructing short proofs of some sequences of tautologies and for proving a p-simulation of one proof system by another. Both are based on the close relation between proof systems and first-order theories.

Given a first-order sentence B valid in all finite structures, in Section 8.2 we shall define a sequence of tautologies $\langle B \rangle_n$ that express that B is valid in all structures of cardinality n. When we want to construct short propositional proofs of these tautologies, a natural idea is to first prove, in some first-order theory, the universal statement that B is valid in all finite structures and then translate such a universal proof into a sequence of propositional proofs for the individual tautologies $\langle B \rangle_n$. The simpler the first-order theory is, the weaker the propositional proof system needs to be. This is analogous, to an extent, to having a Turing machine that decides a language $L \subseteq \{0, 1\}^*$ and constructing from it a sequence of circuits C_n computing the characteristic functions of L restricted to $\{0, 1\}^n$, as in Theorem 1.4.2, the size of C_n being bounded in terms of the time complexity of the machine. Not all sequences of circuits can be constructed in this way and similarly not all short proofs for a sequence of tautologies need to appear in this way either. In particular, if the sequence of the tautologies is not a translation of one first-order sentence then this method does not apply.

The construction of p-simulations using first-order theories is more universal and, with few exceptions, all known p-simulations can be obtained easily in this way. It is based on an analogy with a classical result of mathematical logic that the consistency of a theory implies – over a fixed weak theory – all its universal consequences (modulo some technicalities).

The first and still the key instance of this use of first-order theories in proof complexity was described by Cook [149]. A similar correspondence between a bounded arithmetic theory of Parikh [379] and constant-depth Frege systems was later rediscovered in the language of model theory by Paris and Wilkie [389] and by Ajtai [5]. A general correspondence between first-order theories and proof systems was subsequently developed by Krajíček and Pudlák [316] and linked there with some fundamental questions of proof complexity and mathematical logic (we shall see some examples in Chapter 21). The classes of theories most convenient for this correspondence were introduced by Buss [106], who discovered also another crucial link of his theories to computational complexity via witnessing theorems.

8

Basic Example of the Correspondence between Theories and Proof Systems

Although the first example of the correspondence between theories and proof systems was given by Cook [149] we shall start with the correspondence of Paris and Wilkie [389] and Ajtai [5] between a theory of Parikh [379] and constant-depth Frege systems. It is formally easier and thus illustrates, in my view, the basic ideas underlying this correspondence more clearly.

8.1 Parikh's Theory $I\Delta_0$

The first bounded arithmetic theory, and still the most intuitive in my view, was the theory $I\Delta_0$ defined by Parikh [379]. The term *bounded arithmetic* is used informally and generally refers to theories of arithmetic axiomatized by (the universal closures of) bounded formulas, often instances of the induction scheme IND for a class of bounded formulas in some specific language. The language of the theory $I\Delta_0$ is the language of Peano arithmetic L_{PA}, which we used when discussing the example of the proof system P_{PA} in Section 7.4:

$$0,\ 1,\ +,\ \cdot,\ \leq .$$

The theory $I\Delta_0$ is axiomatized as PA by Robinson's arithmetic Q and by the axiom scheme IND but restricted only to Δ_0-formulas A (see Section 7.4 for the definitions of these notions). Note that the instances (7.4.2) of the IND scheme are not bounded formulas even for a Δ_0-formula $A(x)$. But, for bounded A, (7.4.2) can be rewritten as such:

$$\neg A(0)\ \vee\ (\exists y \leq x, y < x \wedge A(y) \wedge \neg A(y+1))\ \vee\ A(x). \tag{8.1.1}$$

(We implicitly assume that all variables free in (8.1.1) are universally quantified.) The IND axioms of this new form are (taken together) equivalent over Robinson's Q to the original IND axioms.

Parikh's [379] aim was to formalize the informal notion of feasible numbers, that small numbers are feasible and are closed under addition and multiplication but not necessarily under exponentiation. From this perspective, checking the truth value of an instance of a Δ_0-formula for feasible numbers involves only feasible numbers, as all quantifiers in the formula are bounded by a term in the language and terms are constructed from feasible operations.

Bounded formulas are also significant from the computational complexity theory point of view: they define sets in the linear-time hierarchy and these are equivalent to the rudimentary sets; see Section 8.7. Paris and Wilkie [391] realized that it is useful to add an axiom that they called Ω_1, which states that

$$\forall x \exists y, \quad y = x^{|x|}.$$

The function $x \to x^{|x|}$ was denoted $\omega_1(x)$ by them. Here $|x|$ is the bit length of x; it equals $\lceil \log(x+1) \rceil$. In fact it requires a significant amount of work to show that the graph of this function is definable by an L_{PA}-formula and even more work to show that the definition can be bounded. The reason to include Ω_1 is that when formalizing polynomial-time algorithms and proving that they are well defined it is necessary to talk about strings of size (i.e. the bit length of the numbers coding them) polynomial in the size of the input string. Similarly, when formalizing parts of logical syntax one needs to talk about substitutions but substituting a string (a formula or a term) for a symbol (a variable) in another string extends the size non-linearly. All functions bounded by an L_{PA}-term grow linearly with length: for each L_{PA}-term $t(\overline{x})$ have

$$|t(\overline{x})| \leq O(\sum_i |x_i|), \tag{8.1.2}$$

and that prevents the definability of $\omega_1(x)$, by the following theorem.

Theorem 8.1.1 (Parikh's theorem [379]) *Let $A(\overline{x}, y)$ be a Δ_0-formula and assume that $I\Delta_0$ proves*

$$\forall \overline{x} \exists y A(\overline{x}, y).$$

Then there is an L_{PA}-term $t(\overline{x})$ such that $I\Delta_0$ proves even

$$\forall \overline{x} \exists y \leq t(\overline{x}) A(\overline{x}, y)\ .$$

This theorem has a simple model-theoretic proof (Section 8.3) and it is also a consequence of Herbrand's theorem (Section 12.2).

When defining his theories, Buss [106] took a more direct route and added to the language the function symbols for $|x|$ and for (a function similar to) $\omega_1(x)$ (and, in fact, others also), and he included in the theories natural axioms about them. This allowed him to decompose the class of bounded formulas into a natural hierarchy whose levels define predicates in the corresponding levels of the polynomial-time hierarchy and to use that to define natural subtheories of his bounded arithmetic. We shall discuss this in detail in Chapter 9.

There are two main open problems concerning the theory $I\Delta_0$.

Problem 8.1.2 (The finite axiomatizability problem) *Is $I\Delta_0$ finitely axiomatizable?*

Problem 8.1.3 (The Δ_0-PHP problem) *Does $I\Delta_0$ prove the Δ_0-PHP principle?*

The latter problem asks whether the theory $I\Delta_0$ proves, for each bounded formula $A(x,y)$, that for no z does A define a graph of an injective function from $[z+1]$ into $[z]$. That is, can $I\Delta_0$ prove the formula PHP(R) stating that, when $A(x,y)$ is substituted for $R(x,y)$, one of the following three formulas must fail:

1. $\forall x \leq z+1 \exists y \leq z R(x,y)$;
2. $\forall x \leq z+1 \forall y \neq y' \leq z(\neg R(x,y) \vee \neg R(x,y'))$;
3. $\forall x \neq x' \leq z+1 \forall y \leq z(\neg R(x,y) \vee \neg R(x',y))$.

More is known about the finite axiomatizability problem (Section 8.7). Concerning the Δ_0-PHP problem, Paris and Wilkie [389] formulated its weaker version. Add to the language L_{PA} a new binary relation symbol $R(x,y)$. Let $\Delta_0(R)$ be the class of bounded formulas in this extended language and define a theory $I\Delta_0(R)$ to be Robinson's Q with the induction scheme IND accepted for all $\Delta_0(R)$-formulas. The modified problem is a relativization of the Δ_0-PHP problem:

- *Does $I\Delta_0(R)$ prove* PHP(R)*?*

An affirmative answer to this problem would clearly imply that $I\Delta_0$ proves the Δ_0-PHP principle (just substitute everywhere in the proof the formula A for the symbol R). But the relativized problem was solved with a negative answer. The solution followed a reduction of the negative solution, due to Paris and Wilkie [389], to a statement about Frege systems in the DeMorgan language:

- *Assume that there are no $d,c \geq 2$ such that for all $n \geq 1$ there are depth-d, size-n^c Frege proofs of the* PHP$_n$ *tautology from* (1.5.1). *Then* PHP(R) *is not provable in* $I\Delta_0(R)$.

We shall prove the reduction in the next section and the statement about propositional proofs of PHP$_n$ in Chapter 15.

8.2 The Paris–Wilkie Translation $\langle \ldots \rangle$

Extend the language L_{PA} by any countable number of relation symbols and constants. Such an extended language will be denoted by $L_{PA}(\alpha)$, the symbol α indicating that the language contains some unspecified number of new relation symbols and constants. For the presentation in this section we stick to the language $L_{PA}(R)$ extending L_{PA} by one of the binary relation symbols $R(x,y)$ that we used earlier. But it will be clear that this is without loss of generality and that the construction applies to any $L_{PA}(\alpha)$.

The **Paris–Wilkie translation** assigns to any $\Delta_0(R)$-formula $A(x_1,\ldots,x_k)$ and any $n_1,\ldots,n_k \geq 0$ a DeMorgan propositional formula

$$\langle A(\overline{x}) \rangle_{n_1,\ldots,n_k}$$

defined by induction on the logical complexity of A as follows.

1. If B is one of the atomic formulas $t(\overline{x}) = s(\overline{x})$ or $t(\overline{x}) \leq s(\overline{x})$, with t and s terms, then

$$\langle B \rangle_{n_1,\dots,n_k} := \begin{cases} 1 & \text{if } B(n_1,,\dots,n_k) \text{ is true,} \\ 0 & \text{otherwise.} \end{cases}$$

2. If B is the atomic formula $\mathrm{R}(t(\overline{x}), s(\overline{x}))$ and i and j are the values of the terms $t(\overline{x})$ and $s(\overline{x})$ for $\overline{x} := \overline{n}$, respectively, then put

$$\langle B \rangle_{n_1,\dots,n_k} := r_{ij},$$

where r_{ij} are propositional atoms.
3. $\langle \dots \rangle$ commutes with $\neg, \vee, \wedge$.
4. If $A(\overline{x}) = \exists y \leq t(\overline{x}) B(\overline{x}, y)$ then

$$\langle A(\overline{x}) \rangle_{n_1,\dots,n_k} \ := \ \bigvee_{m \leq t(\overline{n})} \langle B(\overline{x}, y) \rangle_{n_1,\dots,n_k,m}.$$

5. If $A(\overline{x}) = \forall y \leq t(\overline{x}) B(\overline{x}, y)$ then

$$\langle A(\overline{x}) \rangle_{n_1,\dots,n_k} \ := \ \bigwedge_{m \leq t(\overline{n})} \langle B(\overline{x}, y) \rangle_{n_1,\dots,n_k,m}.$$

The notation $\langle \dots \rangle_{n_1,\dots,n_k}$ is very close to the notation $\langle \dots \rangle_{n,\mathbf{A}}$ from Section 1.2, the reason being that the translations are essentially the same. Namely, the truth value of a $\Delta_0(R)$-formula for the instances $\overline{n}$ of its variables is determined in the initial interval of $(\mathbf{N}, R)$ with a universe of the form $[0, s(\overline{n})]$, where $s(\overline{x})$ is some term (essentially the composition of the terms bounding the individual quantifiers in the formula).

Lemma 8.2.1 *Let $A(\overline{x})$ be a $\Delta_0(R)$-formula. Then there are $c, d \geq 1$ such that, for all $n_1, \dots, n_k$,*

- $|\langle A(\overline{x}) \rangle_{n_1,\dots,n_k}| \leq (n_1 + \dots + n_k + 2)^c$,
- $\mathrm{dp}(\langle A(\overline{x}) \rangle_{n_1,\dots,n_k}) \leq d$.

Moreover, $A(n_1, \dots, n_k)$ is true for all interpretations of R if and only if $\langle A(\overline{x}) \rangle_{n_1,\dots,n_k} \in$ TAUT.

We note that if we used a language L richer than L_{PA}, in particular if we included $\omega_1(x)$, a similar lemma still holds but we would need to modify the upper bound on the size of the translation. Namely, let $f(x)$ be a non-decreasing function such that for any L-term $t(x_1, \dots, x_k)$ there is some $c \geq 1$ such that it holds that

$$t(x_1, \dots, x_k) \leq f^{(c)}(x_1 + \dots + x_k + 2) \tag{8.2.1}$$

($f^{(c)}$ denotes the c-fold iterate of f). Then the size estimate in the lemma would be

- $|\langle A(\overline{x}) \rangle_{n_1,\dots,n_k}| \leq f^{(c)}(n_1 + \dots + n_k + 2)$, for some $c \geq 1$

(the depth estimate does not change). In particular, by adding $\omega_1(x)$ to L_{PA} the size estimate changes to

$$\omega_1(x)^{(c)} = 2^{|x|^{(2^c)}}.$$

The key result about the Paris–Wilkie translation is the following.

Theorem 8.2.2 (Paris and Wilkie [389]) *Let F be a Frege system in the DeMorgan language. Let $A(\overline{x})$ be a $\Delta_0(R)$-formula and assume that $I\Delta_0(R)$ proves $\forall \overline{x} A(\overline{x})$. Then there are $c, d \geq 1$ such that for each k-tuple $n_1, \dots, n_k$ there is an F-proof $\pi_{n_1,\dots,n_k}$ of $\langle A(\overline{x})\rangle_{n_1,\dots,n_k}$ such that*

$$\mathrm{dp}(\pi_{n_1,\dots,n_k}) \leq d \quad \textit{and} \quad |\pi_{n_1,\dots,n_k}| \leq (n_1 + \cdots + n_k + 2)^c.$$

There are several proofs of this theorem, each appealing to a different intuition and using either proof theory or model theory. A proof-theoretic argument, often branded as the most intuitive, exploits a simple informal idea. Assume that we can find a proof π of $A(\overline{x})$ in the theory such that only $\Delta_0(R)$-formulas appear in π. Then we can apply the $\langle \dots \rangle$ translation step by step and derive the required tautology $\langle A(\overline{x})\rangle_{n_1,\dots,n_k}$ from the $\langle \dots \rangle$ translation of the axioms of $I\Delta_0(R)$, prove these translations of the axioms separately by short Frege proofs of bounded depth and then combine the proofs to get a proof of the formula. To see how a proof of the translation of an axiom may look, consider the IND scheme as in (8.1.1) for a $\Delta_0(R)$-formula $B(x)$ and $x := m$:

$$\neg B(0) \ \vee \ (\exists y \leq m, y < m \wedge B(y) \wedge \neg B(y+1)) \ \vee \ B(m).$$

Denoting $\beta_n := \langle B(x)\rangle_n$, clearly

$$\left[\beta_0 \wedge \bigwedge_{n<m} (\beta_n \to \beta_{n+1})\right] \to \beta_m$$

has a Frege proof of depth bounded by $\mathrm{dp}(\beta_m) + O(1)$ and size $O(m|\beta_m|)$.

The technical problem with this approach is that in order to find a proof π that uses only bounded formulas one needs to introduce into first-order logic new quantifier inference rules that deal with bounded quantifiers. For example, in the sequent calculus formalism we may modify the first-order sequent calculus LK to a calculus called LKB having new rules for introducing bounded quantifiers such as

$$\frac{y \leq t, \Gamma \longrightarrow \Delta, A(y)}{\Gamma \longrightarrow \Delta, \forall y \leq t A(y)}$$

(with y not appearing in Γ, Δ). Further, for proof-theoretic reasons, one also needs to replace the IND axiom scheme by the **IND inference rule**

$$\frac{B(y), \Gamma \longrightarrow \Delta, B(y+1)}{B(0), \Gamma \longrightarrow \Delta, B(x)}$$

(with x and y not occurring in Γ, Δ), and then establish a form of cut elimination that implies the key *subformula property*: every bounded formula provable in $I\Delta_0(R)$ has a proof in which all formulas in all sequents are substitution instances of subformulas of the end-formula or of an induction formula, and hence bounded themselves. This approach is well presented in the literature (Section 8.7) and we shall not work it out here again.

An alternative proof-theoretic argument is based on Herbrand's theorem. We shall use Herbrand's theorem to establish a similar result for translations from Cook's theory PV and from Buss's first-order theories in Chapter 12. The advantage of this approach is, as we shall see, that Herbrand's theorem itself has a simple model-theoretic proof and does not need any involved proof-theoretic construction; the whole argument is elementary. However, to use it for $I\Delta_0(R)$ one needs first to apply a Skolemization of the theory and then to reinterpret the Skolem functions in the propositional setting; both steps are elementary but tedious and they steal some of the appeal of the technical simplicity.

The original proof of Paris and Wilkie [389] was model-theoretic and it is, in my view, very intuitive and easy to modify to other theories and proof systems. However, to understand its technical details presupposes some familiarity with basic mathematical logic, at the level of understanding the Henkin construction underlying the standard proof of the completeness theorem for first-order logic. Nevertheless, I shall give a fairly complete outline of the argument in the next section, as the model-theoretic viewpoint motivates various other ideas that we shall discuss in later parts of the book. A full proof is given in [278].

8.3 Models of $I\Delta_0(R)$

Let L be any language extending the arithmetic language L_{PA} in such a way that each symbol in L has a *standard* interpretation over the set of natural numbers $\mathbf{N}$, this is the **standard model**. For example, we may add to L_{PA} the symbols for the functions $|x|$ or $\omega_1(x)$ discussed earlier in this chapter.

Let $\text{Th}_L(\mathbf{N})$ be the set of all L-sentences true in $\mathbf{N}$. For e, a new constant not in L, consider the language $L(e) := L \cup \{e\}$ and an $L(e)$-theory consisting of

1. $\text{Th}_L(\mathbf{N})$,
2. $e > s_n$ for each numeral s_n for $n \geq 1$ (Section 7.4).

Any finite number of sentences from item 2 is consistent with $\text{Th}_L(\mathbf{N})$ (just interpret e in the standard model by a large enough number) and hence, by the compactness of first-order logic, the theory has a model, say $\mathbf{M}$. The above axioms about e force e to be in $\mathbf{M}$ interpreted by an element that is larger than s_n, for all $n \in \mathbf{N}$. An L-structure with this property is called a **non-standard model**, in this case of $\text{Th}_L(\mathbf{N})$.

The model $\mathbf{M}$ has a number of specific L-substructures called cuts. A **cut** in $\mathbf{M}$ is any $I \subseteq \mathbf{M}$ such that

- I contains all ($\mathbf{M}$-interpretations of) constants from L (so not necessarily e) and is closed under all L-functions,

- I is closed downwards, so that $a < b \wedge b \in I$ implies $a \in I$ for all $a, b \in \mathbf{M}$.

For example, the standard model $\mathbf{N}$ is itself the smallest cut in any non-standard model. Somewhat more interesting is the following cut:

$$I_e := \{a \in \mathbf{M} \mid \text{ there exists an } L\text{-term } t(x) \text{ such that } \mathbf{M} \models a \leq t(e)\}.$$

We can use the cut I_e to prove Parikh's theorem.

Proof of Theorem 8.1.1 Assume that $I\Delta_0$ proves for no L_{PA}-term $t(x)$ the sentence

$$\forall x \exists y \leq t(x) A(x, y).$$

As each of the terms $t_1(x), \ldots, t_k(x)$ is majorized by $t(x) := t_1(x) + \cdots + t_k(x)$, this hypothesis implies that for no L_{PA}-terms $t_1(x), \ldots, t_k(x)$ does $I\Delta_0$ prove the disjunction

$$\bigvee_{i \leq k} \forall x \exists y \leq t_i(x) A(x, y).$$

Consider a theory T in the language L_{PA} with a new constant symbol e added to it and with the axioms

1. $I\Delta_0$,
2. (as item 2 above) $e > s_n$, for each numeral s_n for $n \geq 1$,
3. $\neg \exists y \leq t(e) A(e, y)$, for each L_{PA}-term $t(x)$.

The hypothesis and the compactness theorem imply that T is consistent and hence has a model: let $\mathbf{M}$ be such a model. It is non-standard (owing to item 2) and we can consider the cut I_e. By the axioms in item 3,

$$I_e \models \neg \exists y A(e, y).$$

The proof is concluded by the following two simple but key claims.

Claim 1 For all Δ_0-formulas $B(z_1, \ldots, z_k)$ and any $a_1, \ldots, a_k \in I_e$:

$$I_e \models B(a_1, \ldots, a_k) \quad \text{if and only if} \quad \mathbf{M} \models B(a_1, \ldots, a_k).$$

Claim 1 is established by induction on the number of logical connectives and quantifiers in B, noting that the downward closure of cuts implies that the bounded quantifiers are defined in the same way in $\mathbf{M}$ and in I_e.

Claim 2 $I_e \models I\Delta_0$.

Let $B(x)$ be a Δ_0-formula (possibly with parameters from I_e) and assume that

$$I_e \models B(0) \wedge \neg B(m)$$

for some $m \in I_e$. By Claim 1 we also have

$$\mathbf{M} \models B(0) \wedge \neg B(m)$$

and, as **M** is a model of $I\Delta_0$, there must be an $n < m$ in **M** such that

$$\mathbf{M} \models B(n) \wedge \neg B(n+1).$$

By the downward closure of I_e we also have $n \in I_e$ and by Claim 1

$$I_e \models B(n) \wedge \neg B(n+1).$$

This proves the theorem. □

Note that the same argument applies also to a theory extending $I\Delta_0$ by accepting the IND axioms for all $\Delta_0(L)$-formulas, i.e. bounded formulas in the extended language L considered above.

We shall now outline the following

Proof of Theorem 8.2.2 Let $A(x)$ be a $\Delta_0(R)$-formula (we take for notational simplicity without loss of generality a formula with one free variable x only) and assume that $I\Delta_0(R)$ proves $\forall x A(x)$. By Lemma 8.2.1 there are $c_0, d_0 \geq 1$ such that, for all n,

$$|\langle A(x)\rangle_n| \leq (n+2)^{c_0} \quad \text{and} \quad \mathrm{dp}(\langle A(x)\rangle_n) \leq d_0.$$

Assume for the sake of contradiction that the conclusion of the theorem fails:

- *There are no $c, d \geq 1$ such that for each $n \in \mathbf{N}$ there is a proof π_n of $\langle A(x)\rangle_n$ in a Frege system in the DeMorgan language such that*

$$dp(\pi_n) \leq d \quad \text{and} \quad |\pi_n| \leq (n+2)^c\,. \tag{8.3.1}$$

Consider now a theory T in the language L_{PA} augmented by a new constant e and with the axioms:

1. $\mathrm{Th}_{L_{PA}}(\mathbf{N})$,
2. $e > s_n$, for each numeral s_n for $n \geq 1$,
3. for each $c, d \geq 1$ the axiom:
   There is no proof π of $\langle A(x)\rangle_e$ in a Frege system in the DeMorgan language such that

$$\mathrm{dp}(\pi) \leq d \quad \text{and} \quad |\pi| \leq (e+2)^c. \tag{8.3.2}$$

Formulating the axioms in item 3 in the language of arithmetic presupposes some knowledge of how the logical syntax is formalized in PA; this is a standard mathematical logic background, which we shall not review here.

By the hypothesis (8.3.1) and using compactness we deduce that T is consistent and thus has a model **M**. Moreover, we may assume that **M** is countable (another standard mathematical logic background idea). Our strategy is to show that we can interpret the relation symbol R over the cut I_e by some relation R_G such that

$$(I_e, R_G) \models I\Delta_0(R_G) + \neg A(e), \tag{8.3.3}$$

and thus contradict the hypothesis of the theorem. To do so we shall construct a bounded elementary diagram G of the required structure (I_e, R_G), which is a set of

$\Delta_0(R, I_e)$-sentences in the language $L_{PA}(R, I_e)$. This language is $L_{PA}(R)$ augmented by names for all elements of I_e, such that it holds that:

1. $\neg A(e) \in G$;
2. for each $\Delta_0(R, I_e)$-sentence B, exactly one of $B, \neg B$ is in G;
3. for each $\Delta_0(R, I_e)$-sentence $B \vee C$, $B \vee C \in G$ if and only if $B \in G$ or $C \in G$;
4. for each $\Delta_0(R, I_e)$-sentence $B \wedge C$, $B \wedge C \in G$ if and only if $B \in G$ and $C \in G$;
5. for each $\Delta_0(R, I_e)$-sentence $\exists y \leq t\, B(y)$,
   $\exists y \leq tB(y) \in G$ if and only if $a \leq t \wedge B(a) \in G$ for some $a \in I_e$;
6. for each $\Delta_0(R, I_e)$-sentence $\forall y \leq t\, B(y)$,
   $\forall y \leq tB(y) \in G$ if and only if $\neg a \leq t \vee B(a) \in G$ for all $a \in I_e$;
7. for each $\Delta_0(R, I_e)$-formula $B(y)$ with one free variable y and for each $m \in I_e$,
   (a) $\neg B(0) \in G$,
   (b) or $B(m) \in G$,
   (c) or $B(a) \wedge \neg B(a+1) \in G$ for some $a < m$.

Having such a set G define, for all $a, b \in I_e$,

$$R_G(a, b) \quad \text{if and only if} \quad R(a, b) \in G.$$

The following simple claim, proved by induction on the logical complexity of B, implies that (I_e, R_G) has the required properties.

Claim *For each $\Delta_0(R, I_e)$-sentence B, $(I_e, R_G) \models B$ if and only if $B \in G$.*

The key maneuver is to construct a suitable set G on the *propositional level*, working with the propositional translations of $\Delta(R, I_e)$-sentences instead of the sentences themselves. Let $F \subseteq \mathbf{M}$ be the set of all elements of $\mathbf{M}$ that are, in the sense of $\mathbf{M}$, propositional formulas α in the DeMorgan language satisfying:

(i) α is built from the atoms that occur in $\langle A(x) \rangle_e$;
(ii) $\text{dp}(\alpha) \leq d$ for some $d \in \mathbf{N}$;
(iii) $|\alpha| \leq e^c$ for some $c \in \mathbf{N}$.

This set is not definable in $\mathbf{M}$ (otherwise one could define the numbers in $\mathbf{N}$ as the depths of the formulas in F). As $\mathbf{M}$ is countable, so also is F; assume that $\alpha_0, \alpha_1, \ldots$ enumerates its elements such that each $\alpha \in F$ appears infinitely often, and also enumerate all pairs consisting of a $\Delta_0(R, I_e)$-formula with one free variable and an element of I_e i.e. $(B_0(y), m_0), (B_1(y), m_1), \ldots$

For a set $S \subseteq F$ definable in $\mathbf{M}$ and for $t \in \mathbf{M}$, we say that S is *t-consistent* if and only if in $\mathbf{M}$ there is no Frege proof ρ, of the propositional constant 0 from some formulas in S, such that

$$|\rho| \leq e^t \quad \text{and} \quad \text{dp}(\rho) \leq t.$$

We want to define an increasing chain $S_0 \subseteq S_1 \subseteq \cdots \subseteq F$ of sets definable in $\mathbf{M}$ and satisfying the following conditions:

1. $\neg\langle A\rangle_e \in S_0$;
2. if $k = 2i$ then exactly one of $\alpha_i, \neg\alpha_i$ is in S_k;
3. if $k = 2i$ and if $\alpha_i = \bigvee_j \beta_j$ and $\alpha_i \in S_{k-1}$ then one $\beta_j \in S_k$;
4. if $k = 2i$ and if $\alpha_i = \bigwedge_j \beta_j$ and $\alpha_i \in S_{k-1}$ then all the $\beta_j \in S_k$;
5. if $k = 2i + 1$ and $\beta_j = \langle B_i(y)\rangle_j$ for all $j \leq m_i$ then
   (a) $\neg\beta_0 \in S_k$,
   (b) or $\beta_{m_i} \in S_k$,
   (c) or $\beta_j \wedge \beta_{j+1} \in S_k$ for some $j < m_i$;
6. there is a $t \in \mathbf{M} \setminus \mathbf{N}$ such that S_k is t-consistent.

By the hypothesis (8.3.2) the set consisting of $\neg\langle A\rangle_e$ is t-consistent for some non-standard t and hence either $S_0 := \{\neg\langle A\rangle_e, \alpha_0\}$ or $S_0 := \{\neg\langle A\rangle_e, \neg\alpha_0\}$ is $(t-1)$-consistent.

Let us give two examples of how S_k is defined, the other cases being analogous to these two. Assume that $k = 2i$, $\alpha_i = \bigvee_{j\leq m} \beta_j \in S_{k-1}$ ($m \in I_e$) and S_{k-1} is t-consistent. We want to find β_j that we can add to S_{k-1} such that it stays $(t/2)$-consistent. Assume further for the sake of contradiction that for all j there are Frege proofs ρ_j of 0 from $S_{k-1} \cup \{\beta_j\}$ that are of depth $\leq t/2$ and size $\leq e^{t/2}$. Then we can combine these $m \leq e^{t/3}$ proofs into a proof of 0 from $S_{k-1} \cup \{\alpha_i\} = S_{k-1}$ of size $\leq e^t$ and depth $\leq t/2 + O(1) < t$. This contradicts the t-consistency of S_{k-1}.

For the second example, consider item 5, related to induction: $k = 2i + 1$ and $\beta_j = \langle B_i(j)\rangle$ for all $j \leq m_i$ and S_{k-1} is t-consistent. If adding any of the $m_i + 1$ formulas

$$\neg\beta_0,\ \beta_{m_i},\ \beta_j \wedge \beta_{j+1} \quad \text{for some} \quad j < m_i$$

results in all cases in a $t/2$-inconsistent theory then we can combine the individual inconsistency proofs into a proof of 0 from S_{k-1} with parameters (the depth and size) contradicting its t-consistency.

Having the chain $S_0 \subseteq S_1 \subseteq \cdots$ put an $\Delta(R, I_e)$-sentence B into G if and only if $\langle B\rangle \in \bigcup_{k\geq 0} S_k$. Then G satisfies all seven requirements imposed on it above. This proves the theorem. □

8.4 Soundness of AC^0-Frege Systems

The soundness of a proof system P is the item in Definition 1.5.1 asserting that any formula with a P-proof is a tautology. We shall now consider how to formalize this statement in the language L_{PA} augmented by several relation symbols that will encode a formula, a proof and an evaluation. We shall treat first the simplest case: resolution refutations of sets of clauses.

A clause C in n atoms $p_1, \ldots, p_n$ can be identified with a subset $C' \subseteq [2n]$:

$$p_i \in C \quad \text{if and only if} \quad i \in C' \ \text{and} \ \neg p_i \in C \quad \text{if and only if} \quad n + i \in C'.$$

A set of m clauses $\mathcal{C} = \{C_1, \ldots, C_m\}$ in n atoms can then be represented by a binary relation $F \subseteq [m] \times [2n]$:

$$(j, i) \in F \quad \text{if and only if} \quad i \in C_j'.$$

The same set is represented by a number of relations F depending on the ordering of the clauses. A resolution refutation $D_1, \ldots, D_k$ of $\mathcal{C}$ can be analogously represented by a binary relation $P \subseteq [k] \times [2n]$, and the condition that it is a refutation of $\mathcal{C}$ can be represented by a $\Delta_0(P, F)$-formula $Prf_R(x, y, z)$ with x standing for the number of variables (above, it was n), y for m and z for k:

$$(\forall 1 \leq j \leq z\, [Init(j) \vee (\exists 1 \leq j_1, j_2 < j \exists 1 \leq i \leq x\, Infer(j, j_1, j_2, i))])$$

$$\wedge\ (\forall 1 \leq i \leq x\, \neg P(z, i) \wedge \neg P(z, i + x))$$

where $Init(j)$ formalizes the fact that $D_j \in \mathcal{C}$:

$$\exists 1 \leq t \leq y \forall 1 \leq i \leq 2x\, P(j, i) \equiv F(t, i)$$

and $Infer(j, j_1, j_2, i)$ formalizes the fact that D_j was inferred from D_{j_1}, D_{j_2} by resolving the variable p_i. Thus $Infer(j, j_1, j_2, i)$ is the conjunction of the following formulas:

- $\neg P(j, i) \wedge \neg P(j, x + i)$;
- $[\forall 1 \leq t \leq 2x\, (t \neq i \wedge t \neq x + i) \to P(j, t) \equiv (P(j_1, t) \vee P(j_2, t))]$;
- $[(P(j_1, i) \wedge P(j_2, i + x)) \vee (P(j_1, i + x) \wedge P(j_2, i))]$.

It is easy to modify the definition of Prf_R to capture the provability predicate for R_w or for some other variants of R.

A truth evaluation $\overline{a} \in \{0, 1\}^n$ of n atoms can be identified with a subset $E \subseteq [n]$:

$$i \in E \quad \text{if and only if} \quad a_i = 1.$$

Now let $Sat_2(x, y)$ be the following $\Delta_0(E, F)$-formula (with x again standing for n and y for m) formalizing the fact that E satisfies the formula F:

$$\forall 1 \leq j \leq y\, \exists 1 \leq i \leq x\, [(E(i) \wedge F(j, i)) \vee (\neg E(i) \wedge F(j, i + x))]. \tag{8.4.1}$$

The soundness of resolution can be then formalized by the following $\Delta_0(E, F, P)$-formula, often denoted by $Ref_R(x, y, z)$ and called, in accordance with the terminology in mathematical logic, a **reflection principle** for resolution:

$$Prf_R(x, y, z, P, F) \to \neg Sat_2(x, y, E, F)$$

(for clarity we list explicitly the relations on which the subformulas depend). The negation sign in front of the Sat_2 subformula is there because we are talking about refutations and not about proofs, and the reason for the subscript 2 will become clear shortly.

Lemma 8.4.1 *$I\Delta_0(E, F, P)$ proves the formula $Ref_R(x, y, z)$.*

Proof Assuming $Prf_R(x,y,z,P,F) \wedge Sat_2(x,y,E,F)$, prove the following by induction on $u = 1,\ldots,z$:

$$\forall t \leq u \exists 1 \leq i \leq x\, [(E(i) \wedge P(t,i)) \vee (\neg E(i) \wedge P(t,i+x))].$$

The formula formalizes that E satisfies the first u clauses in the refutation P. For $t = z$ this contradicts the conjunct in Prf_R stating that the last clause is empty. □

We can use the encoding of the depth d DeMorgan formulas described in the proof of Lemma 1.4.4 and write analogously, for each fixed $d \geq 1$,

- a $\Delta_0(F)$-formula $Fla_d(x,y)$ formalizing that F *encodes a depth* $\leq d$ *DeMorgan formula with* $\leq x$ *atoms and of size* $\leq y$;
- a $\Delta_0(F,P)$-formula $Prf_d(x,y,z)$ formalizing that P *encodes a depth* $\leq d$ *Frege proof of a formula F satisfying* $Fla_d(x,y)$;
- a $\Delta_0(E,F)$-formula $Sat_d(x,y)$ formalizing that E *is a truth assignment to the* $\leq x$ *atoms that satisfies the formula F, which itself obeys* $Fla_d(x,y)$;

and the reflection principle $Ref_d(x,y,z)$ expressing the soundness of the depth-d subsystem of a Frege system with respect to proofs of depth $\leq d$ formulas:

$$Prf_d(x,y,z,P,F) \rightarrow Sat_d(x,y,E,F). \tag{8.4.2}$$

The proof of the following lemma is analogous to the proof of Lemma 8.4.1 proving by induction on $u \leq z$ that the first u steps in the F_d-proof P are all satisfied by the assignment E.

Lemma 8.4.2 *$I\Delta_0(E,F,P)$ proves the formula $Ref_d(x,y,z)$.*

It is natural to ask how strong the proof systems are for which we may carry out this argument in $I\Delta_0(E,F,P)$. The provability predicate for any proof system Q is – by the definition – polynomial-time decidable and hence, in particular, is in the class NP. By Fagin's theorem 1.2.2, every NP-relation is $s\Sigma^1_1$-definable. This means that there are some relation symbols $G_1,\ldots,G_t$ (of some arities) on $[z]$ and a $\Delta_0(P,F,G_1,\ldots,G_t)$-formula

$$Prf^0_Q(x,y,z,P,F,G_1,\ldots,G_t)$$

such that, for

$$Prf_Q(x,y,z,P,F) := \exists G_1,\ldots,G_t Prf^0_Q(x,y,z,P,F,G_1,\ldots,G_t)$$

and, for all $x := n, y := m, z := s$, every DeMorgan formula F with $\leq n$ atoms and size $\leq m$ and every relation P it holds that

$$P \text{ is a } Q\text{-proof of } F \text{ of size} \leq s \quad \text{if and only if} \quad Prf_Q(n,m,s,P,F).$$

A general proof system Q proves all DeMorgan tautologies and not just DNFs or bounded-depth formulas. To formulate the reflection principle in full generality requires a formalization of the satisfiability relation for general formulas. This relation is not definable by a Δ_0-formula but it is in coNP and hence definable by a strict Π^1_1-formula. Let $Sat(x,y)$ be a strict Π^1_1-formula of the form

$$Sat(x,y,E,F) := \forall H_1,\dots,H_r Sat^0(x,y,E,F,H_1,\dots,H_r), \tag{8.4.3}$$

where Sat^0 is a $\Delta_0(E,F,H_1,\dots,H_r)$-formula and, for $x := n, y := m$, every DeMorgan formula F with $\leq n$ atoms and size $\leq m$ and set E it holds that

$$E \text{ is a satisfying assignment for } F \quad \text{if and only if} \quad Sat(n,m,E,F).$$

A more general reflection principle for Q, denoted by Ref_Q, is a bounded formula of the same form as before:

$$Prf^0_Q(x,y,z,P,F,G_1,\dots,G_t) \rightarrow Sat^0(x,y,E,F,H_1,\dots,H_r). \tag{8.4.4}$$

We could encode each tuple of the relations G_i and H_i into just one tuple, but it is often more transparent to have separate relations with various roles in the formalization. As we do not know in advance how many relations G_i and H_i we may want to use, it is useful to talk here generically about a $\Delta_0(\alpha)$-formula, like that we used before, in the generic language $L(\alpha)$. Saying that we have a $\Delta_0(\alpha)$-formula $A(E,F,P)$ is meant to stress that E,F,P are prominent relations possibly occurring in A but that there may be other relation symbols as well.

To express the soundness of a proof system just with respect to proofs of formulas of depth at most d, we do not need to use a coNP definition of the satisfiability and may write it using the bounded formula Sat_d used earlier:

$$Prf^0_Q(x,y,z,P,F,G_1,\dots,G_t) \rightarrow Sat_d(x,y,E,F).$$

Denote this formula by $Ref_{Q,d}$.

Note that the freedom in choosing any $s\Sigma^1_1$-definition of Prf_Q works for us: we can incorporate into the relations $G_1,\dots,G_t$ any information useful for the proof of the soundness of a proof system (for an example see Section 11.7). The next theorem shows that even this does not help to prove the soundness of more proof systems in $I\Delta_0(\alpha)$ and that Lemma 8.4.2 cannot be generalized.

Theorem 8.4.3 *Let Q be an arbitrary proof system and let Prf_Q be an arbitrary* NP*-definition of its provability predicate of the form as above.*

For every $d_0 \geq 1$, if $I\Delta_0(\alpha)$ proves Ref_{Q,d_0}, so that

$$Prf^0_Q(x,y,z,P,F,G_1,\dots,G_t) \rightarrow Sat_{d_0}(x,y,E,F), \tag{8.4.5}$$

then there exists $d \geq d_0$ such that F_d p-simulates Q with respect to proofs of depth $\leq d_0$ formulas.

Proof From the hypothesis of the theorem it follows by Theorem 8.2.2 that for some $c,d \geq 1$ the propositional translations $\langle \dots \rangle_{n,m,s}$ of (8.4.5) have size $(n+m+s)^c$ Frege proofs $\pi_{n,m,s}$ of depth $\leq d$.

Assume that Π is a Q-proof of a formula φ in n variables, $|\Pi| \leq s$ and $|\varphi| \leq m$. Let P and F be the relations encoding Π and φ, respectively, and let $G_1,\dots,G_t$ be some relations witnessing the existential quantifiers in $Prf_Q(n,m,s,P,F)$. Let ρ be a size $(n+m+s)^c \le O(s^c)$ F_d-proof (assuming without loss of generality that $n \le m \le s$) of

$$\langle Prf_Q^0(x,y,z) \;\to\; Sat_{d_0}(x,y)\rangle_{n,m,s}, \tag{8.4.6}$$

which is just

$$\langle Prf_Q^0(x,y,z)\rangle_{n,m,s} \to \langle Sat_{d_0}(x,y)\rangle_{n,m}. \tag{8.4.7}$$

Substituting for the atoms in $\langle Prf_Q^0(x,y,z)\rangle_{n,m,s}$ the bits defined by the relations $P, F, G_1, \dots, G_t$, the formula becomes a sentence which is true and has an F_d-proof of size polynomial in the size of the sentence, i.e. in s. Hence we get from ρ a size-$O(s^c)$ F_d-proof ρ' of

$$\langle Sat_{d_0}(x,y)\rangle_{n,m}(\overline{p},\overline{q}/F),$$

where $\overline{p} = p_1, \dots, p_n$ are the atoms corresponding to an unknown evaluation E and where we substitute for the m atoms $\overline{q} = q_1, \dots, q_m$ corresponding to F the bits defining the formula φ.

The formula φ has n atoms and we can write them using $\overline{p}$ as $\varphi(\overline{p})$.

Claim *There are size-$m^{O(1)}$ F_d-proofs of*

$$\langle Sat_{d_0}(x,y,E,F)\rangle_{n,m}(\overline{p},\overline{q}/F) \;\to\; \varphi(\overline{p}). \tag{8.4.8}$$

In fact, there is a p-time algorithm that upon receiving as inputs $1^{(n)}, 1^{(m)}, \varphi$ constructs such a proof.

To prove (8.4.8) one constructs by induction on the depth of φ a proof of a stronger statement:

$$\langle Sat_{d_0}(x,y,E,F)\rangle_{n,m}(\overline{p},\overline{q}/F) \;\equiv\; \varphi(\overline{p}). \tag{8.4.9}$$

For this we need a faithful formalization of Sat_d such that it is easy to prove that an assignment satisfies a conjunction (resp. disjunction) if and only if it satisfies both (resp. at least one) conjunct (resp. disjunct), and that it satisfies the negation of a formula if and only if it does not satisfy the formula itself.

The argument so far shows that F_d simulates Q. To get a p-simulation one needs that the proofs of (8.4.6) provided by Theorem 8.2.2 can be constructed by a p-time algorithm from the inputs $1^{(n)}, 1^{(m)}, 1^{(s)}$. The model-theoretic proof for Theorem 8.2.2 does not establish this, but the proof-theoretic proofs that we mentioned do. We shall verify this in Section 12.3 when we describe a propositional simulation of Cook's theory PV by *EF* based on Herbrand's theorem. □

One may try to strengthen the theorem and to formulate it for the full reflection Ref_Q from (8.4.4):

$$Prf_Q^0(x,y,z,P,F,G_1,\dots,G_t) \;\to\; Sat^0(x,y,E,F,H_1,\dots,H_r)$$

using the $s\Pi_1^1$-definition of the satisfiability for all formulas, not just those of bounded depth. The same argument as before will give us a short proof of

$$\langle Sat^0(x,y,E,F,H_1,\ldots,H_r)\rangle_{n,m}(\overline{p},\overline{q}/F,H_1,\ldots,H_r).$$

However, to derive from this the formula $\varphi(\overline{p})$ we must be able to *define* in terms of E (i.e. of $\overline{p}$) some assignments to $H_1(\overline{p}),\ldots,H_r(\overline{p})$ such that we can even prove that

$$\langle Sat^0(x,y,E,F,H_1,\ldots,H_r)\rangle_{n,m}(\overline{p},\overline{q}/F,H_1/H_1(\overline{p}),\ldots,H_r/H_r(\overline{p}))\ .$$

Typically the witnesses H_i represent some canonical computation of the value of F on $\overline{p}$ and hence we need to define the computation of general formulas. This is not possible by a $\Delta_0(\alpha)$-formula or by a depth d formula; see Section 8.7.

8.5 Ajtai's Argument

Ajtai [5] found an argument that at the same time solved the problem about the provability of PHP(R) in $I\Delta_0(R)$ and gave a lower bound for constant-depth Frege proofs of PHP_n; see Section 8.1. We now describe the structure of his argument.

Let $\mathbf{M}$ be a countable non-standard model of $\mathrm{Th}_L(\mathbf{N})$ as in Section 8.3 and let $n \in \mathbf{M}\setminus\mathbf{N}$ be its non-standard element. Define a particular cut

$$I_n := \{m \in \mathbf{M} \mid m \le n^c \text{ for some } c \in \mathbf{N}\}.$$

Theorem 8.5.1 (Ajtai [5]) *Let $\mathbf{M}$, n and I_n be as above. Assume that $k \in \mathbf{N}$ and that $P \subseteq [n]^k$ is a k-relation on $[n]$ with $P \in \mathbf{M}$.*

Then there exists $R \subseteq [n+1]\times[n]$ such that

$$(I_n, P, R) \models I\Delta_0(P,R) \wedge \neg\mathrm{PHP}(n,R).$$

That is, R is a bijection between $[n+1]$ and $[n]$ in the structure (I_n,P,R).

We shall outline a proof of this theorem in Section 20.3 (Lemma 20.3.2); a full proof is given in [278]. Now we shall derive from it the solutions to the two problems mentioned at the beginning of the section. The first solution is obvious.

Corollary 8.5.2 *The theory $I\Delta_0(R)$ does not prove* PHP(R).

The second solution will require a short argument.

Corollary 8.5.3 *For each $d, c \ge 2$, there are depth d, size ℓ^c Frege proofs of* PHP_ℓ *(the tautology from* (1.5.1)*) for finitely many $\ell \ge 1$ only.*

Proof Assume for the sake of contradiction that for some $d, c \ge 2$ there are depth d Frege proofs of PHP_ℓ of size $\le \ell^c$, for unboundedly many ℓ. The same statement must be true in $\mathbf{M}$ (it is a model of $\mathrm{Th}_L(\mathbf{N})$), and hence for some non-standard n there is a depth d, size n^c Frege proof Π of PHP_n in $\mathbf{M}$. Such a proof can be coded by some k-ary relation P on $[n]$, with $k = c + O(d)$: each of at most n^c lines in Π

will correspond to an $O(d)$-ary relation $P_{\bar{i}}$ defined as $(i_1,\dots,i_c,\bar{y}) \in P$ for specific $\bar{i} = (i_1,\dots,i_c) \in [n]^c$, where $P_{\bar{i}}$ defines a depth-d formula as in Section 8.4.

Let $R \subseteq [n+1] \times [n]$ be the relation provided by Ajtai's theorem 8.5.1 (and we identify it with a unary subset of $[n^2+n]$ to conform with the earlier set-up). By Lemma 8.4.2, depth-d Frege proofs are sound in (I_n, P, R) as formalized by the formulas Prf_d from (8.4.2). The antecedent $Prf_d(n^2+n, m, n^c, P, F)$ of the specific instance of Ref_d holds true there, where n^2+n is the number of atoms in PHP_n, $m = O(n^3)$ is the size of PHP_n and F is a Δ_0-definable relation encoding the formula PHP_n.

Hence in (I_n, P, R) the instance $Sat_d(n^2+n, m, R, F)$ of the succedent of (8.4.2) must hold as well. But R violates PHP_n and, as in the claim establishing (8.4.8), this contradicts $Sat_d(n^2+n, m, R, F)$. □

8.6 The Correspondence – A Summary

Let us review the main points of the correspondence between $I\Delta_0(\alpha)$ and AC^0-Frege systems established in this chapter. We shall formulate them in general terms because variants of the correspondence will be shown later to hold for a number of specific natural pairs

a theory T and a proof system P

(and, in fact, the correspondence can be defined for general theories and proof systems). Our formulation here will be precise but necessarily somewhat informal: we shall talk about a pair T, P although at this point we have just one example. We will formulate general prerequisites and then derive their consequences by the arguments presented so far. Thus for any pair T, P for which we will be able to establish the prerequisites, we will automatically get the consequences.

We allow that the first-order part of the language of T extends by a finite number of symbols L_{PA} (the first-order part of the language of $I\Delta_0(\alpha)$), assuming that true atomic sentences in the extended language are p-time recognizable. This is needed for the first condition in (A) below. We also assume that the L_{PA}-consequences of T are true in $\mathbf{N}$.

The **prerequisites** for the correspondence are as follows.

(A) *The translation $\langle \dots \rangle$ of $\Delta_0(\alpha)$-formulas (in the language of T) $A(x)$ into sequences of propositional formulas $\langle A\rangle_n$, $n \geq 1$, satisfies:*

- *(i) there is a p-time algorithm that computes the formula $\langle A\rangle_n$ from $1^{(n)}$;*
- *(ii) $|\langle A\rangle_n| = n^{O(1)}$ and $dp(\langle A\rangle_n) = O(1)$;*
- *(iii) the sentence $\forall x A(x)$ is true for all interpretations of the relation symbols if and only if all the $\langle A\rangle_n$ are tautologies.*

(B) *If the theory T proves $\forall x A(x)$ then there are P-proofs of $\langle A \rangle_n$ of size $n^{O(1)}$. Moreover, there is a p-time algorithm that constructs a P-proof of $\langle A \rangle_n$ from $1^{(n)}$.*

Note: Following the discussion after Lemma 8.2.1, if the language of T contains functions other than linearly growing in the length and $f(x)$ is a non-decreasing function from (8.2.1) then (B) may hold with size bound to the P-proofs (and time bound to the algorithm) of the form $f^{(c)}(n)$, for some $c \geq 1$.

(C) *The theory T proves the reflection principle $Ref_{P,d}$ for P-proofs of depth-d formulas, $d \geq 2$, for some $s\Sigma_1^1$-formalization of the provability predicate Prf_P of P.*

Note: The formula $Ref_{P,d}$ is defined in an analogous way to Ref_d in Section 8.4. Using the limited extension from Chapter 5 we could restrict to proofs of DNF-formulas (and $Ref_{P,2}$) only but this would alter the definition of bounded-depth systems (as the extension atoms themselves stand for formulas of non-zero depth).

Let us point out that if we wanted to use the full reflection principle with some $s\Pi_1^1$-definition of the satisfiability *Sat* as in (8.4.3) we would also require:

(D) *The theory T proves that every formula can be evaluated on every truth assignment; that is, there exist $H_1, \ldots, H_r$ witnessing Sat^0.*

It is convenient to introduce some suitable terminology:

- P **p-simulates** T if and only if the conditions (A) and (B) are satisfied;
- P **corresponds to** T if and only if P p-simulates T and the condition (C) holds as well.

The **consequences** of the correspondence now follow.

Using only these conditions on T and P we have proved:

(1) Upper bounds *To establish a polynomial upper bound for P-proofs of $\langle A \rangle_n$, $n \geq 1$, it suffices to prove $\forall x A(x)$ in T. This uses (A) and (B).*

(2) p-Simulations for proof systems *To establish the existence of a p-simulation of a proof system Q by P with respect to depth $\leq d$ formulas, it suffices to prove in T the reflection principle $Ref_{Q,d}$ for some $s\Sigma_1^1$-definition of the provability predicate of Q. This uses (A), (B) and (C).*

This is based on the argument underlying Theorem 8.4.3.

(3) Lower bounds *To establish a super-polynomial lower bound for P-proofs of $\langle A \rangle_n$, $n \geq 1$, for a $\Delta_0(\alpha)$-formula $A(x)$, it suffices to construct for any non-standard model* **M** *of* $\mathrm{Th}_{L_{PA}}(\mathbf{N})$ *and any non-standard $e \in$* **M** *an interpretation α_G of all relations in α on the cut I_e such that*

$$(I_e, \alpha_G) \models T + \neg A(e).$$

This is obtained via the Ajtai argument and uses (A), (B) and (C).

We shall prove one more consequence that has not been mentioned yet. Let TAUT_2 be the set of tautological DNF formulas; it is coNP-complete. We take TAUT_2 in place of TAUT in order to be able to use the simple $\Delta_0(E,F)$-formula Sat_2 and to avoid the requirement (D), which is needed when an $s\Pi_1^1$-definition of the satisfiability relation for general formulas is used. We say that

- T **proves** NP $=$ coNP if and only if there is a $s\Sigma_1^1$-formula $\Phi(F,y)$ such that T proves

$$\Phi(F,y) \equiv \forall E, x\, Sat_2(x,y,E,F). \tag{8.6.1}$$

The consequence that we want to state is:

(4) The consistency of NP $\neq$ coNP *A super-polynomial lower bound for P-proofs of any sequence of tautologies τ_n, $n \geq 1$, implies that T does not prove* NP $=$ coNP.

Proof Assume that $\Phi(y,F)$ is of the form $\exists G \subseteq [0,y^c]W(y,F,G)$, with W a $\Delta_0(F,G)$-formula, and that T proves (8.6.1). This means that we can interpret W as defining the provability predicate of some proof system Q. In addition, such a Q is p-bounded, and T proves $Ref_{Q,2}$. Hence, by (2), P p-simulates Q and is therefore also p-bounded. That contradicts the hypothesis that we have a super-polynomial lower bound for P. □

8.7 Bibliographical and Other Remarks

Sec. 8.1 Parikh [379] considered also the feasibility predicate $F(x)$, a new symbol added to L_{PA} and allowed in bounded formulas for which instances of the IND scheme are adopted ($\Delta_0(F)$-IND). Using this predicate it is possible to formalize the intuitive properties of the feasibility as

$$F(0) \;\wedge\; F(1) \;\wedge\; [(F(x) \wedge y \leq x) \rightarrow F(y)]$$

and

$$(F(x) \wedge F(y)) \;\rightarrow\; (F(x+y) \wedge F(x \cdot y))\,.$$

One can then add a new axiom stating that some very large number is not feasible. Buss [118] discussed this issue and gave further references.

The Δ_0-definable sets were defined (under the name *constructive arithmetic sets*) by Smullyan [476]. They are exactly those sets that belong to the linear time hierarchy *LinH* of Wrathall [508] and they are also exactly the *rudimentary predicates* of Smullyan [476] (Wrathall [508] proved that rudimentary sets coincide with *LinH* and Bennett [65] proved that they also coincide with Δ_0-definable sets). Ritchie [449] showed that the Δ_0-sets are included in linear space and Nepomnjascij [365] complemented this by showing that sets decidable with a simultaneous time–space restriction TimeSpace$(n^{O(1)}, n^{1-\Omega(1)})$ are also in Δ_0. In particular, sets decidable in logarithmic space are Δ_0-definable.

The theory $I\Delta_0$ was studied and much developed by Paris and Wilkie and their students in a series of works [380, 381, 382, 171, 383, 385, 387, 386, 388, 389,

390, 391, 392]. The work of Paris and Wilkie [391] circulated in the early 1980s (the main results were reported in [387]) although it was published only in 1987, two years after Buss's Ph.D. thesis [106]. Bennett [65] proved that the graph of the exponentiation $x^y = z$ is definable in the language of rudimentary sets using only bounded quantifiers (in the sense of that language, see [278]). Other Δ_0-definitions of the exponentiation graph were given by Dimitracopoulos and Paris [170] and by Pudlák [407].

The finite axiomatizability problem is due to Paris and Wilkie and the Δ_0-PHP problem to Macintyre. The original formulation of the Δ_0-PHP problem in [389] left out the second clause, that R ought to be a graph of a *function* and allowed also multi-functions. But, having a Δ_0-formula A satisfying the other two conditions, the formula

$$A'(x,y) := A(x,y) \wedge \forall t < y \neg A(x,t)$$

will satisfy also the missing function condition and it is also Δ_0.

Paris and Wilkie [387] showed that if $I\Delta_0 + \Omega_1$ proves that the polynomial-time hierarchy collapses then it is finitely axiomatizable, and this was complemented by Krajíček, Pudlák and Takeuti, who proved that $I\Delta_0 + \Omega_1$ is not finitely axiomatizable unless the polynomial-time hierarchy collapses. This was subsequently strengthened to *provable* collapse by Buss [110] and Zambella [512]. Paris, Wilkie and Woods [392] proved an interesting relation between the two problems. Let Δ_0-WPHP denote Δ_0-PHP with the term $z+1$ bounding x replaced by z^2; i.e. it refers to maps from $[z^2]$ into $[z]$. Paris, Wilkie and Woods [392] proved that if $I\Delta_0$ does not prove Δ_0-WPHP then it is not finitely axiomatizable. More details can be found in [278].

Secs. 8.2 and 8.3 Paris and Wilkie [389] called the statement about the tautologies PHP_n, to which they reduced the relativized PHP-problem the *Cook–Reckhow conjecture* although such a conjecture is to be found in Cook and Reckhow [156] only implicitly. They did not mention explicitly a bound d to the depth but it was clear in their construction that d can be bounded. A bound to the depth is actually crucial; without it the tautologies PHP_n do have polynomial-size Frege proofs, as shown by Buss [108] (see also Section 11.3).

Buss [106] introduced the bounded quantifier version LKB of the first-order sequent calculus LK. Cook and Nguyen [155] worked out the proof of Theorem 8.2.2 in detail using this approach, including the required cut-elimination result. Details of the original Paris and Wilkie [389] proof of Theorem 8.2.2 – only sketched here – can be found in [278]. Buss [113] is a useful survey and Buss [119] uses the Paris–Wilkie propositional translation to prove some witnessing theorems.

Sec. 8.4 The undefinability of computations of general formulas in $I\Delta_0(\alpha)$ and AC^0-Frege systems follows from the fact that general formulas are not equivalent to AC^0-formulas, as obtained by Ajtai [4] and Furst, Saxe and Sipser [192].

Secs. 8.5 and 8.6 Ajtai's paper [5] appeared to me at the time as a miracle. I think that there are about two dozen papers in proof complexity to which one can trace the origins of all the main ideas and this could be perhaps compressed into a dozen if forced by the circumstances of being alone on a desert island. But if I would have to select one it would be this paper of Ajtai. Not only did it prove one of the key results in proof complexity and introduce a new method that has been used ever since, but it opened completely new vistas, showing that proof complexity is a part of a much larger picture and that it does not need to be just a finitary proof theory.

Linking $I\Delta_0(R)$ with the AC^0-Frege systems is strictly speaking not an example of the relation of a theory T to a proof system P, as AC^0-Frege systems refer to a collection of proof systems and not just one. I think this informal formulation does not lead to any confusion, as the theorems clearly link proofs in T to specific F_d. However, the reader who has concerns may consult Ghasemloo [199], who defined *proof complexity classes* and discussed these issues.

9

The Two Worlds of Bounded Arithmetic

The framework in which we formulated the basic example of the relation between a proof system and a theory in Chapter 8 is not the only possibility. One can choose between the second-order set-up (as in Chapter 8) or the first-order set-up, which we yet have to discuss. There is a basic dichotomy regarding which objects of the theory represent binary strings of lengths polynomial in n: either relations on $[n]$, in which case we speak of a two-sorted (or, somewhat loosely, of a second-order) formalization; or numbers of bit-length $n^{O(1)}$ (i.e. having absolute value $2^{n^{O(1)}}$), in which case it is a one-sorted (or first-order) formalization.

In addition there is some leeway in the language for either formalization that one may choose; this depends on whether one studies only links to proof systems as we are doing (in which case the choice of language is far more flexible) or to other topics of bounded arithmetic, e.g. witnessing theorems (and then particulars of the language are often crucial). Stemming from that is a certain fuzziness in the definitions of basic bounded arithmetic theories: authors often tweak the original definitions a bit to fit better their particular choice of language and the specific goal of their investigation. This has the effect that some authors, not hesitant to write expository texts or lecture notes while avoiding the labor of learning the details, present bounded arithmetic in an off-hand and mathematically incorrect way.

An alternative is that authors choose one particular set-up and the language to go with it, and one particular way to formulate theories and propositional translations, and work it out with precision. This is of course preferable to the former approach but it shies away from the key issue that the fundamental ideas underlying the particular set-ups, formalizations and links of theories to proof systems are really the same. When the reader understands them then he or she will have no problem in moving smoothly among the various specific formalizations, in order to use a suitable one for the problem at hand, and to compare results formulated via different formalizations. In this chapter we aim at conveying this understanding; we outline some constructions and proofs and hope that the reader resurfaces at the end of the chapter with an understanding of how it all works and that the formal aspects of the relation between bounded arithmetic and proof systems are less crucial than he or she might have expected.

9.1 The Languages of Bounded Arithmetic

For many bounded arithmetic theories it is useful to expand the language of arithmetic L_{PA} that we used in Chapter 8 by several new function symbols. As mentioned above, because we are concerned here just with the correspondence of bounded arithmetic to propositional proof systems, we are more free to choose which symbols we include. Atomic sentences involving only numbers translate into propositional constants, and the only issue is how fast-growing are the functions defined by the terms in the language. All the functions we add to form the language L_{BA} will have linear growth in the length, as defined in (8.1.2). In addition, all will be p-time computable (this will be important when defining the $\| \dots \|$ translation in Chapter 12).

Let L_{BA}, the **language of bounded arithmetic**, be L_{PA} to which we add:

- the unary function symbols $\lfloor x/2 \rfloor$, $|x|$ and $len(x)$;
- the binary function symbol $(x)_i$; and
- the binary relation symbol $bit(i, x)$.

The canonical interpretation of $\lfloor x/2 \rfloor$ in $\mathbf{N}$ is obvious, $|x|$ is the bit-length of x and $|x| := \lceil \log(x+1) \rceil$, as discussed already in Section 8.1. The relation $bit(i, x)$ holds if and only if the term 2^i occurs in the unique expression of x as a sum of different powers of 2.

All theories will contain a finite set of axioms about the symbols in L_{BA}. This set, denoted by BASIC, will contain groups of axioms 1–5, to be listed now, and further groups 6–8, listed below:

1. Robinson's arithmetic Q;
2. the axioms in the language L_{PA} for the non-negative parts of discretely ordered commutative rings;
3. $|0| = 0 \wedge |1| = 1$ and
   - $x \neq 0 \rightarrow (|\underline{2}x| = |x| + 1 \wedge |\underline{2}x + 1| = |x| + 1)$,
   - $x \leq y \rightarrow |x| \leq |y|$;
4. $x \neq 0 \rightarrow ||\lfloor x/2 \rfloor| + 1 = |x|$ and $x = \lfloor y/2 \rfloor \equiv [y = \underline{2}x \vee y = \underline{2}x + 1]$;
5. $bit(z, x) \rightarrow z \leq |x|$ and
   - $bit(0, x) \equiv (\exists y \leq x x = \underline{2}y + 1)$,
   - $bit(z, x) \equiv bit(z+1, \underline{2}x)$,
   - $(\forall z \leq |x| + |y| \neg bit(z, x) \vee \neg bit(z, y)) \rightarrow (bit(u, x+y) \equiv (bit(u, x) \vee bit(u, y)))$,
   - $\forall z \leq |x| + |y| \; bit(z, x) \equiv bit(z, y) \rightarrow x = y$.

The remaining two functions, $len(x)$ and $(x)_i$, are linked to the way in which numbers encode finite sequences of numbers. Of course, we can define a simple **pairing function**

$$\langle x, y \rangle := \frac{(x+y)(x+y+1)}{2} + x, \tag{9.1.1}$$

which is a bijection between $\mathbf{N} \times \mathbf{N}$ and $\mathbf{N}$, with the projections simply definable with a bounded graph:

$$p_1(z) = x \quad \text{if and only if} \quad \exists y \le z\ \langle x, y\rangle = z$$
$$\text{if and only if} \quad \forall u, v \le z \langle u, v\rangle = z \ \rightarrow\ u = x,$$

and similarly for $p_2(z) = y$. Iterating this, we can encode k-tuples of numbers for any fixed $k \ge 2$. But we want to encode sequences of numbers of variable length as well.

There is no unique way to do that. We shall outline a way that is simple and that will motivate the axioms about $len(x)$ and $(x)_i$ that we put below into BASIC. Let $a_0, \ldots, a_{t-1}$ be a sequence of numbers. Encode it by a number w such that, for all i, j,

$$bit(\langle i, j\rangle, w) \quad \text{if and only if} \quad bit(j, a_i). \tag{9.1.2}$$

For such a w, we define $len(w) := t$ and $(w)_i := a_i$. We cannot use the formula (9.1.2) as an axiom, since we cannot quantify the sequence of a_i. But we can formulate a few axioms that will allow us to prove that sequences can be coded (Lemma 9.3.2). We include the following axioms, all satisfied by the construction just described, in BASIC:

6. $\exists w\ len(w) = 0$;
7. $len(w) \le |w|$ and $(w)_i \le w$;
8. $\forall w, a \exists w' \le (wa)^{10}\ len(w') = len(w) + 1\ \wedge\ (\forall i < len(w)\ (w)_i = (w')_i)\ \wedge\ (w')_{len(w)} = a$.

If L is L_{PA} or L_{BA} and f is a new relation symbol, $L(f)$ will denote $L \cup \{f\}$ and BASIC(f) will denote BASIC to which has been added possibly some axioms about f. The notation $L(\alpha)$ will denote L to which has been added variable symbols for bounded sets, as we did in Section 8.4 for $L_{PA}(\alpha)$, and also a binary relation symbol,

- the **element-hood relation** $x \in X$ between a number x and a set X.

We are adding only sets and not relations of a higher arity, as a k-ary relation can be represented by a set of k-tuples. This somewhat simplifies the syntax as we do not need to attach a specific arity to each relation symbol.

There will be also a set of axioms that BASIC(α) uses: it expands BASIC by two more axioms,

- the **extensionality axiom** $X = Y \equiv [\forall z\ z \in X \equiv z \in Y]$,
- the **set-boundedness axiom** $\exists y \forall x\ x \in X \rightarrow x \le y$,

where X, Y are set variables.

Finally, $L(\alpha, f)$ denotes the language expanded in both ways simultaneously, and similarly for BASIC(α, f).

9.2 One or Two Sorts

In Chapter 8 we worked with natural numbers $\mathbf{N}$ or, more generally, with L_{PA}-structures $\mathbf{M}$. Objects of prime interest such as formulas, evaluations, proofs, graphs of functions etc. were represented by relations. A k-ary relation on $[n]$ represented a binary string of length n^k. Using the pairing function from the last section, we can define a bijection between $\mathbf{N}^k$ and $\mathbf{N}$ and use it to represent k-relations on $[n]$ by unary relations (i.e. sets) on $[n^k]$. It is thus natural to expand the number structure $\mathbf{N}$ by all finite sets representing in this way all possible finite strings. We will add to $\mathbf{N}$ the class $\mathcal{P}_b(\mathbf{N})$ of *finite subsets* of $\mathbf{N}$.

The resulting structure $\mathbf{N}^2$ has a two-sorted universe

$$(\mathbf{N}, \mathcal{P}_b(\mathbf{N})),$$

$\mathbf{N}$ being the **number sort** and $\mathcal{P}_b(\mathbf{N})$ being the **set sort**. Unless stated otherwise, we consider $\mathbf{N}^2$ as an $L_{BA}(\alpha)$-structure: the first-order part is interpreted in the canonical way; the set variables range over $\mathcal{P}_b(\mathbf{N})$. We shall reserve the lower case letters $x, y, \ldots$ for the number-sort variables and the capital letters $X, Y, \ldots$ for the set-sort variables, and similarly $a, b, \ldots$ and $A, B, \ldots$ for elements of the structure of the respective sort.

The structure $\mathbf{N}^2$ is the **standard model** in this two-sorted set-up. Analogously as before we can also consider non-standard models. A general $L_{BA}(\alpha)$-structure $\mathbf{M}^2$ has a universe consisting of two sorts, the number sort $\mathbf{M}_1$ and the set sort $\mathbf{M}_2$, written as

$$(\mathbf{M}_1, \mathbf{M}_2),$$

and interprets L_{BA} on $\mathbf{M}_1$ and $x \in X$ as a relation on $\mathbf{M}_1 \times \mathbf{M}_2$. As before we consider the equality $=$ to be a part of the underlying logic and it is always interpreted as true equality. To maintain some coherent mental picture of $L_{BA}(\alpha)$-structures we shall generally consider only those satisfying BASIC(α).

Every $A \in \mathbf{M}_2$ can be identified with a subset of $\mathbf{M}_1$ consisting of those $a \in \mathbf{M}_1$ for which $\mathbf{M}^2 \models a \in A$, and then we can interpret the $\in$ symbol as the true membership relation. We use the index b in the notation $\mathcal{P}_b(\mathbf{N})$ as referring to *bounded* rather than finite sets because, for a non-standard $L_{BA}(\alpha)$-structure, we can thus identify $\mathbf{M}_2$ with a *subclass* of the class of bounded subsets of $\mathbf{M}_1$:

$$\mathcal{P}_b(\mathbf{M}_1) := \{A \subseteq \mathbf{M}_1 \mid \exists b \in \mathbf{M}_1 \forall a \in A \; a \leq b\}.$$

Note that bounded subsets of $\mathbf{M}_1$ need not be finite (e.g. $\mathbf{N}$ is bounded by any non-standard element but it is not finite).

It is customary to call this set-up informally *second-order*, referring to the fact that we have not only elements (objects of first order) but also sets (objects of second order). However, this is a misnomer; we shall always use only first-order logic. This means that we interpret the set-sort elements as first-order elements of a particular sort, and quantification over the set sort does not have a priori anything to do with the quantification over subsets of the number sort. The chief reason for this is that first-order logic has the fundamental properties of being *complete* and *compact* while, when using second-order logic one can quickly run into properties of structures independent of ZFC. As a famous example, consider the set of reals $\mathbf{R}$ with a predicate for **N**. If we could quantify over functions and sets we could formulate the continuum hypothesis as

$$\forall X \subseteq \mathbf{R}\, (\exists f : \mathbf{N} \to_{\text{onto}} X) \vee (\exists g : X \to_{\text{onto}} \mathbf{R}).$$

The two-sorted set-up is useful when the theories we want to consider do not involve symbols for functions mapping sets to sets. Sets are used to represent binary strings and thus such functions represent functions from strings to strings. However, sometimes these functions may be definable by formulas using bounded number-sort quantifiers but no set-sort quantifiers. In a complexity-theoretic description these are the AC^0-functions, and adding new types of bounded number quantifiers (Section 10.1) also the $AC^0[m]$- and the TC^0-functions (Section 10.2). Using $L_{BA}(\alpha)$ in such a case is something like using a magnifying glass: we can see small details of the translations, for example, the exact depths of the formulas appearing in proofs. But for stronger theories which do operate with functions on strings this loses its appeal and becomes more of a syntactic nuisance. In the magnifying glass analogy: using it to see stars only blurs the picture. For the stronger theories it is, I think, more natural to use the one-sorted set-up in which all objects are numbers. We shall discuss it now.

The language in the one-sorted set-up that we shall use will be $L_{BA}(\#)$ and it expands L_{BA} by

- a binary function symbol $x\#y$,

whose canonical interpretation on **N** is $x\#y := 2^{|x|\cdot|y|}$. Note that we have left out the function symbol $\omega_1(x)$ appearing Section 8.1; the function is now represented by the term $x\#x$. We shall always consider only $L_{BA}(\#)$-structures obeying axioms in BASIC(#), which expands BASIC by the following axioms:

- $|x\#y| = (|x| \cdot |y|) + 1$,
- $0\#x = 1$,
- $x\#y = y\#x$,
- $|y| = |z| \to x\#y = x\#z$,
- $|x| = |y| + |z| \to x\#u = (y\#u) \cdot (z\#u)$,
- $x \neq 0 \to \left[1\#(\underline{2}x) = \underline{2}(1\#x) \wedge 1\#(\underline{2}x + 1) = \underline{2}(1\#x)\right]$.

We will now show that a one-sorted structure determines a natural two-sorted structure and in Section 9.4 we give a complementary construction, of a one-sorted structure from a two-sorted structure.

Let **K** be an $L_{BA}(\#)$-structure. Define an $L_{BA}(\alpha)$-structure

$$\mathbf{M}(\mathbf{K}) = (\mathbf{M}_1(\mathbf{K}), \mathbf{M}_2(\mathbf{K}))$$

by:

- $\mathbf{M}_1(\mathbf{K}) := \{a \in \mathbf{K} \mid \exists b \in \mathbf{K}\ \mathbf{K} \models a = |b|\}$;
- $\mathbf{M}_2(\mathbf{K}) := \mathbf{K}$;

- the interpretation of L_{BA} on $\mathbf{M}_1(\mathbf{K})$ is inherited from $\mathbf{K}$;
- for $a \in \mathbf{M}_1(\mathbf{K})$ and $A \in \mathbf{M}_2(\mathbf{K})$,

$$\mathbf{M}(\mathbf{K}) \models a \in A \text{ if and only if } \mathbf{K} \models bit(a, A).$$

Lemma 9.2.1 *Assume that* $\mathbf{K}$ *satisfies* BASIC*(#). Then* $\mathbf{M}(\mathbf{K})$ *is well defined and satisfies* BASIC*(α).*

Proof We need to verify that $\mathbf{M}_1(\mathbf{K})$ is closed under the L_{BA}-functions: as they are all of linear growth and $\mathbf{M}_1(\mathbf{K})$ is closed downwards it suffices to verify that it is closed under multiplication. But by BASIC(#), we have $|x\#y| \leq |x| \cdot |y| + 1$.

For the extensionality, we need to verify that in $\mathbf{K}$

$$(\forall z(bit(z,x) \equiv bit(z,y))) \rightarrow x = y.$$

However, this is a consequence of two axioms of BASIC(#) (from one axiom we can bound z by $|x| + |y|$ and use the other to derive the extensionality).

Finally, the set-boundedness axiom follows from the axiom $bit(z,x) \rightarrow z \leq |x|$. □

Now we turn the tables and start with an $L_{BA}(\alpha)$-structure $\mathbf{M}$, and define from it an $L_{BA}(\#)$-structure $\mathbf{K}(\mathbf{M})$. Its universe is the set sort $\mathbf{M}_2$ of $\mathbf{M}$. We think of $A \in \mathbf{M}_2$ as representing the number $nb(A) := \sum_{a \in A} 2^a$. The interpretation of the language requires that $\mathbf{M}$ satisfies more than BASIC(α). We first define the interpretation and then discuss what we need from $\mathbf{M}$.

The following symbols of $L_{BA}(\#)$ on $\mathbf{K}(\mathbf{M})$ are interpreted as follows:

- 0 is the empty set $\emptyset$ and 1 is $\{0\}$;
- $nb(A) \leq nb(B)$ if and only if $A = B$ *or the largest* a *such that* $a \in A \triangle B$ *($\triangle$ is the symmetric difference) satisfies* $a \in B$;
- $\lfloor nb(A)/2 \rfloor = nb(B)$, where $A \neq \emptyset$ and $B := A \setminus \{\max(A)\}$, and $\lfloor nb(\emptyset)/2 \rfloor = \emptyset$;
- $|nb(A)| := \max(A) + 1$ and $|nb(\emptyset)| := 0$;
- $nb(A)\#nb(B) = nb(C)$, where $C := \{\max(A) \cdot \max(B)\}$;
- $bit(nb(A), nb(B))$ if and only if $\exists a\, nb(A) = a \wedge a \in B$, where the formula $nb(A) = a$ abbreviates $\forall i \leq |a|\; bit(i,a) \equiv i \in A$.

A sequence of numbers $nb(A_0), \ldots, nb(A_t)$ (assume without loss of generality that the sets A_i are non-empty for a technical reason) is coded by $nb(F)$ where, for all $i = nb(I)$ and $j = nb(J)$,

$$\langle i,j \rangle \in F \quad \text{if and only if} \quad j \in A_i.$$

Then:

- $len(nb(F)) = \min\{i \mid \forall j \langle i,j \rangle \notin F\}$;
- $(nb(F))_i = nb(A_i)$.

It remains to define the interpretations of $+$ and $\cdot$. In these definitions we will need to show that the maxima used above do actually exist, and also some form of induction will be needed to establish the axioms of BASIC(#) for this interpretation. Addition and multiplication are in fact the hardest to define and need even more induction. We shall do this in detail in Section 9.4.

9.3 Buss's Theories

We now present several examples of theories in the languages $L_{BA}(\alpha)$, $L_{BA}(\#)$ and $L_{BA}(\#, \alpha)$. They are collectively called **bounded arithmetic** theories because in each of them the IND axiom scheme is accepted only for some class of bounded formulas. We need to start by describing some classes of these formulas (we add some more in Chapter 10) and we will treat the two-sorted case first.

The class of $L_{BA}(\alpha)$-formulas with all number quantifiers bounded and with no set-sort quantifiers is denoted by $\Sigma_0^{1,b}$. The theory $I\Sigma_0^{1,b}$ is axiomatized by BASIC(α) and the IND scheme, (8.1.1),

$$\neg A(0) \ \vee \ (\exists y \leq x, y < x \wedge A(y) \wedge \neg A(y+1)) \ \vee \ A(x)$$

for all $\Sigma_0^{1,b}$-formulas $A(x)$. The theory V_1^0 expands $I\Sigma_0^{1,b}$ by also having

- the **bounded comprehension axiom** $\Sigma_0^{1,b}$-CA, for all $\Sigma_0^{1,b}$-formulas $B(x)$,

$$\forall x \exists X \forall y \leq x, \quad y \in X \equiv B(y).$$

The theory V_1^0 is a conservative extension of $I\Sigma_0^{1,b}$ with respect to $\Sigma_0^{1,b}$ -consequences: if C is a $\Sigma_0^{1,b}$-formula and V_1^0 proves its universal closure, so does $I\Sigma_0^{1,b}$. To see this, note that any model $\mathbf{M}$ of $I\Sigma_0^{1,b}$ can be expanded to a model of V_1^0: just add to the set sort $\mathbf{M}_2$ all bounded subsets of the number sort $\mathbf{M}_1$ that are definable by a $\Sigma_0^{1,b}$-formula using parameters from $\mathbf{M}$ (of both the number and the set sort).

Introducing a handy notation $X \leq x$ for the formula $\forall y\, y \in X \to y \leq x$, the set-boundedness axiom can be stated simply as $\forall X \exists x X \leq x$. This notation allows us to introduce two **bounded set quantifiers** of the form

$$\exists X \leq t \quad \text{and} \quad \forall X \leq t,$$

where t is a term, and also two new classes of bounded formulas. The classes $\Sigma_1^{1,b}$ and $\Pi_1^{1,b}$ are the smallest classes of $L_{BA}(\alpha)$-formulas such that:

1. $\Sigma_1^{1,b} \supseteq \Sigma_0^{1,b}$ and $\Pi_1^{1,b} \supseteq \Sigma_0^{1,b}$;
2. both $\Sigma_1^{1,b}$ and $\Pi_1^{1,b}$ are closed under $\vee$ and $\wedge$;
3. the negation of a formula from $\Sigma_1^{1,b}$ is in $\Pi_1^{1,b}$ and vice versa;
4. if $A \in \Sigma_1^{1,b}$ then also $\exists X \leq tA \in \Sigma_1^{1,b}$;
5. if $A \in \Pi_1^{1,b}$ then also $\forall X \leq tA \in \Pi_1^{1,b}$.

Strict $\Sigma_1^{1,b}$- and $\Pi_1^{1,b}$-formulas are those having all set-sort quantifiers at the beginning. In particular, a strict $\Sigma_1^{1,b}$-formula looks like

$$\exists X \leq t_1 \ldots \exists X \leq t_k A,$$

where $A \in \Sigma_0^{1,b}$. These two classes of strict formulas are denoted by $s\Sigma_1^{1,b}$ and $s\Pi_1^{1,b}$, respectively. The theory V_1^1 extends V_1^0 by accepting the IND scheme for all $\Sigma_1^{1,b}$-formulas.

We now turn to theories in the language $L_{BA}(\#)$. The class of bounded formulas in $L_{BA}(\#)$ is denoted by Σ_∞^b. As we do not have the set sort (there are not even free set-sort variables), this class is different from the class $\Sigma_0^{1,b}$. Note that $\Sigma_\infty^b \supseteq \Delta_0$ as $L_{BA}(\#) \supseteq L_{PA}$.

The theory analogous to $I\Delta_0$ in this language is T_2: it is axiomatized by BASIC(#) together with the IND scheme for all Σ_∞^b-formulas. It was a significant insight of Buss [106] that the IND scheme itself can be fruitfully modified. He considered two such modifications:

$$\text{LIND}, \ \neg A(0) \ \vee \ (\exists y < |x|, \ A(y) \wedge \neg A(y+1)) \ \vee \ A(|x|)$$

and

$$\text{PIND}, \ \neg A(0) \ \vee \ (\exists y < x, \ A(\lfloor y/2 \rfloor) \wedge \neg A(y)) \ \vee \ A(x).$$

The acronym LIND stands for *length induction* and PIND for *polynomial induction*. The theory S_2 is defined in the same way as T_2 but accepts the LIND scheme instead of the IND scheme for all Σ_∞^b-formulas.

The class Σ_∞^b can be naturally stratified into levels and these levels in turn determine natural subtheories T_2^i and S_2^i of T_2 and S_2, respectively. Call the bounded quantifiers $\exists y \leq t$ and $\forall y \leq t$ **sharply bounded** if and only if the bounding term t has the form $|s|$, where s is another term. Using this notion define the subclasses Σ_i^b and Π_i^b of Σ_∞^b analogously to the classes $\Sigma_1^{1,b}$ and $\Pi_1^{1,b}$:

- the class $\Sigma_0^b = \Pi_0^b$ is the class of bounded $L_{BA}(\#)$-formulas with all quantifiers sharply bounded;
- the classes Σ_i^b and Π_i^b are the smallest classes of $L_{BA}(\#)$-formulas such that, for $i \geq 1$,
  1. $\Sigma_i^b \cap \Pi_i^b \supseteq \Sigma_{i-1}^b \cup \Pi_{i-1}^b$,
  2. both Σ_i^b and Π_i^b are closed under $\vee$ and $\wedge$,
  3. the negation of a formula from Σ_i^b is in Π_i^b and vice versa,
  4. both Σ_i^b and Π_i^b are closed under sharply bounded quantifiers,
  5. if $A \in \Sigma_i^b$ then also $\exists y \leq tA \in \Sigma_i^b$,
  6. if $A \in \Pi_i^b$ then also $\forall y \leq tA \in \Pi_i^b$.

As above, **strict** Σ_i^b- and Π_i^b-formulas are those having all quantifiers not sharply bounded at the beginning. For example, a strict Σ_1^b-formula has the form

$$\exists y_1 \leq t_1 \ldots \exists y_k \leq t_k A,$$

where A is sharply bounded. The strict classes will be denoted by $s\Sigma_i^b$ and $s\Pi_i^b$, respectively.

Using these classes of formulas we can define natural subtheories of T_2 and S_2; for $i \geq 0$,

- the theory T_2^i is the subtheory of T_2 accepting instances of the IND scheme only for Σ_i^b-formulas,
- the theory S_2^i is the subtheory of S_2 accepting instances of the LIND scheme only for Σ_i^b-formulas.

The basic relations among these theories are summarized in the following statement.

Theorem 9.3.1 (Buss [106, 109]) *For $i \geq 1$, it holds that:*

(i) the four theories axiomatized over BASIC*(#) by one of the axiom schemes*

$$\Sigma_i^b\text{-LIND},\ \Sigma_i^b\text{-PIND},\ \Pi_i^b\text{-PIND},\ \Pi_i^b\text{-LIND}$$

are all equivalent;

(ii) the theory $S_2^{i+1}(\alpha)$ proves all axioms of $T_2^i(\alpha)$ and is $\forall\Sigma_{i+2}^b(\alpha)$-conservative over $T_2^i(\alpha)$.

The **Γ-conservativity** of one theory over another means that the latter theory proves all sentences in the class Γ that the former theory proves.

We will use a variant of S_2^1 to prove a lemma about the coding of sequences. Let $S_1^1(\alpha)$ be a theory in $L_{BA}(\alpha)$ containing BASIC(α) and accepting the LIND scheme for $\Sigma_1^b(\alpha)$-formulas in the language, i.e. the # function symbol is omitted. Note that terms in this restricted language have linear growth.

Lemma 9.3.2 *The theory $S_1^1(F)$ proves that*

- *If $|c| = 10|a||b|$ and $F \subseteq [|a|] \times [b]$ is a graph of a function f: $[|a|] \rightarrow [b]$ then there exists $w \leq c$ such that*

$$len(w) = |a| \ \wedge\ \forall i < |a|\ (w)_i = f(i).$$

Proof The parameter c serves only as a bound to w; if we had # we could use the term $(a\#b)^{10}$ instead. But the present formulation implies better estimates of the growth rate of terms later on.

By induction on $t = 1, \ldots, |a|$ (i.e. by LIND), prove that such a code w_t exists for $f \upharpoonright [t]$. The axioms about $(w)_i$ imply that one can put a bound on the length of w_t of the form $|w_t| = 10t|b|$. □

We conclude this section by defining three more theories: V_2^1, U_1^1 and U_2^1. The theory V_2^1 is defined as V_1^1 but in the language $L_{BA}(\alpha, \#)$ and using BASIC(α, #) as the basic axioms. In particular, we also allow in formulas the function symbol $x\#y$ and hence the terms can no longer be bounded above by a polynomial, as in the case of V^1_1. The theory U^1_2 is defined as V^1_2 but using the LIND scheme for all $\Sigma^{1,b}_1$-formulas instead of the IND scheme, and the theory U^1_1 is its subtheory using only the language $L_{BA}(\alpha)$ (i.e. no function symbol #).

9.4 From Two Sorts to One and Back

We are going to complete the construction of the model $\mathbf{K}(\mathbf{M})$ from the end of Section 9.2 and show that the pair of constructions from that section

$$\mathbf{M} \longmapsto \mathbf{K}(\mathbf{M}) \quad \text{and} \quad \mathbf{K} \longmapsto \mathbf{M}(\mathbf{K})$$

establish a correspondence between certain pairs of two-sorted and one-sorted theories. We will demonstrate this for the most interesting example (in which the technical details are also easy) and give further references in Section 9.5.

In the rest of the section, $\mathbf{M}$ is a two-sorted $L_{BA}(\alpha)$-structure satisfying BASIC(α) and $\mathbf{K}$ is a one-sorted $L_{BA}(\#)$-structure satisfying BASIC(#).

Theorem 9.4.1 *If* $\mathbf{M} \models V^1_1$ *then* $\mathbf{K}(\mathbf{M}) \models S^1_2$ *and vice versa; if* $\mathbf{K} \models S^1_2$ *then* $\mathbf{M}(\mathbf{K}) \models V^1_1$.

Proof We start with the construction $\mathbf{K} \longmapsto \mathbf{M}(\mathbf{K})$ as that was fully defined in Section 9.2 and we have already verified that BASIC(α) holds in $\mathbf{M}(\mathbf{K})$ (Lemma 9.2.1). It remains to check the bounded comprehension axiom, $\Sigma^{1,b}_0$-CA, scheme of V^0_1 and the $\Sigma^{1,b}_1$-IND scheme that V^1_1 adds on top of it.

Let $B(x)$ be a $\Sigma^{1,b}_0$-formula with parameters from $\mathbf{M}(\mathbf{K})$ and let $a \in \mathbf{M}_1(\mathbf{K})$, i.e. $\mathbf{K} \models a = |a^*|$ for some $a^* \in \mathbf{K}$. To verify the instance of CA,

$$\exists X \forall y \leq a\ y \in X \equiv B(y),$$

note that a witness X may be required to satisfy without loss of generality $X \leq a$ and so we can rewrite the statement as an $L_{BA}(\#)$-sentence

$$\exists x \leq a^* \forall y \leq |a^*|\ bit(y,x) \equiv \tilde{B}(y), \tag{9.4.1}$$

where $\tilde{B}$ is the translation of B defined by the interpretation of $L_{BA}(\alpha)$ in $\mathbf{K}$. As B has no set-sort quantifiers the formula $\tilde{B}$ is sharply bounded. Hence (9.4.1) holds as a Σ^b_1-formula and can be proved by Σ^b_1-LIND with the induction parameter t in

$$\exists x \leq a^* \forall y \leq t\ bit(y,x) \equiv \tilde{B}(y),$$

and hence $\mathbf{M}(\mathbf{K})$ satisfies the $\Sigma^{1,b}_0$-CA scheme.

The $\Sigma^{1,b}_0$-IND scheme is verified analogously, noting that:

- a $\Sigma^{1,b}_1$-formula over $\mathbf{M}(\mathbf{K})$ is translated into a Σ^b_1-formula over $\mathbf{K}$;
- IND in $\mathbf{M}(\mathbf{K})$ up to a translates into LIND in $\mathbf{K}$ up to $|a^*|$.

For the dual construction $\mathbf{M} \longmapsto \mathbf{K}(\mathbf{M})$, we first need to complete the definition of the interpretation of $L_{BA}(\#)$ in $L_{BA}(\alpha)$ by proving that $\max(X)$ exists in $\mathbf{M}$ (we used that in Section 9.2) and also defining the interpretation of addition and multiplication.

Claim 1 $\mathbf{M} \models \forall X \exists x \in X \forall y \in X\, y \leq x$.

To see this, given $A \in \mathbf{M}_2$ take $a \in \mathbf{M}_1$ such that $A \leq a$ (by the set-boundedness axiom) and then apply the **least number principle** (LNP):

$$\Psi(a) \to \exists x \leq a\, (\Psi(x) \wedge \forall y < x\, \neg\Psi(y))$$

to the $\Sigma_0^{1,b}$-formula

$$\Psi(x) := \forall z \leq a\, z \in X \to z \leq x.$$

In general, the LNP for any $\Psi(x)$ follows by IND: assume that the LNP fails and apply the IND for the induction parameter t to the formula

$$\exists x \leq a\, (\Psi(x) \wedge x + t \leq a),$$

to get a contradiction. As our Ψ is $\Sigma_0^{1,b}$, $\Sigma_0^{1,b}$-IND suffices.

To define addition on $\mathbf{K}(\mathbf{M})$ we define in terms of X, Y, Z the graph of addition $nb(X) + nb(Y) = nb(Z)$ and prove that, for all X, Y, such a Z always exists in $\mathbf{M}$. Let $\Phi(X, Y, Z, W, w)$ be a $\Sigma_0^{1,b}$-formula that is the conjunction of the following conditions:

- $\max(X) \leq w \wedge \max(Y) \leq w \wedge \max(Z) \leq w + 1 \wedge \max(W) \leq w + 1$;
- $(Z_0 \equiv X_0 \oplus Y_0) \wedge (W_0 \equiv X_0 \wedge Y_0)$;
- $\forall u \leq w\, Z_{u+1} \equiv X_{u+1} \oplus Y_{u+1} \oplus W_u \wedge W_{u+1} \equiv \mathrm{TH}_{3,2}(X_{u+1}, Y_{u+1}, W_u)$.

Here X_u abbreviates the formula $u \in X$ (and similarly for Y, Z, W) and the propositional connectives $\oplus$ and $\mathrm{TH}_{3,2}$ are defined as in Section 1.1. The set W represents the string of carried numbers in the summation of $nb(X)$ and $nb(Y)$.

Claim 2 V_1^1 proves that

(i) $\forall X, Y \leq w \exists Z, W \leq w + 1\, \Phi(X, Y, Z, W, w)$;
(ii) $\forall X, Y \leq w \forall Z, Z', W, W' \leq w + 1\, (\Phi(X, Y, Z, W, w) \wedge \Phi(X, Y, Z', W', w)) \to (Z = Z' \wedge W = W')$.

Item (i) is proved by $\Sigma_1^{1,b}$-IND on w; for item (ii) $\Sigma_0^{1,b}$-IND suffices.

In this situation we say that the formula $\exists W \leq w + 1\, \Phi$ defining the graph of addition is $\Delta_1^{1,b}$ **provably** in V_1^1: it is itself $\Sigma_1^{1,b}$ but it is provably (in the theory) equivalent also to a $\Pi_1^{1,b}$-formula

$$\forall Z', W' \leq w + 1\, \Phi(X, Y, Z', W', w) \to Z = Z'.$$

This has the advantage that we can use Φ in a formula in both positive and negative occurrences without leaving the class $\Sigma_1^{1,b}$. The axioms in BASIC(#) concerning addition are proved by induction on w.

Multiplication is interpreted analogously, its graph being defined by a formula that is provably $\Delta_1^{1,b}$ in V_1^1, formalizing the long table multiplication. To write this down in detail is rather tedious but it brings no unwanted surprise complications and to check all axioms in BASIC(#) is straightforward using IND. It suffices to have IND for $\Sigma_1^{1,b}$-formulas, thanks to having the provably $\Delta_1^{1,b}$-definitions of the graphs of the functions in **M**. The details can be found in [274]. □

9.5 Bibliographical and Other Remarks

Secs. 9.1, 9.2 and 9.3 Buss's theories were introduced in his Ph.D. thesis [106], where he proved a number of fundamental results, often in the form of witnessing theorems. We shall touch upon this in Section 19.3. He also wrote several beautiful expositions of various relations between bounded arithmetic theories, propositional proof systems and computational complexity classes; see [116, 113, 119]. These topics are also treated in detail in [278].

Cook and Nguyen [155] developed in detail several two-sorted theories, and their propositional translations and also proved the relevant witnessing theorems. In the two-sorted set-up they used a function symbol $|X|$ for a function assigning to a set the least upper bound on its elements (or 0 if it is empty). (Buss [106] used instead set variables of the form X^t, where t is a term bounding a priori all elements of any set for which the variable stands). This is useful in the primarily proof-theoretic investigations of [106, 155] but for our arguments it is redundant.

The example formulation of the continuum hypothesis was taken from Scott [463]. An alternative to the language of arithmetic is the language of rudimentary sets of Smullyan [476] and Bennett [65] and it is, in fact, more suited for formalizations of logical syntax and of basic notions of computability.

The language $L_{BA}(\#)$ expands the language of Buss's first-order theories by the symbols for $bit(i,x)$, $len(x)$ and $(x)_i$; this is for convenience and it is harmless, as we discuss only propositional translations. These functions are Δ_1^b-definable provably in S_2^1 (see [106, 278]). The index 2 in the names of the theories S_2^i and T_2^i refers to the presence of the symbol # in $L_{BA}(\#)$; if that were left out we would indicate this by using the index 1 and would name the theories T_1^i and S_1^i. On the other hand, one can add a certain function (called $\omega_2(x)$) not majorized by any $L_{BA}(\#)$-term that would be indicated by the index 3, and we can go on and introduce still faster functions (but all below exponentiation); see [278].

Sec. 9.4 The mutual interpretations of a two-sorted structure in an $L_{BA}(\#)$-structure and vice versa, and Theorem 9.4.1 in particular, are from [274]. The mutual interpretations of the theories that this yields are often called the **RSUV isomorphism**, referring to the title of Takeuti's paper [487] where he treated the interpretations proof-theoretically. This was done independently by Razborov [432]. Other examples of pairs of corresponding two-sorted and one-sorted theories can be found in these papers or in [278].

10

Up to EF via the $\langle \dots \rangle$ Translation

In this chapter we shall consider several proof systems of a logical nature and we shall define theories that correspond to them in the sense of Section 8.6. There we showed that AC^0-Frege systems and $I\Delta_0(\alpha)$ (i.e. $I\Sigma_0^{1,b}$) satisfy the conditions (A)–(C) listed there and correspond to each other. Whenever such a correspondence exists between a proof system P and a theory T it has several important consequences for them, listed as (1)–(4) in Section 8.6, which we get then for free, meaning by arguments *identical* to those in Section 8.6.

10.1 $AC^0[m]$-Frege Systems

Let $m \geq 2$ be fixed. A Frege proof system $F(MOD_m)$ results from a Frege system F in the DeMorgan language on extending the language by the connectives $MOD_{m,i}$ from (1.1.3), $0 \leq i < m$, of an unbounded arity, and adding to F the following axiom schemes involving them:

- $MOD_{m,0}(\emptyset)$;
- $\neg MOD_{m,i}(\emptyset)$ for $1 \leq i < m$;
- $MOD_{m,i}(\overline{A}, B) \equiv [(MOD_{m,i}(\overline{A}) \wedge \neg B) \vee (MOD_{m,i-1}(\overline{A}, B) \wedge B)]$,
  where we put $0 - 1 := m - 1$ and $\overline{A}$ stands for a list $A_1, \dots, A_k$ of formulas (possibly empty).

Let $F_d(MOD_m)$ be the subsystems allowing only formulas of depth at most d, where using a MOD_m-connective increases the depth of a formula by 1 (similarly to the use of $\bigvee$ or $\bigwedge$).

When $m = 2$ it is customary to use, instead of $MOD_{m,1}$, the (unbounded) parity connective $\bigoplus$, to define $MOD_{m,0}$ as $\neg \bigoplus$ and to denote the proof system by $F(\oplus)$.

To define a suitable theory, we take $I\Sigma_0^{1,b}$ and add to it a new type of quantifier, a **modular counting quantifier** denoted by Q_m. It is bounded, and $Q_m y \leq t A(y)$ holds if and only if the number of $b \leq t$ for which $A(b)$ holds is not divisible by m. For $m \geq 3$, it is more convenient also to introduce the companion quantifiers $Q_{m,i}$,

$0 \leq i < m$, which express that the number of the witnesses $b \leq t$ is congruent to i modulo m. In the rest of the section we will keep to the notationally simplest case, $m = 2$, and will write just Q_2 for $Q_{2,1}$.

The logical axioms for handling these quantifiers shadow the above rules for the parity connective $\bigoplus$ (and for $\mathrm{MOD}_{m,i}$ in general). In particular, the following are additional logical axioms:

- $\forall y \leq t \neg A(y) \rightarrow \neg Q_2 y \leq t\, A(y)$;
- $A(0) \rightarrow Q_2 y \leq 0 A(y)$;
- $A(t+1) \equiv Q_2 y \leq t A(y) \rightarrow \neg Q_2 y \leq t+1\, A(y)$;
- $A(t+1) \not\equiv Q_2 y \leq t A(y) \rightarrow Q_2 y \leq t+1\, A(y)$.

Let us denote by $Q_2 \Sigma_0^{1,b}$ the class of bounded $L_{BA}(\alpha)$-formulas allowing the new Q_2 quantifier.

Lemma 10.1.1 *For all $d \geq 2$, there is a $Q_2 \Sigma_0^{1,b}$-formula $Sat_d(\oplus)(x, y, E, F)$ defining the satisfiability relation for $F(\oplus)$-formulas of depth at most d.*

We need to extend the $\langle \dots \rangle$-translation to handle the Q_2 quantifier. Set:

- $\langle Q_2 y \leq t(\overline{x})\, B(\overline{x}, y)\rangle_{\overline{n}} := \bigoplus(\langle B(\overline{x}, y)\rangle_{\overline{n},0}, \dots, \langle B(\overline{x}, y)\rangle_{\overline{n},m})$,
  where $m = t(\overline{n})$.

The following lemma is analogous to Lemma 8.2.1.

Lemma 10.1.2 *Let $A(\overline{x})$ be a $Q_2 \Sigma_0^{1,b}$-formula. Then there are $c, d \geq 1$ such that, for all $\overline{n} = (n_1, \dots, n_k)$,*

- $|\langle A\rangle_{\overline{n}}| \leq (n_1 + \cdots + n_k + 2)^c$,
- $\mathrm{dp}(\langle A\rangle_{\overline{n}}) \leq d$.

Moreover, $A(n_1, \dots, n_k)$ is true for all interpretations of the set-sort variables in A if and only if $\langle A\rangle_{\overline{n}} \in$ TAUT.

Finally, let $IQ_2 \Sigma_0^{1,b}$ be the theory defined as $I\Sigma_0^{1,b}$ but using the IND scheme for all $Q_2 \Sigma_0^{1,b}$-formulas. Now we are ready to state the correspondence.

Theorem 10.1.3 *$\mathrm{AC}^0(\oplus)$-Frege systems and the theory $IQ_2 \Sigma_0^{1,b}$ satisfy the conditions (A), (B) and (C) from Section 8.6 and correspond to each other.*

Proof The translation $\langle \dots \rangle$ extended above to handle the Q_2 quantifier satisfies (A) by Lemma 10.1.2. To verify (B) we need to explain how to handle the parity connective $\bigoplus$ in the construction of the chain of sets S_k and of the set G in Section 8.3 in the proof of Theorem 8.2.2.

In the construction of the chain we simply incorporate steps that will guarantee that the eventual structure will satisfy the axioms for handling the Q_2 quantifier. Thus we allow in the set F used in the construction all propositional formulas involving the $\bigoplus$ connective, and we add the following requirements:

- the formula $\neg \bigoplus(\emptyset) \in S_k$ for some k;
- if $\alpha \in S_k$ and $\alpha = \bigoplus(\overline{\beta}, \gamma)$ then some later S_ℓ, $\ell > k$, contains at least one of the formulas

$$\left(\bigoplus(\overline{\beta}) \wedge \neg\gamma\right) \quad \text{or} \quad \left(\neg \bigoplus(\overline{\beta}) \wedge \gamma\right).$$

Finally, the reflection principle for $F_d(\oplus)$ is proved by induction on the number of steps in a proof, using the $Q_2\Sigma_0^{1,b}$-formula $Sat_d(\oplus)$ from Lemma 10.1.1. □

A polynomial f over $\mathbf{F}_2$ is equal to 0 if and only if $\neg \bigoplus(C_1, \ldots, C_m)$ is true, where C_i is the conjunction of variables in the ith monomial of f. Similarly, polynomials over $\mathbf{F}_p$ are expressible using $\mathrm{MOD}_{p,i}$ connectives. Polynomials over finite prime fields can thus be viewed as depth-2 formulas in the language of AC$^0[p]$-Frege systems. Let us conclude by stating for the record an obvious polynomial simulation.

Lemma 10.1.4 *For p a prime,* AC$^0[p]$*-Frege systems p-simulate the polynomial calculus* PC$/\mathbf{F}_p$ *with respect to refutations of sets of polynomials.*

10.2 TC^0-Frege Systems

In this section we consider a particular extension of any Frege system in the DeMorgan language by new **counting connectives** $C_{n,k}$, for $0 \leq k \leq n$, whose meaning is given by

$$C_{n,k}(p_1, \ldots, p_n) = 1 \quad \text{if and only if} \quad \textit{the number of } p_i = 1 \textit{ is exactly } k,$$

and by the axiom schemes

- $C_{1,1}(p) \equiv p$,
- $C_{n,0}(p_1, \ldots, p_n) \equiv \bigwedge_{i \leq n} \neg p_i$,
- $C_{n+1,k+1}(\overline{p}, q) \equiv [(C_{n,k}(\overline{p}) \wedge q) \vee (C_{n,k+1}(\overline{p}) \wedge \neg q)]$, for all $k < n$,
- $C_{n+1,n+1}(\overline{p}, q) \equiv C_{n,n}(\overline{p}) \wedge q$.

The system will be denoted by FC for **Frege with counting**, and FC_d will denote the subsystem allowing in proofs only formulas of depth at most d, where using a $C_{n,k}$-connective increases the depth of a subformula by 1. Note that the $C_{n,k}$ connectives are easily definable using the threshold connectives from Section 1.1 and vice versa:

$$C_{n,k}(\overline{p}) \equiv (\mathrm{TH}_{n,k}(\overline{p}) \wedge \neg \mathrm{TH}_{n,k+1}(\overline{p})) \quad \text{and} \quad \mathrm{TH}_{n,k}(\overline{p}) \equiv \bigvee_{k \leq \ell \leq n} C_{n,\ell}(\overline{p}).$$

The FC_d systems thus serve as examples of TC^0-Frege systems.

There are various ways to define a theory corresponding to FC_d (see Section 10.6 for other variants) and we shall follow one not treated in the literature yet. As in the case of $IQ_m\Sigma_0^{1,b}$ we define the theory by adding to the logic language a new type of quantifier, a **counting quantifier**, denoted by $\exists^{=}$. We shall use it only as bounded:

- $Q^{=s} y \le tA(y)$ *holds* if and only if *the cardinality of* $\{b \le t \mid A(b)\}$ *is exactly* s.

Here s, t are terms not involving y. We also need to put forward some logical axioms for handling this new quantifier; they are analogous to the axioms about the $C_{n,k}$ connectives formulated above:

- $\forall y \le t \, \neg A(y) \to \exists^{=0} y \le t \, A(y)$;
- $(A(t+1) \wedge \exists^{=s} y \le t \, A(y)) \to \exists^{=s+1} y \le t+1 \, A(y)$;
- $(\neg A(t+1) \wedge \exists^{=s} y \le t \, A(y)) \to \exists^{=s} y \le t+1 \, A(y)$.

The class of bounded $L_{BA}(\alpha)$-formulas allowing the new $\exists^{=}$ quantifier will be denoted by $\exists^{=}\Sigma_0^{1,b}$.

The next two lemmas are analogous to corresponding lemmas in Section 10.1.

Lemma 10.2.1 *For all $d \ge 2$, there is a $\exists^{=}\Sigma_0^{1,b}$-formula $Sat_d(C)(x, y, E, F)$ defining the satisfiability relation for FC-formulas of depth at most d.*

The $\langle \dots \rangle$-translation can be extended to the new class of formulas straightforwardly, using the connectives $C_{n,k}$:

- $\langle \exists^{=s} y \le t \, B(\overline{x}, y) \rangle_{\overline{n}} \; := \; C_{m+1,k}(\langle B(\overline{x}, y) \rangle_{\overline{n},0}, \dots, \langle B(\overline{x}, y) \rangle_{\overline{n},m})$,

where $s = s(\overline{x})$ and $t = t(\overline{x})$ are two terms, $k = s(\overline{n})$ and $m = t(\overline{n})$.

Lemma 10.2.2 *Let $A(\overline{x})$ be a $\exists^{=}\Sigma_0^{1,b}$-formula. Then there are $c, d \ge 1$ such that, for all $\overline{n} = (n_1, \dots, n_k)$,*

- $|\langle A \rangle_{\overline{n}}| \le (n_1 + \dots + n_k + 2)^c$,
- $\mathrm{dp}(\langle A \rangle_{\overline{n}}) \le d$.

Moreover, $A(n_1, \dots, n_k)$ is true for all interpretations of the set-sort variables in A if and only if $\langle A \rangle_{\overline{n}}$ is a tautology.

The theory that we shall use in the correspondence will be denoted by $I\exists^{=}\Sigma_0^{1,b}$, and it is defined as $I\Sigma_0^{1,b}$ but allows the IND scheme for all $\exists^{=}\Sigma_0^{1,b}$-formulas.

Theorem 10.2.3 *FC_d-Frege systems, $d \ge 1$, and the theory $I\exists^{=}\Sigma_0^{1,b}$ satisfy the conditions (A), (B) and (C) from Section 8.6 and correspond to each other.*

Proof The translation $\langle \dots \rangle$ extended above to handle the bounded $\exists^{=}$ quantifier satisfies (A) by Lemma 10.2.2. The condition (B) is verified in a way similar to that in the proof of Theorem 10.1.3: we add to the requirements on the chain of sets S_k in the proof of Theorem 8.2.2 the following clauses, aimed at guaranteeing that the eventual structure will satisfy the axioms about the bounded $\exists^{=}$ quantifier. In the set F we allow formulas in the language of FC and add the following requirements:

- the formula $C_{0,0}(\emptyset) \in S_k$, for some k;

- each formula $C_{n,0}(p_1, \ldots, p_n) \equiv \bigwedge_{i \le n} \neg p_i$ is in S_k, for some k;
- if $\alpha \in S_k$ and $\alpha = C_{n+1,k+1}(\overline{\beta}, \gamma)$ then some later S_ℓ, $\ell > k$, contains at least one of the formulas

$$(C_{n,k}(\overline{\beta}) \wedge \gamma) \quad \text{or} \quad (C_{n,k+1}(\overline{\beta}) \wedge \neg\gamma).$$

The reflection principle for FC_d is proved by induction on the number of steps in a proof, using the $\exists^{=}\Sigma_0^{1,b}$-formula $Sat_d(C)$ from Lemma 10.2.1. □

Considering the theory using IND for $\exists^{=}\Sigma_0^{1,b}$-formulas makes it simpler to p-simulate it by TC^0-Frege systems. At the same time, however, it makes it harder to formalize some proofs in such a theory. As in the theory V_1^0, it is sometimes useful (see the examples in Section 11.7) to talk directly about sets encoding natural numbers. Having a model of the theory $I\exists^{=}\Sigma_0^{1,b}$ we can expand it by all subsets of all intervals $[0, n]$ definable by an $\exists^{=}\Sigma_0^{1,b}$-formula. Such an expansion will always satisfy the theory VTC^0 in the two-sorted language that extends both $I\exists^{=}\Sigma_0^{1,b}$ and V_1^0 and also contains the following axioms:

- a bounded CA for all $\exists^{=}\Sigma_0^{1,b}$-formulas;
- every subset of every interval can be enumerated in increasing order (see the next section for a formalization).

The theory VTC^0 is $\exists^{=}\Sigma_0^{1,b}$-conservative over $I\exists^{=}\Sigma_0^{1,b}$.

The circuit class TC^0 is strong enough to contain the definition of the ordering of two n-bit numbers, the definition of the sum of a list of m natural numbers of bit size $\le n$ and also the definition of the product of two such numbers. In addition, the theory VTC^0 is strong enough to prove the basic properties of these arithmetic operations.

Lemma 10.2.4 *Theory* VTC^0 *proves that:*

(i) if $Y = (X_1, \ldots, X_m)$ is a list of m numbers $X_i \le n$ then there exists $Z \le n|m|$ encoding their sum;
(ii) if $X, Y \le n$ are two numbers then there is a $Z \le 2n$ encoding their product.

In addition, the theory proves that

(iii) the definitions of the ordering, the iterated sum and the product satisfy the axioms of the non-negative parts of discretely ordered commutative rings.

10.3 Extended Frege Systems

The proof system EF and the theory V_1^1, which we shall discuss in this section, have been introduced already (Sections 2.4 and 9.3). We will use the satisfiability predicate for all formulas, not only for formulas of a bounded depth, and we need to be more specific about how the *Sat* formula is defined than just appealing to the general form (8.4.3).

Define the formula $Sat(x, y, E, F)$ by:

$$\exists G \leq y\, Sat^0(x, y, E, F, G),$$

where Sat^0 formalizes the following:

- *F is a DeMorgan formula of size $\leq y$ with $\leq x$ atoms, E is a truth assignment to the atoms and G is an assignment of truth values to all subformulas of F that equals E on the atoms, is locally correct (i.e. it respects the connectives) and gives the whole formula the truth value* 1.

Recall the subtheory U_1^1 of V_1^1 from Section 9.3. Let $Enum(X, x, Y)$ be a $\Sigma_0^{1,b}$-formula formalizing that $Y \subseteq [x] \times [x]$ is the graph of a partial function that enumerates the elements of $X \cap [x]$ in an increasing order. That is, if we denote the function by f, it holds that

- $X = \emptyset$ if and only if $f = \emptyset$,
- $X \neq \emptyset \rightarrow f(0) = \min X$,
- $\forall y < x\, X \cap (f(y), x] \neq \emptyset \rightarrow f(y+1) = \min(X \cap (f(y), x])$,

where writing $f(y)$ means implicitly that $f(y)$ is defined.

Lemma 10.3.1 *The theory U_1^1 proves that every set can be enumerated:*

$$\forall x, X \leq x \exists Y \leq x^2\, Enum(X, x, Y).$$

In addition, the theory $I\Sigma_0^{1,b}$ proves that such an enumerating function is unique.

Proof The formula $\exists Y \leq x^2\, Enum(X, x, Y)$ is $\Sigma_1^{1,b}$ and the statement can be obviously proved by induction on x. But in U_1^1 we have only LIND and we need to organize the induction more economically. Writing $Enum(X, u, v, f)$ for

- *f enumerates in increasing order the set $X \cap [u, v)$,*

we prove first that if we can enumerate a set on all intervals of size t then we can also enumerate it on all intervals of size $2t$,

$$(\forall u \leq x \exists f_u \leq x^2\, Enum(X, u, u+t, f_u))$$
$$\rightarrow \forall v \leq x \exists g_v \leq x^2\, Enum(X, v, v+2t, g_v),$$

by putting

$$z \leq t \rightarrow g_u(z) := f_u(z) \quad \text{and} \quad t < z \leq 2t \rightarrow g_u(z) := f_{u+t}(z).$$

Using this, we get the statement by induction up to $|x|$, i.e. by LIND.

The uniqueness of the enumerating function is proved by induction on r up to x, but only for a $\Sigma_0^{1,b}$-formula:

$$\forall z \leq r\, f_1(z) = f_2(z),$$

where f_1, f_2 are two enumerating functions. Hence it is derivable already in $I\Sigma_0^{1,b}$. □

Lemma 10.3.2 *The theory U_1^1 proves that*

$$\forall x, y, E, F\, \exists G \leq y Sat^0(x, y, E, F, G),$$

and the theory V_1^0 proves that such a G is unique:

$$(G \leq y \wedge G' \leq y \wedge Sat^0(x, y, E, F, G) \wedge Sat^0(x, y, E, F, G')) \rightarrow G = G'.$$

Proof Work in U_1^1 and assume that we have a formula $F \leq m$ and an evaluation $E \leq n$ of its atoms. By induction on $t = 0, 1, \ldots, \log_{3/2}(m)$, prove that

$$\forall z \leq m\ Size(F_z) \leq (3/2)^t \rightarrow \exists H \leq y Sat^0(x, y, E, F_z, H),$$

where F_z denotes the subformula of F with top node z and $Size(F_z)$ is its size. The former is $\Sigma_0^{1,b}$-definable; the latter is $\Delta_1^{1,b}$-definable by Lemma 10.3.1.

The induction step is easy using the trick from Spira's lemma 1.1.4, itself easy to prove in U_1^1 as we can count in the theory by Lemma 10.3.1.

The uniqueness is proved in V_1^0 by induction up to m, using that the evaluations have to obey the same local conditions. □

Theorem 10.3.3 *The extended Frege system EF and the theory V_1^1 satisfy the conditions (A), (B) and (C) and, with the formula Sat, also the condition (D), from Section 8.6. In particular, they correspond to each other.*

Proof (A) is obvious and (D) is provided by Lemma 10.3.2. The reflection principles $Ref_{EF,d}$ are proved using the *Sat* formula: by induction on the number of steps in an *EF*-proof P of a formula F and an assignment E to its atoms prove that

- *there exists an assignment E^{ext} to all extension axioms occurring in the first t steps of P such that the first t steps in P are satisfied by $E \cup E^{ext}$ (in the sense of the Sat formula).*

This is a $\Sigma_1^{1,b}$-formula (as *Sat* is) and $\Sigma_1^{1,b}$-IND implies that the last formula F also has to be satisfied by E.

The remaining condition (B), the p-simulation of V_1^1 by *EF*, can be obtained by an extension of the model-theoretic argument used in earlier sections. But perhaps this would stretch the demands on the logical background of the reader too far. The details can be found in [278, Sec. 9.4]. Instead we shall derive this simulation from the relation of V_1^1 to S_2^1 (Theorem 9.4.1) and a self-contained argument for S_2^1 using the Herbrand theorem which we shall present in Section 12.3. □

10.4 Frege Systems

To define a theory corresponding to Frege systems (in the sense that the conditions (A)–(D) from Section 8.6 hold) is rather delicate, and there are no natural examples. The qualification *natural* is subjective and some (maybe even most) experts would disagree with me that the known theories for F are *unnatural*. Let me clarify why I think so.

For the correspondence between a theory T and a proof system P, we require a T that formalizes smoothly the reasoning and arguments in combinatorics, algebra and logic. Only then may we hope that any argument (using concepts of some restricted logical or computational complexity accessible to T) that we may use to prove a combinatorial principle or the soundness of a proof system can be formalized in T, and we can thus obtain the consequences (1) and (2) from Section 8.6. Mathematical logic identifies the prevailing modes of reasoning and formalizes them using axiom schemes like induction, the least number principle, the collection scheme (i.e. replacement) or various choice principles, to name a few. It is then natural in my view to strive to have T axiomatized using principles of this kind rather than by some new ad hoc axioms.

Furthermore, the requirements (A)–(C) from Section 8.6 posed on T only concern its $\forall\Sigma_0^{1,b}$ consequences, and in the case of (D) its $\forall\Sigma_1^{1,b}$ consequences. Thus we can have two theories T and T', both corresponding to P, while T is much stronger than T' and proves more statements of higher quantifier complexity. That is desirable, as it means that we have more room to formalize intuitive arguments.

This is the reason why we have linked the $AC^0[m]$- and the TC^0-Frege systems with theories involving new quantifiers, that are natural from a logical point of view. If we just want to have some theory corresponding to F, we could take

$$V_1^0 \cup \{Ref_{F,d} \mid d \geq 1\}$$

or, less directly,

$$V_1^0 \cup \{\forall x, y, E, F\, \exists G \leq y Sat^0(x, y, E, F, G)\},$$

where the extra axiom is the formula from Lemma 10.3.2.

Instead of these choices, we will link to the F theory U_1^1, although it gives only a quasi-polynomial simulation. I think that is a small price to pay for the natural character of U_1^1.

Theorem 10.4.1 *The Frege systems F and the theory U_1^1 satisfy the conditions (A)–(D) of Section 8.6, using the formula Sat from Lemma 10.3.2 in the condition (D), and with (B) weakened to the following:*

(B′) *if U_1^1 proves $\forall x A(x)$, $A \in \Sigma_0^{1,b}$, then there are F-proofs of $\langle A \rangle_n$ of size $n^{(\log n)^{O(1)}}$. Moreover, these proofs can be constructed by an algorithm running in quasi-polynomial time.*

(We may formulate (B′) by saying that F quasi-polynomially simulates U_1^1.)

Proof As in Theorem 10.3.3, (A) is obvious and (D) is provided by Lemma 10.3.2. The reflection principles $Ref_{F,d}$ are proved as before using the *Sat* formula, but now we do not need to existentially quantify the extension variables. Let F_P be the conjunction of all the formulas in a proof P of F. By Lemma 10.3.2 a (unique) evaluation G of F_P on E exists. Then, by induction on the number of steps in P, show that G evaluates each step in P (a subformula of F_P) to 1: this uses Sat^0 and hence $\Sigma_0^{1,b}$-IND suffices.

For a model-theoretic proof of the condition (B′), we refer the reader to [278, Sec. 9.4]. □

Let us point out here informally where the bound in (B′) changes from polynomial to quasi-polynomial. In the argument the key step is to guarantee the induction for $\Sigma_1^{1,b}$-formulas (say, induction for $\exists Y \leq mB(x, Y)$ on $[0, m]$), and it comes down to two cases:

- either there is some $b < m$ such that we can enforce that $\exists Y \leq mB(b, Y)$ is true while $\exists Y \leq mB(b + 1, Y)$ fails; or
- there are circuits $D_0^b, \ldots, D_m^b$ that compute a witness for $\exists Y \leq mB(b + 1, Y)$ from a witness to $\exists Y \leq mB(b, Y)$, for all $b < m$.

Composing the circuits in the second item for $b = 0, \ldots, m - 1$ yields a circuit that, from a witness for $\exists Y \leq mB(0, Y)$, computes a witness for $\exists Y \leq mB(m, Y)$. The composite circuit will have size polynomial in max $|D_i^b|$ and m, and its depth will be larger than the depth of the individual D_i^b by a multiplicative factor m. In the case of LIND this is on the order of $\log n$ and it looks as though we get $O(\log n)$ depth circuits, i.e. polynomial-size formulas. However, when the LIND axioms used in a proof are nested, the depth of the circuits D_i^b may already be $\log n$ or more and we get, in general, circuits of depth $(\log n)^{O(1)}$ only.

This discussion leads to the following lemma describing a formal system p-simulated by Frege systems. It is not natural in the sense discussed above but it nevertheless offers, I think, the best formal system corresponding to F. When talking about a first-order proof in a theory we can assume without loss of generality that it is tree-like. For a tree-like proof in U_1^1 we define that it has **no nested applications of the** LIND **rule** if and only if no conclusion of a LIND inference is a not necessarily immediate ancestor (Section 3.1) of a hypothesis of another LIND inference. In other words, LIND inferences are arranged in the proof tree in a way such that none is above another.

Lemma 10.4.2 *A formal system allowing only U_1^1 proofs with no nested applications of the* LIND *rule corresponds to Frege system.*

In particular, if U_1^1 proves $\forall x A(x)$, $A \in \Sigma_0^{1,b}$, by a proof in which there are no nested LIND *inferences then there are F-proofs of $\langle A \rangle_n$ of size $n^{O(1)}$. Moreover, these proofs can be constructed by an algorithm running in polynomial time.*

On the other hand, U_1^1 proves the soundness of Frege proofs (item (C)) and the existence of an evaluation of a formula (item (D)) by proofs with no nested LIND *inferences.*

10.5 R-Like Proof Systems

Earlier we linked AC^0-Frege systems with the theory $I\Sigma_0^{1,b}$, and $AC^0[m]$- and TC^0-Frege systems with the extensions of this theory by new quantifiers. We could have used a stronger theory, V_1^0, expanding $I\Sigma_0^{1,b}$ by instances of the comprehension axiom CA (Section 9.3), which is quite convenient for formalizing some arguments. The reason why we have not done so is that by restricting further the complexity of formulas in the IND scheme, from the full class $\Sigma_0^{1,b}$ to one of its subclasses, we may obtain theories corresponding to fragments of the proof systems defined by bounding the depth by a specific value. In the presence of CA this is not possible: by composing instances of CA for (#-free) $\Sigma_1^b(\alpha)$-formulas we may deduce instances of CA for all $\Sigma_0^{1,b}$-formulas.

In this section we concentrate on the proof systems R, R(log) and R*(log) and the theories $T_1^1(\alpha)$ and $T_1^2(\alpha)$ (in Section 10.6 we give references for more theories and proof systems).

Because these R-like proof systems are designed primarily to prove DNF-formulas, i.e. to refute CNF-formulas, we need to be careful about the syntactic form of the bounded arithmetic formulas we want to translate. A **basic formula** is an atomic formula or the negation of an atomic formula. A **DNF$_1$-formula** is a formula in the language $L_{BA}(\alpha)$ that is built from basic formulas by

- first applying any number of conjunctions and bounded universal quantifiers,
- then applying any number of disjunctions and bounded existential quantifiers.

An example is the formula PHP(R) from Section 8.1 when written in prenex form:

$$\exists x \leq z + 1 \exists u \neq v \leq z + 1 \exists w \leq z \, [(\forall y \leq z \neg R(x,y)) \vee (R(u,w) \wedge R(v,w))].$$

Lemma 10.5.1 *Let T and P be any of the following pairs of a theory and a proof system:*

- $T = T_1^1(\alpha)$ *and* $P = \text{R}^*(\log)$ *or* $P = \text{R}$*;*
- $T = T_1^2(\alpha)$ *and* $P = \text{R}(\log)$.

Then the pair T and P satisfies the conditions (A) and (B) from Section 8.6, with (B) restricted to DNF$_1$*-formulas only.*

Further, if T is $T_2^1(\alpha)$ *or* $T_2^2(\alpha)$*, respectively (i.e. they allow the # function), the bound in (B) is quasi-polynomial:* $n^{(\log n)^{O(1)}}$.

Proof Let us just sketch the proof (see Section 10.6 for references with detailed proofs).

A $T_1^2(\alpha)$-proof in the sequent calculus formalism can be assumed to use only $\Sigma_2^b(\alpha)$-formulas. Each such formula translates into a Σ-depth-2 formula and the whole arithmetic proof translates into a sequence of polynomial-size *tree-like* $\text{LK}^*_{2+1/2}$-proofs. In addition, because we are translating a fixed arithmetic proof, there is a constant $c \geq 1$ such that each of these propositional proofs will have at most c formulas in a sequent.

Hence by Lemma 3.4.7 we can transform (by a p-time algorithm) such a proof into an $\text{LK}^*_{1+1/2}$-proof that is still tree-like and further into an $\text{LK}_{1/2}$-proof. But $\text{LK}_{1/2} = \text{R}(\log)$.

For the case $T^1_1(\alpha)$, we start one Σ-depth level lower and, using Lemma 3.4.7, we get an $LK^*_{1/2}$-proof, i.e. an $R^*(\log)$-proof. The R-proof is then obtained via Lemma 5.7.2. □

In Section 8.4 we defined the satisfiability relation for CNF-formulas by the formula Sat_2 in (8.4.1), and we used it to formalize the reflection principle Ref_R for R. We shall use these formulas below. In Lemma 10.5.1 we used $T^1_1(\alpha)$ for R theory as we had its definition ready. For the purpose of establishing the condition (C) from Section 8.6 we need to strengthen the theory a bit.

The class of formulas $\forall^{\leq}_1 \bigwedge(\vee)$ will look a bit ad hoc at first. The formulas in this class are built

- from basic formulas or from disjunctions of two basic formulas;
- by arbitrary conjunctions and bounded universal quantifications.

Let $\forall^{\leq}_1 \bigwedge(\vee)$-LNP be the least number principle from Section 9.4 for all these formulas.

Lemma 10.5.2 *Let T and P be any of the following pairs consisting of a theory and a proof system:*

- $T = T^1_1(\alpha) + \forall^{\leq}_1 \bigwedge(\vee)$–LNP *and* $P = \mathrm{R}$;
- $T = T^2_1(\alpha)$ *and* $P = \mathrm{R}(\log)$.

Then the pair T and P satisfies the condition (C) from Section 8.6 for the Sat_2 formula only (i.e. $d = 2$).

Proof Consider the first pair (the argument for the second is analogous). Assume $\neg Sat_2(n, m, E, F)$, i.e. that the truth assignment E for n atoms satisfies all the clauses of F. The last clause of an R-refutation P is empty and hence not satisfied by E, and by the LNP we can find the first clause D in P that is not satisfied by E: note that this is expressed by a $\forall^{\leq}_1 \bigwedge(\vee)$-formula:

$$\forall 1 \leq i \leq n\, (i \notin E \vee i \notin D) \wedge (n + i \notin E \vee n + i \notin D). \tag{10.5.1}$$

Finding D, however, leads to an immediate contradiction with the correctness of the resolution rule. □

The disjunctions of two basic formulas that we allowed above and used in (10.5.1) spoil the translation of $T^1_1(\alpha) + \forall^{\leq}_1 \bigwedge(\vee)$-LNP into R: one gets proofs in R(2) of Section 5.7. Similarly, the formula $\neg Sat_2$ from (8.4.1) is not a DNF_1-formula, but this is the case only because of the conjunctions of two basic formulas in Sat_2.

However, there is a way to circumvent this technical issue and establish a correspondence. The idea is that, in using the reflection principle to obtain the consequences of (A)–(C) in Section 8.6, we apply it to a specific proof, i.e. one for which we can introduce a name into the first-order language. Then the atomic formulas concerning it are translated by propositional constants, and hence a disjunction of two formulas where one talks about the proof translates into one literal only.

We shall formulate this as a separate lemma using model-theoretic language. The class $\forall_1^{\leq} \bigwedge$ is defined as $\forall_1^{\leq} \bigwedge(\vee)$ but the formulas are built only from basic formulas, i.e. no disjunctions of two basic formulas are allowed. Analogously to the earlier theory, $\forall_1^{\leq} \bigwedge$-LNP is the least number principle for all these formulas.

Let $\mathbf{M}$ be an arbitrary countable model of true arithmetic in a language expanding $L_{BA}(\alpha)$ by the names for any finite or countable set of relations on the standard model $\mathbf{N}$ (one will encode a proof later on). Let $n \in \mathbf{M}$ be any non-standard element and let $\mathbf{M}_n$ be the cut

$$\{u \in M \mid u < n^k, \quad \text{for some } k \in \mathbf{N}\}$$

with structure inherited from $\mathbf{M}$. Denote the language of $\mathbf{M}_n$ by L_n and note that $I\Delta_0(L_n)$ holds in the cut.

Lemma 10.5.3 *Let T and P be one of the following pairs of a theory and a proof system:*

- $T = T_1^1(L_n, \alpha) + \forall_1^{\leq b} \bigwedge$-LNP *and* $P = \mathrm{R}$;
- $T = T_1^2(L_n, \alpha)$ *and* $P = \mathrm{R}(\log)$.

Let $A(x)$ be a DNF_1*-formula. Then, for an arbitrary structure $\mathbf{M}_n$ defined as above the following two statements are equivalent:*

(i) the tautologies $\langle A \rangle_\ell$, $\ell \geq 1$, have no polynomial-size P-proofs;
(ii) there is an expansion of $\mathbf{M}_n$ to a model $(\mathbf{M}_n, \alpha)$ of T in which $\langle A \rangle_n$ fails.

Proof The proof shadows Ajtai's argument from Section 8.5. We shall consider the first pair, $T = T_1^1(L_n, \alpha) + \forall_1^{\leq} \bigwedge$-LNP and $P = \mathrm{R}$; the case of the second pair is analogous.

Assume first that the $\langle A \rangle_\ell$ do have polynomial-size R-proofs. Our language has a name for a relation $P(\overline{x}, y)$ such that, for each $\ell \geq 1$, $P(\overline{x}, \ell)$ defines a relation on $[\ell]^t$, for some fixed $t \geq 1$, that encodes an R-proof of $\langle A \rangle_\ell$. Hence $P(\overline{x}, n)$ encodes in $\mathbf{M}_n$ an R-proof of $\langle A \rangle_n$. As T proves the soundness of P, no expansion in item (ii) can exist.

We explain the opposite implication using a model-theoretic argument similar to that in Chapter 8, as mentioned in the proof of Theorem 8.2.2 in particular. Let F be the set of all clauses (to be denoted by C) formed from the literals occurring in $\langle A \rangle_n$, together with all negations of such clauses (to be denoted by $\neg C$). We shall construct a set $G \subseteq F$ such that:

1. the clauses of $\langle A \rangle_n$ are in G;
2. C or $\neg C$ is in G, for any $C \in F$;
3. if $C \in G$ then $\{\ell\} \in G$ for *some* $\ell \in C$;
4. if $\neg C \in G$, $\{\neg \ell\} \in G$ for *all* $\ell \in C$;
5. if the sequence of clauses $\langle C_0, \dots, C_t \rangle$ from F is defined in $\mathbf{M}$ by an L_n-relation symbol, for $t \in M_n$, then

- either there is minimum $i \leq t$ such that $\neg C_i \in G$, or
- $\{C_0, \ldots, C_t\} \subseteq G$;

6. there is no R-refutation Π definable in $\mathbf{M}$ by an L_n-relation such that all initial clauses in Π are from G.

The set G is built in countably many steps as before, arranging the closure conditions. At the beginning we know (by the hypothesis) that $\langle A \rangle_n$ has no R-refutation in M_n.

Finally, note that $\langle A \rangle_n$ fails because of the first condition and that the expansion is a model of T by the condition in item 5: it implies that the expansion will satisfy $\forall_1^{\leq} \bigwedge$-LNP. □

10.6 Bibliographical and Other Remarks

Sec. 10.1 The proof system and the theory in the section follow [278]. Bounded formulas with modular counting quantifiers were studied by Paris and Wilkie [389].

Sec. 10.2 The proof system FC is from [277]. It is linked in [277, 278] with the theory $(I\Sigma_0^{1,b})^{count}$, an extension of $I\Sigma_0^{1,b}$ by an axiom stating that for every $\Sigma_0^{1,b}$-formula $A(\overline{x}, y)$ and every z there is a function $f(\overline{x}, u)$ such that, for $\overline{n}$, m and some t,

$$f(\overline{n}, 0), f(\overline{n}, 1), \ldots, f(\overline{n}, t)$$

enumerates in increasing order the set $\{b \leq m \mid A(\overline{a}, b)\}$. The theory VTC^0 as defined here contains more axioms than is needed for Lemma 10.2.4 (in particular, one does not need the new $\exists^=$ quantifier); see Cook and Nguyen [155, Chapter IX] for several ways to define VTC^0.

An alternative proof system PTK in the sequent calculus formalism that is p-equivalent to FC was defined by Buss and Clote [121]; see Clote and Kranakis [145, 5.6.6] or Cook and Nguyen [155] for details and its properties.

Secs. 10.3 and 10.4 The (quasi-polynomial) correspondence between U_1^1 and F is from [277, 278]. Cook and Nguyen [155] defined several theories (VNC^1 being a prominent example) giving the *polynomial* correspondence but using an ad hoc axiom stating that the function evaluating *balanced monotone* formulas (an NC^1-complete function under AC^0-reductions of Buss [107]) is well defined. A related theory was defined first by Arai [23]. It is easy to see that U_1^1 restricted to proofs with un-nested applications of the LIND rule proves all these theories.

Sec. 10.5 The results for R, R(log) and R*(log) are from [288]; the simulation of R*(log) by R follows by the same argument as in [276]. Lauria [336] wrote a stand-alone proof for Lemma 5.7.3. The translation of R*(log)-refutations of narrow clauses into a narrow R-refutation was done by Buss, Kolodziejczyk and Thapen [128] (in the proof of Proposition 19 there). They also proved that random resolution, RR (Section 7.6), quasi-polynomially simulates the theory $T_2^1(\alpha) + \text{dWPHP}(\text{PV}, \alpha)$, where the additional axiom formalizes the principle that no p-time function f (represented by a function symbol from L_{PV}) with an oracle access to α can define a map from $[x]$ onto $[2x]$. This theory lies between the theories APC_1 and APC_2 of Section 12.6.

Beckmann, Pudlák and Thapen [62, Appendix A] treat proof-theoretically the more general situation discussed in Section 10.5, when any first-order sentences are also allowed in the place of basic formulas in the definition of the DNF_1-formulas.

We shall consider in Section 13.1 a class of formulas (to be called DNF_2-formulas), extending the class of DNF_1-formulas, which can still be meaningfully translated into propositional CNF formulas.

10.6.1 Further Remarks

The definition of $F(MOD_m)$ was taken from [278]. The modular counting quantifiers Q_m were considered by Paris and Wilkie [389].

To further illustrate the discussion about possible *natural choices* for the theories T in Section 10.4, let us explain another way to define a theory for the $AC^0[2]$-Frege systems, used by Cook and Nguyen [155]. If we can compute the parity (using the quantifier Q_2) then we can define, for each set, a function counting the parity of the set on initial intervals. In fact, this may be done uniformly in parameters, and it leads to the following notion. Let $R(\overline{x}, y)$ be a relation. A relation $S(\overline{x}, z)$ is a **2-cover** of R (this terminology is from [304, Sec. 13.2]) if and only if, for all $a \leq n$ and all parameters $\overline{b}$ from $[n]$,

$$S(\overline{b}, a) \quad \text{if and only if} \quad Q_2 y \leq a\, R(\overline{b}, y).$$

This equivalence allows to define S in terms of R using Q_2, *but also the other way around*: $Q_2 y \leq z\, R(\overline{b}, z)$ is equivalent to $S(\overline{b}, z)$. Hence, instead of using the parity quantifier we could add an axiom, as Cook and Nguyen [155] do, that each relation has a 2-cover on each interval, where the 2-cover is defined by a few natural conditions on how it changes its truth value locally.

More generally, given a circuit class $\mathcal{C}$, Cook and Nguyen [155] selected a function F: $\{0,1\}^* \rightarrow \{0,1\}^*$ that is computed by circuits from the class $\mathcal{C}$ and is AC^0-complete for $\mathcal{C}$: all other functions computed in $\mathcal{C}$ can be retrieved from F via AC^0-circuits (cf. [155] for the technical details). Then they established the correspondence between $\mathcal{C}$-Frege systems and a theory extending V_1^0 by an axiom

$$\forall X \leq x \exists Y \leq t(x)\, F(X) = Y.$$

The reader will find a number of examples of this construction in [155].

There are other pairs T, P for which a correspondence has been established and used fruitfully. Perhaps the most important of those omitted in this chapter are the correspondences between the subtheories $T_1^i(\alpha)$ of $T_1(\alpha)$ and AC^0-Frege systems of a specific depth. The proof systems occurring in this correspondence are better defined using the sequent calculus LK and the notion of Σ-depth from Section 3.4: $T_1^{i+2}(\alpha)$ corresponds to the system $LK_{i+1/2}$. This is from [276, 278]. Recall from Section 3.4 that R(log) is $LK_{1/2}$.

11

Examples of Upper Bounds and p-Simulations

In this chapter we use the correspondences between theories and proof systems established earlier to prove several length-of-proof upper bounds and p-simulations.

11.1 Modular Counting Principles in $AC^0[m]$-Frege

Fix $m \geq 2$. We shall consider m-**partitions** R of sets $[n]$: these are partitions that satisfy the condition that

- *all blocks of R have the same size m with possibly one exceptional block – the* **remainder block** *– whose size may be strictly between* 0 *and m.*

The size of the remainder block will be denoted by $rem(R)$, and we set $rem(R) := 0$ if R has no exceptional block (i.e. if all blocks have the size m).

A **modular counting principle** formalizes that no set $[n]$ with $m|n$ admits an m-partition R with $0 < rem(R)$. We shall denote the bounded formula formalizing it by $Count^m(x, R)$, x being the variable for n. Its exact form is not important as long as it is a faithful formalization, and we leave its choice to the reader.

Lemma 11.1.1 *Let $m \geq 2$. Then $IQ_m\Delta_0(R)$ proves $Count^m(x, R)$.*

Proof Work in $IQ_m\Delta_0(R)$ and assume $m|n$. Assume also without loss of generality that 1 is in the remainder R-block and $0 < rem(R) = i < m$. For $1 \leq t \leq n$, define $A_t \subseteq [n]$ by

$$i \in A_t \quad \text{if and only if} \quad i \in [n] \wedge \exists j \in [t]\ R(i,j).$$

Prove by induction on t that

$$Q_{m,i}x \in [n]\ x \in A_t.$$

As $A_n = [n]$, m cannot divide n, which is a contradiction. □

We shall be more specific about the propositional formulation of the principle than we were for the first-order formulation. For a fixed $m \geq 2$ and any $n \geq m$, let $Count_n^m$ be the propositional formula built from atoms r_e, one for each m-element subset $e \subseteq [n]$ (and hence we have at least one, as $n \geq m$) that is a disjunction of the following formulas:

- $\neg r_e \vee \neg r_f$, for each pair e,f of **incompatible blocks** e,f (denoted by $e \perp f$): $e \neq f \wedge e \cap f \neq \emptyset$;
- $\bigvee_{e:i\in e} r_e$, for each $i \in [n]$.

Lemma 11.1.1 and Theorem 10.1.3 yields the following statement.

Corollary 11.1.2 *The formulas $Count_n^m$ have polynomial-size $F_d(m)$-proofs, for some fixed $d \geq 2$.*

11.2 PHP in TC^0-Frege

Let $\mathrm{PHP}(z, R)$ be the formula from Sections 8.1 and 10.5 to which we have added a condition that R is a function; it is the disjunction of the formulas

- $\exists x \leq z + 1 \forall y \leq z \neg R(x, y)$,
- $\exists u \neq v \leq z + 1 \exists w \leq z\, R(u, w) \wedge R(v, w)$,
- $\exists w \leq z + 1 \exists u \neq v \leq z\, R(w, u) \wedge R(w, u)$.

We may prove by induction on $t = 0, \ldots, z + 1$ that

$$|rng(R \cap (\{0, \ldots, t\} \times \{0, \ldots, z\}))| = t + 1$$

but also that

$$|\{0, \ldots, z\}| = z + 1.$$

These statements are easily formalized using the $\exists^=$ quantifier:

$$\exists^{=t+1} y \leq z \exists x \leq t\, R(x, y) \quad \text{and} \quad \exists^{=z+1} u \leq z\, u = u.$$

They can be proved by induction in $I\exists^=\Delta_0(R)$.

The propositional translation PHP_n of $\mathrm{PHP}(z - 1, R)$ is the formula (1.5.1):

$$\neg \left[\bigwedge_i \bigvee_j p_{ij} \wedge \bigwedge_i \bigwedge_{j \neq j'} (\neg p_{ij} \vee \neg p_{ij'}) \wedge \bigwedge_{i \neq i'} \bigwedge_j (\neg p_{ij} \vee \neg p_{i'j}) \right]. \tag{11.2.1}$$

(The use of $z - 1$ is due to the discrepancy between the sets $\{0, \ldots, n\}$ and $[n]$ used in the arithmetic and the propositional formulas, respectively.)

Hence by Theorem 10.2.3 we get

Lemma 11.2.1 *The* PHP_n *formulas have polynomial-size* TC^0*-Frege proofs.*

11.3 Simulation of TC^0-Frege by Frege

In this section we want to show that Frege systems p-simulate the theory $I\exists^=\Sigma_0^{1,b}$, and obtain the simulation of TC^0-Frege systems by Frege systems as a corollary;

consequence (2) in Section 8.6. By Lemma 10.3.1 we know that in U_1^1 we can define functions enumerating sets in increasing order, and hence we can also count the cardinalities of sets. By Theorem 10.4.1, U_1^1 proves the soundness of FC and that Frege systems simulate U_1^1 (and thus also FC and TC^0-Frege systems), but unfortunately by *quasi-polynomial-size* proofs only. We shall show in this section that Frege systems can define counting functions and prove their main properties in polynomial size, and thus a p-simulation of FC actually exists.

We need to define by some DeMorgan propositional formulas the connectives $C_{n,k}(p_1,\ldots,p_n)$ from Section 10.2. The definition of the counting used in Lemma 10.3.1 yields only quasi-polynomial-size formulas. To improve upon this, Buss [108] used **carry–save addition**. Using it one may compute the sum of n n-bit numbers by a propositional formula (using the bits of the numbers as atoms) of size polynomial in n. A special case will yield the definition of $C_{n,k}$: interpret $p_1,\ldots,p_n$ as n one-bit numbers and use the $\log n$ bits $r_0,\ldots,r_{|n|-1}$ of the output of the carry–save addition formula to define the value of $\sum_i p_i$. Then we have

$$C_{n,k}(p_1,\ldots,p_n) \equiv \bigwedge_{j\in K_0} \neg r_j \wedge \bigwedge_{j\in K_1} r_j,$$

where K_0 and K_1 are the positions j where the bit of value k is 0 or 1, respectively. The construction makes it clear that the inductive properties of these connectives forming the axioms of FC in Section 10.2 have polynomial-size Frege proofs.

We are going to define a number of propositional formulas in atoms $p_1,\ldots,p_n$ that will represent various numbers (carries and partial sums) of the bit-length up to $|n|$ or define operations on them. For

$$u = 0,\ldots,|n|-1 \quad \text{and} \quad v = 0,\ldots,n/2^u - 1,$$

let a_{uv} and b_{uv} be abbreviations for the $|n|$-tuples of formulas a_{uv}^w and b_{uv}^w, respectively, for $w = 0,\ldots |n|-1$.

These formulas will be defined inductively, using the $|n|$-tuples of formulas

$$\overline{A}(\overline{q},\overline{r},\overline{s},\overline{t}) \quad \text{and} \quad \overline{B}(\overline{q},\overline{r},\overline{s},\overline{t}),$$

where each $\overline{q},\overline{r},\overline{s},\overline{t}$ is itself an $|n|$-tuple of atoms that define the carries and partial sums in the carry–save algorithm for addition. Omitting, for better readability, the overbars on A and B, these formulas are defined as

$$A := A_0(A_0(\overline{q},\overline{r},\overline{s}), B_0(\overline{q},\overline{r},\overline{s}), \overline{t})$$

and

$$B := B_0(A_0(\overline{q},\overline{r},\overline{s}), B_0(\overline{q},\overline{r},\overline{s}), \overline{t}),$$

where the (tuples of the) formulas A_0 and B_0 represent the sums modulo 2 and the carries, respectively, of the following three numbers:

$$A_0^w(\overline{q},\overline{r},\overline{s}) := q^w \bigoplus r^w \bigoplus s^w,$$
$$B_0^0(\overline{q},\overline{r},\overline{s}) := 0,$$

and, for $0 < w < |n|$,

$$B_0^w(\overline{q},\overline{r},\overline{s}) := (q^{w-1} \wedge r^{w-1}) \vee (q^{w-1} \wedge s^{w-1}) \vee (r^{w-1} \wedge s^{w-1}).$$

Note that the sum of the numbers $\overline{q},\overline{r},\overline{s}$ modulo $2^{|n|}$ is equal to the sum of A_0 and B_0. We define

$$a_{uv}^w = \begin{cases} p_v & \text{if } w = 0, \\ 0 & \text{if } w > 0, \end{cases}$$

put $b_{0v} := \overline{0}$ and, using the $|n|$-tuples of the formulas A and B (still written without the over-line), define

$$a_{u+1,v} := A(a_{u,2v}, b_{u,2v}, a_{u,2v+1}, b_{u,2v+1})$$

and

$$b_{u+1,v} := B(a_{u,2v}, b_{u,2v}, a_{u,2v+1}, b_{u,2v+1}).$$

Take a DNF-formula $C(\overline{e},\overline{f})$, $\overline{e},\overline{f}$ being $|n|$-tuples of atoms, that defines the sum of the two numbers determined by $\overline{e}$ and $\overline{f}$. Such a DNF-formula has size $2^{O(|N|)} = n^{O(1)}$ and it exists by Lemma 1.1.2. Then Frege systems can prove the usual properties of addition by verifying them for all assignments.

Using C we define:

$$c_{uv} := C(a_{uv}, b_{uv}).$$

These are the bits of $\sum_{v2^u \leq i < (v+1)2^u} p_i$ and, in particular, $c_{|n|-1,0}$ are the bits of the sum $\sum_i p_i$. Hence

$$r_k := c_{|n|-1,0}(p_1, \ldots, p_{k-1}, 0, \ldots, 0).$$

Using the properties of C we can verify in Frege systems that

$$p_k \to C(r_k, 0 \ldots 0, 1) = r_{k+1} \quad \text{and} \quad \neg p_k \to r_k = r_{k+1},$$

which follow by induction on u in

$$p_k \equiv (C(c_{uv}(p_k/0),\ 0 \ldots 0, 1) = c_{uv}),$$

where v is chosen to be the unique number such that $k \in [v2^u, (v+1)2^u)$.

The construction readily yields the following statement.

Lemma 11.3.1 *The size of the formulas $C_{n,k}$ is polynomial in n and the axioms of FC from Section 10.2 have polynomial-size Frege proofs.*

Corollary 11.3.2 *The proof system F polynomially simulates FC- and hence also* TC^0*-Frege systems.*

Lemma 11.2.1 then implies

Corollary 11.3.3 (Buss [108]) *The* PHP_n*-formulas* (11.2.1) *have polynomial-size Frege proofs.*

11.4 WPHP in AC^0-Frege

The $WPHP(s, t, R)$-formula is a variant of the $PHP(z, R)$-formula from Section 11.2, where we have changed the bounds to the quantifiers. The formula is the disjunction of the following three formulas:

- $\exists x \leq s \forall y \leq t \neg R(x, y)$;
- $\exists u \neq v \leq s \exists w \leq t\ (R(u, w) \wedge R(v, w))$;
- $\exists w \leq s \exists u \neq v \leq t\ (R(w, u) \wedge R(w, u))$.

We shall consider primarily its two versions

$$WPHP(2z, z, R) \quad \text{and} \quad WPHP(z^2, z, R).$$

The variant WPHP formalizes the **weak** PHP; the qualification *weak* refers to the fact that now s can be much larger than t rather than just equal to $t + 1$. Hence the principle is weaker and it ought to be easier to prove it.

We shall denote the propositional $\langle \dots \rangle$ translations of the two versions by

$$WPHP_n^{2n} \quad \text{and} \quad WPHP_n^{n^2},$$

respectively.

Lemma 11.4.1 *There is a relation S definable from a relation R by an* L_{BA}*-formula (i.e. one not including #) that is* $\Delta_1^b(R)$ *in* $S_1^1(R)$ *and such that* $S_1^1(R)$ *proves the following:*

$$(|m| = |n|^2 \wedge \neg WPHP(2n, n, R)) \rightarrow \neg WPHP(n^2, n, S).$$

Proof Work in $S_1^1(R)$ and assume that for some m, n it holds that $|m| = |n|^2$; now assume for the sake of contradiction that $WPHP(2n, n, R)$ fails.

Let F: $[2n] \rightarrow [n]$ be the function whose graph is R. When we use it in a term, say $t(F(x))$, an atomic formula like $t(F(x)) = s$ can be expressed by an L_{BA}-formula in a $\Delta_1^b(R)$ way:

$$\begin{aligned} t(F(x)) = s \quad &\text{if and only if} \quad \exists y \leq n\ (R(x, y) \wedge t(y) = s) \\ &\text{if and only if} \quad \forall y \leq n\ (R(x, y) \rightarrow t(y) = s). \end{aligned}$$

Hence we can use F in formulas without altering the $\Sigma_i^b(R)$-complexity of formulas $(i \geq 1)$.

Using F define the function F_2: $[4n] \to [n]$ by

$$F_2(x) := \begin{cases} F(F(x)) & \text{if } x \in [2n], \\ F(n + F(x)) & \text{if } x \in [4n] \setminus [2n]. \end{cases}$$

Repeating this construction for $t = 3, \ldots, |n|$ gives us the functions $F_t : [2^t n] \to [n]$ and eventually

$$F_{|n|}\colon [n^2] \to [n] .$$

The formal definitions of the functions F_t involve coding-length-t sequences of numbers from $[n]$. By the coding lemma 9.3.2 this can be done in $S^1_1(\alpha)$ if a number with bit-size $t|n| \le |n|^2$ exists; this is the role of m provided by the hypothesis of our lemma.

It is easy to prove that all the F_t are injective if F is. □

Let us recall, before the next lemma, the **Cantor diagonal argument** showing that the power set $\mathcal{P}(I)$ of a non-empty set I has a larger cardinality than the set itself. Assume for the sake of contradiction that

$$g\colon \mathcal{P}(I) \to I$$

is an injective function. Taking the *diagonal* set

$$D := \{i \in I \mid i \in rng(g) \wedge i \notin g^{(-1)}(i)\},$$

we get a contradiction: for $d := g(D)$;

$$d \in D \quad \text{if and only if} \quad d \notin D.$$

The condition $i \in rng(g)$ in the definition of D is there in order that $g^{(-1)}(i)$ is well defined and in the argument below, it would cause an increase in the quantifier complexity of the formulas used (in the induction, in particular). To avoid that, we shall assume first that g is also surjective (i.e. it is a bijection) and discuss the general case later.

Let *onto*WPHP(s,t,R) be the formula for the disjunction of WPHP(s,t,R) with the negation of the surjectivity condition

- $\exists y \le t \forall x \le s \neg R(x,y)$.

That is, an R violating the principle is the graph of a bijection between $[s]$ and $[t]$. The propositional translation of the formula for $s = n^2$ and $t = n$ will be denoted by

$$onto\text{WPHP}^{n^2}_n .$$

Our aim now is to prove the *onto*WPHP(z^2, z, R) principle in bounded arithmetic. Let us first explain the construction and the argument and only then worry about how it formalizes.

Assume that *onto*WPHP(n^2, n, R) fails and denote by G the function whose graph is R; it is a bijection

$$G\colon [n^2] \to [n] .$$

Thinking of a natural bijection between $[n^2]$ and $[n] \times [n]$ (similar to the bijection (9.1.1) between $\mathbf{N} \times \mathbf{N}$ and $\mathbf{N}$) we shall write G as a binary function $G(x, y)$ defined on $[n] \times [n]$, and define the inverse function $G^{(-1)}(z)$ for $z \in [n]$ as

$$(P_0(z), P_1(z)),$$

where

$$P_0(z) = x \quad \text{if and only if} \quad \exists y \in [n] G(x, y) = z$$
$$\text{if and only if} \quad \forall u, v \in [n](G(u, v) = z \to u = x),$$

and analogously for P_1. Note that this is a $\Delta_1^b(R)$-definition in L_{BA} (no #). We shall now iterate G in a similar way to F in the proof of Lemma 11.4.1 but this time faster. Set $G_1 := G$ and for $2 \le t \le |n|$ define the maps $G_t := [n]^{2^t} \to [n]$ by

$$G_t := (a_1, \dots, a_{2^t}) \in [n]^{2^t} \to G(b_0, b_1) \in [n],$$

where

$$b_0 := G_{t-1}(a_1, \dots, a_{2^{(t-1)}}) \quad \text{and} \quad b_1 := G_{t-1}(a_{2^{t-1}+1}, \dots, a_{2^t}) .$$

For $t = |n|$, this gives a bijection $G_{|n|}$ between $[n]^n$ and $[n]$, to which we can apply the Cantor diagonal argument and get a contradiction.

The first issue to tackle in the formalization is how to represent the elements of $[n]^n$. Coding them by numbers would require numbers of size n^n, and that would destroy the hope of getting from the arithmetic proof short propositional proofs. This issue is solved by thinking of the elements of $[n]^n$ as being represented by oracles for functions $\mu\colon [n] \to [n]$, $\mu(i)$ being the ith coordinate of the element in $[n] \times \cdots \times [n]$ (n-times) that μ represents.

The second issue is that when we represent the elements of $[n]^n$ as oracles we cannot talk about the function $G_{|n|}$, as we are considering only functions whose arguments are numbers, not sets or functions. The solution to this is that instead of talking about the map $G_{|n|}$ we shall talk about the values of the inverse function. That is, we shall define functions g_t on $[n] \times \{0, 1\}^t$ as follows: for $a \in [n]$ and $w \in \{0, 1\}^t$ put

$$g_t(a, w) := \text{ the } w\text{-th coordinate of the sequence } G_t^{(-1)}(a) \in [n]^{2^t},$$

where we think of $\{0, 1\}^t$ as being ordered lexicographically and representing $[2^t]$ ordered in the usual way. Using such maps we can represent the statement that G_t maps μ to a by saying that

$$\forall w \in \{0, 1\}^t \; \mu(w) = g_t(a, w).$$

Using $g := g_{|n|}$, we can then simulate the diagonal argument. Namely, define $\delta: [n] \to \{0, 1\}$ by

$$\delta(b) = 1 \quad \text{if and only if} \quad g(b, b) = 0.$$

If we had $d \in [n]$ to which $G_{|n|}$ maps δ then we would have

$$\delta(d) = 1 \quad \text{if and only if} \quad \delta(d) = 0,$$

getting a contradiction as before. The following lemma summarizes what we need to prove in order to be able to formalize this construction and the argument.

Lemma 11.4.2 *Assume that $|m| = |n|^2$ and that onto*WPHP(n^2, n, R) *fails. Let G_t and g_t be the functions defined earlier. Then (see Section 9.3):*

(i) $S^1_1(R)$ can $\Delta^b_1(R)$-define the functions g_t by L_{BA}-formulas (i.e. no #);
(ii) $S^3_1(R, \mu)$ proves that

$$\exists a \in [n] \forall w \in \{0, 1\}^{|n|}\ \mu(w) = g_{|n|}(a, w).$$

Proof For item (i), define $g_t: [n] \times \{0, 1\}^t \to [n]$ as follows: $g_t(a, w) = c$ if and only if

$$\exists u, v \le n^{10(t+1)}\ len(u) = len(v) = t + 1 \wedge (u)_0 = a \wedge (u)_t = c \wedge$$

$$[\forall i < t\, (w_{i+1} = 0 \to G(u_{i+1}, v_{i+1}) = u_i) \wedge (w_{i+1} = 1 \to G(v_{i+1}, u_{i+1}) = u_i)]\,.$$

The sequences u and v are unique, so the definition can be written also in a $\Delta^b_1(R, \mu)$ way.

For item (ii), prove by induction on $s = 1, \ldots, |n|$ (i.e. by LIND) that

$$\forall w \in \{0, 1\}^{|n|-s} \exists b \in [n] \forall v \in \{0, 1\}^s\ g_s(b, v) = \mu(w \frown v)$$

($w \frown v$ is the concatenation of the words w and v). The induction step from s to $s+1$ amounts to combining the two codes by one application of G.

The formula to which we apply the induction is $\Pi^b_3(\alpha)$ and, by Theorem 9.3.1, $\Pi^b_3(\alpha)$-LIND is the same as $S^3_1(\alpha)$. □

Hence the diagonal argument can be applied in $S^3_1(R)$ and, together with Lemma 11.4.1 (the construction there preserves the surjectivity), it yields

Corollary 11.4.3 (Paris, Wilkie and Woods [392]) *Theory $S^3_1(R)$ proves that*

$$|u| = |z|^2 \to \ onto\text{WPHP}(2z, z, R) \wedge onto\text{WPHP}(z^2, z, R).$$

We would like to get from the arithmetical proofs short proofs in R(log). We know how to translate proofs in $T^2_1(R)$ into R(log) (Lemma 10.5.1) but here we have proofs in $S^3_1(R)$. If we could use the # function then this would be simple: it is known that $S^3_2(R)$ is $\forall\Sigma^b_2(R)$-conservative over $T^2_2(R)$ (the conservativity argument needs the # function), and hence a proof of WPHP in $S^3_1(R) \subseteq S^3_2(R)$ implies that the statement is also provable in $T^2_2(R)$. But that ruins the upper bound as one gets only $n^{(\log n)^{O(1)}}$

bounds on the R(log) proofs, while we are aiming at a better, $n^{O(\log n)}$, bound. What we need is simply a proof of WPHP (assuming that $n^{\log n}$ exists, i.e. assuming that, for some m, $|m| = |n|^2$) in $T^2_1(R)$. The proof of the following lemma has a lot in common with how the conservativity of $S^3_2(R)$ over $T^2_2(R)$ is actually proved.

Lemma 11.4.4 *Theory $T^2_1(R)$ proves:*

$$|u| = |z|^2 \rightarrow \ onto\text{WPHP}(2z, z, R) \wedge onto\text{WPHP}(z^2, z, R).$$

Proof By (a statement completely analogous to Lemma 11.4.1 for surjective maps) it suffices to prove WPHP(n^2, n, R), assuming that $|m| = |n|^2$. Assume that f: $[n] \rightarrow [n^2]$ is the bijection given by R (it is convenient to have it from $[n]$ onto $[n^2]$ rather than the other way around). Denote $k := |n|$ and, in fact, assume without loss of generality that $n = 2^k$, identify $[n]$ with $\{0, 1\}^k$ and also identify $[n^2]$ with $[n] \times [n]$.

We shall write $f(x) = (f_0(x), f_1(x))$, with $f_i(x) \in \{0, 1\}^k$. When $b = (b_1, b_2, \dots) \in \{0, 1\}^*$, f_b stands for the composition $\cdots \circ f_{b_2} \circ f_{b_1}$ and it is a bijection on $[n]$.

As before, consider a complete binary tree of height k. Its leaves are addressed by paths (strings from $\{0, 1\}^k$) through the tree and its inner nodes are addressed by strings from $\{0, 1\}^{<k}$. Let μ be an oracle giving to each leaf a label from $[n]$.

For $w \in \{0, 1\}^j$ and $u \in [n]$, let *Correct*(u, w) be the $\Pi^b_1(\mu)$ formula

$$\forall v \in \{0, 1\}^{k-j} f_v(u) = \mu(w \frown v).$$

The universal quantifier is bounded by $n^{\log n}$, i.e. by m, and the formulas formalize that u is the label that is assigned to the node in T addressed by w when the labeling of the leaves is compressed by f, as in the proof of Lemma 11.4.2.

Now define the formula *Neil*(w), formalizing a property of the strings $w \in \{0, 1\}^k$, as follows:

$$\exists u_1, \dots, u_k \forall i < k \, [w_i = 1 \rightarrow Correct(u_i, w_1 \dots w_i 0)].$$

In words, at every node $w_1 \dots w_i$ where w continues to the right-hand child, u_{i+1} is a correct label for the left-hand child. The existential quantifier is again bounded by $n^{\log n}$.

Then induction on b can be used to show that *Neil*(w) holds for the string $1 \dots 1$, and a label for the root of T obtained. The induction step splits into two cases:

- if w ends in 0, i.e. it has the form $w'0$, then for *Neil*$(w'1)$ we add the label $\mu(w'0)$;
- if w ends in 1, i.e. it has the form $w'01 \dots 1$, then we combine $\mu(w)$ with the labels for $w'00$, $w'010$, etc. guaranteed by *Neil*(w) to get the label for $w'0$, which is what we need for $w + 1 = w'10 \dots 0$.

Note that *Neil*(w) is a $\Sigma^b_2(R)$-formula and hence the argument is indeed within $T^2_1(R)$. □

Lemmas 10.5.1 and 11.4.4 then entail the required upper bound.

Corollary 11.4.5 *The propositional formulas*

$$onto\mathrm{WPHP}_n^{2n} \text{ and } onto\mathrm{WPHP}_n^{n^2}$$

have size $n^{O(\log n)}$ proofs in R(log).

This upper bound to R(log)-proofs of the WPHP with n^2 can be significantly improved. Let $|x|^{(k)}$ denotes the k-times iterated length function $|\ldots|x|\ldots|$, and similarly let $\log^{(k)} n$ denote the k-times iterated logarithm.

Theorem 11.4.6 (essentially Paris, Wilkie and Woods [392]) *For any $k \geq 1$, $S_1^3(R)$ proves that*

$$|u| = |z| \cdot |z|^{(k)} \rightarrow onto\,\mathrm{WPHP}(z^2, z, R).$$

The propositional formulas onto $\mathrm{WPHP}_n^{n^2}$ *have size-$n^{O(\log^{(k)} n)}$ proofs in* R(log).

Proof Assume that $z = n$ and that we have m such that $|m| = |n| \cdot |n|^{(2)}$. By the construction above we can define functions G_t for $t \leq \log\log n$ only. But such a $G_{||n||}$ codes sequences (of numbers from $[n]$) of length $\log n$ and hence we can use this function to encode the sequences of length up to $\log n$ needed in the definition of $G_{|n|}$.

If only $|n| \cdot |n|^{(3)}$ exists then we use it to define $G_{|||n|||}$; using that we define $G_{||n||}$, and eventually $G_{|n|}$. A similar procedure is used for other fixed k. All maps are $\Delta_1^b(R)$-definable and so we stay with induction in the same theory. □

We need another argument to get rid of the surjectivity assumption.

Theorem 11.4.7 (essentially Maciel, Pitassi and Woods [348]) *$S_1^3(R)$ proves that*

$$|u| = |z|^2 \rightarrow \mathrm{WPHP}(z^2, z, R) \wedge \mathrm{WPHP}(2z, z, R).$$

Both the propositional formulas WPHP_n^{2n} *and* $\mathrm{WPHP}_n^{n^2}$ *have size $n^{(\log n)^{O(1)}}$ proofs in* R(log).

Proof By Lemma 11.4.1 it suffices to prove the weaker principle $\mathrm{WPHP}(z^2, z, R)$ in $S_1^3(R)$. Therefore work in $S_1^3(R)$ and put $z := n$.

For a sequence w from $[n]^{\leq |n|}$ (finite sequences of elements from $[n]$ having length at most $|n|$), define the relations

$$R(w, x, y) \subseteq [n^2] \times [n/2^\ell]$$

by induction on the length ℓ of the sequence w as follows:

- $R(\emptyset, x, y) := R(x, y)$;
- for $i < n$, $R(w \frown i, x, y) :=$

$$(\exists\, z < nR(x, z) \wedge R(w, in + z, y)) \wedge y < \frac{n}{2^{\ell+1}};$$

- $R(w \frown n, x, y) :=$

$$\exists\, z < nR(x, z) \wedge (\exists\, u \in [zn, (z+1)n),\ R(w, u\frac{n}{2^{\ell+1}} + y)) \wedge y < \frac{n}{2^{\ell+1}}.$$

By induction on ℓ, prove that

$$\neg\text{WPHP}(n^2, n, R) \rightarrow \exists w \in [n]^\ell \; \neg\text{WPHP}(n^2, \frac{n}{2^\ell}, R(w, x, y)).$$

This is straightforward, as any $S \subseteq [n^2] \times [n/2^\ell]$ either maps an interval of the form $[in, (i+1)n)$ into $[0, n/2^{\ell+1})$ or each such interval contains an element z that is mapped (i.e. has at least one S-value in) to $[n/2^{\ell+1}, n/2^\ell)$.

All the relations $R(w \frown i, x, y)$ and $R(w \frown n, x, y,)$ are $\Sigma_1^b(R)$-definable by an L_{BA} formula, and hence the witnesses to the Σ_1^b-quantifiers can be pulled together for $\ell = O(\log n)$ (via Lemma 9.3.2) to get a $\Sigma_1^b(R)$-definition of the *ternary* relation $R(w, x, y)$ for $|w| \leq |n|$.

For a $\Sigma_1^b(R)$-definable relation, S is the formula $\text{WPHP}(z^2, z, S)$ in $\Pi_2^b(R)$ and $\Sigma_3^b(R)$-LIND is enough to carry through the induction. For $\ell = |n|$ we get an obvious contradiction.

Now applying the argument via conservativity explained before Lemma 11.4.4 (the $\Sigma_3^b(R)$-conservativity of $\Sigma_3^b(R)$-LIND over $\Sigma_2^b(R)$-LIND), the proof can be carried in $T_2^2(R)$. The second part of the theorem then follows from Lemma 10.5.1. □

The following theorem was proved via an argument that is analogous to Nepomnjascij's theorem. We shall need it in Section 14.5 but will not prove it here. The second part follows from Theorem 8.2.2.

Theorem 11.4.8 (Paris and Wilkie [389]) *For all $k \geq 1$, $I\Delta_0(R)$ proves* $\text{WPHP}(|n|^k + 1, |n|^k, R)$. *Hence for all $k \geq 1$ there are $e, c \geq 1$ such that $\neg\text{PHP}_{(\log n)^k}$ has an $\text{LK}_{e+1/2}$-refutation of size at most n^c.*

11.5 Simulations by *EF*

We first use bounded arithmetic to prove Lemma 2.4.4, whose proof we postponed until now. Let us state it again.

Lemma 11.5.1 *$EF \geq_p SF$.*

Proof By Theorem 10.3.3 and consequence (2) from Section 8.6 it suffices to prove in V_1^1 the soundness of *SF*. That is easy, however; if P is a size m *SF*-proof of the formula F,

$$Prf_{SF}(n, m, F, P),$$

prove by induction on $t \leq m$ that

- *the first t steps in P are all tautologies.*

As the property that a formula H is a tautology is $\Pi_1^{1,b}$-definable,

$$\forall E \leq n \; G \leq m \; Sat^0(n, m, E, G, F)$$

(the universal set-sort quantifier ranges over all possible truth assignments and evaluations of the formula), $\Pi_1^{1,b}$-IND suffices. But V_1^1 proves all instances of the $\Pi_1^{1,b}$-IND scheme: induction on $[0,s]$ for $A(x) \in \Pi_1^{1,b}$ follows from induction for $\neg A(s - y) \in \Sigma_1^{1,b}$ on the same interval. □

An analogous argument applies to the system S^2F, the relativized Frege system with second-order substitution from Section 4.2. By Lemma 4.2.2 a relativized formula C is satisfiable if and only if the set $S(C)$ of ordinary propositional formulas, constructed there by a p-time algorithm from C, is satisfiable. Hence the logical validity of relativized formulas is also $\Pi_1^{1,b}$-definable.

Lemma 11.5.2 *$EF \geq_p S^2F$ and hence EF, SF and S^2F are all p-equivalent.*

We now turn to the implicit proof systems from Section 7.3. In Lemma 7.3.4 we stated the p-equivalence of ER and $[\mathrm{R}, \mathrm{R}^*]$ but proved there only $ER \leq_p [R, R^*]$. Now we prove the opposite simulation.

Lemma 11.5.3 *$ER \geq_p [\mathrm{R}, \mathrm{R}^*]$ and hence $ER \equiv_p [\mathrm{R}, \mathrm{R}^*]$.*

Proof Work in V_1^1 and assume that (P, C) is a size s $[\mathrm{R}, \mathrm{R}^*]$-refutation of a CNF-formula F, of size m and built from n atoms. By Lemma 7.3.1 we may assume that C is a circuit that on every assignment computes a clause over the $2n$ literals (together with information on how it was inferred) and that P is an R-proof of the statement that these clauses, when ordered by the lexicographic ordering of the input strings, form an R*-refutation of F.

We identify the strings from $\{0,1\}^{\leq t}$ with the nodes in the complete binary tree of depth t. Without loss of generality we may assume that the inputs to C come from $\{0,1\}^{\leq t}$, $t \leq s$, and that $C(w)$ for $w \in \{0,1\}^{\leq t}$ is the clause sitting at node w in the tree. Hence C computes from the empty word the empty clause, and from a word of length t some initial clause, i.e. some clause of F.

Let E be any truth assignment.

Claim *There exists a path Q in the proof tree from the root to a leaf such that E falsifies all clauses on Q.*

By induction on r, prove that there exists such a *partial* path of length r; this is a $\Sigma_1^{1,b}$-statement and hence V_1^1 proves an induction for it. The induction step follows from the correctness of P.

Having such a Q, we get a clause in F (the clause labeling the leaf in Q) that is falsified by E; hence no E satisfies F, and P is sound. □

Note that if we wanted to prove analogously the soundness of $[R, R]$ we would need to find the lexicographically first $Q \in \{0,1\}^t$ leading to a clause falsified by E. Such a principle is not apparently available in V_1^1 and seems to need $\Sigma_2^{1,b}$-IND.

11.6 Subexponential Simulations by AC^0-Frege Systems

This section is devoted to proofs of the two statements in Theorem 2.5.6 about subexponential simulations of TC^0-Frege systems and of Frege systems by AC^0-Frege systems. Let us recall the statement of the theorem:

- *Let P be a TC^0-Frege system or a Frege system in the DeMorgan language and $d_0 \geq 0$. Then for every $\epsilon > 0$ there is a $d \geq d_0$ such that, for any formula A of depth* $dp(A) \leq d_0$, *it holds that*

$$\mathbf{s}_{F_d}(A) \leq 2^{\mathbf{s}_P(A)^\epsilon}.$$

The case of TC^0-Frege systems (the first statement in Theorem 2.5.6) is much easier and uses the fact underlying Theorem 11.4.8 that for any fixed $t \geq 1$ there are $\Delta_0(\alpha)$-definitions $\Phi_t(\alpha, a)$ of the counting functions for $\alpha \subseteq [0, (\log a)^t]$ such that $I\Delta_0(\alpha)$ proves their properties. Hence there are also size $n^{O(1)}$ AC^0-definitions of the connectives $TH_{(\log n)^t, k}$, whose properties can prove an AC^0-Frege system in size $n^{O(1)}$.

Given a TC^0-Frege proof of size s, for any $k \geq 1$ we can write $s = (\log m)^k$ and simulate the proof by an AC^0-Frege proof of size $m^{O(1)} = 2^{O(s^{1/k})}$. As $k \geq 1$ can be arbitrarily large, the statement follows.

Our strategy for proving the second statement in Theorem 2.5.6, about unrestricted Frege systems (tacitly in the DeMorgan language), is to prove in the theory $T_1(\alpha)$ (i.e. with no # function) for all constants $c \geq 1$ the reflection principle for Frege proofs P satisfying the following conditions on the size $|P| = s$ and the logical depth $\ell dp(P) = \ell$:

$$s \leq |b|^c \quad \text{and} \quad \ell \leq c\|b\|. \tag{11.6.1}$$

The parameter b serves only as a bound and it will play a role in the formalization of the argument in bounded arithmetic.

By the balancing lemma 2.3.3 every Frege proof of size s_0 of a formula of logical depth ℓ_0 can be transformed by a p-time function into a proof of the same formula that has size $s \leq s_0^{O(1)}$ and logical depth $\ell \leq O(\log s_0) + \ell_0$. Further, a depth-$d_0$ formula of size at most s_0 has logical depth at most $d_0(\log s_0)$. Putting this together: if a depth-d_0 formula A has a Frege proof of size $s_0 \leq |b|^{c_0}$, then it also has a proof of size $s = |b|^{O(c_0)}$ and logical depth $\ell \leq O(\log s_0) + \ell dp(A) \leq O((c_0 + d_0)||b||)$, and such a proof will satisfy (11.6.1) for some $c \geq 1$.

The structure of the proof of the theorem is then the following. Assuming that A has a Frege proof of size s_0, it has also a proof of size $s = s_0^{O(1)}$ using only balanced formulas (in the above sense) of logical depth ℓ. For an arbitrary $c \geq 1$ we can write $s = |b|^c$ and $\ell \leq c\|b\|$ for some parameter b. The provability of the reflection principle for such proofs in $T_1(\alpha)$ yields, via Theorem 8.2.2, F_d-proofs of size $b^{O(1)}$ of its $\langle \dots \rangle$ translation for some $d \geq d_0$, which further yields an F_d-Frege proof of A of size $b^{O(1)} = 2^{O(s_0^{1/c})}$ via consequence (2) in Section 8.6. As $c \geq 1$ was arbitrary, the theorem follows. We note here that the construction will yield a bound for d as well: $d = O(c)$.

It remains to prove the reflection principle for proofs satisfying (11.6.1). Our strategy is the same as in earlier proofs of other reflection principles: find a definition of the satisfaction relation for the formulas that may occur in a proof and then use it to prove (by induction on the number of steps in the proof) that it proves a tautology.

Formulas, proofs and evaluations are represented by sets F, P, E, as in Section 8.4, and we shall use the formulas $Prf_F(x, y, z, P, F)$ (expressing the provability relations for formulas F of size $\leq y$, i.e. $F \leq y$) with x atoms and proofs of size $\leq z$ (i.e. $P \leq z$), and $Sat_d(x, y, E, F)$ or $Sat(x, y, E, F)$ (expressing the satisfiability for formulas F and evaluations E of x atoms (i.e. $E \leq x$). In $T_1(\alpha)$ we cannot quantify sets and hence we could not express claims such as *an evaluation of a formula exists*. The idea is that, whenever we would like to quantify a set, we first encode it by a number (via the coding Lemma 9.3.2) and then quantify this number. The parameter b plays a crucial role in this.

Let us now proceed formally. Let $P \leq s$ be a Frege proof of a depth-d_0 formula F (so that $F \leq s$ too) and let the logical depth of P be $\ell := \ell\text{dp}(P)$ and assume that s and ℓ satisfy (11.6.1) for some parameter b and constant c. We are going to define for $t = 1, 2, \ldots$ functions $Eval_t(b, E, F)$ that will compute, for formulas F satisfying

$$\ell\text{dp}(F) \leq t\|b\| \quad \text{and hence} \quad E \leq |b|^t, \quad F \leq |b|^{t+1},$$

the truth value of F under the assignment E. We shall show that it can be defined by a bounded formula and that $T_1(\alpha)$ can prove that $Eval_t$ respects the logical connectives; thus

$$Eval_t(b, E, F) = 1 - Eval_t(b, E, \neg F)$$

and

$$Eval_t(b, E, F) = Eval_t(b, E, F_1) \circ Eval_t(b, E, F_2),$$

where $F = F_1 \circ F_2$ and $\circ = \vee, \wedge$. We denote the formula by *Eval* rather than by *Sat* because of the additional parameter b.

To define $Eval_1(b, E, F)$ we work in $T_1(\alpha)$. Using the coding lemma 9.3.2 we find a number w_0 coding the assignment E to the atoms of F. There are at most $2^{\ell\text{dp}(F)} \leq |b|$ such atoms, so $len(w_0) \leq |b|$ and $w_0 \leq b^{O(1)}$. Further, by induction on the size of F, i.e. up to $2|b| = |b^2|$, prove that there exists w, $len(w) \leq 2|b|$, the evaluation of subformulas of F extending the evaluation w_0 of the atoms. Hence we take for the definition of $Eval_1(b, E, F) = v$ the formalization of the following:

- there exist $w_0, w \leq b^{O(1)}$ such that
  - w_0 codes the evaluation E of the atoms of F,
  - w is the evaluation of F extending w_0,
  - the output value of w is v.

This is a $\Sigma_1^b(\alpha)$-formula (but it is not strict).

For $t > 1$, assume that we have a definition of the graph of $Eval_{t-1}$, and define $Eval_t$ as follows. Find formulas $A(p_1, \ldots, p_{|b|})$ and B_i for $i \leq |b|$ such that

$$F = A(B_1, \ldots, B_{|b|}), \quad \ell\mathrm{dp}(A) \leq \|b\|, \quad \forall i \leq |b| \; \ell\mathrm{dp}(B_i) \leq (t-1)\|b\|. \quad (11.6.2)$$

Then we use $Eval_{t-1}$ to prove (using the coding lemma 9.3.2 again) that there exist w_0, $len(w_0) \leq |b|$, coding the values of $Eval_{t-1}(b, E, B_i)$ for $i \leq |b|$, and next we use $Eval_1$ to compute the value $Eval_t(b, E, F) := Eval_1(b, w_0, A)$. The formula defining $Eval_t(b, E, F) = v$ thus has the following form:

- there exists a sequence u, $len(u) \leq |b|$, of vertices in the tree of F, such that if we define B_i to be the subformula of F whose top node is the ith vertex in u, and the formula A to be the top part of F ending with vertices in u, the inequalities (11.6.2) hold, and
- there exists $w_0 \leq b^{O(1)}$ such that
  - for all $i \leq |b|$, the ith bit of w_0 codes the value of $Eval_{t-1}(b, B_i, F)$,
  - w is the evaluation of A extending w_0,
  - the output value of w is v.

Proving that $Eval_t$ respects the logical connectives is easy as it involves only the formula $Eval_1$ applied to the top part of F. Subsequently we can prove the soundness of a proof satisfying (11.6.1). Let us summarize this formally.

Lemma 11.6.1 *For any $c \geq 1$, $T_1(\alpha)$ proves the soundness of Frege proofs satisfying* (11.6.1).

This concludes the proof of Theorem 2.5.6.

Let us remark that the definitions of the graphs of all $Eval_t$ are $\Sigma_1^b(\alpha)$ and, in fact, provably $\Delta_1^b(\alpha)$ as the evaluations are unique. However, they are not *strict* $\Sigma_1^b(\alpha)$ because without the # function we cannot move a sharply bounded universal quantifier to a position behind a bounded existential quantifier. If we had a strict definition, the d in the theorem could be taken to be d_0, but that is impossible (as follows e.g. from Corollary 11.3.3 and also from Theorem 15.3.1 below).

11.7 Simulations of Algebro-Geometric Systems by Logical Systems

Our first example is a partial p-simulation of the **Nullstellensatz proof system** $\mathrm{NS}/\mathbf{F}_p$ over a finite prime field $\mathbf{F}_p$ in an AC^0-Frege system augmented by instances of the counting modulo-p principle $Count^p$ (Section 11.1) as new axioms. The qualification *partial* refers to the fact that the simulation does not apply to all sequences of systems $\mathcal{F}_n$ of polynomial equations in n variables over $\mathbf{F}_p$ but only to those obeying the following

size restriction *There is a constant $c \geq 1$ such that, for all $n \geq 1$, each polynomial in $\mathcal{F}_n$ has at most c monomials.*

The simulation will apply, in fact, to a larger class of polynomial systems; we shall discuss this after the construction. We shall present the construction for $p = 2$, as that is syntactically the easiest (the coefficients are represented by truth values) and explain at the end the difference for a general p.

Let $d \geq 0$ and $n \geq 1$. A monomial of degree d in n variables, say $x_{i_1} \cdots x_{i_d}$, can be represented by a number w coding the sequence $i_1, \ldots, i_d$, i.e. satisfying, by the coding lemma 9.3.2,

$$len(w) = d \quad \text{and}, \quad \forall j < d, \quad (w)_j = i_{j+1} \quad \text{and} \quad w \leq n^{10d}. \tag{11.7.1}$$

An alternative way (in the second-order world) would be to represent $x_{i_1} \cdots x_{i_d}$ by a set W such that

$$W \subseteq [n] \quad \text{and}, \quad \forall j \in [n], \quad j \in W \equiv j \in \{i_1, \ldots, i_d\}.$$

This looks more economical, as we do not use any numbers with value $\geq n^d$, but in fact such numbers can be avoided only when we are talking about one monomial. For the purpose of the proof system NS (and PC), we use a dense representation of polynomials, i.e. we list the coefficients of all monomials up to some fixed degree d (i.e. there are $\binom{n+d}{d} = n^{O(d)}$ of them). A polynomial would be thus represented by a binary relation on $[n^{O(d)}] \times [n]$, and the number n^d is needed anyway.

The Boolean axioms $x_i^2 - x_i = 0$ obey the size restriction and we shall assume that the polynomials $x_i^2 - x_i$ are included in all the systems $\mathcal{F}_n$. Assume we have an NS/$\mathbf{F}_2$-refutation P of $\mathcal{F} = \{f_1, \ldots, f_m\}$:

$$P: \sum_{i \in [m]} h_i f_i = 1, \tag{11.7.2}$$

where the degree of each $h_i f_i$ is at most d. We assume that $\mathcal{F}$ is represented by a binary relation $F \subseteq [m] \times [n^{10d}]$ such that

- the monomial represented by w appears in f_i with coefficient 1 if and only if $(i, w) \in F$.

We now take advantage of the fact that in formulating the reflection principle we are allowed to use *any* NP-definition of the provability predicate and, in particular, we can incorporate into the NP-witness any additional information about the proof that may help us in proving the reflection principle. In our case we shall represent the refutation P by two relations, H and R. The first, H, represents (11.7.2) in a straightforward way, and it is defined as follows:

- $H \subseteq [m] \times [n^{10d}]$ and a monomial w appears in h_i with coefficient 1 if and only if $(i, w) \in H$.

The second, R, contains the auxiliary information that we shall use later in the soundness proof. To define R we first need to define a ternary relation $Mon_P \subseteq$

$[m] \times [n^{10d}] \times [n^{10d}]$ by:

$$(i, w_1, w_2) \in Mon_P \quad \text{if and only if} \quad (i, w_1) \in H \wedge (i, w_2) \in F.$$

Then we can state the following.

- The relation R is a 2-partition of Mon_P with remainder $rem(R) = 1$ (see Section 11.1 for the definition) and such that
  - each block consists of pairs defining the same monomial, that is, two triples (i, u_1, u_2) and (j, v_1, v_2) such that $u_1 u_2 = v_1 v_2$,
  - the unique exceptional block of size 1 contains a pair (i, w_1, w_2) giving the constant monomial 1, i.e. $w_1 = w_2 = 1$.

A partition R satisfying the condition exists if and only if (11.7.2) is valid in $\mathbf{F}_2[\overline{x}]$: all monomials on the left-hand side have to cancel out except one constant term 1.

Now assume that $E \leq n$ is an $\mathbf{F}_2$-assignment to the n variables such that $f_i(E) = 0$ for all $i \in [m]$. Put

$$Mon_P(E) \ := \ \{(i, u, v) \in Mon_P \mid uv \ \text{ evaluates to one under } E\ \}.$$

On $Mon_P(E)$ we can define two 2-partitions:

- R_E, which is R restricted to $Mon_P(E)$;
- S_E, defined as follows:
  - For each $i \in [m]$, define a total pairing on the monomials of f_i that acquire the value 1 under E. (Here we use the size restriction: because of that we can define one such pairing S_i using definition by cases; there is no need to define the parity of any set.)
  - S_E pairs together two elements $(i, u, v), (i, u, w) \in Mon_P(E)$ for which S_i pairs v and w.

Lemma 11.7.1 *If E satisfies all the equations $f_i = 0$, $i \in [m]$, then $Count_t^2(T)$ is violated for some even number t and some relation T definable by a bounded formula from E, F, H, R.*

Proof Let $Mon_P \subseteq [s]$ for some $s \leq mn^{20d}$. Hence we also have $Mon_P(E) \subseteq [s]$. Take $t := 2s$ and define a 2-pairing T on $[t]$ as follows:

- if $w \in [s] \setminus Mon_P(E)$, pair w with $w + s$;
- on $[s] \cap Mon_P(E)$ define T as R_E;
- for $w_1, w_2 \in [s] \cap Mon_P(E)$, pair together $s + w_1$ and $s + w_2$ if and only if w_1, w_2 are paired by S_E.

The relation T is definable from R_E and S_E and hence from E, F, H, R, and it is easy to see that it is a 2-pairing with remainder $rem(T) = 1$. This contradicts the $Count^2$ principle. □

Corollary 11.7.2 *$I\Delta(\alpha) + Count^2$ proves the soundness of* NS/$\mathbf{F}_2$ *with respect to the refutation of all polynomial systems obeying the size restriction.*

Consequence (2) from Section 8.6 then yields

Corollary 11.7.3 *There is a $d_0 \geq 2$ such that the following holds. Assume that $\mathcal{F}$ is a polynomial system with m n-variable polynomials containing all the Boolean axioms and obeying the size restriction. Assume that $\mathcal{F}$ has an* NS/$\mathbf{F}_2$*-refutation of degree d.*

Then there is an F_{d_0}-refutation of $\mathcal{F}$ from instances of the $Count^2$ principles of total size at most $mn^{O(d)}$.

For NS over $\mathbf{F}_p$, we just need to consider p-partitions and the principle $Count^p$; otherwise the argument is the same. Let us also note that the size restriction condition is only used to obtain some $\Delta_0(\alpha)$-definitions of the partitions S_i. Such a definition may exist even if a polynomial has an unbounded number of monomials. For example, we can formulate the PHP principle using polynomial equations as the following unsatisfiable system, which we shall denote by $\neg\mathrm{PHP}_n$:

1. $\sum_j x_{ij} - 1 = 0$ for each $i \in [n+1]$;
2. $x_{ij}x_{ij'} = 0$ for all $i \in [n+1]$ and $j \neq j' \in [n]$;
3. $x_{ij}x_{i'j} = 0$ for all $i \neq i' \in [n+1]$ and $j \in [n]$.

While the polynomials in item 1 do not obey the size restriction, it is simple to prove in $I\Delta_0(\alpha)$ that if E satisfies all the equations in item 2 then exactly one x_{ij} in the polynomial $\sum_j x_{ij} - 1$ acquires the value 1. In other words, $Mon_P(E)$ will contain exactly two non-zero monomials in each such equation, and that suffices to define the relations S_i.

11.7.1 Cutting Planes

Now we turn to another example, simulations of the cutting planes proof system CP and its extension LK(CP) by Frege systems.

Let $\mathcal{C}$: $C_1, \ldots, C_m$ be clauses in n variables and let $L_1, \ldots, L_m$ be the integer linear inequalities that represent these clauses in the sense of (6.0.2). Think of each inequality as $n+2$ integers: the coefficients of the variables plus two absolute terms, one on the left-hand side and one on the right-hand side.

A CP-refutation of a set of initial inequalities $\mathcal{C}$ with k steps and of total size s can be represented by a ternary relation $P \subseteq [k] \times [n+2] \times [s]$:

$(i, j, t) \in P$ if and only if the jth coefficient in the ith step in P has the tth bit 1.

Here s is also the bound on the bit-size of the coefficients, and we can assume without loss of generality that one bit signals whether the coefficient is positive or negative.

To evaluate an inequality in the refutation thus means adding together $\leq s+1$ integers of bit-size at most s and comparing the result with another bit-size s integer. By the construction in Section 11.3 this can be done in F by counting formulas of size $s^{O(1)}$, and the arithmetic properties needed to prove the soundness of CP inference rules using such counting formulas can also be proved in polynomial size.

In fact, already by Lemma 10.2.4 TC^0-Frege systems suffice. So a TC^0-Frege system can prove the soundness of the refutation step by step and in size $s^{O(1)}$. Then the consequence (2) from Section 8.6 again yields a simulation.

Lemma 11.7.4 *Assume that $\mathcal{C}$: $C_1, \ldots, C_m$ are clauses in n variables and L_i, $i \in [m]$, are the integer linear inequalities representing the clauses C_i in the sense of (6.0.2). Assume further that the inequalities $L_1, \ldots, L_m$ have a CP-refutation with k steps and of size s.*

Then $\mathcal{C}$ has a TC^0*-Frege refutation of size $s^{O(1)}$.*

One can formulate a somewhat better estimate using an a priori estimate of the size of the coefficients.

Lemma 11.7.5 *Assume that $L_1, \ldots, L_m$ are inequalities in n variables such that the bit-size of every coefficient in any L_i is at most ℓ. Assume further that the system has a CP-refutation with k steps.*

Then there is another CP*-refutation of the same system that has $O(k^3(\ell + \log m))$ steps, and the bit-size of all coefficients in the refutation is bounded above by $O(k + \ell + \log m)$.*

This lemma allows one to express the size of the constructed TC^0-Frege proof as

$$O(k^3(\ell + \log m)[(n+2)(k + \ell + \log m)]^{O(1)}) = (n(k + \ell + \log m))^{O(1)}.$$

In Frege systems or even in TC^0-Frege systems we have no problem in simulating logical reasoning on top of CP.

Corollary 11.7.6 TC^0*-Frege systems p-simulate the system* R(CP) *and Frege systems p-simulate the system* LK(CP) *from Section 7.1.*

11.7.2 Semi-Algebraic Calculus

The next example concerns the proof system SAC from Section 6.4. Lemma 10.2.4 gave us a theory VTC^0 corresponding to TC^0-Frege systems that can handle the arithmetic operations on natural numbers, hence also on the integers **Z** and on the rationals **Q**. In particular, for every $d \geq 1$ there is a bounded formula $Eval_d(F, x, X, y, Y)$ in the language of VTC^0 such that VTC^0 proves that if F is (the dense notation for) a degree $\leq d$ polynomial in x variables and X is a list of x rational numbers $X_i \leq y$ (i.e. the bit-size is $\leq y$) then there exists a number Y that is the value of F on the assignment X_i. In addition, by the third item in Lemma 10.2.4, VTC^0 proves the usual properties of the evaluation of polynomials. Note that we have to restrict to a fixed d because VTC^0 is not known to be able to define an iterated product.

Lemma 11.7.7 *The theory* VTC^0 *proves:*

(i) *for every $d \geq 1$, the soundness of* SAC_d*-proofs in which only rational coefficients occur with respect to assignments from* **Q***;*

(ii) *the soundness of* SAC*-proofs in which only rational coefficients occur (no degree restriction) for Boolean assignments.*

Proof For the first statement, use the formula $Eval_d$ to prove by induction on the number of steps in an SAC proof of degree $\leq d$ that all lines are true.

The second statement follows by noting that a product of any number of Boolean inputs is expressible by the conjunction and hence the evaluation of polynomials on Boolean assignments reduces to the iterated sum only. □

Note that the restriction in the lemma to proofs in which only rational numbers occur is not severe, as only such numbers can be represented by a finite string and we are primarily interested in Cook–Reckhow proof systems.

We conclude by stating without proof two statements about the polynomial calculus, which are proved by arguments analogous to those discussed above. We need to be able to add a set of numbers and to multiply two numbers; in the first case modulo a prime and in the second case in the rationals (which can be reduced to integers).

Lemma 11.7.8

(i) *Let p be a prime. The systems* $\mathrm{PC}/\mathbf{F}_p$ *(Section 6.2 and* $\mathrm{R}(\mathrm{PC}/\mathbf{F}_p)$ *(Section 7.1) can be p-simulated by an* $\mathrm{AC}^0[p]$*-Frege system.*
(ii) *The systems* $\mathrm{PC}/\mathbf{Q}$ *and* $\mathrm{R}(\mathrm{PC}/\mathbf{Q})$ *can be p-simulated by* TC^0*-Frege systems.*

11.8 Bibliographical and Other Remarks

Sec. 11.1 Modular counting principles were considered first in our context by Ajtai [6, 7] and subsequently by Beame *et al.* [52], Riis [447], Buss *et al.* [124] and others (Section 15.7).

Sec. 11.2 I am not aware of a reference for Lemma 11.2.1.

Sec. 11.3 The polynomial-size simulation of counting in Frege systems is due to Buss [108]; the quasi-polynomial construction via U_1^1 is from [278]. The particular definition of the connectives $C_{n,k}$ via the DeMorgan formulas follows closely the construction in [278, Chapter 13], including the notation. A slightly different presentation is given in Clote and Kranakis [155]. The idea of carry–save addition goes back to von Neumann [501].

Short Frege proofs for a number of formulas can be obtained by first constructing a short TC^0-Frege proof using one of the theories that the system p-simulates and then applying the p-simulation from Corollary 11.3.2. For example, quasi-polynomial-size Frege proofs of the Kneser–Lovász theorem by Buss *et al.* [2] can be obtained in this manner (it uses only basic counting).

Unfortunately, one cannot just dismiss the various theories for Frege systems that we branded as *unnatural* in Sections 10.4 and 10.6 and say that you can always find a short TC^0-Frege proof (using one of the *natural* theories). For example, the original Cook and Reckhow [156] proof of PHP_n formalizes in U_1^1 and one gets quasi-polynomial-size Frege proofs, but the same argument does not seem to yield short TC^0-Frege proofs as well. The formalization of a proof of Frankl's theorem by Buss *et al.* [3] is another such example: while it works in U_1^1 and gives quasi-polynomial-size Frege proofs, the existence of short TC^0-Frege proofs is open.

Sec. 11.4 The first proof of the WPHP in $T_2(R)$ was due to Paris, Wilkie and Woods [392]. Statements 11.4.1–11.4.3 and 11.4.6 are based on their construction (we have followed [278] to an extent). The proof of Lemma 11.4.4 is due to Thapen. If the additional condition of surjectivity is left out, the theory $S_1^3(R)$ has to be replaced by $S_1^4(R)$ and the proof system R(log) by the Σ-depth 1 system $LK_{3/2}$ (Section 3.4).

The idea of the proof of Theorem 11.4.7, giving a sharper result, is due to Maciel, Pitassi and Woods [348]. They presented the proof combinatorially using only propositional logic; the proof via the formalization in bounded arithmetic is from my unpublished notes for the Prague–San Diego email seminar that we ran in the 1990s. It is open whether there are also R(log)-proofs of $WPHP_n^{n^2}$ of sizes $n^{\log^{(k)} n}$ for any $k \geq 1$, as there are for the surjective version, by Theorem 11.4.6. By not applying the conservativity at the end of the proof of Theorem 11.4.7 we get size $n^{O(\log n)}$ proofs but in $LK_{3/2}$.

Theorem 11.4.8 is a consequence of [389, Theorem 7] which concerned the WPHP for all Δ_0-relations. The uniformity of the argument giving the oracle version was verified in Impagliazzo and Krajíček [242, Fact 0.4]. Let us note that Paris and Wilkie showed also that $T_1(R)$ proves the ordinary (i.e. not the weak) PHP, but for small numbers: no relation R can be the graph of an injective function of $[|z|^k]$ into $[|z|^k - 1]$, for any fixed $k \geq 1$.

Sec. 11.5 Lemma 11.5.1 was first noted by Dowd [172].

Sec. 11.6 This section is based on Muller's [359] proof of the second part of Theorem 2.5.6.

Sec. 11.7 The simulation of NS by AC^0-Frege systems augmented by counting principles was pointed out (using a different argument) by Impagliazzo and Segerlind [248].

The p-simulation of CP by F goes back to Goerdt [202]; I have not found a reference for the p-simulation of CP by TC^0-Frege systems. Lemma 11.7.5 is due to Buss and Clote [121]; it is also well presented in Clote and Kranakis [145, Theorem 5.6.6 and Corollary 5.6.2].

The p-simulation of SAC (resp SAC_d) by TC^0-Frege systems in Lemma 11.7.7 appears to be new. In connection with the restriction to proofs in which only rational numbers occur, it may be interesting to note an example, found by Scheiderer [460], of a degree-4 polynomial with integer coefficients which is a sum of squares of polynomials over **R** but not of polynomials over **Q**.

11.8.1 Further Remarks

Cook and Soltys [479, 158] defined three theories of increasing strengths (LA, LAP and ∀LAP) in order to study the proof complexity of various linear algebra facts (e.g. the Caley–Hamilton theorem). This is partially motivated by Cook's question whether there are p-size Frege proofs of the formulas translating the commutativity of matrix inverses, $AB = I_n \rightarrow BA = I_n$. Subsequently Cook and Fontes [151] introduced theories corresponding (in the sense of the witnessing theorems) to the complexity classes $\oplus L$ and DET, whose complete problems include determinants over $\mathbf{F}_2$ and $\mathbf{Z}$, respectively. These theories interpret LAP. Some of the theories were separated by formulas in a higher quantifier complexity than zero by Soltys and Thapen [480].

Fernandes and Ferreira [180] formalized systematically the part of analysis in weak theories related to polynomial time and Ferreira and Ferreira [182] extended this by using theories formalizing counting (see also the survey in Fernandes, Ferreira and Ferreira [181]). The propositional applications of these formalizations have not yet been studied.

12
Beyond EF via the $\| \dots \|$ Translation

12.1 Cook's Theory PV and Buss's Theory S^1_2

There have been suggestions in the past to study feasible computations via a theory having function symbols for all p-time functions and axiomatized by all true universal statements (some workers have even suggested all true statements as axioms – but then why bother with a theory?). This is a rather unnatural idea, because, while the motivation is computational feasibility, even the language of the theory is not feasible in the sense that there is no natural way to associate its symbols with specific functions. Further, the set of true universal statements (even in a finite sub-language containing only a few suitable functions) is not even algorithmically decidable. This is analogous to Trakhtenbrot's theorem [490] that the set of sentences (in a rich enough finite language) valid in all finite structures is Π^0_1-complete. A subtler approach is to take a language whose symbols correspond to polynomial-time *algorithms* in some programming scheme. However, the problem with the undecidability of the set of true universal statements will still persist.

The theory that we introduce in this section, Cook's theory PV, has the additional feature that we accept as axioms only identities that are used to define new symbols from old ones (i.e. we omit the set of all true universal statements). There is one more important feature of the theory, which we shall discuss later.

Cook [149] based his theory on Cobham's theorem [146], which characterizes the class of p-time functions in a particular way not involving a machine model but using instead the scheme of **limited recursion on notation** (LRN).

Let $s_0(y)$ and $s_1(y)$ be two functions defined as

$$s_0(y) := 2y \quad \text{and} \quad s_1(y) := 2y + 1.$$

That is, they add one bit, 0 or 1 respectively, at the end of the string (represented by the number y). If g, h_0, h_1, ℓ are functions, a function f is said to be obtained from the 4-tuple by the LRN if it holds that

1. $f(\overline{x}, 0) = g(\overline{x})$,
2. $f(\overline{x}, s_i(y)) = h_i(\overline{x}, y, f(\overline{x}, y))$, for $i = 0, 1$,

3. $f(\overline{x}, y) \leq \ell(\overline{x}, y)$.

Theorem 12.1.1 (Cobham's theorem [146]) *The class of polynomial-time functions is the smallest class of functions that contains the constant 0 and the functions $s_0(y), s_1(y)$ and $x\#y$ and is closed under:*

- *the permutation and renaming of variables,*
- *the composition of functions,*
- *the limited recursion on notation.*

The definition of the theory PV (for polynomially verifiable) is rather complex, as its language, axioms and derivations are introduced simultaneously and in infinitely many steps (according to a notion of a rank). The theory is equational, i.e. its statements only assert that two terms are equal. Its definition is as follows.

1. The **function symbols of rank** 0 (the initial functions) are: the constant 0, the unary function symbols $s_0(y), s_1(y)$, the binary function symbol # used in Theorem 12.1.1, a new unary function symbol $Tr(x)$ and two new binary function symbols $x \frown y$ and $Less(x, y)$.

   (The symbol $\frown$ denotes concatenation; $Less(x, y)$ is a function that subtracts $|y|$ rightmost bits from x and $Tr(x)$ truncates x by one bit.)
2. The **defining equations of rank** 0 are:
   - $Tr(0) = 0$ and $Tr(s_i(x)) = x$, for $i = 0, 1$;
   - $x \frown 0 = x$ and $x \frown (s_i(y)) = s_i(x \frown y)$, for $i = 0, 1$;
   - $x\#0 = 0$ and $x\#s_i(y) = x \frown (x\#y)$, for $i = 0, 1$;
   - $Less(x, 0) = x$ and $Less(x, s_i(y)) = Tr(Less(x, y))$, for $i = 0, 1$.
3. The **inference rules** are:

   (a)
$$\frac{t = u}{u = t} \quad \frac{t = u \quad u = v}{t = v} \quad \text{and} \quad \frac{t_1 = u_1, \dots, t_k = u_k}{f(t_1, \dots, t_k) = f(u_1, \dots, u_k)},$$

   (b)
$$\frac{t = u}{t(x/v) = u(x/v)}.$$

   (c) Let $E_1, \dots, E_6$ be two sets of copies of the equations 1–3 from the definition of the LRN: three for f_1 and three for f_2, each in place of f. Then:
$$\frac{E_1, \dots, E_6}{f_1(\overline{x}, y) = f_2(\overline{x}, y)}.$$

4. A PV-**derivation** is a sequence of equalities $E_1, \dots, E_t$ in which each E_i is either a defining equation or is derived from some earlier equations by one of the PV-rules. A derivation has rank $\leq k$ if all the function symbols in it and all the defining equations in it are of rank at most k.
5. For a term t built from function symbols of rank $\leq k$, f_t is a function symbol of rank $k + 1$ and $f_t = t$ is a **defining equation of rank** $k + 1$.

6. For all function symbols $g, h_0, h_1, \ell_0, \ell_1$ maximum rank k and PV-derivations π_i, $i = 0, 1$, of rank k of the equations

$$Less(h_i(\overline{x}, y, z), z \frown \ell_i(\overline{x}, y)) = 0, \tag{12.1.1}$$

we introduce a new function symbol,

$$f = f_{<g, h_0, h_1, \ell_0, \ell_1, \pi_0, \pi_1>}.$$

It has **rank** $k + 1$, and the systems 1 and 2 before Theorem 12.1.1 defining f from g, h_i by the LRN are the **defining equations of rank** $k + 1$.

We shall denote the **language of** PV by L_{PV}.

The following lemma asserts that every p-time function can be represented by a PV function symbol and that PV can efficiently evaluate such functions. This is not obvious because of the requirement that before we can introduce into L_{PV} a symbol for a function defined by the LRN we must have a proof of the bounds (12.1.1). The proof of the lemma relies on an analysis of the proof of Theorem 12.1.1 (see [146]). Recall the dyadic numerals $\underline{n}$ from Section 7.4.

Lemma 12.1.2 *Let $F(x_1, \ldots, x_k)$ be any polynomial time function. Then there is a* PV*-function symbol $f(x_1, \ldots, x_k)$ such that, for every $m_1, \ldots, m_k$ and $n = F(m_1, \ldots, m_k)$,* PV *proves that*

$$f(\underline{m}_1, \ldots, \underline{m}_k) = \underline{n}.$$

Cook [149] considered an extension of the theory that allows propositional reasoning (the formal system was called PV1) and it was further extended to an ordinary first-order theory, PV_1, in [323]. All axioms of PV_1 are universal sentences and contain all equations $t = u$ provable in PV. In addition, PV_1 has new axioms replacing the induction axioms for open formulas. In particular, for any open formula $A(x)$, a new function symbol $h(b, u)$ is introduced by the conditions following:

1. $h(b, 0) = (0, b)$;
2. for $h(b, \lfloor u/2 \rfloor) = (x, y)$ and $u > 0$, put

$$h(b, u) := \begin{cases} (\lceil \frac{x+y}{2} \rceil, y) & \text{if } \lceil \frac{x+y}{2} \rceil < y \wedge A(\lceil \frac{x+y}{2} \rceil), \\ (x, \lceil \frac{x+y}{2} \rceil) & \text{if } x < \lceil \frac{x+y}{2} \rceil \wedge \neg A(\lceil \frac{x+y}{2} \rceil), \\ (x, y) & \text{otherwise.} \end{cases}$$

The function h simulates the binary search for a witness for the induction axiom for A. We include in PV_1 as an axiom the following (universal) formula:

$$(A(0) \wedge \neg A(b) \wedge h(b, b) = (x, y)) \rightarrow (x + 1 = y \wedge A(x) \wedge \neg A(y)).$$

To link PV with Buss's theories, denote by $S^1_2(\mathrm{PV})$ the theory defined in the same way as S^1_2 but in L_{PV}: it extends

$$\mathrm{BASIC}(L_{\mathrm{PV}}) := \mathrm{BASIC}(\#) \cup \{\text{all defining equations of PV}\}$$

by $\Sigma_1^b(L_{PV})$-LIND.

Theorem 12.1.3 (Buss [106]) *The theory $S_2^1(\mathrm{PV})$ is $\forall\Sigma_1^b(L_{PV})$-conservative over* PV_1.

The proof of the theorem relies on Buss's witnessing theorem and we shall not present it here (see [278]).

12.2 Herbrand's Theorem

Herbrand's theorem is a statement about witnessing existential quantifiers in logically valid first-order formulas of a certain syntactic form. Let L be an arbitrary first-order language (possibly empty) and let $A(x, y)$ be a quantifier-free L-formula. The simplest form of Herbrand's theorem says that if $\forall x \exists y A(x, y)$ is logically valid then there are terms $t_1(x), \dots, t_k(x)$ such that

$$A(x, t_1(x)) \vee \cdots \vee A(x, t_k(x)) \tag{12.2.1}$$

is already logically valid. Note that even if $L = \emptyset$ we have terms: the variables. This can be interpreted as saying that we can compute a witness for y from a given argument x by one of the k terms but not necessarily the same term for all x.

Many results in proof theory have a simple rudimentary version and also a number of more or less (often more rather than less) technically complicated stronger variants. Herbrand's theorem is no exception. The technically more difficult versions are, for example, those describing how to find the terms t_i from *any* first-order proof of $\forall x \exists y A(x, y)$ or formulations for formulas A that are not quantifier-free or are not even in a prenex form (the most cumbersome variant). Fortunately, we will not need these difficult results but only a slight extension of the above informal formulation.

The key difference of the variant we need from the one given above is that the *logical validity* of (12.2.1) is replaced by *propositional validity*. We define a quantifier-free L-formula B to be **propositionally valid** if and only if any assignment of propositional truth values 0, 1 to atomic formulas in B evaluates the whole formula to 1. The only requirement in this assignment is that the same atomic formulas get the same value, but the assignment has a priori no connection with the Tarski truth definition in some L-structure.

Not all logically valid formulas are also propositionally valid. For example, none of the equality axioms

$$x = x, \quad x = y \rightarrow y = x, \quad (x = y \wedge y = z) \rightarrow x = z \tag{12.2.2}$$

is propositionally valid, and neither are the equality axioms for a relation symbol $R(\overline{x})$ or for a function symbol $f(\overline{x})$ from L:

$$Eq_R(\overline{x}, \overline{y}) : \ \bigwedge_i x_i = y_i \ \rightarrow \ R(\overline{x}) \equiv R(\overline{y}) \tag{12.2.3}$$

and

$$Eq_f(\overline{x},\overline{y}) : \bigwedge_i x_i = y_i \rightarrow f(\overline{x}) = f(\overline{y}). \tag{12.2.4}$$

For example, you can give to $x = y$ the value 1 and to $y = x$ the value 0, or to all $x_i = y_i$ the values 1 and to $f(\overline{x}) = f(\overline{y})$ the value 0.

One advantage of Herbrand's theorem over the cut-elimination procedure is that one can give its *complete* proof using only the compactness of propositional logic.

Theorem 12.2.1 (Herbrand's theorem) *Let L be an arbitrary first-order language and let $A(\overline{x},\overline{y})$ be a quantifier-free formula. Assume that $\forall\overline{x}\exists\overline{y}A(\overline{x},\overline{y})$ is logically valid.*

Then there are, for $e \geq 0$ and $k \geq 1$,

- *equality axioms $Eq_j(\overline{u},\overline{v})$, $j \leq e$, of the form (12.2.2), (12.2.3) or (12.2.4) for some symbols of L,*
- *tuples of terms $\overline{r}^i_j(\overline{x})$, $\overline{s}^i_j(\overline{x})$ and $\overline{t}^i(\overline{x})$, for $j \leq e$, $v \leq a$ and $i \leq k$,*

such that

$$(\bigvee_{i,j} \neg Eq_j(\overline{r}^i_j(\overline{x}), \overline{s}^i_j(\overline{x}))) \vee \bigvee_i A(\overline{x}, \overline{t}^i(\overline{x})) \tag{12.2.5}$$

is propositionally valid.

Proof Assume for the sake of contradiction that the conclusion of the theorem is not true. Consider a theory T consisting of

- all instances of all the equality axioms $Eq(\overline{u},\overline{v})$ of the three forms (12.2.2), (12.2.3) and (12.2.4) for all symbols of L, and, for all tuples of terms $\overline{r}(\overline{x}), \overline{s}(\overline{x})$,

$$Eq(\overline{r}(\overline{x}), \overline{s}(\overline{x})),$$

- all instances of $\neg A$ for all tuples of terms $\overline{t}(\overline{x})$,

$$\neg A(\overline{x}, \overline{t}(\overline{x})).$$

Claim 1 T is propositionally satisfiable. That is, it is possible to assign to all atomic formulas occurring in T propositional truth values such that all formulas in T become satisfied.

This is the place where we will use the compactness of propositional logic. If T is not propositionally satisfiable, already some finite $T_0 \subseteq T$ is not. But that would mean that a disjunction of the negations of formulas in T_0 is propositionally valid. But such a disjunction is of the form (12.2.5), contradicting our assumption.

Now we define a first-order L-structure in which the sentence $\forall\overline{x}\exists\overline{y}A(\overline{x},\overline{y})$ fails. Let h be a truth assignment to atomic formulas occurring in T that makes all formulas in T true. Let A be the set of all L-terms $w(\overline{x})$. On A define the relation

$$u \sim v \quad \text{if and only if} \quad h(u = v) = 1 .$$

Owing to the (instances of the) equality axioms (12.2.2) in T it is an equivalence relation. In fact, owing to the (instances of the) equality axioms (12.2.3) and (12.2.4) it is a congruence relation for all symbols of L: the axioms Eq_R and Eq_f hold with $\sim$ in place of $=$.

This determines an L-structure $\mathbf{B}$ with the universe B consisting of all $\sim$-blocks $[u]$ of $u \in A$, $B := A/\sim$, and with L interpreted on B via T:

$$\mathbf{B} \models R([u_1], \ldots, [u_n]) \quad \text{if and only if} \quad h(R(u_1, \ldots, u_n)) = 1$$

and analogously for all function symbols f.

Claim 2 For all quantifier-free L-formulas $C(z_1, \ldots, z_n)$ and all $[u_1], \ldots, [u_n] \in B$, we have

$$\mathbf{B} \models C([u_1], \ldots, [u_n]) \quad \text{if and only if} \quad h(C(u_1, \ldots, u_n)) = 1.$$

The claim is readily established by the logical complexity of C and it implies that

$$\mathbf{B} \models \neg\exists\overline{y}A([x_1], \ldots, y_1, \ldots),$$

contradicting the hypothesis of the theorem. □

From Theorem 12.2.1, we get, without any additional effort, a similar statement for the consequences of **universal theories**, theories all of whose axioms are universal sentences of the form

$$\forall\overline{z}B(\overline{z}),$$

where B is quantifier-free.

Corollary 12.2.2 *Let L be an arbitrary first-order language and let T be a universal L-theory. Let $A(\overline{x}, \overline{y})$ be a quantifier-free formula and assume that $\forall\overline{x}\exists\overline{y}A(\overline{x}, \overline{y})$ is provable in T, i.e. that it is valid in all models of T.*

Then there are $e, a \geq 0$, $k \geq 1$,

- *equality axioms $Eq_j(\overline{u}, \overline{v})$, $j \leq e$, of the form* (12.2.2), (12.2.3) *or* (12.2.4) *for some symbols of L,*
- *axioms $\forall\overline{z}B_u(\overline{z}) \in T$, $u \leq a$,*
- *tuples of terms $\overline{r}^i_j(\overline{x})$, $\overline{s}^i_j(\overline{x})$ and $\overline{w}^i_v(\overline{x})$ and $\overline{t}^i(\overline{x})$, for $j \leq e$ and $i \leq k$,*

such that

$$(\bigvee_{i,v} \neg B_v(\overline{w}^i_v(\overline{x}))) \vee (\bigvee_{i,j} \neg Eq_j(\overline{r}^i_j(\overline{x}), \overline{s}^i_j(\overline{x}))) \vee \bigvee_i A(\overline{x}, \overline{t}^i(\overline{x})) \tag{12.2.6}$$

is propositionally valid.

Proof If T proves the formula, already a finite number of axioms $\forall\overline{z}B_v(\overline{z})$, $v \leq a$, from T suffices. Apply Theorem 12.2.1 to the formula

$$\forall\overline{x}\exists\overline{y}, \overline{z}_1, \ldots, \overline{z}_a\, A(\overline{x}, \overline{y}) \vee \bigvee_v \neg B_v(\overline{z}_v).$$ □

We have formulated a version of Herbrand's theorem having propositional validity because that is what we shall use in proving simulations of theories by proof systems. But now we formulate two corollaries of the theorem just in terms of first-order provability in universal theories. These serve as witnessing theorems, and they will be used in some arguments later on.

For simplicity of notation we shall consider just single quantifiers rather than blocks of similar quantifiers (this is without loss of generality).

Corollary 12.2.3 *Let T be a universal theory in a language L and let $\forall x \exists y A(x, y)$, where A is quantifier-free, be provable in T.*

Then there are $k \geq 1$ and L-terms $t_i(x)$, $i \leq k$, such that T proves

$$\bigvee_{i \leq k} A(x, t_i(x)). \tag{12.2.7}$$

Proof First-order logic includes the equality axioms and hence T proves that all instances of all such axioms, as well as of its own axioms, are true. What remains from the disjunction (12.2.6) in Corollary 12.2.2 is given by (12.2.7). □

Let us now assume that our formula is more complex than just $\forall\exists$, say it is a $\forall\exists\forall$-formula, i.e. a formula of the form

$$\forall x \exists y \forall z D(x, y, z), \tag{12.2.8}$$

with D quantifier-free. Let $h(x, y)$ be a new binary function symbol *not* in L. It is often called a **Herbrand function**. Then (12.2.8) is logically valid if and only if

$$\forall x \exists y D(x, y, h(x, y)) \tag{12.2.9}$$

is logically valid; in fact, a theory T in the language L proves (12.2.8) if and only if it proves (12.2.9). It is clear that the validity of the former in an L-structure implies the validity of the latter. But the opposite is also true in the following sense: if (12.2.8) were not true then there would be an a in the structure such that for each b we can find a c there such that $\neg D(a, b, c)$; hence, taking for $h(a, b)$ one such c, interprets the Herbrand function in a way such that (12.2.9) fails.

Combining this reasoning with Corollary 12.2.3 yields the next statement.

Corollary 12.2.4 (The KPT theorem [323]) *Let T be a universal theory in a language L and let $\forall x \exists y \forall z D(x, y, z)$ be provable in T where D is quantifier-free.*

Then there are $k \geq 1$ and L-terms

$$t_1(x), t_2(x, z_1), \ldots, t_k(x, z_1, \ldots, z_{k-1})$$

such that T proves

$$D(x, t_1(x), z_1) \vee D(x, t_2(x, z_1), z_2) \vee \cdots \vee D(x, t_k(x, z_1, \ldots, z_{k-1}), z_k). \tag{12.2.10}$$

Proof Think of T as a theory in the language $L \cup \{h\}$, with h the symbol for the Herbrand function corresponding to the formula. Then the hypothesis of the theorem implies by Herbrand's theorem that T proves (12.2.9) and, hence, a disjunction of the form

$$\bigvee_{i \le k} D(x, t'_i(x), h(x, t'_i(x))). \tag{12.2.11}$$

Modify this disjunction as follows. Find a subterm s occurring in (12.2.11) that starts with the symbol h and that has the maximum size among all such terms. It must be one of the terms $h(x, t'_i(x))$ sitting at a position z in one of the disjuncts of (12.2.11); say it is $h(x, t'_k(x))$. Replace all its occurrences in the disjunction by a new variable z_k. This maneuver clearly preserves the validity on all structures for $L \cup \{h\}$ that are models of T because we can interpret h arbitrarily. Note that by choosing the maximum-size subterm we know that it does not occur in any t'_i with $i < k$.

Now choose the next to maximum size subterm s of the required form and replace it everywhere by z_{k-1}. The subterm s is either in t'_k in which case we have just simplified t'_k but have not changed anything else, or it may be one of the $h(x, t'_i(x))$, say $h(x, t'_{k-1}(x))$. The subterm s then does not occur in any t'_i for $i < k - 1$ but it may still occur in t'_k. This will transform t'_k into a term $t''_k(x, z_{k-1})$ that may depend also on z_{k-1}. Hence the last two disjuncts on the disjunction will look like

$$\cdots \vee D(x, t'_{k-1}(x), z_{k-1}) \vee D(x, t''_k(x, z_{k-1}), z_k)$$

for some term t''_k with the variables shown.

Repeat this process as long as there is any occurrence of the symbol h. □

There is a nice interpretation of the disjunction (12.2.3) in terms of a two-player game, the so-called **Student–Teacher game**. Assume that

$$\forall x \exists y \forall z D(x, y, z)$$

is valid in an L-structure (this is usually applied to the standard model, so we may consider that the formula is true). Consider a game between Student and Teacher proceeding in rounds. They both receive some $a \in \{0, 1\}^*$, and the task of Student is to find $b \in \{0, 1\}^*$ such that $\forall z D(a, b, z)$ is true. They play as follows.

- In the first round Student produces a candidate solution b_1. If $\forall z D(a, b_1, z)$ is true then Teacher says so. Otherwise she gives Student a *counter-example*: some $c_1 \in \{0, 1\}^*$ such that $\neg D(a, b_1, c_1)$ holds.
- Generally, before the ith round, $i \ge 2$, Student has suggested solutions $b_1, \ldots, b_{i-1}$ and has received counter-examples $c_1, \ldots, c_{i-1}$. He sends a new candidate solution b_i and Teacher either accepts it or sends her counter-example.

The play may continue for a fixed number of rounds or for an unlimited number, as prearranged. Student wins if and only if he finds a valid solution.

Assume that the disjunction (12.2.11) is valid. Student may use the terms t_i as his strategy: in the first round he sends

$$b_1 := t_1(a_1).$$

If that is incorrect and he gets c_1 as a counter-example, he sends

$$b_2 := t_2(a, c_1)$$

in the second round and similarly in the later rounds. But, because (12.2.11) is valid in the structure, in at most the kth round his answer must be correct. Hence we get

Corollary 12.2.5 *Let T be a universal theory in a language L and let $\forall x \exists y \forall z$ $D(x,y,z)$, where D is quantifier-free, be provable in T. Let $\mathbf{M}$ be any model of T.*

Then there are $k \geq 1$ and L-terms $t_i(x, z_1, \dots, z_{i-1})$, $i \leq k$, such that Student has a winning strategy for the Student–Teacher game, associated with the above formula over $\mathbf{M}$, such that he wins in at most k rounds for every a. Moreover, his strategy is computed by the terms $t_1, \dots, t_k$ as described above.

12.3 The $\|\dots\|$ Translation

In Section 1.4 we presented the set of clauses $\text{Def}_C(\overline{x}, \overline{y})$ that define a circuit C: the set is satisfied if and only if $\overline{y}$ is the computation of C on an input $\overline{x}$. We considered there only circuits with one output, the last y-bit. Now we need to extend this notation to circuits which output multiple bits (i.e. strings). It will also be convenient to consider circuits with multiple string inputs (rather than combining one-string inputs). By $\text{Def}_C(\overline{x}_1, \dots, \overline{x}_r; \overline{y}, \overline{z})$ we denote the set of clauses whose conjunction means that $\overline{y}$ is the computation of C on the inputs $\overline{x}_i$ with output string $\overline{z}$.

The following statement formalizes in resolution the fact that the computations of circuits are uniquely determined by the inputs; it is easily proved by induction on the size of the circuit.

Lemma 12.3.1 *Let C be a size s circuit. Then there are size $O(s)$ resolution derivations of*

$$y_j \equiv u_j \quad \textit{and} \quad z_i \equiv v_i, \textit{ for all } i,j,$$

from the initial clauses

$$\text{Def}_C(\overline{x}_1, \dots, \overline{x}_r; \overline{y}, \overline{z}) \cup \text{Def}_C(\overline{x}_1, \dots, \overline{x}_r; , \overline{u}, \overline{v}).$$

Recall that $L_{BA}(\text{PV})$ is the language L_{BA} of bounded arithmetic augmented by all function symbols of L_{PV}; in particular, # is among them. Our aim is to define for all sharply bounded (i.e. Σ_0^b) $L_{BA}(\text{PV})$-formulas $A(x_1, \dots, x_k)$ a sequence of propositional formulas

$$\|A(x_1, \dots, x_k)\|^{n_1, \dots, n_k}$$

with the property that the formula is a tautology if and only if

$$\forall x_1 (|x_1| = n_1) \dots \forall x_k (|x_k| = n_k)\, A(x_1, \dots, x_k)$$

is true in $\mathbf{N}$.

There is one relation symbol in $L_{BA}(\text{PV})$ (besides the equality symbol $=$, which is always present): the ordering $x \leq y$. To avoid dealing with this extra case below we shall assume without loss of generality that $\leq$ is represented by a binary function $\leq (x, y)$ that is equal to 1 if $x \leq y$ and to 0 otherwise. So, without loss of generality, we can assume that we have only function symbols in $L_{BA}(\text{PV})$.

Every function symbol $f(x_1, \dots, x_r)$ from $L_{BA}(\text{PV})$ is interpreted in $\mathbf{N}$ by a polynomial-time function and, in fact, the introduction of the symbols into the language associates with every f a specific p-time algorithm computing it. In particular, with f is also associated a canonical sequence of circuits

$$C^f_{n_1,\dots,n_r}(\overline{x}_1, \dots, \overline{x}_r),$$

which computes f on inputs $\overline{x}_1, \dots, \overline{x}_r$ of lengths $n_1, \dots, n_r$, respectively. The length $|f(x_1, \dots, x_r)|$ of the function value may differ even if the lengths of the inputs are fixed, but $C^f_{n_1,\dots,n_r}$ has a (large enough) fixed number of output bits depending polynomially on just $n_1, \dots, n_r$ and allowing the expression of any possible value (under the input-length restriction).

Let $t(x_1, \dots, x_k)$ be an $L_{BA}(\text{PV})$-term. We associate with it a circuit $C^t_{n_1,\dots,n_k}$ by induction on the size of t, as follows.

- If t is just one function symbol f, $C^t_{n_1,\dots,n_k} := C^f_{n_1,\dots,n_k}$.
- If $t = f(s_1(x_1, \dots, x_k), \dots, s_r(x_1, \dots, x_k))$ and m_j are the output lengths of the circuits $C^{s_j}_{n_1,\dots,n_k}$ computing the terms s_j, then

$$C^t_{n_1,\dots,n_k}(\overline{x}_1, \dots, \overline{x}_k) :=$$

$$C^f_{m_1,\dots,m_r}(C^{s_1}_{n_1,\dots,n_k}(\overline{x}_1, \dots, \overline{x}_k), \dots, C^{s_r}_{n_1,\dots,n_k}(\overline{x}_1, \dots, \overline{x}_k)).$$

That is, the defining clauses are

$$\bigcup_{j \leq r} \text{Def}_{C^{s_j}}(\overline{x}_1, \dots, \overline{x}_k; \overline{y}_j, \overline{z}_j) \cup \text{Def}_{C^f}(\overline{z}_1, \dots, \overline{z}_r; \overline{u}, \overline{v})$$

(we omit the indices referring to the input sizes here in order to simplify the notation).

With this notation we define the $\| \dots \|$ **translation** by induction on the logical complexity of an open $L_{BA}(\text{PV})$-formula as follows.

1. If $A(x_1, \dots, x_k)$ is an atomic formula of the form

$$t(x_1, \dots, x_k) = s(x_1, \dots, x_k)$$

then set

$$\|A(x_1, \dots, x_k)\|^{n_1,\dots,n_k} :=$$

$$\bigwedge \text{Def}_{C^t_{n_1,\dots,n_k}}(\overline{x}_1, \dots, \overline{x}_k; \overline{y}, \overline{z}) \wedge \bigwedge \text{Def}_{C^s_{n_1,\dots,n_k}}(\overline{x}_1, \dots, \overline{x}_k; \overline{u}, \overline{v})$$

$$\to \bigwedge_i z_i \equiv v_i.$$

2. If $A(x_1, \dots, x_k)$ is the negation of an atomic formula

$$t(x_1, \dots, x_k) \neq s(x_1, \dots, x_k)$$

then set

$$\|A(x_1, \dots, x_k)\|^{n_1, \dots, n_k} :=$$

$$\bigwedge \mathrm{Def}_{C^t_{n_1, \dots, n_k}}(\overline{x}_1, \dots, \overline{x}_k; \overline{y}, \overline{z}) \wedge \bigwedge \mathrm{Def}_{C^s_{n_1, \dots, n_k}}(\overline{x}_1, \dots, \overline{x}_k; \overline{u}, \overline{v})$$

$$\to \bigvee_i z_i \not\equiv v_i.$$

3. The translation $\| \dots \|$ commutes with $\vee$ and $\wedge$.
4. If A is not in negation normal form (i.e. there are negations that are not in front of atomic formulas) then first apply the DeMorgan rules to put A into an equivalent negation normal form A' and then set $\|A\| := \|A'\|$.

Lemma 12.3.2 *For all open $L_{BA}(\mathrm{PV})$ formulas $A(x_1, \dots, x_k)$ and all $n_1, \dots, n_k$, the size of $\|A(x_1, \dots, x_k)\|^{n_1, \dots, n_k}$ is $\leq (n_1 + \cdots + n_k + 2)^c$, where c depends only on A.*

In fact, $\|A(x_1, \dots, x_k)\|^{n_1, \dots, n_k}$ can be constructed by a p-time algorithm from the inputs $1^{(n_1)}, \dots, 1^{(n_k)}$.

The following statement is a consequence of Lemma 12.3.1.

Lemma 12.3.3 *Let Eq_f be the equality axiom* (12.2.4) *for a symbol $f(x_1, \dots, x_k)$ from $L_{BA}(\mathrm{PV})$. Then the formulas*

$$\|Eq_f\|^{n_1, \dots, n_k, n_1, \dots, n_k}$$

have polynomial-size resolution derivations, meaning that all $z_i \equiv w_i$ are derived from the clauses of

$$\mathrm{Def}_{C^f}(\overline{x}_1, \dots, \overline{x}_k; \overline{y}, \overline{z}) \cup \mathrm{Def}_{C^f}(\overline{u}_1, \dots, \overline{u}_k; \overline{v}, \overline{w}) \cup \{x_{ij} \equiv u_{ij} \mid i, j\}.$$

(We have omitted the length bounds for better readability.)

Lemma 12.3.4 *Let $A(x_1, \dots, x_k)$ be any axiom (i.e. a defining equation) of* PV_1, *as introduced in Section 12.1. Then the formulas*

$$\|A\|^{n_1, \dots, n_k}$$

have size-$(n_1 + \cdots + n_k + 2)^c$ ER-proofs. In fact, such proofs can be constructed by a p-time algorithm from the inputs $1^{(n_1)}, \dots, 1^{(n_k)}$.

Proof The key ingredient of the argument is the specific polynomial-time algorithms associated with the function symbols from $L_{BA}(\mathrm{PV})$, as they are introduced via composition and limited recursion on notation. In the case of a composition, the circuit for the composed function is built from the circuits for the starting functions as in the definition of the circuits C^t for the terms t above. In the case of LRN they are obtained by composing the circuits for the starting functions in exactly the same way as for the recursion on notation.

Proving the axioms is simple for the initial functions (i.e. the defining equations of rank 0) as well as for the defining equations corresponding to a composition of functions or to LRN. The remaining axioms corresponding to item 3(c) in the definition of PV are proved by invoking Lemma 12.3.1.

The derivations of the axioms described above are, in fact, all within R: from the clauses of one or more sets Def_C of clauses defining various circuits we derive some statement. The reason why the whole derivation of $\|A\|^{n_1,\dots,n_k}$ is not in R as well is that the proof may contain axioms about functions not occurring in A and hence some sets Def_C for those functions are not among such sets occurring in $\|A\|^{n_1,\dots,n_k}$. That is, these sets act as extension axioms in the eventual ER derivation. □

12.4 PV, S^1_2 and ER

In this section we want to establish a correspondence between $T = \mathrm{PV}_1$ or $T = S^1_2$ and $P = ER$ using the $\| \dots \|$ translation, that is similar to the correspondence in Section 8.6. The soundness of a proof system P can be formulated as a $\forall\Pi^b_1\, L_{BA}(\mathrm{PV})$-formula, which we shall denote by RFN_P,

$$P(y,x) \rightarrow Sat(x,z),$$

where $P(y,x)$ is the function symbol for the characteristic function of the provability relation of P (Definition 1.5.1) and $Sat(x,z)$ is the function symbol for the characteristic function of the satisfiability relation.

Lemma 12.4.1 *Both* PV_1 *and* S^1_2 *prove the soundness of ER.*

Proof By Theorem 12.1.3, S^1_2 is $\forall\Sigma^b_1$-conservative over PV_1 and it is even the case that the formulas RFN_P are Π^b_1; hence it suffices to prove RFN_{ER} in S^1_2.

Work in S^1_2 and assume π that is an ER-proof of a formula τ. By induction on the number of steps in π (i.e. by LIND) prove that, for any given truth assignment $\overline{a} \in \{0,1\}^n$ to the atoms of τ, there exists an assignment $\overline{b} \in \{0,1\}^*$ to the extension atoms that satisfies the extension axioms in π. Then prove that all steps in π are true under $\overline{a}, \overline{b}$ (again by LIND), i.e. $\tau(\overline{a})$ is also true. □

We are ready for the key simulation theorem.

Theorem 12.4.2 (Cook [149]) *The proof systems* PV_1 *(or* S^1_2*) and ER correspond to each other in the following sense:*

(i) PV_1 *(or* S^1_2*) proves the soundness of ER, i.e.* RFN_{ER}.

(ii) *If* $A(x)$ *is an open* $L_{BA}(\mathrm{PV})$*-formula and* PV_1 *or* S^1_2 *prove* $\forall x A(x)$ *then there are polynomial-size ER-proofs of the formulas* $\|A\|^n$.

In fact, such proofs can be constructed by a p-time algorithm from $1^{(n)}$.

Proof The first part is provided by Lemma 12.4.1. For the second part we may assume, by the conservativity theorem 12.1.3, that PV_1 proves $A(x)$. As A is open, we may apply Herbrand's theorem 12.2.1 and its corollary 12.2.2.

The disjunction (12.2.6) in the corollary is a propositional tautology and thus has a fixed-size resolution proof. When we translate the disjunction by $\| \dots \|^n$ we are using this fixed proof as a template to get a derivation σ of the translation, of size proportional to the size of the translation. Because there is no existential quantifier in $A(x)$, the disjunction (12.2.6) contains term instances only in the negations of either the equality axioms or of the axioms of PV_1. But both are refutable in *ER* by polynomial-size (in fact, p-time constructible) derivations using Lemmas 12.3.3 and 12.3.4. Composing these refutations with σ we obtain the required *ER* proof of $\|A\|^n$. □

In accordance with the earlier terminology for the $\langle \dots \rangle$ translation we shall describe the situation in the second item of Theorem 12.4.2 by saying that PV_1 and S^1_2 **p-simulate** *ER*. More generally, for a theory in the language $L_{BA}(\mathrm{PV})$, P **p-simulates** T if, whenever T proves an open formula A, P admits p-size proofs of the formulas $\|A\|^n$ and these proofs can be constructed by a p-time algorithm from $1^{(n)}$.

The correspondence has consequences the same as consequences (1)–(4) in Section 8.6 in the case of the $\langle \dots \rangle$ translation. Let us now state it for the record, in order to be able to refer to it later. The arguments are identical to those in Section 8.6.

Corollary 12.4.3 *Assume that* T *is an* $L_{BA}(\mathrm{PV})$*-theory and* P *is a proof system such that*

- T *proves* RFN_P,
- P *p-simulates* T.

Then the following hold:

(1) *the provability in* T *implies polynomial upper bounds in* P;
(2) *for* P*-simulations, if* T *proves* RFN_Q *for some proof system* Q *then* P *p-simulates* Q;

(we have left out the third consequence from Section 8.6, linking lower bounds and the model theory of T*; this will be discussed in Part IV)*

(4) *the existence of super-polynomial lower bounds for* P*-proofs for any sequence of tautologies implies that* $\mathrm{NP} \neq \mathrm{coNP}$ *is consistent with* T.

Let us mention a variant of the $\| \dots \|$ translation aimed at proof systems below *ER*. If the language of T contains only function symbols corresponding to some natural circuit subclass C of $\mathrm{P}/poly$, and closed under composition, then all circuits whose defining sets Def_C occur in the propositional derivations will be from $\mathcal{C}$. Hence the derivations of the translations we construct will be not only in *ER* (i.e. in the circuit Frege proof system *CF* of Section 7.2) but, in fact, by unwinding the extension axioms we obtain proofs in AC^0-Frege systems or NC^1-Frege systems, and in similar subsystems of *CF*, as determined by $\mathcal{C}$.

12.5 G, G_i and G_i^*

We shall now extend the translation from open formulas to bounded $L_{BA}(\mathrm{PV})$-formulas using the quantified propositional formulas Σ^q_∞ from Section 4.1. The restriction to open formulas in the negation normal form in the last section (and the somewhat ad hoc treatment of basic formulas) was imposed in order to keep the propositional reasoning within resolution. Now we shall use (fragments of) the quantified propositional system G and we can thus drop this requirement.

For a formula $A \in \Sigma^b_\infty$, we need the following definitions.

1. If $A(x_1, \ldots, x_k)$ is an atomic formula of the form

$$t(x_1, \ldots, x_k) = s(x_1, \ldots, x_k)$$

then put

$$\|A(x_1, \ldots, x_k)\|^{n_1,\ldots,n_k} :=$$

$$\exists \overline{y}, \overline{z}, \overline{u}, \overline{v} \bigwedge \mathrm{Def}_{C^t_{n_1,\ldots,n_k}}(\overline{x}_1, \ldots, \overline{x}_k; \overline{y}, \overline{z})$$

$$\wedge \bigwedge \mathrm{Def}_{C^s_{n_1,\ldots,n_k}}(\overline{x}_1, \ldots, \overline{x}_k; \overline{u}, \overline{v}) \wedge \bigwedge_i z_i \equiv v_i.$$

2. The translation $\|\ldots\|$ commutes with $\neg$, $\vee$ and $\wedge$.
3. If $A(x_1, \ldots, x_k) = \exists x_{k+1} \leq |t(x_1, \ldots, x_k)| B(x_1, \ldots, x_k, x_{k+1})$ and the circuit $C^{|t|}$ computing $|t|$ on inputs of lengths $n_1, \ldots, n_k$ outputs $m = O(\log(n_1 + \cdots + n_k + 2))$ bits then put:

$$\|A(x_1, \ldots, x_k)\|^{n_1,\ldots,n_k} :=$$

$$\bigvee_{b \in \{0,1\}^m} \|x_{k+1} \leq |t(x_1, \ldots, x_k)| \wedge B(x_1, \ldots, x_k, x_{k+1})\|^{n_1,\ldots,n_k,m}(\overline{x}_{k+1}/b).$$

4. If $A(x_1, \ldots, x_k) = \forall x_{k+1} \leq |t(x_1, \ldots, x_k)| B(x_1, \ldots, x_k, x_{k+1})$ and m is as above then put:

$$\|A(x_1, \ldots, x_k)\|^{n_1,\ldots,n_k} :=$$

$$\bigwedge_{b \in \{0,1\}^m} \|x_{k+1} \leq |t(x_1, \ldots, x_k)| \rightarrow B(x_1, \ldots, x_k, x_{k+1})\|^{n_1,\ldots,n_k,m}(\overline{x}_{k+1}/b).$$

5. If $A(x_1, \dots, x_k) = \exists x_{k+1} \leq t(x_1, \dots, x_k) B(x_1, \dots, x_k, x_{k+1})$ and the circuit C^t computing t on inputs of lengths $n_1, \dots, n_k$ outputs m bits then put:

$$\|A(x_1, \dots, x_k)\|^{n_1, \dots, n_k} :=$$

$$\exists y_1, \dots, y_m \, \|x_{k+1} \leq t(x_1, \dots, x_k) \wedge B(x_1, \dots, x_k, x_{k+1})\|^{n_1, \dots, n_k, m}(\overline{x}_{k+1}/\overline{y}).$$

6. If $A(x_1, \dots, x_k) = \forall x_{k+1} \leq t(x_1, \dots, x_k) B(x_1, \dots, x_k, x_{k+1})$ and m is as above then put:

$$\|A(x_1, \dots, x_k)\|^{n_1, \dots, n_k} :=$$

$$\forall y_1, \dots, y_m \, \|x_{k+1} \leq t(x_1, \dots, x_k) \to B(x_1, \dots, x_k, x_{k+1})\|^{n_1, \dots, n_k, m}(\overline{x}_{k+1}/\overline{y}).$$

This definition differs from the earlier one in Section 12.3 for basic formulas, and hence we need to establish that it yields equivalent formulas whose equivalence has a short proof.

Lemma 12.5.1 *Let A be an atomic L_{BA}(PV)-formula. Denote by $\|A\|_1^n$ its translation from Section 12.3 and by $\|A\|_2^n$ the translation into a Σ_1^q-formula defined above. Then it holds that*

- *there are polynomial-size G_1^*-proofs of the implications*

$$\|A\|_1^n \to \|A\|_2^n \quad \textit{and} \quad \|A\|_2^n \to \|A\|_1^n,$$

- *$\|A\|_1^n$ is Δ_1^q-provable in G_1^*. That is, there is a Π_1^q-formula B such that G_1^* proves in polynomial size the equivalence*

$$\|A\|_1^n \equiv B.$$

Proof The first item follows because, for any circuit C, G_1^* can prove in polynomial size that

$$\exists \overline{y}, \overline{z} \bigwedge \mathrm{Def}_C(\overline{x}; \overline{y}, \overline{z}).$$

The second item follows from the uniqueness of the witnesses $\overline{y}, \overline{z}$ (Lemma 12.3.1). □

Hence, as long as we work in a proof system containing G_1^*, the original and the new definitions of the translation are equivalent. The second item of Lemma 12.5.1 implies the quantifier complexity estimate in the next statement.

Lemma 12.5.2 *Let $A \in \Sigma_i^b$. Then there is a p-time algorithm that constructs the formula $\|A\|^n$ from $1^{(n)}$. In particular, the size of the formula is $n^{O(1)}$.*

Further, if $i \geq 1$ then $\|A\|^n \in \Sigma_i^q$ and if $i = 0$ then $\|A\|^n$ is Δ_1^q-provable in G_1^* by size $n^{O(1)}$ proofs.

Let $i \geq 0$ and let P be G or one of its subsystems G_j and G_j^* for $j \geq i$. The formula $i\text{RFN}_P$ is the formalization of the soundness of P with respect to the proofs of Σ_i^q-formulas:

$$P(y, x) \rightarrow Sat_i(x, z),$$

where Sat_i is a Σ_i^b $L_{BA}(\text{PV})$-formula defining the satisfiability relation for Σ_i^q-formulas.

Lemma 12.5.3 *For $i \geq 1$, T_2^i proves $i\text{RFN}_{G_i}$ and S_2^i proves $i\text{RFN}_{G_i^*}$; also, U_2^1 proves $i\text{RFN}_G$ for all $i \geq 1$.*

Proof All three propositions are proved by utilizing a suitable definition of the satisfiability relation for sequents consisting of Σ_i^q-formulas in the first two cases, and of arbitrary Σ_∞^q-formulas in the last case. We outline the proof of the first case and will give references for the other cases in Section 12.9.

The satisfiability relation for a Σ_i^q-formula, $i \geq 1$, is definable by a Σ_i^b-formula and hence the satisfiability relation for sequents consisting of such formulas is Σ_{i+1}^b-definable. Given a G_i-proof π and a truth assignment to the atoms in the end-sequent, prove by induction on the number of steps in π that all sequents in π are satisfied by the assignment (i.e. also the end sequent). This needs Σ_{i+1}^b-LIND, i.e. S_2^{i+1}, and hence by Theorem 9.3.1 T_2^i suffices. □

Theorem 12.5.4 (Krajíček and Pudlák [318], Krajíček and Takeuti [326]) *For $i \geq 1$ and a Σ_i^b-formula $C(x)$:*

(1) if T_2^i proves $\forall x C(x)$ then there are size $n^{O(1)}$ G_i-proofs of $\|C\|^n$, and if S_2^i proves $\forall x C(x)$ then there are size $n^{O(1)}$ G_i^*-proofs of $\|C\|^n$;
(2) if U_2^1 proves $\forall x C(x)$ then there are size $n^{O(1)}$ G-proofs of $\|C\|^n$.

Moreover, the proofs can be constructed by a p-time algorithm from the input $1^{(n)}$.

Proof The original proofs (Section 12.9) used the sequent calculus formulations of the bounded arithmetic theories involved and a suitable form of cut elimination. Here we shall give an idea of how these statements could be proved using Herbrand's theorem, as for Theorem 12.4.2. We shall outline a proof of a weaker statement,

- G *simulates* T_2,

as that is free of various technical complications.

The idea of the argument is to Skolemize T_2 by introducing new Skolem functions and adding new Skolem axioms. The result will be a universal theory $Sk[T_2]$ in a richer language $L_{BA}(\text{PV})^{Sk}$ proving all axioms of T_2, and we will then apply the earlier argument to this theory.

For an $L_{BA}(\text{PV})$-formula $A(x)$ of the form $\exists y \leq t(x) B(x, y)$, with B open, introduce a new **Skolem function** symbol f_A and a new **Skolem axiom**

$$AxSk[A]\colon (y \leq t(x) \wedge B(x, y)) \rightarrow (f_A(x) \leq t(x) \wedge B(x, f_A(x))).$$

Note that it is an open formula, and that modulo this axiom is $A(x)$ equivalent to

$$Sk_A\colon f_A(x) \leq t(x) \wedge B(x, f_A(x)).$$

Repeat this process ad infinitum, transforming all bounded formulas into an equivalent open formula in the language $L_{BA}(\mathrm{PV})^{Sk}$, i.e. $L_{BA}(\mathrm{PV})$ augmented by the new function symbols.

Let $Sk[T_2]$ be the theory

$$\mathrm{PV}_1 \cup \{AxSk[A] \mid \text{ all } A\} \cup \{Sk_B \mid B \text{ a } T_2 \text{ axiom}\}.$$

It is a universal theory, which proves all axioms of T_2, and hence we may apply the argument of Theorem 12.4.2. But we need to explain how we shall translate into propositional logic the terms involving the new Skolem function symbols. This cannot be done via circuits, as the functions are not in general p-time computable (they are computable only somewhere in the polynomial-time hierarchy).

For $n := |x|$ and m bounding the length $|t(x)|$ for $|x| \leq n$, define the translation $\|f_A(x) = y\|^{n,m}$ as

$$\|y \leq t(x) \wedge B(x,y) \wedge \forall z < y \neg(z \leq t(x) \wedge B(x,z))\|^{n,m}$$

$$\vee \|y = 0 \wedge \forall y \leq t(x) \neg B(x,y)\|^{n,m} .$$

In other words, we interpret f_A as the least witness for the existential quantifier if the least witness exists, and by 0 otherwise.

It is then tedious but not difficult to verify that G can prove in polynomial size all the axioms of $Sk[T_2]$. □

Analogously with the earlier terminology we shall say, for the situation described by Theorem 12.5.4 that P **p-simulates the Σ_i^b-consequences of** T.

Corollary 12.5.5 *For $i \geq 1$, let T and P be any of the pairs*

- $T = T_2^i$ *and* $P = \mathrm{G}_i$,
- $T = S_2^i$ *and* $P = \mathrm{G}_i^*$,
- $T = U_2^1$ *and* $P = \mathrm{G}$.

Then T and P correspond to each other: T proves $i\mathrm{RFN}_P$ and P p-simulates the Σ_i^b-consequences of T.

We use the consequences of the correspondence, now standard, to establish a simulation result for the systems G_i and G_i^*.

Lemma 12.5.6 *For $i \geq 1$, G_i and G_{i+1}^*, p-simulate each other with respect to proofs of Σ_i^q-sequents.*

Also, ER p-simulates G_1^* with respect to proofs of quantifier-free formulas.

Proof The construction of a p-simulation of G_i by G_{i+1}^* is completely analogous to the construction transforming Frege proofs into a tree-like form in the proof of

Theorem 2.2.1. But we can also see this via the correspondence in Corollary 12.5.5: S_2^{i+1} contains T_2^i by Theorem 9.3.1 and hence it also proves $i\mathrm{RFN}_{G_i}$ and thus G^*_{i+1} p-simulates G_i with respect to all proofs (tacitly of Σ^q_i-sequents).

The same argument works also for the opposite simulation: by Theorem 9.3.1 T_2^i proves all Σ^q_{i+1}-consequences of S_2^{i+1}. Hence it also proves $i\mathrm{RFN}_{G^*_{i+1}}$, and the p-simulation by G_i follows. □

This correspondence can also be used to prove the following statement (we shall omit the proof; see Section 12.9).

Lemma 12.5.7 *Let $i \geq 1$ and assume that G_i proves all Σ^q_1-tautologies by polynomial-size proofs. Then it also proves all Σ^q_i-tautologies by polynomial size proofs. The same holds for G^*_i and for G.*

12.6 Jeřábek's Theories APC_1 and APC_2

We saw in Chapter 11 that having the possibility of counting in a theory allows one to formalize various intuitive combinatorial arguments. A number of constructions in combinatorics and in complexity theory use, in fact, only the ability to count approximately, with not too big an error. The cardinality $|X|$ of a set $X \subseteq [a]$ may be determined approximately, with an error a^ϵ estimated in terms of the size of the ambient interval or, even more precisely, with a possibly much smaller error $|X|^\epsilon$.

Jeřábek [257, 259] showed that such approximate counting can be defined in bounded arithmetic using only a form of the weak PHP, which we shall call **dual** WPHP. Let g be a function symbol. The dual WPHP for g, denoted by $d\mathrm{WPHP}(g)$, is a formula formalizing the statement that g cannot be a surjective function from $\{0,1\}^n$ onto $\{0,1\}^{n+1}$ ($n(1+1/\log n)$ is used in [257, 259]):

$$\forall z\, \exists y(|y| = |z| + 1)\forall x(|x| = |z|)\; g(x) \neq y\,.$$

Here the role of n is taken by $|z|$ and of $\{0,1\}^n$ by $\{x \mid |x| = |z|\}$.

The theory APC_1 is PV_1-augmented by the axioms

$$d\mathrm{WPHP}(g), \qquad \text{for all PV function symbols } g.$$

The Theory APC_2 is a relativization of APC_1 by an NP-oracle: it is defined as APC_1 but at one level higher in the Σ^b_i-hierarchy. It is the theory $T^1_2(\mathrm{PV})$ together with all instances $\mathrm{dWPHP}(g)$ for all functions g that are Σ^b_2-definable in $T^1_2(\mathrm{PV})$. The first theory allows counting approximately with an error a^ϵ, the second with a smaller error $|X|^\epsilon$. We shall not discuss here how this is done.

Before the next statement recall the WPHP Frege proof system WF from Section 7.2.

Lemma 12.6.1 *The theory APC_1 proves the soundness of WF, RFN_{WF}.*

Proof By Lemma 12.4.1, PV_1 proves the soundness of *EF*. We need only to see how we can prove that, using instances of the extra *WF* axiom, (7.2.3) is sound. However, that was noted in Lemma 7.2.3 to follow from a suitable instance of the dWPHP (7.2.2). □

The next statement is the harder part of the correspondence at which we are aiming.

Lemma 12.6.2 *For an open $L_{BA}(\mathrm{PV})$-formula $A(x)$, if APC_1 proves $\forall x A(x)$ then there are size $n^{O(1)}$ WF-proofs of $\|A\|^n$, and the proofs can be constructed by a p-time algorithm from the input $1^{(n)}$.*

That is, WF p-simulates APC_1.

Proof In order to be able to use Herbrand's theorem we need to replace APC_1 by a universal theory. This is done by a Skolemization of the dWPHP axioms. For each such axiom

$$\mathrm{dWPHP}(g)\colon \forall z\, \exists y(|y| = |z| + 1)\forall x(|x| = |z|)\; g(x) \neq y,$$

introduce a new function symbol G and an axiom

$$d\mathrm{WPHP}(g, G)\colon \forall z \forall x(|x| = |z|)\; (|G(z)| = |z| + 1 \wedge g(x) \neq G(z))\ .$$

Claim *Each axiom* $\mathrm{dWPHP}(g)$ *follows from* $\mathrm{dWPHP}(g, G)$.

Hence, if APC_1 proves A, so does the theory

$$T := \mathrm{PV}_1 \cup \{d\mathrm{WPHP}(g, G) \mid g \text{ is a PV function symbol}\}\ .$$

This is a universal theory and hence we can apply to it the same construction as in the proof of Theorem 12.4.2. However, we have to explain how we translate terms involving the new function symbols G. These are not necessarily computable by polynomial-size circuits. So, whenever we get to the stage of translating a term of the form $G(t(\overline{x}))$ and we have a circuit C^t with m output bits corresponding to $t(\overline{x})$ for inputs of bit size n, we simply invoke the extra rule of *WF* corresponding to C^t, (7.2.3), and use the m-tuple $\overline{r}$ introduced there as the translation of the term. We use the same tuple $\overline{r}$ for all instances of G with the same input length. □

Corollary 12.6.3 APC_1 *and WF correspond to each other.*

Lemma 12.6.4 *T^2_2 proves RFN_{WF} and hence G_2 p-simulates WF.*

Proof The dual WPHP, $\mathrm{dWPHP}(g)$, for the parameter z is an instance of the ordinary $\mathrm{WPHP}(a^2, a, R)$ with $a := 2^{|z|}$ and $R(u, v)$ that holds for $u \in [2a]$ and $v \in [a]$ if and only if $g(v) = u$.

By Theorem 11.4.7, $T^2_2(R)$ proves $\mathrm{WPHP}(2a, a, R)$ and hence $T^2_2(\mathrm{PV})$ proves all the APC_1 axioms $d\mathrm{WPHP}(g)$. The rest follows from Theorem 12.5.4 and Corollary 12.5.5. □

12.7 Higher Up: V_2^1 and *iER*

We now give an example of how the correspondence can be fruitfully used for proof systems still stronger than *ER* by proving Lemma 7.3.3, that implicit *ER* from Section 7.3 p-simulates the quantified propositional calculus *G*.

By Lemma 7.3.1 we may work with an f-implicit version of *iER*. Let τ be a set of clauses in the variables $p_1, \dots, p_n$. By Lemma 5.8.4 we may assume that the clauses in an *ER*-refutation contain at most $2n$ literals (the bound to the width of the clauses in τ). Recall that a refutation π of τ in *iER* is a pair (α, β), where β is a circuit with inputs $i_1, \dots, i_k$ that outputs clauses such that the sequence

$$\beta(\overline{0}), \dots, \beta(\overline{i}), \dots, \beta(\overline{1}),$$

with $\{0,1\}^k$ ordered lexicographically, is an *ER*-refutation of τ; $\beta(\overline{i})$ also contains information about the particular inference that was used in the derivation of the $\overline{i}$th clause (or about which extension axiom was introduced). The string α is an *ER*-proof of the tautology $Correct_\beta(x_1, \dots, x_k)$ (Section 7.3) formalizing that β describes a valid *ER*-refutation.

We are in the regime of the $\| \dots \|$ translation and hence strings such as formulas, circuits, proofs, assignments etc. are coded by numbers. Having also sets in V_2^1 will give us an extra edge for proving various statements. The following lemma quantifies the amount of edge. The formal argument is similar to (but different from) the argument underlying Theorem 9.4.1. There we established a link between models of V_1^1 and of S_2^1 (the RSUV isomorphism). Here we shall show another link, this time between V_2^1 and a formal system defined using S_2^1. The RSUV isomorphism translates Σ_i^b-formulas over a model of S_2^1 into $\Sigma_i^{1,b}$-formulas over a particular model of V_1^1, and vice versa. The link in the next lemma does not change the formulas but relates instead the validity of a Σ_i^b-formula in a model of V_2^1 to the validity of the *same* formula in a particular model of S_2^1.

Lemma 12.7.1 *Let $A(x)$ be a Σ_∞^b-formula. Then V_2^1 proves $A(x)$ if and only if there is an $L_{BA}(PV)$-term $t(x)$ such that S_2^1 proves*

$$|y| \geq t(x) \rightarrow A(x). \tag{12.7.1}$$

Proof We shall prove just the "only-if" part as that is what we shall use (Section 12.9). The idea is that subsets of $[t(a)]$ can act as strings (i.e. numbers) of bit size $t(a)$.

Assume that (12.7.1) fails for all terms $t(x)$. By the compactness of first-order logic there is a model $\mathbf{M}$ of S_2^1 with $a, b \in \mathbf{M}$ such that $A(a)$ fails in $\mathbf{M}$ but $|b| \geq t(a)$ holds for all terms $t(x)$. Take the cut

$$J := \{c \in \mathbf{M} \mid c \leq t(a) \text{ for some term } t(x)\}$$

and equip it with the set sort consisting of those subsets of J coded by some number in $\mathbf{M}$. It is easy to see that the resulting structure satisfies V_2^1. □

Lemma 12.7.2 *V^1_2 proves the soundness of iER, i.e. the formula* RFN_{iER}.

Proof Let $\pi = (\alpha, \beta)$ be an *iER*-refutation of τ with parameters as above, and let $s = |\pi|$. In particular, the output string of β has length at most $|\beta| \leq s$.

Let K, N, S be numbers such that $|K| = k$, $|N| = n$ and $|S| = s$ (they are bounded by the number coding π). We identify K and N with $\{0,1\}^k$ and $\{0,1\}^n$, respectively.

Claim 1 V^1_2 proves that there is a relation $F \subseteq \{0,1\}^k \times S$ such that, for each $\bar{i} \in \{0,1\}^k$, there is a unique $w \leq S$ such that $F(\bar{i}, w)$. Such a w is the clause coded by $\beta(\bar{i})$.

This is proved by induction on the length of the derivation described by β, i.e. by induction up to K. The induction statement is $\Sigma^{1,b}_1$ and hence V^1_2 suffices.

A similar induction argument establishes the next claim. We also use that, because we have the proof α, we know that in V^1_2 (as S^1_2 already proves the soundness of *ER*) F is indeed an *ER* refutation.

Claim 2 V^1_2 proves the following. Let $\omega \leq N$ be an assignment to the variables of τ and assume that ω satisfies all the clauses of τ. Then there exists a set coding an assignment to the extension atoms introduced in the refutation F such that all the clauses in F are satisfied.

This is a contradiction and hence τ cannot be satisfied. □

Lemma 12.7.3 *iER p-simulates V^1_2: if V^1_2 proves an open $L_{BA}(\mathrm{PV})$-formula $A(x)$ then there are size $n^{O(1)}$ iER proofs of $\|A\|^n$. Moreover, these proofs can be constructed by a polynomial-time algorithm from $1^{(n)}$.*

Proof By Lemma 12.7.1 the hypothesis of the lemma implies that, for some term $t(x)$, (12.7.1) holds. By the conservativity (Theorem 9.3.1) of $S^1_2(\mathrm{PV})$ over PV_1, PV_1 also proves (12.7.1).

Now apply an argument using Herbrand's theorem as in Theorem 12.4.2. We need only to verify that the resulting proof is very uniform, so that it can be described by a small circuit. This stems from the following claim.

Claim *Let $t(\bar{x})$ be an $L_{BA}(\mathrm{PV})$-term. The circuit C^t described in the definition of the $\|\ldots\|$-translation computing t for inputs of length $M = 2^m$ can be bit-wise described by a circuit D^t with $O(m)$ inputs and of size $m^{O(1)}$.*

The translation $\|\ldots\|^n$ of the disjunction (12.2.6) in the construction of the propositional proof results in circuits of size 2^m for $m = n^{O(1)}$. However, by the claim, these circuits (as well as the proofs of the axiom instance) can be described by circuits of size $n^{O(1)}$. □

Lemmas 12.7.2 and 12.7.3 imply

Corollary 12.7.4 *V^1_2 and iER correspond to each other.*

The following was stated without proof in Lemma 7.3.3.

Corollary 12.7.5 *iER p-simulates* G.

Proof By Lemma 12.7.3, *iER* p-simulates V^1_2, which contains U^1_2 and hence, by Lemma 12.5.3, proves 0RFN_G. □

12.8 Limits of the Correspondence

Are there any limits to the correspondence between theories and proof systems? If we mean by that question how high we can get, i.e. how strong are the theories and proof systems for which we can find a counterpart, the answer is *arbitrarily high*. But if we mean how deeply does the correspondence run and how wide is it, i.e. what aspects of a theory and a proof system are related, then there are limits.

Let us address the former, easier, question first. Let S be an arbitrary theory; we shall assume for simplicity that its language contains $L_{BA}(\text{PV})$ and $S \supseteq \text{PV}_1$. Further, we shall assume that:

- the set of axioms of S is decidable in polynomial time; and
- all universal $L_{BA}(\text{PV})$-sentences provable in S are true in $\mathbf{N}$.

For example, we can extend the language of the set theory ZFC by $L_{BA}(\text{PV})$ and add to the ZFC axioms all the PV_1 axioms and an axiom defining the realm of these axioms to be the natural numbers as defined in ZFC.

The first list item implies that there is an open $L_{BA}(\text{PV})$-formula $Prf_S(y,x)$ formalizing that *y is an S-proof of the formula x*, and we also have the $L_{BA}(\text{PV})$ open formula $Sat(z,x)$ used earlier, defining propositional satisfiability: z satisfies the propositional formula x. Recall that $Taut(x) := \forall z \leq x Sat(z,x)$. By the second list item, S then proves $Taut(\ulcorner \tau \urcorner)$ if and only if $\tau \in \text{TAUT}$. Here $\ulcorner \tau \urcorner$ is the closed term representing τ as defined in Section 7.4.

Define the theory $T := \text{PV}_1 + Con_S$, where Con_S is the universal $L_{BA}(\text{PV})$-formula

$$\forall y\, \neg Prf_S(y, \ulcorner 0 \neq 0 \urcorner)$$

and, analogously to P_{PA} in Section 7.4, define the proof system P_S by

$$P(\pi, \tau) \quad \text{if and only if} \quad \pi \text{ is an } S\text{-proof of } Taut(\ulcorner \tau \urcorner).$$

The next lemma is a standard fact of mathematical logic, that Con_S implies, over a fixed weak theory (PV_1 suffices), all universal consequences of S (subject to some technical conditions, satisfied by our S).

Lemma 12.8.1 *The theory T proves all universal $L_{BA}(\text{PV})$-formulas provable in S.*

Thus T is – for the purpose of links to propositional logic – as strong as S.

Lemma 12.8.2 *The theories T and P correspond to each other:*

- *T proves RFN_P; and*

- *P p-simulates T.*

Proof For the first item, note that if an assignment falsifies a formula τ then S can prove it and thus also prove that τ is not a tautology. Moreover, T proves the following fact about S as well:

$$\neg Sat(z,x) \to \exists v Prf_S(v, \ulcorner \neg Taut(\ulcorner \tau \urcorner)\urcorner).$$

Hence RFN_P follows in T from Con_S.

The second item rests on the existence (and their p-time constructibility) of S-proofs of

$$Taut(\ulcorner \| Con_S \|^n \urcorner)$$

of size $n^{O(1)}$. These do indeed exist and their construction stems from the construction underlying the upper bound in Theorem 21.3.1. We shall not present it here. □

Note that, in particular, there is a proof system that p-simulates ZFC (+ PV_1).

Let us now turn to the second question, of how deep and wide is the correspondence. We have presented the examples of corresponding pairs in a fairly pedestrian manner, stating simple technical requirements and their consequences (Section 8.6). The way in which some people think about the correspondence and refer to it is, however, more general and unfortunately also more informal. In particular, it often encompasses more topics in computational complexity.

A theory T (think primarily of theories akin to those discussed earlier) may correspond to a proof system P_T, which could be a Frege-style system restricted to using only formulas from a certain circuit class. However, T itself can define sets from some complexity class and, in particular, may assume induction axioms only for formulas defining sets from a particular complexity class. There are variants of induction schemes, though (we have used IND and LIND) and variants of languages, and it turns out to be more natural to associate with T the class F_T of all multi-functions with NP graphs and of polynomial growth that are definable in T. Such multi-functions are called **total** NP **search problems** and they are simply binary relations R on $\{0,1\}^*$ that can be defined as follows: for an NP binary relation $R_0(x,y)$ and a constant $c \geq 1$,

$$R(x,y) \quad \text{if and only if} \quad |y| \leq |x|^c \wedge R_0(x,y)$$

and it holds that

$$\forall x \exists y R(x,y).$$

The witness y is not required to be unique, so it is a *multi*-function.

A formula corresponding to NP is $\Sigma_1^{1,b}$ in the two-sorted set-up or Σ_1^b in the one-sorted set-up. We say that R is definable in T if, for some NP formula θ defining R_0, T proves that

$$\exists y(|y| \leq |x|^c)\, \theta(x,y).$$

Thus we associate with T two objects,

$$F_T \leftarrow T \rightarrow P_T,$$

and we may think of this picture as also linking F_T with P_T. This is often fruitful, especially when we are considering theories and proof systems of some restricted form, for example, subsystems of V_1^1 and subsystems of *EF*. Examples of similar triples include:

$$\text{p-time functions} \leftarrow \text{PV}_1 \rightarrow EF,$$

$$\text{AC}^0\text{- functions} \leftarrow V_1^0 \rightarrow \text{AC}^0\text{-}F,$$

$$\text{PLS (see Section 13.7)} \leftarrow T_2^1 \rightarrow G_1,$$

$$\text{PSPACE} \leftarrow U_2^1 \rightarrow \text{G},$$

$$\text{EXPTIME} \leftarrow V_2^1 \rightarrow iER.$$

However, if we are aiming to understand general proof systems that are possibly very strong, these well-behaved examples may be misleading. The discrepancy stems from the fact that the right-hand side relation

$$T \rightarrow P_T$$

depends on the universal consequences, or, more generally, the $\forall\Pi_1^b$-consequences, of T, while the left-hand side

$$F_T \leftarrow T$$

depends on the $\forall\Sigma_1^b$-consequences of T.

It may happen that $F_T = F_{T'}$ while $P_T \not\equiv_p P_{T'}$. In the two sorted set-up, a good example is $T = I\Sigma_0^{1,b}$ and $T' = I\Sigma_0^{1,b} + Count^2$ (Section 11.1). In both cases the class of multi-functions is the class of AC^0-functions, while the corresponding proof systems are AC^0-F and also AC^0-F augmented by instances of $\langle Count^2 \rangle$ as new axioms, respectively. The latter proof system is provably stronger than the former (Chapter 15).

More generally, F_T is not influenced by adding to T some new true universal axioms (or $\forall\Pi_1^b$ axioms, in general), while this is, of course, key from the propositional logic point of view as these are the formulas that give rise to sequences of tautologies.

It may also happen that $F_T \neq F_{T'}$ while $P_T = P_{T'}$, although in the next example $P_{T'}$ simulates T' but maybe does not p-simulate it. Let $R(y)$ be a Σ_1^b $L_{BA}(\text{PV})$-formula defining the set of primes. Take $T := \text{PV}_1$ and

$$T' := \text{PV}_1 + \forall x \geq 2 \exists y (|y| = |x|)\, R(y).$$

It is unknown (and unlikely) that $T = T'$ (Section 12.9), but *EF* simulates T': the witnesses y depend only on the length of x and hence *EF* can use a sequence of *advice* witnesses a_n, such that $R(a_n) \wedge |a_n| = n$, as values of a Skolem function witnessing the extra axiom. But it is also unknown whether any such sequence $\{a_n\}_n$ is p-time constructible, and so it may be that $F_T \neq F_{T'}$.

Another issue refers to the subtle point of how the propositional translations are defined. Assume that we have a Π_1^b-formula $A(x)$:

$$\exists i \leq |x| \forall z \leq x\, B(x, i, z), \tag{12.8.1}$$

with B an open $L_{BA}(\mathrm{PV})$-formula. Then $\|A\|^n$ is

$$\bigvee_{i \leq n} \|z \leq x \rightarrow B(x, i, z)\|^n. \tag{12.8.2}$$

Now assume that we have proved (12.8.2) in a proof system P, and we know in T that P is sound. Then we can deduce that (12.8.2) is a tautology, which means that it is true for all assignments to its atoms. Formally,

$$\forall x(|x| = n)\forall w(|w| \leq n^{10})\exists i \leq n\; C(w) \rightarrow ((w)_i \leq x \rightarrow B(x, i, (w)_i)) \tag{12.8.3}$$

is true, where $C(w)$ formalizes the statement that

- w is a sequence of $\leq n$ bit strings $(w)_0, \ldots, (w)_n$.

However, to obtain from the validity of (12.8.3) the validity of the original formula (12.8.1) requires a switch from sharply bounded quantifier to a bounded one. This can be done with the help of the so-called **bounded collection axioms** (equivalently, **replacement axioms**)

$$\forall j \leq |x| \exists y \leq x D(x, j, y) \;\rightarrow\; \exists w(|w| \leq |x|^{10})\forall j \leq |x| (w)_j \leq x \wedge D(x, j, (w)_j).$$

Nevertheless, this axiom scheme is probably not provable from any set of true universal $L_{BA}(\mathrm{PV})$-formulas and, in particular, not in PV_1 (Section 12.9).

These remarks are not meant to question the splendid usefulness of the correspondence between theories and proof systems, as has been established in a number of cases. Rather, I want to point out that there are limits to what one can derive from it and that to get an insight about what happens beyond the limits may be a key to understanding strong proof systems.

12.9 Bibliographical and Other Remarks

Sec. 12.1 Cobham's theorem [146] built on an earlier work by Bennett [65]. The class of polynomial-time functions also has a *finite basis*: it can be generated from a finite subset by composition only. See Muchnik [357] and Jones and Matiyasevich [262]. More details about PV are given in [278].

The theory PV_1 was defined by Krajíček, Pudlák and Takeuti [323]. Theorem 12.1.3 probably cannot be improved; the theory $S_2^1(\mathrm{PV})$ is axiomatized by $\forall\Sigma_2^b(L_{PV})$-sentences and hence $\forall\Sigma_2^b(L_{PV})$-conservativity would imply that $S_2^1(\mathrm{PV}) = \mathrm{PV}_1$, which would in turn imply that $\mathrm{NP} \subseteq \mathrm{P}/poly$ by [323].

Sec. 12.2 Herbrand's theorem [226] appears abundantly in the logic literature; see e.g. [470, 278]. In the context of bounded arithmetic Corollary 12.2.4 is often called the KPT theorem, after Krajíček, Pudlák and Takeuti [323], who formulated it and first applied it in that context. The Student–Teacher game was defined by Krajíček, Pudlák and Sgall [322], who also proved some hierarchy theorems with respect to the number of rounds in the game.

Sec. 12.3 The translation $\| \dots \|$ for atomic (and hence open) $L_{BA}(\mathrm{PV})$-formulas is essentially that of Cook [149].

Sec. 12.4 The proof of the simulation theorem 12.4.2 via Herbrand's theorem is new; the original proof was for the equational version of PV only.

Sec. 12.5 The extension of the $\| \dots \|$ translation to all bounded formulas is from Krajíček and Pudlák [318], as are the results about the correspondence (i.e. the soundness and the simulations) involving G_i or G_i^*. The results about the correspondence of U_2^1 and G are from Krajíček and Takeuti [326]. Dowd [172] established earlier a relation between G and bounded arithmetic for PSPACE but his translation was different (the quantifier complexity of the propositional translation grew with the space bounds). In both cases the original arguments used the sequent calculus formalization of bounded arithmetic theories and a suitable cut-elimination theorem. Lemmas 12.5.6 and 12.5.7 are from Krajíček and Pudlák [318]; see also [278] (the proof of Lemma 12.5.7 can be found there). All this material is also treated via the sequent calculus formalism in [278].

Sec. 12.6 A form of approximate counting in the bounded arithmetic $I\Delta_0$ was described by Paris and Wilkie [389]. Jeřábek's approximate counting in APC_1 and APC_2 is to be found in [257, 259]. The link between dWPHP and the proof system *WF* is from his Ph.D. thesis [256]. He also studied mutual provability relations for various variants of the (W)PHP principles in [258]. Sometimes the dual WPHP is referred to as *surjective* (and other variants as *injective* and *bijective*). Analogously to Lemma 12.6.4 it can be shown that G_3 simulates proofs in APC_2. We note here (but give an argument for it in Section 19.4) that it is unlikely that *EF* simulates *WF* (see also [291]).

Sec. 12.7 The content of Section 12.7 is from [293]. Because *iER* corresponds to V_2^1 and G to U_2^1 and the former theory is conjectured to prove more Π_1^b-statements than the latter, it appears unlikely that Lemma 7.3.3 (Corollary 12.7.5) can be strengthened to p-equivalence. We note that it is known (unconditionally) that V_2^1 proves more Π_1^b statements than S_2^1 and, in particular, that S_2^1 does not prove RFN_{iER}; see [293]. Unfortunately, this does not seem to imply that *ER* does not simulate *iER* (but it gives certain evidence for it). The formal system used in Lemma 12.7.1 is called $S_2^1 + 1 - Exp$ in [274, 278].

Sec. 12.8 The requirements on S in Section 12.8 can be weakened substantially; see Paris and Wilkie [390], Pudlák [408, 409, 410] and Krajíček and Pudlák [316]. Lemma 12.8.1 can be found in, for example, Smorynski [475]. The existence of short S-proofs of the formulas in the proof of Lemma 12.8.2 follows from [409, 410]; see Chapter 21.

NP search problems will be discussed in more detail in Section 19.3. Cook and Nguyen [155] gave numerous examples of the pairs T and F_T. The infinitude of primes was derived over T_2 from WPHP by Paris, Wilkie and Woods [392], extending earlier work of Woods [507]. The conditional unprovability of the bounded collection scheme in PV_1 was proved by Cook and Thapen [159].

Cook's construction relating PV_1 to ER (Theorem 12.4.2) can be smoothly extended to any consistent r.e. theory T extending PV_1, offering an alternative to the construction from Section 12.8. Namely, let T_1 be the universal $L_{BA}(PV)$-consequences of T; it is r.e. theory and it is easy to find p-time $T^* \subseteq T_1$ axiomatizing it over PV_1. Define $\|T^*\|$ to be the set of all $\|B\|^n$ translations of all $\forall x B \in T^*$. Define a proof system $P(T)$ to operate as ER but using also all substitution instances off all formulas in $\|T^*\|$ as extra axioms. For $k \geq 1$, let $P_k(T)$ be its subsystem using only those axioms $B \in T^*$ having size $|B| \leq k$.

Claim 1: *If $\forall x A(x) \in T_1$ then there is $k \geq 1$ such that formulas $\|A\|^n$ have $P_k(T)$-proofs of size polynomial in n and p-time constructible from $1^{(n)}$.*

The hypothesis of the claim and Herbrand's theorem (Corollary 12.2.3) implies that $A(x)$ follows in PV_1 from finitely many formulas $B_i(t_i(x))$, where $B_i \in T^*$ and where t_i are PV-terms. The claim follows from Theorem 12.4.2, using the new axioms from $\|T^*\|$ corresponding to formulas B_i.

Claim 2: *For all $k \geq 1$, T proves $RFN_{P_k(T)}$.*

This follows as each universal sentence $\forall x B(x)$ implies over PV_1 that all formulas $\|B\|^n$ are tautologies.

If T^* can be chosen finite then one of the systems $P_k(T)$ corresponds to T (on the other hand, if any Q corresponds to T then T_1 is finitely axiomatized over PV_1 by RFN_Q). If T^* is infinite, then the whole collection of systems $P_k(T)_k$ corresponds to T in the sense of the claims. These collections are analogous to collections of bounded-depth Frege systems $\{F_d\}_d$ or fragments $\{G_i\}_i$ of G corresponding to $I\Delta_0(\alpha)$ and S_2, respectively.

Part III

Lower Bounds

Part III is devoted to a presentation of some of the main length-of-proof lower bounds that were proved until recently. Lower bounds are considered to be the heart of the subject. One reason is that the fundamental problems from Section 1.5 are formulated using lengths of proofs, and hence proving a lower bound of that type gives the impression that it elucidates the problems. An additional reason is, I think, that what a lower bound is and whether it is stronger than some other lower bound is (mostly) very specific. This gives the whole field a clear sense of purpose and a rule helping to sort out new ideas.

What do we know when we have a super-polynomial lower bound for a proof system Q? Obviously Q cannot be a proof system witnessing that NP $=$ coNP, and if the lower bound is for a sequence of formulas that do have short proofs in some other proof system (this happens whenever the sequence is p-time constructible, see Chapter 21), then Q cannot be optimal either. As the length of a proof is an absolute lower bound on the time required to find it, obtaining proofs in such a proof system takes a long time as well. In other words, we know then that Q does not contradict the expected answers to fundamental problems. Using the correspondence between theories and proof systems from Part II we can show that lower bounds for Q imply, in fact, more specific statements about the problems than meets the eye at first.

Assume for a moment that actually P $=$ NP and that f is a p-time algorithm (represented by its PV symbol) solving SAT: if there is any satisfying assignment y for the atoms in a formula x then $f(x)$ is also a satisfying assignment. Using the formalism of Part II (as found e.g. in Section 8.4) we can write this as a universal statement in the language of PV:

$$Sat(x, y) \;\rightarrow\; Sat(x, f(x)). \tag{III}$$

The mere validity of (III) is not enough; we also need to have its proof in some established theory formalizing mathematical reasoning, e.g. in the set theory ZFC. If the above proof system Q corresponds (in the sense of Part II) to a theory T then we know that T does not prove (III) for any f (see consequence (4) in Section 8.6). In other words, *any* super-polynomial lower bound for Q rules out as p-time SAT algorithms any f whose soundness is provable in T. For some proof systems for which lower bounds are known (and this will be discussed in the coming chapters), the class of SAT algorithms whose soundness is provable in a theory corresponding to the proof system contains all currently studied algorithms and much more.

We have used ZFC as the standard example of a theory formalizing mathematics but for complexity theory a much weaker theory suffices. All the material in any textbook on the subject is provable in bounded arithmetic augmented by exponentiation (the theory is called $I\Delta_0 + Exp$), which is itself a very weak subtheory of Peano arithmetic. In fact, most standard textbook material can be formalized in its fragment $(I\Sigma_0^{1,b})^{count}$ (Section 10.6) and many well-known results are provable even in theories like PV, APC_1 or APC_2, which we discussed in Chapter 12 (we shall return to these issues in Chapter 22). Proving the consistency of the conjecture that $NP \neq coNP$ with one of these theories would go a long way, in my view, as evidence that the conjecture is true. Such a consistency follows from *any* super-polynomial lower bound (i.e. the hard formulas need not encode anything about the P vs. NP problem or circuit lower bounds, etc.) for proof systems corresponding to these theories (e.g. *ER* or *WF*, see Sections 12.4 and 12.6).

A philosophically minded reader may contemplate the following theoretical situation. Assume that the above statement regarding $Sat\ (x, y)$ is true for some p-time algorithm f and is even provable in, say, PA but not in PV. That is, we have a p-time way to handle NP-queries, but p-time concepts themselves are not enough to prove the soundness of the algorithm; we need induction for stronger concepts. In which sense is f then truly feasible? To make the issue more transparent, assume that while (III) cannot be proved in PV (i.e. using induction for P-predicates), it is possible to prove it using induction for NP-predicates (the theory T_2^1 of Sections 10.5 and 12.6). Then it does not seem that we can honestly claim that NP is feasible.

There are essentially two methods for proving lower bounds on size, in the sense that they offer a somewhat general methodology for approaching a lower bound problem and are successful in a number of varied situations. They are the **restriction method** and the method of **feasible interpolation**. The former is presented primarily in Chapters 14 and 15, dealing with lower bounds for constant-depth subsystems of Frege systems and of the sequent calculus. The latter method, feasible interpolation, will be treated in detail in Chapters 17 and 18.

Both methods will also appear in Chapter 13, which presents some lower bound criteria for resolution. Resolution has a singular place among all proof systems. It is the most studied proof system, due partly to its links with SAT solving and partly to its formal simplicity and elegance. A number of methods for proving resolution lower bounds do apply, however, to resolution or to its variants (and sometimes to some algebro-geometric systems) only. We shall treat resolution and related systems in Chapter 13.

Algebro-geometric systems will be treated in Chapters 17 and 18 on feasible interpolation and also in Chapter 16; there we present some lower bounds proved by ad hoc combinatorial arguments.

13

R and R-Like Proof Systems

There have been perhaps more papers published about the proof complexity of resolution than about all the remaining proof complexity topics combined. In this situation one needs to make some choice of the type of results to present. We shall concentrate in this chapter on presenting results that can be branded as *lower bound criteria*, statements that do imply resolution lower bounds (or lower bounds in some related systems) under some reasonably general conditions. The reader who wishes to learn the whole spectrum of results about resolution and who does particularly care about the links with the rest of proof complexity will be better off with some recent survey (cf. Section 13.7). Also, we generally give preference to simplicity and generality over the numerically strongest results (e.g. results that achieve the best lower or upper bound on some parameter).

13.1 Adversary Arguments

We shall consider in this section a first-order sentence Φ in a relational language L (possibly with constants) in the DNF_1-form introduced in Section 10.5. If such a sentence fails in all finite L-structures then the $\langle\Phi\rangle_n$ are, for all $n \geq 1$, unsatisfiable sets of clauses. We shall link the complexity of their $R^*(\log)$-refutations with the existence of an *infinite* model of Φ. Recall that DNF_1-formulas are built from basic formulas by first applying any number of conjunctions and universal quantifiers and then any number of disjunctions and existential quantifiers. The prime example is the pigeonhole principle PHP considered earlier (see Sections 8.1 and 10.5). We want to formalize it here without reference to the ordering of numbers so, instead of the language of arithmetic, we take a language with one binary relation $R(x,y)$ and one constant c and write PHP as the disjunction of the sentences

1. $\exists x \forall y \neq c\ \neg R(x,y)$,
2. $\exists x \exists y \neq c \exists y' \neq c\ (y \neq y' \wedge R(x,y) \wedge R(x,y'))$,
3. $\exists x \exists x' \neq x \exists y \neq c\ (R(x,y) \wedge R(x',y))$.

The restriction to DNF_1-formulas and to relational languages with constants can be removed, and the criterion we shall formulate can be proved in a more general setting. This will be demonstrated in Section 13.2 but for now we shall confine ourselves to the simple set-up in order not to burden with extra syntactic complexities our first encounter with the link to infinite structures.

We shall formulate the criterion first for a computational model based on particular search trees and then derive from it the criterion for R*(log). The computational model for solving the search problem Search($\mathcal{C}$) is from Section 5.2: given an unsatisfiable set of clauses $\mathcal{C}$ and a truth assignment $\overline{a}$, find a clause in $\mathcal{C}$ that is not satisfied by $\overline{a}$. The specific problem Search($\langle \neg\Phi \rangle_n$) determined by a DNF$_1$ sentence Φ in a language L and valid in all finite structures can be restated as follows: given an L-structure on $[n]$, find a clause of $\langle \neg\Phi \rangle_n$ that fails under the assignment determined by the structure.

In Sections 5.2 and 5.3 we considered how to solve Search($\mathcal{C}$) using decision trees or read-once branching programs. The model we shall introduce now generalizes the decision tree model. A DNF-**tree** for solving Search($\mathcal{C}$) is a rooted finite tree T such that:

- each leaf of T is labeled by a clause of $\mathcal{C}$;
- each inner node u is labeled by a DNF-formula $D_1 \vee \cdots \vee D_e$, and the tree branches at u into two subtrees, which are labeled as affirmative and negative.

The tree is a k-DNF **tree** if all DNFs labeling the inner nodes are k-DNF, i.e. $|D_i| \leq k$.

A truth assignment $\overline{a}$ to the atoms of $\mathcal{C}$ determines a path $P(\overline{a})$ in T, and hence the label $T(\overline{a})$ of the unique leaf in $P(\overline{a})$, as follows:

- if $\overline{a}$ satisfies some term D_i in the label u on $P(\overline{a})$ then the path continues via the affirmative subtree;
- otherwise $P(\overline{a})$ continues via the negative subtree.

Let us first tie these trees with R*(log)-refutations or, more generally, with (DNF-R)*-refutations.

Lemma 13.1.1 *Assume* $\mathcal{C}$ *has a tree-like* DNF-*R refutation* π *(i.e. it is a* (DNF-*R*)**-refutation) in which all terms have size at most t, and assume* π *has* $k := \mathbf{k}(\pi)$ *steps.*

*Then there is a t-*DNF *tree T solving* Search($\mathcal{C}$) *whose height is* $O(\log k)$.

Proof Using Spira's lemma 1.1.4 find a DNF-clause E in π such that the number of steps k_1 in the subproof π_1 of π ending with E satisfies $k/3 \leq k_1 \leq (2k)/3$. The DNF-tree T will ask at its root about the validity of E. If the answer is negative, we restrict to π_1; it is a proof of a falsified clause from the initial clauses $\mathcal{C}$. If it is affirmative we restrict to π_2, the rest of π not above E; it is a proof of a falsified clause (the empty one) from the clauses $\mathcal{C} \cup \{E\}$, and E is satisfied. Hence any falsified clause among the initial clauses of π_2 has to be from $\mathcal{C}$.

The tree T is built in the same way until the remaining proofs shrink to one node. That node must be labeled by a falsified clause from $\mathcal{C}$. The height of the tree is bounded above by $\log_{3/2}(k) = O(\log k)$. □

Theorem 13.1.2 *Let q be the maximum arity of a relation symbol occurring in a* DNF$_1$*-sentence Φ. Assume that $\neg\Phi$ has an infinite model.*

*Then any t-*DNF *tree T of height h solving* Search$(\langle\neg\Phi\rangle_n)$ *must satisfy*

$$(h+1)t \geq \frac{n}{q}.$$

Proof For simplicity of presentation we shall prove the theorem for a Φ whose language consists of one q-ary relation symbol R. The general case is completely analogous.

Let $\mathbf{M} = (M, S)$ be an infinite model of $\neg\Phi$, where $S \subseteq M^q$ interprets the symbol R. We shall think of a size n structure for the language as having the universe $[n]$. Consider triples (U, R, h) where:

- $\emptyset \neq U \subseteq [n]$ and $R \subseteq U^q$;
- h is an isomorphism of the structure (U, R) with substructure $\mathbf{M}$.

Let (U, R, h) be such a triple and $r_{i_1,\ldots,i_q}$ an atom of $\langle\Phi\rangle_n$. Now define the forcing relation:

1. (U, R, h) forces $r_{i_1,\ldots,i_q}$ to be true if and only if $\{i_1, \ldots, i_q\} \subseteq U$ and $R(h(i_1), \ldots, h(i_q))$ holds in $\mathbf{M}$;
2. (U, R, h) forces $\neg r_{i_1,\ldots,i_q}$ to be true if and only if $\{i_1, \ldots, i_q\} \subseteq U$ and $\neg R(h(i_1), \ldots, h(i_q))$ holds in $\mathbf{M}$.

We extend this forcing relation also to logical terms and t-DNF clauses by:

3. (U, R, h) forces a logical term D to be true if and only if it forces to be true all literals in D;
4. (U, R, h) forces a t-DNF clause $\bigvee_i D_i$ to be true if and only if it forces some D_i to be true.

Note that we do not define what it means to force a term or a t-DNF to be false.

Let $(\mathcal{R}, \preceq)$ be the partial ordering of all these triples, defined by

$$(U, R, h) \preceq (U', R', h') \quad \text{if and only if} \quad U \subseteq U' \wedge R' \cap U^q = R \wedge h \subseteq h'.$$

Let T be a t-DNF tree of height h solving Search$(\langle\neg\Phi\rangle_n)$. We will define a path $P = (u_1, u_2, \ldots)$ through T, u_1 being the root, and triples (U_ℓ, R_ℓ, h_ℓ) from $\mathcal{R}$ for $\ell = 1, 2, \ldots$ such that:

- $(U_\ell, R_\ell, h_\ell) \preceq (U_{\ell+1}, R_{\ell+1}, h_{\ell+1})$;

- $|U_\ell| \leq \ell qt$;
- the path continues from u_ℓ to the affirmative tree if and only if (U_ℓ, R_ℓ, h_ℓ) forces the t-DNF labeling u_ℓ to be true, and this happens if and only if there is any $(U', R', h') \succeq (U_{\ell-1}, R_{\ell-1}, h_{\ell-1})$ forcing it to be true (or any (U', R', h') if $\ell = 1$).

Assuming that a partial path led us to node u_ℓ, we have $(U_{\ell-1}, R_{\ell-1}, h_{\ell-1})$ satisfying the conditions above and we need to define (U_ℓ, R_ℓ, h_ℓ) (or $\ell = 1$, the node is the root and we need to define (U_1, R_1, h_1)). Assume that u_ℓ is labeled by a t-DNF $E = \bigvee_i D_i$, where each D_i is a $\leq t$-term. Consider two cases:

(a) E can be forced to be true by some $(U', R', h') \succeq (U_{\ell-1}, R_{\ell-1}, h_{\ell-1})$;
(b) case does not hold.

In case (a) it suffices if the triple (U', R', h') forces true at least one term D_i in E, i.e. at most t literals have been not forced true already by $(U_{\ell-1}, R_{\ell-1}, h_{\ell-1})$. The r-atoms in these literals involve at most qt elements of $[n]$ in total. Hence we may assume that $|U'| \leq |U_{\ell-1}| + qt \leq \ell qt$. Select any such (U', R', h') as (U_ℓ, R_ℓ, h_ℓ) and continue the path to the affirmative subtree.

In case (b) put $(U_\ell, R_\ell, h_\ell) := (U_{\ell-1}, R_{\ell-1}, h_{\ell-1})$ and continue the path to the negative subtree.

Assume that the path P reaches a leaf u_ℓ, i.e. $\ell \leq h + 1$, and let $C \in \mathcal{C}$ be the initial clause labeling the leaf. The structure (U_ℓ, R_ℓ) is isomorphic to a finite substructure (M_0, S_0) of (M, S) and Φ fails in (M, S). That means that the (instance of the) existential quantifier that C translates is witnessed in (M, S). Hence we can find a finite substructure of (M, S) extending (M_0, S_0) and containing the witness and use it to define $(U', R', h') \succeq (U_\ell, R_\ell, h_\ell)$, forcing C to be true. But that would contradict the hypothesis that the search tree solves Search$(\langle\neg\Phi\rangle_n)$: answering all queries in T by (U', R', h') would define the same path P and answer the search problem incorrectly by C.

The only way out of this contradiction is that there is no room in $[n] \setminus U_\ell$ to extend h_ℓ. In particular, this implies that

$$n - qt < qt\ell \leq qt(h + 1),$$

and the theorem follows. □

Corollary 13.1.3 *Let q be the maximum arity of a relation symbol occurring in a* DNF$_1$*-sentence Φ. Assume that Φ is valid in all finite structures but that $\neg\Phi$ has an infinite model.*

Then the size s of any R*(log)*-refutation of $\langle\neg\Phi\rangle_n$ must satisfy*

$$s \geq 2^{\Omega((n/q)^{1/2})}.$$

Proof An R*(log)-refutation of size s is a $(\log s)$-DNF refutation and has at most s steps. Lemma 13.1.1 and Theorem 13.1.2 then imply the corollary. □

The class of formulas for which a similar criterion exists can be extended somewhat beyond the DNF_1 class. Let us consider first two examples. The Ramsey theorem for graphs (simple, undirected and without loops) states that any graph on n vertices must have an induced homogeneous subgraph of size at least $\lfloor (\log n)/2 \rfloor$. A homogeneous subgraph is either a **clique**, i.e. a subgraph with all possible edges among its vertices, or an **independent set**, with no edges at all. In the language with a binary relation symbol $R(x, y)$ for the edge relation and the name H for the vertex set of a subgraph, this can be formalized by the formula $\mathrm{RAM}(R, n)$:

$$Graph(R, n) \rightarrow \big[\, \exists H \subseteq [n](|H| = \lfloor (\log n)/2 \rfloor)\ (\forall x, y \in H\, x \neq y \rightarrow R(x, y)) \\ \vee\ \exists H \subseteq [n](|H| = \lfloor (\log n)/2 \rfloor)\ (\forall x, y \in H\, \neg R(x, y))\big],$$

where $Graph(R, n)$ is

$$(\forall x \neg R(x, x)) \wedge (\forall x, y\, R(x, y) \rightarrow R(y, x)).$$

This can be translated into a DNF propositional formula by taking a disjunction over all possible $H \subseteq [n]$; we shall denote the resulting formula by RAM_n and it will feature later on as well. To avoid the nuisance of conditioning everything on R being symmetric and anti-reflexive, we build the formula RAM_n from atoms r_e, where e runs over all $\binom{n}{2}$ possible edges among the vertices of $[n]$. We think of e as being an unordered pair of vertices and so it makes sense to write $e \subseteq H$, i.e. both endpoints of e are in H. The formula RAM_n is then given by

$$\bigvee_H \bigwedge_{e \subseteq H} r_e \vee \bigvee_H \bigwedge_{e \subseteq H} \neg r_e,$$

where H ranges over all subsets of $[n]$ of size $\lfloor (\log n)/2 \rfloor$. Note that the formula is a $(\log n/2)^2$-DNF with $2\binom{n}{\log n} \leq n^{\log n}$ terms.

The second example is similar but it leads to more complex formulas. The **tournament principle** speaks about **tournaments**, which are directed graphs without loops such that between any two different vertices is an edge in exactly one direction:

$$Tour(R, n) \ := \ \forall x \neg R(x, x) \ \wedge \ \forall x, y\, R(x, y) \equiv \neg R(y, x)\ .$$

Think of the edge from i to j as meaning that player i beats player j in some game. The principle says that in any tournament of size n there is a dominating set $D \subseteq [n]$ of size $\lceil \log n \rceil$: every player outside D has lost to some player in D; thus $TOUR(R, n)$ is defined by

$$Tour(R, n) \ \rightarrow \ \exists D \subseteq [n](|D| = \lceil \log n \rceil) \forall y \notin D \exists x \in D\, R(x, y).$$

Using atoms r_{ij} for the (now directed) edges, the propositional translation $TOUR_n$ is given by

$$\bigvee_D \bigwedge_{j \notin D} \bigvee_{i \in D} r_{ij}.$$

This is not a DNF but a formula of Σ-depth 2.

While the negation of RAM_n is a set of clauses, the negation of $TOUR_n$ is a set of $(\log n)$-DNFs and it still makes good sense to talk about their $\mathrm{R}^*(\log)$-refutations.

Both the formulas $\mathrm{RAM}(R,n)$ and $TOUR(R,n)$ are what we shall call $b(n)$-DNF_2 formulas, where $b(n)$ is any function majorized by n. Such formulas Ψ are built from basic formulas by applying first disjunctions and existential quantifiers over elements relativized to some subset X of the universe of size $\leq b(n)$, then conjunctions and universal quantifiers and finally disjunctions and existential quantifiers either over elements or over subsets X of the universe of size $\leq b(n)$. Then $\langle\neg\Psi\rangle_n$ is a set of at most $O(n^{b(n)})$ formulas, each being an $O(b(n))$-DNF.

Now we will generalize Theorem 13.1.2.

Theorem 13.1.4 *Let $b(n) \leq n$ and let Ψ be a $b(n)$-DNF_2 system. Let q be the maximum arity of a relation symbol in Ψ. Assume that there is an infinite structure for the language of Ψ such that no existential set-quantifier in Ψ is witnessed by a finite subset of the structure.*

Then, for any $t \geq b(n)$, any t-DNF tree T of height h solving $\mathrm{Search}(\langle\neg\Psi\rangle_n)$ *must satisfy*

$$(h+1)\cdot t \geq \frac{n}{q}.$$

Proof The proof is the same as that of Theorem 13.1.2. The tree cannot find a falsified initial clause of $\mathrm{Search}(\langle\neg\Psi\rangle_n)$, as that would mean finding a set witnessing one of the existential set-quantifiers and the image of that set under the partial isomorphism h_ℓ would witness the same quantifier in the infinite structure. But that is impossible by the hypothesis. □

We get from Theorem 13.1.4 a corollary similar to Corollary 13.1.3.

Corollary 13.1.5 *Let q be the maximum arity of a relation symbol occurring in a $b(n)$-DNF_2 sentence Ψ. Assume that Ψ is valid in all finite structures but that in some infinite structure no set-quantifier of Ψ is witnessed by a finite set.*

Then the size s of any $\mathrm{R}^*(\log)$*-refutation of $\langle\neg\Psi\rangle_n$ must satisfy*

$$s \geq 2^{\Omega((n/q)^{1/2})}.$$

In particular, this lower bound is true for $TOUR_n$.

The particular case of $TOUR_n$ follows as it is easy to construct an infinite tournament having no finite dominating set. However we cannot use the argument to get a lower bound for $\mathrm{R}^*(\log)$-refutations of RAM_n as well, because any infinite graph has finite (even infinite) homogeneous subgraphs. We shall return to RAM_n in the next section and again in Chapter 15.

We conclude the section by complementing the lower bound criterion by the opposite statement. We state it only for DNF_1 formulas.

Lemma 13.1.6 *Assume that Φ is a* DNF$_1$ *formula that fails in all finite as well as in all infinite structures. Then the CNFs $\langle\neg\Phi\rangle_n$ have size $n^{O(1)}$* R$^*(\log n)$*-refutations as well as size $n^{O(1)}$* R*-refutations.*

Proof The hypothesis of the lemma implies that $\neg\Phi$ is logically valid and hence, by the completeness theorem, provable in the predicate calculus alone. In particular, it is also provable in $T^1_1(\alpha)$ and the lemma follows by Lemma 10.5.1. □

We remark that Lemma 13.1.6 can be improved to get R$^*(O(1))$-refutations: the proof of $\neg\Phi$ in predicate logic uses no $\Sigma_0^{1,b}$-part of the induction formulas (which in Lemma 10.5.1 result in the $\log n$ term); in fact, it uses no induction at all.

13.2 Relativization

In the previous section we confined ourselves for the sake of simplicity to DNF$_1$- and DNF$_2$-formulas that translate into propositional DNFs. However, the construction and the criterion in Theorem 13.1.2 apply to any formula pre-processed with a little bit of logic. Let Φ be any first-order sentence in any language L. Write Φ in a prenex form:

$$\forall x_1 \exists y_1 \ldots \forall x_k \exists y_k\ \varphi(\overline{x}, \overline{y}),$$

where φ is an open formula in conjunctive normal form. Define the **Skolemization** Φ_{Sk} of the formula Φ to be

$$\forall x_1 \ldots \forall x_k\ \varphi(\overline{x}, \ldots, y_i/h_i(x_1, \ldots, x_i), \ldots),$$

where $h_i(x_1, \ldots, x_i)$, $i = 1, \ldots, k$, are *new* function symbols (the so-called **Skolem functions**). Let L_{Sk} be the language of Φ_{Sk}.

Now rewrite Φ_{Sk}, using more universally quantified variables, in a form where all atomic formulas are $R(z_1, \ldots, z_\ell)$, $f(z_1, \ldots, z_\ell) = z_{\ell+1}$ or $u = v$, where $\overline{z}, u, v$ are variables and R and f are ℓ-ary relational and function symbols, respectively. For example, $\neg R(g(x), h(u, v))$ can be written as

$$\forall y, z\ \neg g(x) = y \vee \neg h(u, v) = z \vee \neg R(y, z)$$

(it would be natural to write this formula using implications but in the end we want to get a DNF$_1$-formula and such formulas do not use implications). Let us denote the resulting sentence by Φ_1.

Next we want to get rid of all function symbols: for any $f(z_1, \ldots, z_\ell)$ from L_{Sk} introduce a *new* $(\ell + 1)$-ary relation symbol $F(z_1, \ldots, z_{\ell+1})$ and change all atomic formulas $f(z_1, \ldots, z_\ell) = z_{\ell+1}$ in Φ_1 to $F(z_1, \ldots, z_{\ell+1})$. We shall call the new symbols F **function-graph relation symbols**. The formula Φ_2 is a formula in a *relational* language that we shall denote by L_2 (i.e. it is L_{Sk} with all function symbols f replaced by the associated relation symbols F), obtained by this process from Φ_1 together with the universal closures of the formulas

$$(F(z_1,\ldots,z_\ell,u) \wedge F(z_1,\ldots,z_\ell,v)) \to u = v.$$

Note that it is still a CNF prefixed by universal quantifiers.

Let Φ_3 be Φ_2 in conjunction with all the sentences

$$\forall z_1,\ldots,z_\ell \exists z_{\ell+1}\, F(\overline{z}), \tag{13.2.1}$$

for all the function-graph relation symbols in L_2.

Finally, we arrive at the formula we are aiming at: the **relativization** Φ^{rel} of Φ is a formula constructed as follows. Let $U(x)$ be a new unary relation symbol. The formula Φ^{rel} is the conjunction of the formulas

- $\exists x U(x)$,
- Φ_2^U, which is Φ_2 with all (universal) quantifiers relativized to U, i.e. each $\forall y \ldots$ is replaced by $\forall y\, \neg U(y) \vee \ldots$,
- $\forall z_1,\ldots,z_\ell \exists z_{\ell+1} \bigvee_{i\leq\ell} \neg U(z_i) \vee F(z_1,\ldots,z_\ell,z_{\ell+1})$, for all function-graph relation symbols,
- $\forall z_1,\ldots,z_{\ell+1} \bigvee_{i\leq\ell} \neg U(z_i) \vee \neg F(z_1,\ldots,z_\ell,z_{\ell+1}) \vee U(z_{\ell+1})$, again for all function-graph symbols F.

Recall from Sections 10.5 or 13.1 the notion of a DNF$_1$-formula. The following lemma follows from the construction.

Lemma 13.2.1 *(i)* $\neg\Phi^{rel}$ *is a* DNF$_1$*-formula and its* $\langle\ldots\rangle_n$ *translation yields, for each* $n \geq 1$*, a CNF.*

(ii) Any model of Φ *can be expanded into a model of* Φ_3 *and hence of* Φ^{rel}*, and any model of* Φ^{rel} *contains a substructure (whose universe is the interpretation of* U*) that is a model of* Φ*. In particular, if* Φ *has an infinite model, so does* Φ^{rel}*.*

Theorem 13.2.2 (Dantchev and Riis [164]) *Let* Φ *be an arbitrary first-order sentence and assume that* Φ *has no finite models but that it does have an infinite model.*

Then the $\langle\Phi^{rel}\rangle_n$*,* $n \geq 1$*, are unsatisfiable CNFs of size* $n^{O(1)}$ *and there exists* $\delta > 0$ *depending on* Φ *but not on* n *such that every* R*-refutation of* $\langle\Phi^{rel}\rangle_n$ *has at least* 2^{n^δ} *clauses.*

The non-existence of a finite model implies the unsatisfiability of the resulting CNFs, and we need to prove only the lower bound. This will use the existence of an infinite model. The proof will utilize two lemmas that we shall prove first and it will be summarized at the end of the section.

Think of the formula $\langle\Phi_3\rangle_m$ as having atoms $r_{i_1,\ldots,i_\ell}$ representing atomic sentences with the original relation symbols R of L, and $f_{i_1,\ldots,i_{\ell+1}}$ representing the new function-graph symbols F of L_2. The **support** of an r-atom $r_{i_1,\ldots,i_\ell}$ is $\{i_1,\ldots,i_\ell\} \subseteq [m]$ and that of an f-atom $f_{i_1,\ldots,i_{\ell+1}}$ is also $\{i_1,\ldots,i_\ell\}$ (i.e. $i_{\ell+1}$ is not included). The support of a clause C, $supp(C)$, is the union of the supports of all atoms that occur in C (positively or negatively).

Lemma 13.2.3 *Let $q \geq 1$ be the maximum arity of a relation or a function symbol in L_{Sk} (i.e. it bounds the size of the support of any atom in $\langle \Phi_3 \rangle_m$).*

Then any R-refutation of $\langle \Phi_3 \rangle_m$ must contain a clause C whose support satisfies

$$|supp(C)| \geq \Omega(m^{1/q}) .$$

Proof Let $\mathbf{M}$ be an infinite model of Φ (with universe M) and, by Lemma 13.2.1(ii) consider that its expansion is a model of Φ_3. Let π be an R-refutation of $\langle \Phi_3 \rangle_m$. We shall attempt to construct a sequence of clauses $C_0, C_1, \ldots, C_e$, starting with the empty end-clause $C_0 := \emptyset$ of π, and such that C_{t+1} is one of the two clauses used in π to derive C_t and C_e is an initial clause (i.e. from $\langle \Phi_3 \rangle_m$). The requirements on the sequence will be such that at a certain step the construction cannot be continued, and this will imply the lower bound.

The requirements ask that we should also construct a sequence of partial injective maps $\alpha_t :\subseteq [m] \to M$ having the following properties:

1. $supp(C_t) \subseteq dom(\alpha_t)$,
2. whenever $f_{i_1,\ldots,i_\ell,i_{\ell+1}}$ is a function-graph atom in C_t (and hence $\{i_1, \ldots, i_\ell\} \subseteq supp(C_t)$) then there is a $j \in dom(\alpha_t)$ (not necessarily in $supp(C_t)$ or equal to $i_{\ell+1}$) such that
$$\mathbf{M} \models F(\alpha_t(i_1), \ldots, \alpha_t(i_\ell), \alpha_t(j));$$
3. α_t defines a partial truth-assignment $\hat{\alpha}_t$ by
   - if $r_{i_1,\ldots,i_\ell}$ is a relation atom and $\{i_1, \ldots, i_\ell\} \subseteq dom(\alpha_t)$ then
$$\hat{\alpha}_t(r_{i_1,\ldots,i_\ell}) := \begin{cases} 1 & \text{if } \mathbf{M} \models R(\alpha_t(i_1), \ldots, \alpha_t(i_\ell)), \\ 0 & \text{otherwise,} \end{cases}$$
   - if $f_{i_1,\ldots,i_\ell,i_{\ell+1}}$ is a function-graph atom and $\{i_1, \ldots, i_\ell\} \subseteq dom(\alpha_t)$ then
$$\hat{\alpha}_t(f_{i_1,\ldots,i_\ell,i_{\ell+1}}) := 1, \text{ if } i_{\ell+1} \in dom(\alpha_t) \text{ and } \mathbf{M} \models F(\alpha_t(i_1), \ldots, \alpha_t(i_{\ell+1})),$$
$$\hat{\alpha}_t(f_{i_1,\ldots,i_\ell,i_{\ell+1}}) := 0, \text{ if } i_{\ell+1} \neq j \in dom(\alpha_t) \text{ and } \mathbf{M} \models F(\alpha_t(i_1), \ldots, \alpha_t(i_\ell), \alpha_t(j));$$
4. $\hat{\alpha}_t(C_t) = 0$.

At the beginning we take $\alpha_0 := \emptyset$. Assume that we already have $C_0, \ldots, C_t$ and $\alpha_0, \ldots, \alpha_t$ obeying all conditions above, and assume that C_t was derived as

$$\frac{D \cup \{p\} \quad E \cup \{\neg p\}}{C_t (= D \cup E)},$$

where p is some atom (relation or function-graph). We attempt to define C_{t+1} and α_{t+1} by considering three cases, depending on what is p.

(a) The support of p is included in $dom(\alpha_t)$: then $\hat{\alpha}_t$ decides its truth-value and we put

$$C_{t+1} := \begin{cases} D \cup \{p\} & \text{if } \hat{\alpha}_t(p) = 0, \\ E \cup \{\neg p\} & \text{if } \hat{\alpha}_t(p) = 1, \end{cases}$$

and take for α_{t+1} the $\subseteq$-minimum submap of α_t satisfying 1–3 above (i.e. we forget some values that are not needed for item 4 with respect to C_{t+1}).

(b) The atom p is a relation atom $r_{i_1,\dots,i_\ell}$ and $\{i_1,\dots,i_\ell\} \not\subseteq dom(\alpha_t)$: extend α_t to some β such that $\{i_1,\dots,i_\ell\} \subseteq dom(\beta)$ (there is room for that as M is infinite) and define

$$C_{t+1} := \begin{cases} D \cup \{p\} & \text{if } \hat{\beta}(p) = 0, \\ E \cup \{\neg p\} & \text{if } \hat{\beta}(p) = 1, \end{cases}$$

and again take for α_{t+1} the $\subseteq$-minimum submap of β satisfying 1–3 with respect to C_{t+1}.

(c) (This is the case that will imply the lower bound.) The atom p is a function-graph atom $f_{i_1,\dots,i_\ell,i_{\ell+1}}$ and $\{i_1,\dots,i_\ell\} \not\subseteq dom(\alpha_t)$.

We first extend α_t to some injective β such that $\{i_1,\dots,i_\ell\} \subseteq dom(\beta)$. Then we examine $\mathbf{M}$ and take the unique $u \in M$ such that

$$\mathbf{M} \models F(\beta(i_1),\dots,\beta(i_\ell),u).$$

If $u \in rng(\beta)$, say $\beta(j) = u$, then $\hat{\beta}$ gives a truth-value to the atom and we can define C_{t+1} and α_{t+1} as in (a) and (b).

The *crucial case* is when $u \notin rng(\beta)$. Then – as the map we are constructing ought to be injective – we need to take $j \notin dom(\beta)$, extend β to β' by stipulating $\beta'(j) = u$ and define C_{t+1} and α_{t+1} as before. However, there need not be any $j \notin dom(\beta)$, i.e. it can be the case that $dom(\beta) = [m]$.

Indeed, this situation must occur during the construction of the sequence of C_t and α_t as otherwise we would reach an initial clause C_e and its falsifying assignment $\hat{\alpha}_e$, which copies a substructure of $\mathbf{M}$ (i.e. $rng(\alpha_e)$). But that is impossible as $\mathbf{M} \models \Phi_3$, i.e. all substructures satisfy all clauses.

If α_t is the map in (c) that cannot be extended to any suitable β and β', and $w = |supp(C_t)|$, then the number of j's in $dom(\beta)$ that are *not* in $supp(C_t)$ is bounded above by

$$O(w^q)$$

(recall that q is the maximum arity): there is a constant number of functions in L_1 (the O-constant), and w^q bounds the number of all possible input tuples for which we need to give a value. Hence

$$m \leq O(w^q) + w + q \leq O(w^q)$$

for $q \geq 1, w \geq 2$, where the second summand w is the size of $supp(C_t)$ and the last summand bounds the size of the support of the resolved atom.

This proves the lemma. □

The next lemma will essentially show that if a short R-refutation exists then there is another in which all the clauses have a small support. In it we will use **restrictions** of clauses by a partial truth assignment, as in Section 5.4. In particular, there we defined the restriction of C by setting the value of a literal ℓ to $\epsilon \in \{0,1\}$, $C \restriction \ell = \epsilon$. If ρ is a partial truth assignment then

$$C^{\rho} := (C \restriction p_1 = \rho(p_1)) \restriction p_2 = \rho(p_2) \ldots, \quad \text{where } \{p_1, p_2, \ldots\} \text{ is } dom(\rho).$$

Lemma 13.2.4 *Let σ be an R-refutation of $\langle \Phi^{rel} \rangle_n$ of size $s \leq 2^{n^\delta}$, where $\delta := 1/q$ (q is the maximum arity as before).*

Then there is an m, $n/3 \leq m \leq 2n/3$, and an R_w-refutation of $\langle \Phi_3 \rangle_m$ in which the size of the support of any clause is at most $m^{1/(q+1)}$.

(R_w from Section 5.4 is the extension of R by the weakening rule and by allowing the constant 1.)

Proof Define a partial truth assignment to the atoms of $\langle \Phi^{rel} \rangle_n$ by the following process.

1. Each u_i is given the value 0 or 1, uniformly and independently.

By the Chernoff bound we get

Claim 1 With the probability of failing at most $2^{-\Omega(n)}$, the set

$$U^{\rho} := \{i \in [n] \mid \rho(u_i) = 1\}$$

has size $n/3 \leq |U^{\rho}| \leq 2n/3$.

Let $m := |U^{\rho}|$.

1. Assign to each relation atom $r_{i_1,\ldots,i_\ell}$ such that $\{i_1,\ldots,i_\ell\} \not\subseteq U^{\rho}$ either 0 or 1, uniformly and independently.
2. Assign to each function-graph atom $f_{i_1,\ldots,i_\ell,j}$ the value 0, if $\{i_1,\ldots,i_\ell\} \subseteq U^{\rho}$ and $j \notin U^{\rho}$. If $\{i_1,\ldots,i_\ell\} \not\subseteq U^{\rho}$ and j is arbitrary, assign to $f_{i_1,\ldots,i_\ell,j}$ either 0 or 1, uniformly and independently.

Claim 2 Assume the clause C in σ contains literals $\ell_1,\ldots,\ell_t$ that correspond to relation atoms or to function-graph atoms (i.e. not to U-atoms) and are such that

$$supp(\ell_u) \not\subseteq \bigcup_{v<u} supp(\ell_v),$$

for $u = 2,\ldots,t$. Then

$$Prob_{\rho}[C^{\rho} \neq \{1\}] \leq (3/4)^t .$$

Imagine that ρ has been defined on the literals $\ell_1, \ldots, \ell_{u-1}$ and U^ρ has been defined for the elements in $\bigcup_{v<u} supp(\ell_v)$ only, and not for some

$$i \in supp(\ell_u) \setminus \bigcup_{v<u} supp(\ell_v) \ .$$

Then, with probability $1/2$, we put $u_i = 0$ (i.e. stipulate that $i \notin U^\rho$) and, in that case, with probability $1/2$ we make the literal ℓ_u true. That is, we will fail to reduce C to $\{1\}$ with probability at most $3/4$. These events are independent for $u = 1, \ldots, t$ and the claim follows.

Claim 3 Assume that $w = |supp(C)|$ and $q(t-1) < w$. Then there are literals $\ell_1, \ldots, \ell_t$ satisfying the hypothesis of Claim 2.

The union of the supports of $t-1$ literals has size at most $q(t-1) < w$, so there must be one whose support is not covered yet.

The last two claims imply

Claim 4 If $w = |supp(C)|$ for $C \in \sigma$ then

$$Prob_\rho[C^\rho \neq \{1\}] \leq (3/4)^{w/q} \ .$$

To prove the lemma take $w := (n/3)^{1/(q+1)}$. For a random ρ, $m \geq n/3$ with an exponentially small probability of failing (by Claim 1) and, by Claim 4, σ^ρ fails to have the required properties with probability at most

$$s(3/4)^{w/q} = 2^{n^\delta}(3/4)^{(1/q)(n/3)^{1/(q+1)}}$$

which goes to 0 by a suitable choice of δ. □

Proof of Theorem 13.2.2 An R-refutation of $\langle \Phi^{rel} \rangle_n$ of size $\leq 2^{n^{1/q}}$ yields (via Lemma 13.2.4) an R_w-refutation of $\langle \Phi_3 \rangle_m$ in which the support of every clause is of size less than $m^{1/(q+1)}$, contradicting Lemma 13.2.3. □

Simple examples of Φ to which Theorem 13.2.2 applies include the pigeonhole principle and the modular counting principles from Section 11.1, the usual axioms of dense linear orderings, the usual axioms of non-commutative fields reaching outside combinatorics and many others.

We shall now return to the Ramsey formulas $\mathrm{RAM}(n, R)$ and RAM_n from the previous section and show that the method behind the proof of Theorem 13.2.2 applies to them as well, if the relativization is suitably defined, even if the theorem itself does not. We shall assume that $n = 4^k$ and hence RAM_n says that a graph on n vertices has a homogeneous subgraph of size at least k. We shall stress this set-up by using the notation $\mathrm{RAM}(n, k)$.

The first relativization $\mathrm{RAM}^U(n, k)$ formalizes that for any $U \subseteq [n]$ either the induced subgraph with vertex set U or the induced subgraph with the vertex set $[n] \setminus U$ contains a homogeneous subgraph of size $k-1$. The propositional formulation

$\mathrm{RAM}^U(n,k)$ is a DNF in the variables r_e used already in RAM_n but using also new variables u_i that translate the atomic sentences $i \in U$. It is the disjunction of the following $4\binom{n}{k-1}$ terms of size $k-1+\binom{k}{2}$:

$$\bigwedge_{i\in H} u_i \wedge Cli(H) \quad \text{and} \quad \bigwedge_{i\in H} \neg u_i \wedge Cli(H)$$

and

$$\bigwedge_{i\in H} u_i \wedge Ind(H) \quad \text{and} \quad \bigwedge_{i\in H} \neg u_i \wedge Ind(H)$$

where H ranges over the subsets of $[n]$ of size $k-1$ and we use the abbreviations $Cli(H)$ and $Ind(H)$ for two relativization as follows:

$$Cli(H) := \bigwedge_{e\subseteq H} r_e \quad \text{and} \quad Ind(H) := \bigwedge_{e\subseteq H} \neg r_e.$$

The second relativization is stated in a roundabout way but it is technically a little easier to use. The formula $\mathrm{RAM}^f(n,k)$ says that for any injective function $f : [n/4] \to [n]$ there is an induced homogeneous subgraph of size $k-1$ whose vertex set is included in the range $rng(f)$ of f. This is a true principle as, owing to the injectivity of f, the range $rng(f)$ has size 4^{k-1}.

The propositional version $\mathrm{RAM}^f(n,k)$ again uses the atoms r_e and also new atoms $f_{i,j}$, where $i \in [n/4]$ and $j \in [n]$. The formula is the disjunction of the following terms:

1. $\bigwedge_j \neg f_{i,j}$, for any i;
2. $f_{i,j_1} \wedge f_{i,j_2}$, for any i and $j_1 \neq j_2$;
3. $f_{i_1,j} \wedge f_{i_2,j}$, for any $i_1 \neq i_2$ and j;
4. $f_{i_1,j_1} \wedge \cdots \wedge f_{i_k,j_k} \wedge Cli(\{j_1,\ldots,j_k\})$, for any ordered k-tuples of different elements $i_1,\ldots,i_k \in [n/4]$ and $j_1,\ldots,j_k \in [n]$;
5. $f_{i_1,j_1} \wedge \cdots \wedge f_{i_k,j_k} \wedge Ind(\{j_1,\ldots,j_k\})$, for any ordered k-tuples of different elements $i_1,\ldots,i_k \in [n/4]$ and $j_1,\ldots,j_k \in [n]$.

Here $i, i_1, i_2 \in [n/4]$ and $j, j_1, j_2 \in [n]$. Note that $\mathrm{RAM}^f(n,k)$ has $n^{O(k)}$ terms. Because of item 1 they are not narrow (as all the terms in $\mathrm{RAM}^U(n,k)$ are).

Now we do not have an infinite structure to guide us when forcing a path through a refutation along falsified clauses. But, for each $\ell \gg 0$ we do have a graph that has no homogeneous subgraph of size ℓ and that has at least $2^{\ell/2}$ vertices; this was proved by one of the first probabilistic arguments in combinatorics by Erdös [175]. Denote such a graph by E_ℓ.

For a clause C using the variables r_e of $\mathrm{RAM}(m,\ell)$, denote by $Ver(C)$ all vertices of any edge e for which the atom r_e occurs in C. Clearly the width of C satisfies

$$\mathbf{w}(C) \leq \binom{|Ver(C)|}{2}.$$

We may use the graph E_ℓ in place of the infinite structure $\mathbf{M}$ used earlier, building a path of clauses C and, for each, a graph G_C on $Ver(C)$ and a partial isomorphism between G_C and a subgraph of E_ℓ, such that the evaluation of the atoms in C according to G_C falsifies C. This process goes on as before except that we could now run out of vertices in E_ℓ when trying to evaluate a resolved atom (earlier this could not happen as $\mathbf{M}$ was infinite). However, as long as $|Ver(C)| < |E_\ell| = 2^{\ell/2}$ the process may continue. At the end of the path we reach an initial clause D of RAM(M, ℓ). Such a clause D is falsified if and only if G_D is either a clique or an independent set of size ℓ; but that is impossible as G_D is isomorphic to a subgraph of E_ℓ which does not have sufficiently large homogeneous subgraphs.

To avoid a contradiction, the construction of the path must stop earlier and hence we must encounter a clause with $|Ver(C)| \geq 2^{\ell/2}$. Clearly the width of C satisfies $2\mathbf{w}(C) \geq |Ver(C)|$. This yields the following lemma.

Lemma 13.2.5 *Any R-proof of* RAM(m, ℓ) *must have width at least*

$$\frac{1}{2} 2^{\ell/2}.$$

A lower bound for the width may, in principle, imply a lower bound for the size, by Theorem 5.4.5, but in this case it is too small compared with the number of variables. We shall study how to use the width for the size lower bound in Section 13.3.

Instead we shall use the partial restriction method from the proof of Theorem 13.2.2. Choosing an evaluation of the atoms u_i in RAM^U randomly and uniformly or the atoms f_{ij} in RAM^f randomly among all injective maps from $[n/4]$ into $[n]$, calculations completely analogous to these in the proof of Lemma 13.2.4 yield the following statements.

Lemma 13.2.6 *Let $k \geq 2$ and $n = 4^k$. Assume that there is an R-proof of* RAM$^U(n, k)$ *of size $s \leq 2^{n^{1/11}}$. Then* RAM$(n, k-1)$ *has an R-proof of width at most $n^{1/5}$.*

The same statement is true for the formula RAM$^f(n, k)$.

We may combine the two lemmas (Lemma 13.2.5 plays the role of Lemma 13.2.3 and Lemma 13.2.6 that of Lemma 13.2.4) and obtain, in analogy with Theorem 13.2.2, the following lower bounds.

Theorem 13.2.7 *Let $k \geq 2$ and $n = 4^k$. Then every R-proof of* RAM$^U(n, k)$ *or of* RAM$^f(n, k)$ *must have size at least $\Omega(2^{n^{1/11}})$.*

13.3 Width and Expansion

The arguments in Sections 13.1 and 13.2 are explicitly or implicitly linked with the width of refutations. In this section we shall study how to derive size lower bounds directly from Theorem 5.4.5, i.e. from width lower bounds. There is, however, an a priori problem. A DNF that expresses some combinatorial principle is the translation $\langle\Phi\rangle_n$ of a first-order sentence Φ in a DNF_1 form. For it to be non-trivial the language of Φ must contain at least one non-unary relation symbol. Hence the number of atoms will be at least $\binom{n}{2}$ while the natural parameter of the tautology is n, the size of the universe, and often $O(n)$ will upper-bound the width of an R-refutation of the tautology proceeding by some sort of induction. However, to be able to use Theorem 5.4.5 we would need a width lower bound of at least $n^{1+\Omega(1)}$.

Take, for example, the pigeonhole principle PHP_n formulas (1.5.1). If Φ from Section 13.2 is the PHP principle expressed using a function symbol f and a constant c, where

- *f is an injective map from the universe into itself whose range does not contain c,*

the sentence Φ_3 is the PHP(R)-formula expressed using a relation symbol R and thus PHP_n is $\langle\Phi_3\rangle$. Hence Lemma 13.2.3 yields an $\Omega(n)$ lower bound for R-refutations of PHP_n. However, PHP_n has $(n+1)n$ atoms and Theorem 5.4.5 yields nothing (not to mention that the maximum width of an initial clause is n itself).

One possible way (we shall see another after Theorem 13.3.4) to avoid these two problems, i.e. too-wide initial clauses and too many variables, is to fix a parameter $d \geq 2$ and choose for all $n \gg 1$ in advance a binary relation $E \subseteq [n+1] \times [n]$ such that

- $deg_E(i) \leq d$ for all $i \in [n+1]$,

where $deg_E(i)$ is the number of $j \in [n]$ such that $(i,j) \in E$ (the graph-theoretic degree of the node), and then express the PHP only for $R \subseteq E$. Working directly with the formula PHP_n means that we substitute 0 for all atoms p_{ij} such that $(i,j) \notin E$. The resulting formula, to be denoted by E-PHP_n, has at most dn atoms and size $O(n^3)$ (more precisely $O(ne^2)$ where $e \leq n+1$ is the maximum degree of some $j \in [n]$, often $O(d)$ too), and the maximum width of its clauses is $\mathbf{w}(E\text{-}PHP_n) = d$. Hence if we find a relation E with these properties such that the minimum width of a refutation of E-PHP_n must be $\Omega(n)$ then we obtain a size lower bound $2^{\Omega(n)}$ via Theorem 5.4.5.

We need not expect that refutations of E-PHP_n will be hard for all E. For example, if we choose $E := [n+1] \times \{1\}$ then a constant-size refutation exists. A crucial idea of Ben-Sasson and Wigderson [73] was that graphs E with a suitable *expansion property* will yield instances of the PHP principle requiring refutations of large width.

We shall present the construction and the argument in a way that we can also use with little modification in the next section (Lemmas 13.4.4 and 13.4.5). We will also consider the $WPHP_n^m$ principles for $m > n$.

To define a relation $E \subseteq [m] \times [n]$ we define its **incidence matrix** A as an $m \times n$ matrix A over the two-element field $\mathbf{F}_2$ such that $(i,j) \in E$ if and only if $A(i,j) = 1$. For a parameter $1 \leq \ell \leq n$, call such a matrix **ℓ-sparse** if and only if each row contains at most ℓ non-zero entries. Put $J_i := \{j \in [n] \mid A_{ij} = 1\}$; hence $|J_i| \leq \ell$ if A is ℓ-sparse. A **boundary** of a set of rows $I \subseteq [m]$, denoted by $\partial_A(I)$, is the set of all $j \in [n]$ such that for exactly one $i \in I$ it holds that $A_{ij} = 1$.

Let $1 \leq r \leq m$ and $\epsilon > 0$ be any parameters. The matrix A is an (r, ϵ)-**expander** if and only if for all $I \subseteq [m]$, $|I| \leq r$, we have $|\partial_A(I)| \geq \epsilon\ell|I|$. Expanders simulate, in a sense, matrices with disjoint sets J_i of the maximum size ℓ. In such a case it would hold that $|\partial_A(I)| = \ell|I|$. An (r, ϵ)-expander achieves (as long as $|I| \leq r$) at least an ϵ-percentage of this maximum value.

The following existence statement can be proved by a probabilistic argument. Consider the following random process yielding an ℓ-sparse matrix A. For every $i \in [m]$ and $u \leq \ell$ let $j_{i,u}$ be chosen independently and uniformly at random from $[n]$. Let $J_i \subseteq [n]$ be the set of these values for fixed i, and define $A_{i,j} = 1$ if and only if $j \in J_i$. The following theorem is a special case of [12, Theorem 5.1].

Theorem 13.3.1 (Alekhnovich *et al.* [12, Theorem 5.1]) *For every $\delta > 0$, there is an $\ell \geq 1$ such that for all sufficiently large n the random ℓ-sparse $n^2 \times n$-matrix constructed by the random process above is an $(n^{1-\delta}, 3/4)$-expander with probability approaching* 1.

For a set of rows $I \subseteq [m]$, let $J(I) := \bigcup_{i \in I} J_i$, and let A_I be the $(m-|I|)\times(n-|J(I)|)$-matrix obtained from A by deleting all the rows in I and all the columns in $J(I)$.

Lemma 13.3.2 *Assume that A is an ℓ-sparse $m \times n$-matrix that is an $(r, 3/4)$-expander. For any set of rows $I \subseteq [m]$ of size $|I| \leq r/2$ there is an $\hat{I} \supseteq I$, $|\hat{I}| \leq 2|I|$, such that, for any $i \notin \hat{I}$,*

$$|J_i \setminus \bigcup_{u \in \hat{I}} J_u| \geq \frac{\ell}{2}. \tag{13.3.1}$$

Moreover, for any $\hat{I}$ of size $|\hat{I}| \leq r$ having the property (13.3.1), *$A_{\hat{I}}$ is an $(r, 1/4)$-expander. Furthermore, there exists a smallest (w.r.t. inclusion) such $\hat{I}$.*

Proof Assume that $|I| \leq r/2$. Put $I_0 := I$. Add to I_t in consecutive steps $t = 0, \ldots$ any one row i as long as

$$|J_i \cap \bigcup_{k \in I_t} J_k| > \frac{\ell}{2}. \tag{13.3.2}$$

We claim that this process stops before t reaches $|I|$. Assume not, and let $I' = I_{|I|}$. Then, by (13.3.2), it holds that

$$\partial_A(I') < \ell|I| + (\ell/2)|I| = (3/4)\ell|I'|.$$

This contradicts the expansion property of A, as $|I'| \leq r$.

Let $\hat{I}$ be the last I_t in the process, so that $t < r/2$ and $|\hat{I}| \leq 2|I|$. Clearly $\hat{I}$ has the property (13.3.1). Thus we need only to verify the expansion property of $A_{\hat{I}}$. Let K be a set of $\leq r$ rows in $A_{\hat{I}}$. Then

$$\partial_{A_{\hat{I}}}(K) = \partial_A(K) \setminus \bigcup_{i \in \hat{I}} J_i(A).$$

As for all $i \in K \setminus \hat{I}$ we have $|J_i(A) \cap \bigcup_{k \in \hat{I}} J_k(A)| \leq \ell/2$, this equality implies that

$$|\partial_{A_{\hat{I}}}(K)| \geq |\partial_A(K)| - (\ell/2)|K| \geq (3/4)\ell|K| - (\ell/2)|K| \geq (1/4)|K|.$$

□

Any I satisfying the condition (13.3.1) from Lemma 13.3.2 will be called a **safe set of rows**. The following is the key lemma.

Lemma 13.3.3 *Let $1 \leq \ell \leq n < m \leq n^2$ and assume that $E \subseteq [m] \times [n]$ is a bipartite graph whose incidence matrix is A, an ℓ-sparse $m \times n$-matrix that is an $(r, 3/4)$-expander.*

Then any R*-refutation of E-*WPHP$_n^m$ *must have width at least $r/2$.*

Proof The argument is analogous to similar adversary arguments in Sections 13.1 and 13.2. Assume an R-refutation π of E-WPHP$_n^m$ has width $w < r/2$. We shall construct a path $C_0, C_1, \ldots$ of clauses in π, and partial injective maps $\alpha_0, \alpha_1, \ldots$ for all $\alpha_t \subseteq E$ such that:

- C_0 is the empty end-clause and $\alpha_0 := \emptyset$;
- C_{t+1} is one of the two hypotheses of the inference yielding C_t, for all t;
- $|\alpha_t| \leq 2w$ and α_t forces C_t to be false (as in the adversary arguments before);
- $dom(\alpha_t)$ is a safe set of rows (i.e. of pigeons, see below).

Having C_t, α_t, let p_{uv} be the atom resolved when deriving in π the clause C_t. If $u \in dom(\alpha_t)$, take $\alpha_{t+1} := \alpha_t$ and choose for C_{t+1} the hypothesis of the inference forced to be false by the map.

If $u \notin dom(\alpha_t)$ then, by the definition of safe sets of rows, there is at least one $j \in J_u \setminus \bigcup_{i \in dom(\alpha_t)} J_i$. In particular, $(u,j) \in E$. Add (u,j) to α_t to form the map β, take for C_{t+1} the hypothesis of the inference forced to be false by β and take for $\alpha_{t+1} \subseteq \beta$ some submap of β whose domain is a $\subseteq$-minimal safe set of rows that still forces C_{t+1} to be false.

As no map of size $2w < r \leq n$ can force any initial clause to be false, the width cannot obey the inequality $w < r/2$. □

Theorem 13.3.4 *Let $1 \leq n < m \leq n^2$. Then for any $\delta > 0$ there is a bipartite graph $E \subseteq [m] \times [n]$ such that any* R*-refutation of E-*WPHP$_n^m$ *must have size at least*

$$2^{\Omega(n^{2-2\delta}/m)}.$$

In particular, PHP$_n$ *requires* R*-refutations of size $2^{\Omega(n^{1-o(1)})}$ and* WPHP$_n^m$ *for $m \leq n^{2-\Omega(1)}$ requires* R*-refutations of size $2^{n^{\Omega(1)}}$.*

Proof An ℓ-sparse incidence matrix A defines a bipartite graph $E \subseteq [m] \times [n]$ whose degree for any $i \in [m]$ is bounded by ℓ, and hence the formula E-WPHP$_n^m$ has ℓm atoms and all its clauses have width at most ℓ.

Taking $\delta > 0$, $r := n^{1-\delta}$ and ℓ a large enough constant, Theorems 13.3.1, 5.4.5 and Lemma 13.3.3 yield the lower bound

$$2^{\Omega((r\ell-\ell)^2/\ell m)} \geq 2^{\Omega(n^{2-2\delta}/m)} .$$

The last part of the theorem follows since a lower bound for any E implies a lower bound for the complete graph $[m] \times [n]$. □

Now we present another way to modify the PHP so that a length-of-proof lower bound can be derived from a width lower bound. The modification, **bit** PHP, talks about the hole where pigeon i sits using not an n-tuple of atoms p_{ij} but only logarithmically many atoms. Assume $n = 2^k$. The formula $b\mathrm{PHP}_n$ is built from atoms q_{ij} with $i \in [n+1]$ and $j \in [k]$. For a given i we think of the truth values of the atoms $q_{i1}, \ldots, q_{ik}$ as defining a hole in $\{0,1\}^k$ in which i sits. We thus do not need to stipulate that every i sits somewhere. The negation of $b\mathrm{PHP}_n$ is the set of the following $\binom{n+1}{2}n$ clauses, each of size $2k$, expressing that the hole assignment is injective:

$$\bigvee_{j\in[k]} q_{i_1j}^{1-b_j} \vee \bigvee_{j\in[k]} q_{i_2j}^{1-b_j}, \qquad \text{for each } \overline{b} \in \{0,1\}^k \text{ and } i_1 \neq i_2 \in [n+1].$$

Theorem 13.3.5 *Assume that n is a power of* 2. *Every* R*-refutation of* $b\mathrm{PHP}_n$ *must have width at least* $n+1$ *and hence size at least* $2^{\Omega(n/\log n)}$.

Proof The width lower bound is proved by an adversary argument similar to those we have used already. Let π be an R-refutation of $b\mathrm{PHP}_n$. Walk, in π, from the end-clause towards the initial clauses, at any step keeping a partial injective map $f :\subseteq [n+1] \rightarrow \{0,1\}^k$ modified along the way. The map f held at a given step falsifies all literals in the current clause. In particular, the size of f needs to be at most the size of the clause and so we not only add to f but also delete from f any pairs not needed.

It is possible to maintain such maps f as long as $\mathbf{w}(\pi) \leq n$. That would take us to an initial clause of $b\mathrm{PHP}_n$ but none of these can be falsified in this way (f is injective). Hence $\mathbf{w}(\pi) \geq n+1$. Theorem 5.4.5 then implies a size lower bound of the form

$$2^{\Omega((n+1-2\log n)^2/(n\log n))} \geq 2^{\Omega(n/\log n)}.$$

□

We now mention a prominent class of tautologies, the so-called **Tseitin formulas**, introduced by Tseitin [492]. These formulas or their variants have found almost as many applications in proof complexity as the variants of the PHP.

Let $G = ([n], E)$ be a graph (tacitly simple, undirected and without loops). Let $f\colon [n] \rightarrow \{0,1\}$ be a function assigning to each vertex 0 or 1. Consider the set of equations in $\mathbf{F}_2$ in the variables x_e, where $e = \{i,j\}$ are the edges from E:

$$\bigoplus_{j:\{i,j\}\in E} x_{ij} = f(i), \qquad \text{for each } i \in [n] .$$

If d is the maximum degree of a vertex in G then each such condition can be written as a conjunction of 2^{d-1} d-clauses, resulting in a d-CNF Tseitin formula $TSE_{G,f}$ having at most $n2^{d-1}$ d-clauses in at most $dn/2$ atoms.

In the sum of all equations each edge appears twice and hence the sum is 0 in $\mathbf{F}_2$. Hence we have

Lemma 13.3.6 *Assume that $\sum_{i\in[n]} f(i) = 1$ in $\mathbf{F}_2$. Then $TSE_{G,f}$ is unsatisfiable.*

One can obviously restrict to only certain classes of G, f; it is enough to consider connected graphs G, and one may assume that f is identically equal to 1 and that n is odd. Any formula $TSE_{H,g}$ can be deduced shortly in R* from the *TSE*-formula for one of these special G, f classes.

Strong lower bounds are known for the $TSE_{G,f}$ formulas (and we shall see these in Section 15.4 for AC^0-Frege systems also). We state them without proof, as we shall prove a lower bound for a closely related formula in Lemma 13.4.5 and Corollary 13.4.6.

For a connected graph $G = ([n], E)$, define its **expansion**

$$e(G) := \min \{|E(W, [n] \setminus W)| \mid n/3 \leq |W| \leq (2n)/3\},$$

where $E(W, [n] \setminus W)$ is the set of edges with one end-point in W and the other outside W. It is well known that there exists graphs with a small constant degree and of a high expansion. In particular, there exist the so-called **3-regular expanders**: the degree of each vertex in G is 3 and $e(G) \geq \Omega(n)$ (cf. [73]).

Theorem 13.3.7 (Urquhart [496], Ben-Sasson and Wigderson [73]) *Let $G = ([n], E)$ be a connected graph and $f: [n] \to \{0, 1\}$ be such that $\sum_{i\in[n]} f(i) = 1$ in $\mathbf{F}_2$. Then $\mathbf{w}_R(TSE_{G,f}) \geq e(G)$.*

In particular, if G is a 3-regular expander then $|TSE_{G,f}| = O(n)$ and each R-*refutation of $TSE_{G,f}$ must have size at least $2^{\Omega(n)}$.*

13.4 Random Formulas

It is folklore that randomly chosen formulas share all the natural hardness properties of the formulas that one can specifically tailor to make a particular hardness measure high. This is a bit vague, but in proof complexity and SAT solving it means that if you can establish a lower bound for the size (the height, the number of steps, the degree etc.) or for the time of a SAT algorithm, for a specific sequence of formulas, then the same lower bound is likely to hold (with high probability) for formulas chosen randomly. The phrase *chosen randomly* involves specifying a probability distribution on formulas, and that involves setting various parameters. Some such distributions give the impression of being more *natural* than others, and some are even boldly described by researchers as being *canonical*.

As an example consider CNF-formulas in n atoms. You may pick $0 \leq p \leq 1$ and choose a random CNF by including in it each of a possible 4^n clauses – allowing complementary literals to appear at the same time – with probability p. Or you may a priori disregard clauses containing both an atom and its negation, keeping the remaining 3^n. Or you may select $k \geq 1$ and pick only clauses with k literals (exactly or at most k). Or, instead of using p, you may select in advance $0 \leq m \leq 3^n$ and choose the CNF uniformly at random from all those having exactly m clauses (or k-clauses). The properties of the resulting distributions may be very different. A well-known example is the *satisfiability threshold*. As an illustration we shall formulate a special case.

For the next theorem, denote by $\mathcal{D}^k_{n,m}$ a distribution on k-CNFs with n atoms and m clauses, defined as follows: pick m k-clauses, uniformly and independently from all $\binom{n}{k}2^k$ possible k-clauses, with repetitions of clauses allowed.

Theorem 13.4.1 (Friedgut [191]) *There is a function $n \rightarrow r(n)$ such that, for all $\epsilon > 0$,*

$$Prob_{D \in \mathcal{D}^3_{n,(r(n)-\epsilon)n}}[D \in \mathrm{SAT}] \rightarrow 1 \quad \text{and} \quad Prob_{D \in \mathcal{D}^3_{n,(r(n)+\epsilon)n}}[D \in SAT] \rightarrow 0.$$

It is open whether $r(n)$ converges to some value, but it is expected that it does and that the limit is close to 4.27.

For proof complexity, this means that choosing 3CNFs according to, say, $\mathcal{D}^3_{n,5n}$ is a source of random formulas that, with a high probability, are unsatisfiable, and we may study their proof complexity. An early example was

Theorem 13.4.2 (Chvátal and Szemerédi [141]) *Let $k \geq 3$ and let $c \geq 2^k \ln 2$. Then, as $n \rightarrow \infty$, with probability tending to 1, a random k-CNF chosen according to $\mathcal{D}^k_{n,cn}$ is unsatisfiable but requires* R*-refutations of exponential size $2^{\Omega(n)}$.*

The proof of this theorem illustrates a weak point of random formulas. Lower bounds for them, for a specific proof system or a specific complexity measure, so far have never come first. Instead, a lower bound is proved first for some specific sequence of formulas defined for the purpose, and only later is it verified that the key property of the sequence needed in the argument is also shared by a random sequence chosen from a suitable distribution. Initial discoveries of proof complexity lower bounds seem *always* to rely closely on the meaning of the specific hard formulas used. In this respect random formulas have so far not been very useful in proof complexity. That may change with the discovery of new methods (but not, I think, by formulating any number of new hypotheses).

The resolution lower bound for random formulas was recovered (and strengthened) by a width argument (involving expanding hypergraphs) as well.

Theorem 13.4.3 (Ben-Sasson and Wigderson [73]) *Let $0 < \epsilon \leq 1/2$ be arbitrary and put $m = m(n) := (n^{1/2-\epsilon})n$. Then, with high probability, a random formula*

chosen from $\mathcal{D}^3_{n,m}$ is unsatisfiable but requires R*-refutations of width at least*

$$m^{-2/(1-\epsilon)}n .$$

In particular, it requires refutations of size at least

$$2^{\Omega(m^{-4/(1-\epsilon)}n)} .$$

We will not prove these theorems, as we shall recover some lower bounds for random formulas from a different type of statement that we discuss now (and in more detail in Section 19.4).

Consider an ℓ-sparse $m \times n$ matrix A over the two-element field $\mathbf{F}_2$, as in Section 13.3, and recall that we denoted $J_i := \{j \in [n] \mid A_{ij} = 1\}$. We shall interpret A now as defining a linear map from $\mathbf{F}_2^n$ into $\mathbf{F}_2^m$. In particular, the ith output bit is computed as

$$(A \cdot \overline{x})_i = \sum_j A_{ij}x_j = \bigoplus_{j \in J_i} x_j.$$

Because we are assuming that $m > n$, $rng(A)$ is a proper subset of $\mathbf{F}_2^m$. Take any vector $\overline{b} \in \mathbf{F}_2^m \setminus rng(A)$. Then the linear system

$$A \cdot \overline{x} = \overline{b}$$

has no solution (in $\mathbf{F}_2$) and this unsolvability can be expressed as a tautology, to be denoted by $\tau_{\overline{b}}(A)$, in a DNF as follows:

$$\tau_{\overline{b}}(A) := \bigvee_{i \in [m]} \quad \bigvee_{\epsilon \in \{0,1\}^{|J_i|}, \bigoplus_{j \in J_i} \epsilon_j = 1 - b_i} \quad \bigwedge_{j \in J_i} x_j^{\epsilon_j} .$$

Here we have used the notation x^ϵ from Section 5.4. The formula says that there is some bit i such that the ith bits of $A \cdot \overline{x}$ and $\overline{b}$ differ, which itself is expressed by saying that there are evaluations ϵ_j, of the bits x_j of x that belong to J_i which determine the ith bit of $A \cdot \overline{x}$ as $1 - b_i$, i.e. as different from b_i. Note that the size of the formula is bounded above by $O(m2^\ell \ell)$ and that the clauses of $\neg\tau_{\overline{b}}$ have width $\leq \ell$.

Using the notion of a *safe set of rows* of a matrix A from Section 13.3 we define the following.

1. A partial assignment $\rho :\subseteq \{x_1, \ldots, x_n\} \to \{0, 1\}$ is called **safe** if and only if $dom(\rho) = \bigcup_{i \in I} J_i$ for some safe I. One such I chosen in some canonical way is the support of ρ, denoted by $supp(\rho)$.
2. Let $\overline{b} \in \{0, 1\}^m$. A safe partial assignment ρ with support I is a **safe partial solution** of $A \cdot \overline{x} = \overline{b}$ if and only if $\rho(x_j) = b_i$ for all $J_i \subseteq \bigcup_{u \in I} J_u$, $\bigoplus_{j \in J_i}$.
3. For a safe partial solution ρ with $supp(\rho) = I$, $\overline{b}^\rho$ is the $(m - |I|)$-vector with ith coordinate $b_i \bigoplus \bigoplus_{j \in J_i \cap dom(\rho)} \rho(x_j)$, for i such that $J_i \not\subseteq dom(\rho)$. Denote by x_I the tuple of variables not in $J(I)$.

Clearly, if ρ is a safe solution whose support is I, and ξ is a solution of $A_I \cdot \overline{x}_I = \overline{b}^\rho$, then $\rho \cup \xi$ is a solution of $A \cdot \overline{x} = \overline{b}$.

Lemma 13.4.4 *Assume that $I \subseteq I' \subseteq [m]$ are safe systems and that $|I' \setminus I| \leq r$. Assume further that ρ is a safe assignment such that $supp(\rho) = I$ and that values $c_i \in \{0,1\}$ for $i \in I' \setminus I$ are arbitrary.*

Then the assignment ρ can be extended to a safe assignment ρ' with $supp(\rho') = I'$ such that

$$\bigoplus_{j \in J_i} \rho'(x_j) = c_i,$$

for all $i \in I' \setminus I$.

Proof The map A_I is an $(r, 1/4)$-expander (by Lemma 13.3.2) and the expansion property implies that every subset of $I' \setminus I$ has a non-empty border in A_I. In particular, no subset of $I' \setminus I$ can be a linearly dependent set of rows of A_I. Hence the system

$$\bigoplus_{j \in J_i \setminus dom(\rho)} x_j = c_i \oplus \bigoplus_{j \in J_i \cap dom(\rho)} \rho(x_j)$$

has a solution, say ξ. Put $\rho' := \rho \cup \xi$. □

Lemma 13.4.5 *Assume that A is an ℓ-sparse $m \times n$ matrix that is an $(r, 3/4)$-expander. Let $\overline{b} \notin rng(A)$ be arbitrary.*

Then every R*-proof of $\tau_{\overline{b}}(A)$ must have width at least $\geq r/4$.*

Proof Let π be a resolution refutation of $A \cdot \overline{x} = \overline{b}$, i.e. a proof of $\tau_{\overline{b}}(A)$. Let us denote the width of π by w.

We shall construct a sequence of clauses $C_0, \ldots, C_e$ occurring in π and a sequence of partial safe assignments $\alpha_t :\subseteq \{x_1, \ldots, x_n\} \to \{0,1\}$ for $t = 0, \ldots, e$, such that the following conditions are satisfied:

1. $C_0 := \emptyset$ is the end-clause of π. Each C_{t+1} is a hypothesis of an inference in π yielding C_t, and C_e is an initial clause;
2. if x_j occurs in C_t then $x_j \in dom(\alpha_t)$;
3. C_t is false under the assignment α_t;
4. $|supp(\alpha_t)| \leq 2w$.

Put $\alpha_0 := \emptyset$. Assume we have C_t and α_t and that C_t has been inferred in π by

$$\frac{D_1 \cup \{x_j\} \qquad D_2 \cup \{\neg x_j\}}{C_t (= D_1 \cup D_2)}.$$

Let $I' \supseteq supp(\alpha_t)$ be a minimum safe set with some row containing j. It exists, by Lemma 13.3.2, as long as $|supp(\alpha_t)| + 1 \leq r/2$; since $|supp(\alpha_t)| \leq 2w$ this inequality follows if $w < r/4$.

By Lemma 13.4.4 there is a partial safe solution $\rho' \supseteq \alpha_t$. Take for $\alpha_{t+1} \subseteq \rho'$ a minimum safe assignment obeying condition 2. Finally, take for C_{t+1} the clause among $D_1 \cup \{x_j\}, D_2 \cup \{\neg x_j\}$ made false by α_{t+1}.

Now note that the conditions posed on C_e and α_e lead to a contradiction. Since C_e is an initial clause, α_e makes true its negation, which is one of the conjunctions $\bigwedge_{j\in J_i} x_j^{\epsilon_j}$ in τ_b. In particular, $\oplus_{j\in J_i}\epsilon_j = 1 - b_i$. But that violates the assumption that α_e satisfies all the equations of $A \cdot \overline{x} = \overline{b}$ evaluated by α_e.

We have constructed the sequence under the assumption that $w < r/4$. Hence $w \geq r/4$. □

Corollary 13.4.6 *Assume that A is an ℓ-sparse $m \times n$ matrix that is an $(r, 3/4)$-expander. Let $\overline{b} \notin rng(A)$ be arbitrary. Then every* R*-proof of $\tau_{\overline{b}}(A)$ must have size at least $\geq 2^{\Omega((r/4-\ell)^2/n)}$.*

In particular, for every $\delta > 0$ there is an $\ell \geq 1$ such that, for all sufficiently large n, there exists an ℓ-sparse $n^2 \times n$ matrix A such that $\tau_{\overline{b}}(A)$ requires R*-proofs of size at least $\geq 2^{\Omega(n^{1-\delta})}$.*

Proof Apply Theorem 13.3.1 for $\delta/2$, to get $\ell \geq 1$ and an ℓ-sparse $n^2 \times n$ matrix A which is an $(n^{1-\delta/2}, 3/4)$-expander. By Lemma 13.4.5 every R-proof of $\tau_b(A)$ must have width at least $\Omega(n^{1-\delta/2})$.

From the width–size relation given in Theorem 5.4.5, it follows that the size of any such proof must be at least $\exp(\Omega((n^{1-\delta/2} - \ell)^2/n))$, as ℓ bounds the width of the initial clauses. This is $2^{\Omega(n^{1-\delta})}$. □

Now we return to random formulas. Let C be an ℓ-clause with literals corresponding to the variables x_j from $j \in J \subseteq [n]$ of size ℓ. Let $\overline{a} \in \{0,1\}^n$ be any truth assignment. The condition $C(\overline{a}) = 0$ forces the values of $\overline{a}$ on J and hence also a unique value $b(C) := \bigoplus_{j\in J} a_j$.

Assume now that D is an ℓ-CNF formula with ℓ-clauses C_i, $i \in [m]$, the $J_i \subseteq [n]$ being the (indices of) the variables occurring in C_i. Then D is unsatisfiable if and only if

$$\forall x \in \{0,1\}^n \exists i \in [m]\ C_i(x) = 0,$$

which implies that

$$\forall x \in \{0,1\}^n \exists i \in [m] \bigoplus_{j\in J_i} x_j = b(C_i);$$

this is further equivalent to

$$\forall x \in \{0,1\}^n \exists i \in [m] \bigoplus_{j\in J_i} x_j \neq b_i,$$

where we set $b_i := 1 - b(C_i)$. Put $\overline{b} := (b_1, \ldots, b_m)$.

Let A be the $m \times n$ matrix with ones in the ith row of columns J_i. Then $\overline{b} \notin rng(A)$ and a short refutation of D yields a short refutation of $\tau_{\overline{b}}(A)$ (here we need $\ell = O(\log n)$ or better $\ell = O(1)$ in order to manipulate the DNF expressing the parity $\bigoplus$

of ℓ variables efficiently). If D was random, so is A and, by Theorem 13.3.1, A has with a high probability the expansion property required in Corollary 13.4.6 and so the lower bound applies.

13.5 Interpolation

In Section 5.6 we considered an unsatisfiable set of clauses

$$A_1(\overline{p},\overline{q}),\ldots,A_m(\overline{p},\overline{q}),B_1(\overline{p},\overline{r}),\ldots,B_\ell(\overline{p},\overline{r}) \tag{13.5.1}$$

containing only literals corresponding to the atoms

$$\overline{p} = p_1,\ldots,p_n, \quad \overline{q} = q_1,\ldots,q_s, \quad \text{and } \overline{r} = r_1,\ldots,r_t.$$

We showed that one can extract a circuit $C(\overline{p})$ from an R-refutation of (13.5.1) that interpolates the implication

$$\bigwedge_{i\le m}(\bigvee A_i) \longrightarrow \neg\bigwedge_{j\le \ell}(\bigvee B_j) \tag{13.5.2}$$

and one can give an upper bound on the size of C in terms of the size of the refutation (Theorem 5.6.1).

We shall now reformulate this situation as follows. Consider two sets $U_n, V_n \subseteq \{0,1\}^n$ defined by

$$U_n := \{\overline{a} \in \{0,1\}^n \mid \bigwedge_{i\le m} A_i(\overline{a},\overline{q}) \in \text{SAT}\} \tag{13.5.3}$$

and

$$V_n := \{\overline{a} \in \{0,1\}^n \mid \bigwedge_{j\le \ell} B_j(\overline{a},\overline{r}) \in \text{SAT}\}\,. \tag{13.5.4}$$

Then (13.5.1) is unsatisfiable if and only if U_n and V_n are disjoint,

$$U_n \cap V_n = \emptyset,$$

and C interpolates (13.5.2) if and only if the set

$$W_n := \{\overline{a} \in \{0,1\}^n \mid C(\overline{a}) = 1\}$$

separates U_n from V_n:

$$U_n \subseteq W_n \quad \text{and} \quad W_n \cap V_n = \emptyset.$$

If U_n is closed upwards, i.e. $\overline{a} \in U_n \wedge \overline{a} \le \overline{b}$ implies $\overline{b} \in U_n$, then there is a separating set W_n that is also closed upwards and can be defined by a monotone circuit. The same is true if V is closed downwards (i.e. $\overline{a} \in V_n \wedge \overline{a} \ge \overline{b}$ implies $\overline{b} \in V_n$).

A general way to obtain a sequence of sets U_n and V_n that are defined in this way is to use two disjoint NP sets U and V. We shall formulate this as a lemma for future reference; it follows immediately from the NP-completeness of SAT.

Lemma 13.5.1 *Assume that U and V are two disjoint NP sets. Then, for each $n \geq 1$, there are $m, \ell, t, s \leq n^{O(1)}$ and $m + \ell$ clauses A_i and B_j of the form* (13.5.1) *such that*

$$U_n := U \cap \{0,1\}^n \quad \text{and} \quad V_n := V \cap \{0,1\}^n$$

satisfy (13.5.3) *and* (13.5.4), *respectively.*

In addition, if U is closed upwards then such clauses A_i can be found in which the atoms $\overline{p}$ occur only positively, and if V is closed downwards then such clauses B_j can be found in which the atoms $\overline{p}$ occur only negatively.

Using Theorem 5.6.1 we can now formulate a lower bound criterion.

Theorem 13.5.2 *Let U, V be two disjoint NP-sets and let $s(n)$ be the minimum size that any circuit separating U_n from V_n must have. Then any* R*-refutation of the clauses A_i, B_j satisfying the conditions of Lemma 13.5.1 must have at least $\Omega(s(n)/n)$ clauses.*

If, in addition, U is closed upwards (or V is closed downwards), $s^+(n)$ is the minimum size of a monotone circuit separating U_n from V_n and the clauses A_i are chosen so that all atoms $\overline{p}$ occur only positively in them (or only negatively in all clauses B_j), then any R*-refutation of the clauses A_i, B_j must have at least $\Omega(s^+(n)/n)$ clauses.*

Hence, in order to obtain, say, a super-polynomial lower bound for R via this criterion the only thing we need to do is to take two NP sets that cannot be separated by a set in P/*poly* or the monotone version of this. This is, however, not at all easy because there are no known non-trivial lower bounds on general (non-monotone) circuits. In Chapter 17 we shall consider examples of disjoint NP sets for which the hardness of their separation is a well-established conjecture in computational complexity. Fortunately we are not confined to proving only conditional results: there are strong lower bounds for monotone circuits separating two (monotone) NP sets and luckily we can then use the monotone part of the criterion in Theorem 13.5.2.

We start by defining perhaps the most famous pair of disjoint NP sets where one is closed upwards and the other is closed downwards. As in Section 1.4, where we defined the Boolean function $\text{Clique}_{n,k}$, we shall identify graphs (tacitly undirected, simple and without loops) with strings of length $\binom{n}{2}$. We shall also use the symbol $\binom{n}{2}$ to denote the set of unordered pairs of elements of $[n]$.

For $n, \omega, \xi \geq 1$, define $\text{Clique}_{n,\omega}$ to be the set of graphs on $[n]$ (i.e., the set of strings of length $\binom{n}{2}$) that have a clique of size at least ω, and define $\text{Color}_{n,\xi}$ to be the set of graphs on $[n]$ that are ξ-colorable. The former set is closed upwards and the latter downwards: adding more edges cannot remove a clique and deleting some edges cannot violate a graph coloring. They are defined by the following two sets of clauses.

The set of clauses to be denoted by $\text{Clique}_{n,\omega}(\overline{p}, \overline{q})$ uses $\binom{n}{2}$ atoms p_{ij}, $\{i,j\} \in \binom{n}{2}$, one for each potential edge in a graph on $[n]$, and $\omega \cdot n$ atoms q_{ui}, $u \in [\omega]$ and $i \in [n]$, intended to describe a mapping from $[\omega]$ to $[n]$. It consists of the following clauses:

- $\bigvee_{i \in [n]} q_{ui}$, for all $u \leq \omega$;
- $\neg q_{ui} \vee \neg q_{uj}$, for all $u \in [\omega]$ and $i \neq j \in [n]$;
- $\neg q_{ui} \vee \neg q_{vi}$, for all $u \neq v \in [\omega]$ and $i \in [n]$;
- $\neg q_{ui} \vee \neg q_{vj} \vee p_{ij}$, for all $u \neq v \in [\omega]$ and $\{i,j\} \in \binom{n}{2}$.

The set $\text{Color}_{n,\xi}(\overline{p}, \overline{r})$, using, besides the atoms $\overline{p}$, $n \cdot \xi$ atoms r_{ia}, $i \in [n]$ and $a \in [\xi]$, intended to describe a coloring of the graph. It consists of the following clauses:

- $\bigvee_{a \in [\xi]} r_{ia}$, for all $i \in [n]$;
- $\neg r_{ia} \vee \neg r_{ib}$, for all $a \neq b \in [\xi]$ and $i \in [n]$;
- $\neg r_{ia} \vee \neg r_{ja} \vee \neg p_{ij}$, for all $a \in [\xi]$ and $\{i,j\} \in \binom{n}{2}$.

Truth assignments to atoms $\overline{q}$ satisfying $\text{Clique}_{n,\omega}(\overline{a}, \overline{q})$ can be identified with injective maps from the set $[\omega]$ onto a clique in the graph (determined by) $\overline{a}$, and truth assignments to $\overline{r}$ satisfying $\text{Color}_{n,\xi}(\overline{a}, \overline{r})$ are colorings of the graph $\overline{a}$ by ξ colors. Clearly

$$\text{Clique}_{n,\omega} = \{\overline{a} \mid \text{Clique}_{n,\omega}(\overline{a}, \overline{q}) \in \text{SAT}\}$$

and similarly

$$\text{Color}_{n,\xi} = \{\overline{a} \mid \text{Color}_{n,\xi}(\overline{a}, \overline{r}) \in \text{SAT}\},$$

and these two sets are disjoint as long as $\xi < \omega$. Note also that the atoms $\overline{p}$ occur only positively and negatively in the Clique and Color clauses, respectively.

Here is the circuit lower bound we shall use.

Theorem 13.5.3 (Razborov [430]; improved by Alon and Boppana [18]) *Assume that* $3 \leq \xi < \omega$ *and* $\sqrt{\xi}\omega \leq n/(8 \log n)$. *Then every monotone Boolean circuit* $C(\overline{p})$ *separating* $\text{Clique}_{n,\omega}$ *from* $\text{Color}_{n,\xi}$ *must have size at least* $2^{\Omega(\sqrt{\xi})}$.

We get a resolution lower bound as an immediate corollary of Theorems 13.5.2 and 13.5.3.

Corollary 13.5.4 *There is a constant* $c > 0$ *such that whenever* $3 \leq \xi < \omega$ *and* $\sqrt{\xi}\omega \leq n/(8 \log n)$ *the following lower bound holds.*

Any R-*refutation of*

$$\text{Clique}_{n,\omega} \cup \text{Color}_{n,\xi}$$

must have at least $2^{\Omega(\xi^{1/2})}/n$ *clauses.*

A useful choice of the parameters is

$$\omega := n^{2/3} \quad \text{and} \quad \xi := n^{1/3}$$

as that will relate later the Clique–Color clauses to the weak pigeonhole principle $\text{WPHP}(m^2, m)$, for $m := n^{2/3}$.

13.6 Games

Many concepts and constructions in mathematical logic can be explained in terms of two-player (finite or infinite) games. A classic example is the Ehrenfeucht-Fraissé game characterizing the elementary equivalence of two structures. Model-theoretic constructions often allow a natural game-theoretic formulation; see Hodges [232]. Games are also frequent in computer science and complexity theory and we have used games a few times already as well.

Games are used in proof complexity more as a way to express a certain adversary argument than in a genuine game-theoretic sense. In particular, often one player has a canonical strategy. It may be useful to have a variety of games to choose from when we want to frame an argument in that language. We have abbreviated somewhat the treatment of space lower bounds and to remedy this at least slightly we present here the class of so-called **pebbling games** and related **pebbling formulas** (or **pebbling contradictions** as they are also called). Other games will be mentioned in the next section.

Pebbling arguments have been used in computational complexity theory successfully for various space bounds or time–space tradeoff results, and they provide a well established technique. The rough idea is that one uses pebbles to mark data that ought to be kept in the memory in order to progress with a computation (or with a proof), stipulating strict elementary rules governing moves of the pebbles. The minimum number of pebbles that suffices is the minimum memory requirement. An example of this set-up is the pebbling definition of the space of R*- and R-refutations from Section 5.5.

The pebbling techniques in R were developed originally to tackle a few general questions and have been developed since then into a sophisticated technical tool. The original questions included the optimality of Theorem 5.5.5, i.e. whether it could be the case that the clause space and the width measures essentially coincide (as is the case for the height and the total space by Lemma 5.5.4). If not, then there could still be some tight relation between the size and the clause space that is analogous to Theorem 5.4.5. These problems as well as general questions about the existence of proofs minimizing at the same time two or more of the measures considered were solved. Section 13.7 offers details and the relevant bibliography. Here we shall confine ourselves to explaining the games and the formulas.

We defined in Section 5.5 the pebbling of a directed acyclic graph (a dag) and used it to define the clause space of R-refutations. If we restrict to dags G with just one sink, we can think of G as the underlying graph of a circuit that uses only conjunctions (of an arbitrary arity). The pebbling rule that a node can be pebbled if all its predecessors are pebbled just amounts to the condition that the value 1 passes through the $\bigwedge$ gate. This game is called the **black pebble game**. In actual applications a more general **black–white pebble game** is used. As the name suggests there are two kinds of pebbles, black and white, and some new rules come with them:

1. any node may hold at most one pebble;
2. a white pebble can be put on any non-pebbled node;
3. a black pebble can be put on any non-pebbled source node;
4. a black pebble can be put on a node if all its (immediate) predecessors are pebbled (this incorporates the previous rule);
5. a black pebble can be removed at any time from a node;
6. a white pebble can be removed at any time from a source;
7. a white pebble can be removed from a node if all its (immediate) predecessors are pebbled.

If G is the proof graph of a refutation we consider that a black pebble on a clause certifies that we have already verified that it was correctly derived, and a white pebble signals our *assumption* that the clause was correctly derived.

The **black–white pebbling price** of a graph is the minimum number of black and white pebbles counted together that are needed in some pebbling game that puts a *black* pebble on the sink. The **pebbling time** is the minimum number of steps in a pebbling game that puts a black pebble on the sink.

A **pebbling contradiction** may be constructed from any dag G. Assume that $V(G)$ are its nodes, $w \in V(G)$ is the unique sink and $\emptyset \neq S \subseteq V(G)$ is the non-empty set of source nodes in G. Attach to each $v \in V(G)$ an atom x_v and use these atoms to define the formula Peb_G to be the conjunction of the following clauses:

- $\{x_u\}$ for all $u \in S$;
- $\{\neg x_{u_1}, \ldots, \neg x_{u_t}, x_v\}$ for all $v \in V(G) \setminus S$ whose immediate predecessors are $u_1, \ldots, u_t$;
- $\{\neg x_w\}$ for the sink w.

Note that Peb_G has $|V(G)|+1$ clauses whose maximum width is 1 plus the maximum in-degree of a node in G.

These formulas are not hard for R. The following lemma is analogous to Lemma 12.3.1.

Lemma 13.6.1 *For any dag G with one sink and maximum in-degree d, Peb_G has an* R*-refutation with $O(|V(G)|)$ steps and of width $d+1$.*

In order to use the pebbling contradictions for lower bounds, Ben-Sasson and Nordström [72] combined them with substitutions. In particular, let $d \geq 2$ be a constant and let $f : \{0,1\}^d \to \{0,1\}$ be a suitable Boolean function (in applications the function should have the property that one cannot fix its value by a small partial restriction, e.g. the parity function). Because d is a constant, we can write constant-size DNFs representing both f and its negation $\neg f$ and use these DNFs to substitute $f(y_1, \ldots, y_d)$, in new variables $\bar{y}$, for the positive and negative occurrences of the atoms of the original pebbling formula, thus transforming clauses to clauses. This will increase the size and the width of the formula proportionally.

There are many modifications of this basic idea and a similarly large number of its applications. Rather than giving a sample here, we refer the reader to the next section for references.

13.7 Bibliographical and Other Remarks

Sec. 13.1 is from [278], which expanded Riis [446]. The DNF trees are a special case of the test trees considered in [278]. The Ramsey theorem and the tournament principle can be found in [296]. The dichotomy between polynomial-size and exponential-size proofs as in Corollary 13.1.3 and Lemma 13.1.6 was popularized by Riis [448] under the term *complexity gaps*. Adversary arguments can be rephrased via partial orderings of partial truth assignments (under various names: consistency conditions, dynamic satisfiability, extension systems etc.) We shall expand upon the forcing approach in Section 20.3.

Beyersdorff, Galesi and Lauria [77] gave a game-theoretic characterization of the minimum proof size in R*, using a variant of the Prover–Delayer game (see below). Ben-Sasson, Impagliazzo and Wigderson [70] gave an almost optimal separation of R* from R. Buss and Johannsen [125] studied variants of linear resolution and compared it with regular and general resolution.

Sec. 13.2 is based on my unpublished notes for students and expands a bit on Dantchev and Riis [164]. The idea of relativization in proof complexity and the particular question these authors answered is from [294]. The examples of the relativized Ramsey principles are from [303]. Lemma 13.2.5 is from [288], and improves upon an earlier width lower bound for $RAM(n_r, r)$ with *critical Ramsey* number n_r by Krishnamurthy and Moll [329]. Lemma 13.2.6 and Theorem 13.2.7 are from [303].

Motivation for investigation of the proof complexity of the Ramsey principle (in [288, 303] at least) is the following open problem (see [303]):

Problem 13.7.1 (The R(log) problem) *Find a sequence of* DNF-*formulas (preferably with narrow terms) that have short constant-depth Frege proofs (in the DeMorgan language) but require long proofs in R*(log).

The primary motivation for this problem is to improve upon the known non-conservativity results in bounded arithmetic and the known non-simulation results among bounded-depth Frege systems; this problem is the *base case*. We shall discuss this in some detail when talking about the **depth** d **vs. depth** $d+1$ problem 14.3.2 in Section 14.3. For now let us stress only that, for a solution to have these consequences, the qualification *narrow* should be interpreted as *poly-logarithmic* (in the size of the formula), a *short* proof in AC^0-Frege systems should mean *quasi-polynomial*, and a *long* R(log)-proof ought to be *not quasi-polynomially bounded*. An optimal solution would present kDNFs with constant k having polynomial-size AC^0-Frege proofs but requiring exponential-size R(log)-proofs.

Both relativized Ramsey principles from Section 13.2 are potential candidates, as is the unrelativized Ramsey principle RAM_n; Theorem 13.2.7 gives lower bounds for the relativized principles and Pudlák [420] established an exponential lower bound for R-refutations of RAM_n. Furthermore, all these principles have quasi-polynomial-size proofs in the Σ-depth 1 system $LK_{3/2}$ by the translation of a bounded arithemtic proof of the Ramsey principle due to Pudlák [411], as improved by Aehlig and Beckmann [1] (see [303] for upper bounds for the relativized principles).

From the results for (W)PHP explained in this chapter, PHP principles are not good candidates for solving the R(log)-problem 13.7.1. Atserias, Müller and Oliva [40] defined a relativization of the weak pigeonhole principle formalizing that if at least $2n$ out of n^2 pigeons are mapped into n holes then it cannot be an injective map. They showed that the principle requires slightly super-polynomial R(*id*) proofs (i.e. in DNF-R with no restrictions on the bottom fan-in) but that there are quasi-polynomial-size R(log) proofs. Hence the relativized PHP principle will not help either.

Formulas that are DNF and will separate R(log) from LK_d or $LK_{d+1/2}$ if anything can are the reflection principles for the latter systems: subexponential R(log)-proofs of these formulas would imply a subexponential simulation of the stronger systems (by Section 8.6). However, these are probably hard to use directly. Instead some formulas formalizing in a combinatorially transparent way a principle equivalent to the reflection formulas may be useful. A class of such formulas formalizing the **colored** PLS principle defined by Krajíček, Skelley and Thapen [325] gives an example of how this can be done for R.

Sec. 13.3 presents ideas from Ben-Sasson and Wigderson [73, 66] but instead of their original width-lower-bound argument (modifying Haken's bottleneck counting) we used a construction analogous to the one from [291] as that is used also for Lemma 13.4.5. The idea of restricting R in the PHP to subsets of a suitable relation E was used first in [276, 278], but not for an E of bounded degree and with expansion properties (see Section 14.1 for *modified* WPHP and its use there). The idea of using the combinatorial expansion in proof complexity originated with Ben-Sasson and Wigderson [73].

The definition of expanding matrices is from Alekhnovich *et al.* [12, Definition 2.1]. The proof of Theorem 13.3.1 can be found in [291] and is a special case of the proof of [12, Theorem 5.1]. Lemma 13.3.2 is from [291] and slightly generalizes [12, Lemma 4.6]; the proof here follows that in [291]. Lemma 13.3.3 is due to Ben-Sasson and Wigderson [73, Theorem 4.15]; the proof is new. Theorem 13.3.4 is also from [73]. The special cases for PHP_n and $WPHP_n^m$ are due to Haken [216] and Buss and Turan [133], respectively; they (as well as [73]) get a slightly better bound, with exponent $\Omega(n^2/m)$. We opted to use a formulation of Theorem 13.3.1 that gives an only slightly weaker bound, as the same formulation could be used also in Section 13.4. Eventually Raz [426] proved an exponential bound for arbitrary

m (this was preceded by the same result for regular R by Pitassi and Raz [401]). Razborov [437, 438, 439] further extended these results.

A method representing functions by their bit-graphs was used in bounded arithmetic by Chiari and Krajíček [138] in order to decrease the quantifier complexity of some combinatorial principles (Section 14.6). This corresponds to changing some clauses to narrow ones. The bit PHP was defined by Filmus *et al.* [185], who studied space complexity in PC (Theorem 13.3.5 is surely implicit there).

The second part of Theorem 13.3.7 follows from the width lower bound (due to [73]) via Theorem 5.4.5, but it was proved earlier by Urquhart [496] using a different method.

Sec. 13.4 Statements (as well as their proofs) 13.4.4–13.4.6 are from [291]. Random formulas may enter proof complexity also in a different way, as random axioms. Building on an original proposal by Dantchev, Buss, Kolodziejczyk and Thapen [128, Sec. 5.2], the notion of the δ-**random resolution** (denoted by RR) was defined: a δ-random **refutation distribution** of a set of clauses $\mathcal{C}$ is a probability distribution on pairs $\pi_r, \mathcal{D}_r$ such that π_r is an R-refutation of $\mathcal{C} \cup \mathcal{D}_r$ and such that the distribution $(\mathcal{D}_r)_r$ obeys the condition that any total truth assignment satisfies $\mathcal{D}_r$ with probability at least $1 - \delta$. This is a sound proof system. Buss *et al.* [128] explained its relation to a bounded arithmetic theory (this was the motivation for them to consider the proof system) and Pudlák and Thapen [424] presented variants of the definition and several upper and lower bounds.

Beame, Culberson, Mitchell and Moore [51] investigated the proof complexity of a class of random formulas expressing the k(n)-colorability for random graphs, and Atserias *et al.* [28] investigated random formulas (in regular resolution) expressing the non-existence of a large clique.

Sec. 13.5 is from [280].

Sec. 13.6 Beyersdorff and Kullmann [80] used games to define in a uniform way all proof complexity measures for R-refutations. The games considered by them generalized the Prover–Delayer game of Impagliazzo and Pudlák [245], designed originally for proving lower bounds for R*. The game is played in the same way as the Buss–Pudlák Prover–Liar game (Section 2.2), but Prover may ask only for truth values of atoms. On the other hand Liar has more options: he can either answer (in a way locally consistent with the initial clauses) but he may also defer the answer to Prover. Whenever he does that he scores a point. The existence of a strategy for Liar to score at least p points against every Prover strategy implies a size-2^p lower bound for R*. Galesi and Thapen [193] generalized the Prover–Delayer game in another direction in order to characterize various subsystems of R.

Pudlák [415] considered a restriction of the Buss–Pudlák game (Section 2.2) for resolution that restricted the queries to literals. The minimum length of play between Prover and Liar corresponds to the minimum height of R-refutations (analogously to Lemma 2.2.3). However, he showed that, by considering instead the *memory* that Prover has to use, it is possible to capture the *size* of R-refutations. The idea is that Prover can use an R-refutation π of a CNF as her strategy (as in the proof of Lemma 2.2.3) and at any given point she just needs to remember which clause of π she is in. Hence the memory is at most (and, in fact, is proportional to) the logarithm of the number of clauses in π (as an example Haken's [216] lower bound for PHP was presented in this framework).

The pebbling formulas and the pebbling game were used for lower bounds by Nordström [368] and by Ben-Sasson and Nordström [71] and further improved in Nordström [369, 370]. Nordström [371] is a detailed survey of the topic.

13.7.1 Clause Space Remarks

A lower bound for the width implies lower bounds for the clause space (Theorem 5.5.5) and, with luck, for the size (Theorem 5.4.5). One can ask whether it is possible to separate these measures by finding formulas allowing one to be small but requiring the other to be large, or whether it is possible to minimize two or more of the measures simultaneously. In addition, the space measure comes in (at least) three variants (clause, total, variable). The variety of different results is large and ever growing, and the reader interested in the details should consult some specialized up-to-date survey. Below we mention only a few selected results; the cited works offer further references.

Nordström [368, 369] investigated the space measure and its relations to the width and to the size. In particular, he gave examples for which short or narrow refutations must be spacious. Thapen [489] described a family of narrow CNF-formulas having polynomial-size refutations, but all narrow refutations must have exponential size.

A simple counting argument shows that any formula refutable in R within a width w must have an R-refutation of size $n^{O(w)}$. Atserias, Lauria and Nordström [36] proved that this is tight: they constructed 3CNF formulas with n atoms that do have R-refutations of width w but require them to be of size $n^{\Omega(w)}$.

Bonacina, Galesi and Thapen [89] proved quadratic lower bounds for the total space of refutations of random k-CNFs and also for two variants of the pigeonhole principle. Bonacina [88] then proved a general theorem implying that one can get good lower bounds on the total space from lower bounds on the width. Nordström and Hastad [373] proved a tight lower bound on the space of refutations of the pebbling contradictions by an argument that did not use a width lower bound. Beame, Beck and Impagliazzo [49] proved a size–space tradeoff that applies to a super-linear space.

13.7.2 Links to SAT Solving

We did not discuss results about the proof complexity of resolution motivated by various SAT solving algorithms. Instead we offer now at least some links to a rather vast literature on the subject that can serve the reader as starting points for exploring that area. Nordström [372] is a useful survey.

Atserias, Fichte and Thurley [32] showed that a width-k resolution can be thought of as $n^{O(k)}$ restarts of the unit resolution rule together with learning.

Lower bounds for R* imply via Lemma 5.2.1 lower bounds for the time of DPLL algorithms; these lower bounds are for unsatisfiable formulas. Alekhnovich, Hirsch and Itsykson [13] gave lower bounds for two classes of DPLL algorithms for *satisfiable* formulas as well.

Goerdt [203] demonstrated a super-polynomial separation between regular and unrestricted resolution. This was improved to an exponential separation by Alekhnovich *et al.* [14]. They used the so-called Stone tautologies formalizing a form of induction. Buss and Kolodziejczyk [127] proved that these tautologies have short proofs in *pool resolution* and in an extension of regular tree-like resolution, implying that they do not separate R from DPLL with clause learning.

Kullmann [332] defined a generalization of resolution capturing propositional logic, constraining satisfaction problems and polynomial systems at the same time. He also defined [331] a *generalized ER*, *GER*, a parameterized version of *ER*; *ER* p-simulates GER but Kullmann [331] showed that one can prove strong lower bounds for GER.

Beame *et al.* [54] considered several proof systems corresponding to extensions of DPLL by a formula-caching scheme. In particular, they defined a proof system FC^{W}_{reason} which can polynomially simulate regular resolution and has an exponential speed-up over general resolution on some formulas.

Beame, Kautz and Sabharwal [55] characterized precisely (in a proof complexity sense) extensions of R by various forms of clause learning. Buss, Hoffmann and Johannsen [123] considered characterizations of DPLL with clause learning via tree-like R with lemmas.

Bonacina and Talebanfard [90] considered a proof system weaker than R, but generalizing regular R, and proved a strong exponential lower bound for it, described by them as being consistent with the so-called *exponential time hypothesis* (an assumption used in analyzing some SAT algorithms).

14

$\mathrm{LK}_{d+1/2}$ and Combinatorial Restrictions

In this chapter we prove the exponential separation of $\mathrm{LK}_{d+1/2}$ from $\mathrm{LK}^*_{d+1/2}$ (and, by Lemmas 3.4.2 and 3.4.7, also from $\mathrm{LK}_{d-1/2}$) if $d \geq 1$. In particular, by the simulation of Theorem 3.2.4 we get exponential lower bounds for AC^0-Frege systems. The hard formulas depend on the depth (otherwise we could not get a separation of the subsystems). Formulas that are hard for all constant-depth subsystems will be presented in Chapter 15.

We present the results for the subsystems $\mathrm{LK}_{d+1/2}$ defined by the Σ-depth rather than for the LK_d defined using the depth. The reason is that the bounded arithmetic theories $S^i_1(\alpha)$ and $T^i_1(\alpha)$ from Section 9.3 translate into proof systems defined by the Σ-depth (Section 10.5). The general idea works equally well for the systems LK_d. Using constant-depth subsystems of the sequent calculus LK rather than of a Frege system in the DeMorgan language allows us to use the optimal simulations among them proved in Section 3.4 and thus to prove sharper separations.

An important part of the argument rests upon an analysis of the effect that a restriction (i.e. a partial truth assignment) of a certain type has on a constant-depth formula. The qualification *combinatorial* in the title of the chapter refers to the fact that we shall study how restricted formulas behave with respect to the set of all truth assignments, i.e. semantically. This makes the whole argument conceptually simpler, I think, than the argument we shall study in the next chapter.

14.1 The Lifting Idea

The general idea is fairly simple. Assume that we have two subsystems P and Q of LK of small (Σ-)depths and that we also have a formula $A(\overline{p})$ of small depth that has a short proof in Q but only long proofs in P. Replace each atom p_i by a generic depth d formula written using new sets of atoms $\{s^i_j\}_j$ that are disjoint for all i. Such a generic formula is naturally obtained as the $\langle \dots \rangle$ translation (for some n) of a generic first-order formula of the quantifier complexity d. The most natural such formula is perhaps

$$\forall y_1 \exists y_2 \dots Q_d y_d \, S(y_1, \dots, y_d),$$

with the quantifiers alternating (i.e. $Q_d = \forall$ if and only if d is odd) and with S a d-ary relation symbol. The translation with size parameter n is called complexity theory the depth d **Sipser function**:

$$S^{d,n} := \bigwedge_{j_1} \bigvee_{j_2} \bigwedge_{j_3} \bigvee_{j_4} \dots s_{j_1,\dots,j_d} \tag{14.1.1}$$

with the indices $j_1, \dots, j_d$ ranging over $[n]$.

If we substitute in the Q-proof of $A(\overline{p})$ for each p_i the formula $S_i^{d,n}$ written as (14.1.1) but using the atoms $s^i_{j_1,\dots,j_d}$ for different i then we get a proof of $A(S_1^{d,n}, S_2^{d,n} \dots)$ in $Q + d$, i.e. Q lifted by the (Σ-)depth d. The hard part is to show that there is no short proof of the formula in $P + d$, i.e. P lifted by the depth d. The strategy is to find a restriction σ of the atoms of all $S_i^{d,n}$ which would collapse a short proof π in $P + d$ of $A(S_1^{d,n}, \dots)$ into a short P-proof of the original formula $A(\overline{p})$, which we know does not exist. It is easy to define a σ that collapses each $S_i^{d,n}$ into one atom but we need σ to collapse all depth d subformulas that occur in π. This is more difficult and, in particular, σ will not collapse $A(S_1^{d,n}, \dots)$ into the original formula $A(\overline{p})$ but into a formula that is *close* to it.

The formula A that we shall work with is *onto*$\text{WPHP}_n^{n^2}$ from Corollary 11.4.5 in Section 11.4. For technical reasons, we want all indices to range over the same set $[n]$ and to represent elements of $[n^2]$ by pairs from $[n] \times [n]$. Thus for $\neg$*onto*$\text{WPHP}_n^{n^2}$ we take the following unsatisfiable set of clauses in the atoms $p_{i_1,i_2,j}$:

- $\{p_{i_1,i_2,j} \mid j \in [n]\}$, for all $i_1, i_2 \in [n]$;
- $\{p_{i_1,i_2,j} \mid i_1, i_2 \in [n]\}$, for all $j \in [n]$;
- $\{\neg p_{i_1,i_2,j}, \neg p_{i_1,i_2,j'}\}$, for all $i_1, i_2 \in [n]$ and $j \neq j' \in [n]$;
- $\{\neg p_{i_1,i_2,j}, \neg p_{i'_1,i'_2,j}\}$, for all $(i_1, i_2) \neq (i'_1, i'_2) \in [n] \times [n]$ and $j \in [n]$.

The base proof systems P and Q will be $R^*(\log)$ and $R(\log)$, respectively. By Corollaries 11.4.5 and 13.1.3, $\neg$*onto*$\text{WPHP}_n^{n^2}$ has a size-$n^{O(\log n)}$ $R(\log)$-refutation while each $R^*(\log)$-refutation must be exponentially large, $2^{\Omega(n^{1/2})}$. However, because of the collapsing process we shall need an $R^*(\log)$ lower bound for the formula close to $\text{WPHP}_n^{n^2}$, **modified** WPHP.

Let $E \subseteq [n] \times [n] \to [n]$ and define

$$\neg(E\text{-}onto\text{WPHP}_n^{n^2})$$

to be as $\neg$*onto*$\text{WPHP}_n^{n^2}$ but with all atoms $p_{i_1,i_2,j}$ such that $((i_1, i_2), j) \notin E$ deleted from it. For $(i_1, i_2) \in [n] \times [n]$, denote by $E((i_1, i_2), -)$ the set of all $j \in [n]$ such that $((i_1, i_2), j) \in E$ and analogously by $E(-, j)$ the set of all $(i_1, i_2) \in [n] \times [n]$ such that $((i_1, i_2), j) \in E$.

Lemma 14.1.1 *Assume $E \subseteq [n] \times [n] \to [n]$ is such that:*

(i) for all $(i_1, i_2) \in [n] \times [n]$, $|E((i_1, i_2), -)| \geq n^{1/2}$;
(ii) for all $j \in [n]$, $|E(-, j)| \geq n^{3/2}$.

Then every $R^*(\log n)$*-refutation of* $\neg(E\text{-onto}WPHP^{n^2}_n)$ *must have size at least* $2^{\Omega(n^{1/4})}$.

Proof Use the adversary argument from the proof of Corollary 13.1.3 but build the partial bijections R_ℓ between $[n] \times [n]$ and $[n]$ only within E. This decreases the power $1/2$ in the exponent of $n^{1/2}$ to $1/4$. □

One could start also with the pair R^* and R as the base case. However, by the nature of the collapsing process, σ will transform a proof in $R^* + d$ into an $R^*(\log)$-proof anyway, so there is no advantage in starting with the pair R^*, R.

14.2 Restrictions and Switchings

To make the whole idea work we need a substitution with the properties of σ discussed in the last section. We shall assume that our original formula is built from the atoms $p_{\bar{i}}$ indexed by $\bar{i} \in [n]^k$, for some $k \geq 1$ (we had $k = 3$ for WPHP). For each $d, n \geq 1$ we take the Sipser functions $S^{d,n}_{\bar{i}}$ as in (14.1.1), one for each $\bar{i} \in [n]^k$. The atoms of $S^{d,n}_{\bar{i}}$ are $s^{\bar{i}}_{\bar{j}}$, for all $\bar{j} \in [n]^d$, and their set will be denoted by $Var(S^{d,n}_{\bar{i}})$.

For the next statement, recall from Section 3.3 and 3.4 the definition of the Σ-depth of a formula and of the formula classes $\Sigma^{S,t}_d$ and $\Pi^{S,t}_d$ and define the class Δ^t_1 to consist of formulas that are in both $\Sigma^{S,t}_1$ and $\Pi^{S,t}_1$.

Lemma 14.2.1 *Let* $c, d \geq 1$ *and* $n \gg 1$ *and assume that* $S \leq 2^{n^{1/3}}$ *and* $t = \log S$. *Assume further that:*

- $A_1, \dots, A_s$ *are* $\Sigma^{S,t}_d$*-formulas built from atoms in* $\bigcup_{\bar{i}\in[n]^k} Var(S^{d,n}_{\bar{i}})$ *of total size* $\sum_{u \leq s} |A_u| \leq S$,
- $U_1, \dots, U_r \subseteq [n]^k$ *are arbitrary* $r \leq n^c$ *sets.*

Then there is a substitution

$$\sigma\colon \bigcup_{\bar{i}\in\{0,1\}^k} Var(S^{d,n}_{\bar{i}}) \rightarrow \{0,1\} \cup \{p_{\bar{i}} \mid \bar{i} \in [n]^k\}$$

such that:

(i) *all formulas* $\sigma(A_u)$, $u \leq s$, *are* Δ^t_1;
(ii) *for any* $\bar{i} \in [n]^k$, $\sigma(S^{d,n}_{\bar{i}}) \in \{0, 1, p_{\bar{i}}\}$;
(iii) *for all* U_v, $v \leq r$, *it holds that, for at least* $|U_v|n^{-(1/2)}$ *k-tuples* $\bar{i} \in U_v$, $\sigma(S^{d,n}_{\bar{i}}) = p_{\bar{i}}$.

We shall use this lemma in the next section to prove the lower bound, but let us outline in the rest of this section how it is proved. We do not give full proofs here; see Section 14.6 for references.

Partition the set $\bigcup_{\bar{i}\in[n]^k} Var(S^{d,n}_{\bar{i}})$ into n^{k+d-1} blocks

$$B(\bar{i}, j_1, \ldots j_{d-1}) \ := \ \{s^{\bar{i}}_{j_1,\ldots,j_{d-1},j} \mid j \in [n]\},$$

one for each choice of $\bar{i} \in [n]^k$ and $j_1, \ldots, j_{d-1} \in [n]$.

For $0 < q < 1$, define a probability space $\mathbf{R}^+_{k,d,n}(q)$ of restrictions defined on $\bigcup_{\bar{i}\in[n]^k} Var(S^{d,n}_{\bar{i}})$ as follows (the statement that an atom is set to $*$ means that it is left unchanged):

1. For any $w = (\bar{i}, j_1, \ldots, j_{d-1}) \in [n]^{k+d-1}$, put

$$r_w := \begin{cases} * & \text{with probability } q, \\ 0 & \text{with probability } 1-q. \end{cases}$$

2. For any atom $s \in B(w)$, put

$$\rho(s) := \begin{cases} r_w & \text{with probability } q, \\ 1 & \text{with probability } 1-q. \end{cases}$$

The probability space $\mathbf{R}^-_{k,d,n}(q)$ is defined analogously but exchanging the roles of 0 and 1.

Each $\rho \in \mathbf{R}^+_{k,d,n}(q)$ is further extended to $\hat{\rho}$ in order to replace the original atoms $\bigcup_{\bar{i}\in[n]^k} Var(S^{d,n}_{\bar{i}})$ by the atoms of the Sipser function one level down, i.e. the atoms from $\bigcup_{\bar{i}\in[n]^k} Var(S^{d-1,n}_{\bar{i}})$:

1. for $w = (\bar{i}, j_1, \ldots, j_{d-1}) \in [n]^{k+d-1}$ and $j, j' \in [n]$, let $\tilde{\rho}$ extend $\rho \in \mathbf{R}^+_{k,d,n}(q)$ by assigning the value 1 to any $s^{\bar{i}}_{j_1,\ldots,j_{d-1},j}$ for which ρ is undefined as long as, for some $j' > j$, $\rho(s^{\bar{i}}_{j_1,\ldots,j_{d-1},j'})$ is also undefined;
2. $\hat{\rho}(s^{\bar{i}}_{j_1,\ldots,j_{d-1},j}) := s^{\bar{i}}_{j_1,\ldots,j_{d-1}}$ for all $s^{\bar{i}}_{j_1,\ldots,j_{d-1},j}$ that are not evaluated by $\tilde{\rho}$ to 0 or 1.

For $\rho \in \mathbf{R}^-_{k,d,n}(q)$, we define $\hat{\rho}$ analogously but changing the values of $s^{\bar{i}}_{j_1,\ldots,j_{d-1},j}$ in item 1 to 0 instead of to 1.

The following lemma shows that $\hat{\rho}$ does reduce the Σ-depth of formulas.

Lemma 14.2.2 (Hastad [222]) *Assume that $d, e \geq 1$ and $n \gg 1$, and let $0 < q < 1$. Let C be an $\Sigma^{S,t}_{e+1}$-formula built from the atoms in $\bigcup_{\bar{i}\in[n]^k} Var(S^{d,n}_{\bar{i}})$.*

Then, for ρ chosen from $\mathbf{R}^+_{k,d,n}(q)$ or from $\mathbf{R}^-_{k,d,n}(q)$, the probability that $\hat{\rho}(C)$ is not equivalent to a $\Sigma^{S,t}_e$-formula (with atoms from $\bigcup_{\bar{i}\in[n]^k} Var(S^{d-1,n}_{\bar{i}})$) is at most $(6qt)^t$.

We also require that $\hat{\rho}$ does not collapse the Sipser functions too much, i.e. it should be by at most one level. To be able to formulate the relevant statement we need first to modify their definition a little. For a parameter $\ell \geq 1$, define $T^{d,n,\ell}_{\bar{i}}$ in the same way as $S^{d,n}_{\bar{i}}$ except that the arity of the last $\bigwedge$ (if d is odd) or $\bigvee$ (if d is even) ranges only over

$$\left(\frac{1}{2}\ell n \log n\right)^{1/2}.$$

Note that $T^{d,n,\ell}_{\bar{i}}$ is a substitution instance of $S^{d,n}_{\bar{i}}$ for $\ell \ll n/\log n$ (in the actual construction ℓ is a constant). We shall say that a formula C **contains** formula D if C can be obtained from D by renaming or erasing some variables.

Lemma 14.2.3 (Hastad [222]) *Assume that $d, e, \ell \geq 1$ and $n \gg 1$ and let $\bar{i} \in [n]^k$; denote $T^{d,n,\ell}_{\bar{i}}$ simply by T. Put*

$$q = \left(\frac{2\ell \log n}{n}\right)^{1/2}$$

and assume that $q \leq 1/5$.

(i) If $d \geq 2$ and ρ is chosen from $\mathbf{R}^+_{k,d,n}(q)$ if d is odd or from $\mathbf{R}^-_{k,d,n}(q)$ if it is even, then the probability that $\hat{\rho}(T)$ does not contain $T^{d-1,n,\ell-1}_{\bar{i}}$ is at most

$$\frac{n^{-\ell+d-1}}{3}.$$

(ii) If $d = 1$ and ρ is chosen from $\mathbf{R}^+_{k,d,n}(q)$ then, with probability at least

$$1 - \frac{n^{-\ell+k}}{6},$$

all n^k circuits $T^{1,n,\ell}_{\bar{i}}$ for all $\bar{i} \in [n]^k$ are transformed by $\hat{\rho}$ to $*$ or 0 and, with at least the same probability, at least

$$((\ell - 1) \log n)^{1/2} n^{k-1/2}$$

$*$ symbols are assigned.

The required substitution σ is obtained as a composition of the substitutions $\hat{\rho}$ chosen in an alternating way from $\mathbf{R}^+_{k,d,n}(q)$ and $\mathbf{R}^-_{k,d,n}(q)$ for $d, d-1, \ldots, 1$. The reader can find the details in the references given in Section 14.6.

14.3 Depth Separation

We have everything ready to make the lifting idea work.

Theorem 14.3.1 (Krajíček [276]) *For every $d \geq 0$ and $n \gg 1$, there is a set $Z_{d,n}$ of sequents consisting of formulas of depth at most d and of total size $|Z_{d,n}| = O(n^{3+d})$ such that:*

- *$Z_{d,n}$ can be refuted in $\mathrm{LK}_{d+1/2}$ by a refutation of size $n^{O(\log n)}$;*

- *every refutation of $Z_{d,n}$ in* $\mathrm{LK}^*_{d+1/2}$ *(and thus also in* $\mathrm{LK}_{d-1/2}$ *for* $d \geq 1$ *by Lemmas 3.4.4 and 3.4.7) must have size at least* $2^{\Omega(n^{1/3})}$.

Proof Let A be the set of clauses $\neg onto\mathrm{WPHP}^{n^2}_n$ from Section 14.1 and for all $\bar{i} \in [n]^3$ let $S^{d,n}_{\bar{i}}$ be the Sipser formulas as in (14.1.1), all of which are in disjoint sets of atoms. The set $Z_{d,n}$ consists of the substitution instances of all clauses from $\neg onto\mathrm{WPHP}^{n^2}_n$ with each atom $p_{\bar{i}}$ substituted by $S^{d,n}_{\bar{i}}$.

By Corollary 11.4.5, R(log) refutes $\neg onto\mathrm{WPHP}^{n^2}_n$ in size $n^{O(\log n)}$ and the substitution of the Sipser formulas for all atoms in the refutation yields a refutation in $\mathrm{LK}_{d+1/2}$, still of size $n^{O(\log n)}$ (for any fixed d).

Assume now that $Z_{d,n}$ has a size S refutation π in $\mathrm{LK}^*_{d+1/2}$. Put $t := \log S$ and assume that $S \leq 2^{n^{1/3}}$. Take for the sets U_v in Lemma 14.2.1 all n^3 sets of the form

- $\{(i_1, i_2, j) \mid j \in [n]\}$ for all $i_1, i_2 \in [n]$,
- $\{(i_1, i_2, j) \mid i_1, i_2 \in [n]$ for all $j \in [n]$.

Let the A_i from the lemma be all the formulas in π and let σ be the substitution guaranteed to exist by that lemma.

The substitution transforms π into a (tree-like) refutation of the substitution $\sigma(\neg onto\mathrm{WPHP}^{n^2}_n)$ in which all formulas collapse to Δ^t_1. But $\sigma(\neg onto\mathrm{WPHP}^{n^2}_n)$ is the set $\neg(E\text{-}onto\mathrm{WPHP}^{n^2}_n)$ for

$$E := \{\bar{i} \in [n]^3 \mid \sigma(p_{\bar{i}}) = p_{\bar{i}}\}.$$

By item (iii) of Lemma 14.2.1 the relation $E \subseteq [n]^3$ satisfies the hypothesis of Lemma 14.1.1, and hence $\sigma(\pi)$ (and thus also π) must have size greater than $2^{n^{1/3}}$. □

The following problem goes back to [276] and [278, p. 243].

Problem 14.3.2 (The depth d vs. depth $d+1$ problem) *Is there a constant $c \geq 0$ such that for all $d \geq 0$ and $n \gg 1$ there are sets of clauses $W_{d,n}$ of size $n^{O(1)}$ such that all formulas in $W_{d,n}$ have depth at most c and such that* $\mathrm{LK}_{d+1/2}$ *has more than quasi-polynomial speed-up over* $\mathrm{LK}^*_{d+1/2}$ *on refutations of sets $W_{d,n}$?*

In particular, can one always choose $c = 0$ (i.e. can the proof systems be always separated by refutations of sets of clauses)?

We do not require the sequence $W_{d,n}$ to be uniform in any way: the existence of any sequence implies (cf. Section 8.6) that a suitable reflection principle also separates the proof systems, and these are uniformly constructed formulas.

We shall show in Theorem 14.5.1 that one can get a super-polynomial speed-up with $c = 0$ (i.e. the separation is realized by DNF-formulas). A super-quasi-polynomial speed-up is needed in order to derive from it a non-conservativity result in bounded arithmetic (Section 14.6).

Another open question is whether one can realize the separation using sets of *narrow* clauses of formulas, i.e. whether each sequent in $Z_{d,n}$ can be required to contain at most $O(\log n)$ formulas. This is possible for $d = 0$, i.e. for a separation of R(log) from R*(log) (Section 14.6).

14.4 No-Gap Theorem

It may seem that part of the difficulty in Problem 14.3.2 is the close proximity of the proof systems involved. In this section we shall show that this is not so. We shall formulate the statements for dag-like systems but analogous constructions apply to tree-like systems.

Lemma 14.4.1 *Let $1 \le d < e$ and $c \ge 0$. Let $\mathcal{A}$ be a set of sequents of total size S consisting of formulas that are $\Sigma_c^{S,t}$ for $t := \log S$ (i.e. for $c = 0$ these are small conjunctions or disjunctions).*

Assume that $\mathcal{A}$ has an $\mathrm{LK}_{e+1/2}$ refutation of size s_e but every $\mathrm{LK}_{d+1/2}$ refutation must have size at least s_d. Then there is a set $\mathcal{B}$ of sequents consisting of $\Delta_1^{\log s_e}$ formulas such that:

- *$\mathcal{A} \cup \mathcal{B}$ has an $\mathrm{LK}_{e-1/2}$ refutation of size $O(s_e)$;*
- *every $\mathrm{LK}_{d-1/2}$ refutation of $\mathcal{A} \cup \mathcal{B}$ must have size at least $\Omega(s_d/s_e)$.*

Proof Let π be an $\mathrm{LK}_{e+1/2}$ refutation of $\mathcal{A}$ of size s_e. For every $\Sigma_1^{s_e,t_e}$ subformula $C = \bigvee_i D_i$ occurring in π, where the D_i are of size $\le \log s_e$ conjunctions, introduce a new atom q_C and include in $\mathcal{B}$ all defining clauses of the extension axiom

$$q_C \equiv C.$$

These clauses are

$$\{\neg q_C, D_1, \dots\} \quad \text{and} \quad \{\neg D_i, q_C\} \quad \text{for all } i.$$

For the subformulas $\bigwedge_i D_i$, where the D_i are small disjunctions, do the same for their negation. Note that the total size of $\mathcal{B}$ is at most $O(s_e)$.

Claim 1 If $\mathcal{A} \cup \mathcal{B}$ has a size z $\mathrm{LK}_{d-1/2}$-refutation then $\mathcal{A}$ has a size $O(zs_e)$ $\mathrm{LK}_{d+1/2}$-refutation.

This is obtained by substituting C for each q_C and filling in inferences removing the extension axioms and their uses.

Claim 2 $\mathcal{A} \cup \mathcal{B}$ has a size $O(s_e)$ $\mathrm{LK}_{e-1/2}$-refutation.

Replace each C in π by q_C and use the extension axioms to simulate the inferences in π. □

The idea of the following theorem is that if we can separate $\mathrm{LK}_{d+1/2}$ from $\mathrm{LK}_{e+1/2}$, $d + 2 \le e$, then for some $d \le w < e$ we can also separate $\mathrm{LK}_{w+1/2}$ from

$\mathrm{LK}_{w+3/2}$ and then use Lemma 14.4.1 $(w-d)$ times to pull the separation down to $\mathrm{LK}_{d+1/2}$ and $\mathrm{LK}_{d+3/2}$.

Theorem 14.4.2 (The no-gap theorem, Chiari and Krajíček [139]) *Let $0 \leq c$ and $0 \leq d < e$, and let $f(n), g(n) \geq n$ be two non-decreasing functions on $\mathbf{N}$. Assume that $\mathcal{A}_n$ are sets of sequents consisting of formulas of depth $\leq c$ such that, for $n \gg 1$ it holds that*

- *$\mathcal{A}_n$ has an $\mathrm{LK}_{e+1/2}$-refutation of size $f(n)$,*
- *every $\mathrm{LK}_{d+1/2}$-refutation of $\mathcal{A}_n$ must have size at least $g^{(e-d)}(f(n))$ (the $(e-d)$-th iterate of g on $f(n)$).*

Then there is a constant $0 \leq v < e-d$ and sets $\mathcal{C}_n$ of sequents consisting of formulas of depth $\leq c$ and Δ_1^t formulas, $t := \log(f(n)) + O(1)$, such that, for $n \gg 1$,

- *if m is the minimum size of an $\mathrm{LK}_{d+3/2}$-refutation of $\mathcal{C}_n$ then every $\mathrm{LK}_{d+1/2}$-refutation of $\mathcal{C}_n$ must have size at least $\Omega(g(m)/(m^v))$.*

Proof There is a $d \leq w < e$ such that $\mathcal{A}_n$ has a size $s_{w+1} := g^{(e-w-1)}(f(n))$ $\mathrm{LK}_{w+3/2}$-refutation but every $\mathrm{LK}_{w+1/2}$-refutation must have size at least

$$s_w := g(s_{w+1}) = g^{(e-w)}(f(n)) .$$

Apply Lemma 14.4.1 $v := (w-d)$-times to this pair in place of $[s_e, s_d]$ in the lemma; so, from the initial pair $[s_{w+1}, s_w]$ we get $[O(s_{w+1}), \Omega(s_w/s_{w+1})]$ and eventually $[O(s_{w+1}), \Omega(s_w/(s_{w+1})^v)]$ (with different constants implicit in the O- and Ω- notations, depending on v). □

Let us give a specific example of the theorem aimed at a situation relevant for bounded arithmetic and the R(log) problem 13.7.1 (Sections 13.7 and 14.6).

Corollary 14.4.3 *Let $0 \leq c, d$ and let $\mathcal{A}_n$ be sets of sequents consisting of formulas of depth $\leq c$ such that for $n \gg 1$ it holds that*

- *for some $e \geq d+1$, $\mathcal{A}_n$ has an $\mathrm{LK}_{e+1/2}$-refutation of size $n^{(\log n)^{O(1)}}$ (i.e. quasi-polynomial),*
- *every $\mathrm{LK}_{d+1/2}$-refutation of $\mathcal{A}_n$ must have size at least $n^{(\log n)^{\omega(1)}}$ (i.e. super-quasi-polynomial).*

Then there are sets $\mathcal{C}_n$ of sequents consisting of depth $\leq c$ formulas and Δ_1^t formulas, $t := (\log n)^{O(1)}$, such that, for $n \gg 1$,

- *$\mathcal{C}_n$ has a quasi-polynomial-size $\mathrm{LK}_{d+3/2}$-refutation but every $\mathrm{LK}_{d+1/2}$-refutation of $\mathcal{C}_n$ must have super-quasi-polynomial size.*

Proof Apply Theorem 14.4.2 with $f(n)$ the quasi-polynomial upper bound for $\mathrm{LK}_{e+1/2}$ and $g(n) := n^{\log n}$. □

14.5 DNF Separation

We shall prove now that we can get a super-polynomial separation in the depth-d vs. depth $d+1$ problem 14.3.2. The proof uses a lower bound that we shall prove in the next chapter (Theorem 15.3.1) but the topic fits better in this chapter so we present it here.

Theorem 14.5.1 (Impagliazzo and Krajíček [242]) *For all $d \geq 0$, there are sets of clauses $\mathcal{B}_n$ and $\epsilon > 0$ such that, for all $n \gg 1$, if m is the minimum size of an $\mathrm{LK}_{d+3/2}$-refutation of $\mathcal{B}_n$ then every $\mathrm{LK}_{d+1/2}$-refutation of $\mathcal{B}_n$ must have size at least*

$$m^{(\log m)^{\epsilon}}.$$

Proof The proof is a fairly simple combination of more difficult results. From the next chapter we shall need:

1. Every $\mathrm{LK}_{d+1/2}$-refutation of $\neg\mathrm{PHP}_m$ (the set of clauses expressing the negation of (1.5.1)) must have size at least $2^{m^{10^{-(d+1)}}}$.

This is Theorem 15.3.1; here we have 10 in place of 5 in the theorem because there we work in a language without $\bigwedge$, and its replacement via $\bigvee$ and $\neg$ causes a possible doubling of the depth, and the extra $+1$ in the exponent of 10 arises because we use the Σ-depth here and hence the ordinary depth is possibly higher by 1.

Fix $d \geq 0$. For a parameter $k \geq 1$ whose value we shall fix a bit later, and for $n \gg 1$, put $m(k) := (\log n)^k$. By Theorem 11.4.8:

2. For $n \gg 1$, there are $e_k, c_k \geq 1$ such that $\neg\mathrm{PHP}_{m(k)}$ has an $\mathrm{LK}_{e_k+1/2}$-refutation of size at most $f(n) := n^{c_k}$.

Now pick $k := 2 \times 10^{(d+1)}$. By point 1 any $\mathrm{LK}_{d+1/2}$-refutation of $\neg\mathrm{PHP}_{m(k)}$ must have size at least $n^{\log n}$. Depending on $e - d$ choose $\epsilon > 0$ small enough that for

$$g(n) := n^{(\log n)^{\epsilon}}$$

it holds that

$$g^{(e-d)}(f(n)) \leq n^{\log n}.$$

The no-gap theorem 14.4.2 then yields what we need. □

14.6 Bibliographical and Other Remarks

Secs. 14.1 and 14.2 The lifting idea is from [276], as is the construction of the substitution σ. Lemmas 14.2.2 and 14.2.3 used in the construction were proved by Hastad [222]; see also Yao [510]. Our formulations are close to those used in Buss and Krajíček [130]. A related construction was used for a separation result in bounded arithmetic in [275]. The review [278] offers an exposition of all this material.

Sec. 14.3 Theorem 14.3.1 was the first exponential lower bound for constant-depth Frege systems (the first non-polynomial lower bound was Ajtai's theorem 15.3.1, which we shall prove in the next chapter); it was available as a preprint in mid 1991.

The non-conservativity which would follow from an affirmative solution to Problem 14.3.2 is that of $T_2^{d+2}(\alpha)$ over $S_2^{d+2}(\alpha)$ with respect to $\forall\Sigma^b_{c+2}(\alpha)$ formulas. At present we have such a non-conservativity for all $c = d$ and for $c = 1$ and $d = 1, 2$; see [275], Chiari and Krajíček [138] and Buss and Krajíček [130]. The narrow separation of R(log) from R*(log) corresponds via the $\langle \dots \rangle$ translation to the separation of $T_1^2(\alpha)$ from $S_2^2(\alpha)$ by $\forall\Sigma_1^b(\alpha)$-formulas. This was achieved in Chiari and Krajíček [138] by finding first a $\forall\Sigma_1^b(f)$ separating formula, for f a new function symbol, and then encoding f via its **bit-graph**:

$$\beta(x,j) \quad \text{if and only if} \quad \text{the } j\text{th bit of } f(x) \text{ is } 1.$$

However, we do not have a generalization of the lifting method of Section 14.2 to this situation. The difficulty is that the second argument of β ranges over a substantially smaller set than the first. Chiari and Krajíček [139] found a coherent generalization but not suitable formulas to which it could be applied and which at the same time separate $T_1^2(f)$ from $S_2^2(f)$. Note that such proof complexity separations can be usually turned into a non-reducibility results for the corresponding relativized NP search problems; see Section 19.3.

Secs. 14.4 and 14.5 transformed to propositional logic the constructions from Chiari and Krajíček [139, Sec. 5] and from Impagliazzo and Krajíček [242], respectively, that were formulated for bounded arithmetic theories and non-conservativity relations among them. It was Thapen who pointed out that the argument from [242] should work in the non-uniform (i.e. propositional) case as well, and that it ought to give a super-polynomial DNF separation of constant-depth systems. Skelley and Thapen [473] improved [139] to $\forall\Sigma_1^b(\alpha)$-formulas, and presumably one can derive from their result a new propositional no-gap theorem for sets of *narrow* clauses, that would be analogous to Theorem 14.4.2. Thapen [488] gives candidate formulas based on the so-called *game iteration principle* that could yield a stronger (in particular, non-quasi-polynomial) separation.

15

F_d and Logical Restrictions

This chapter is devoted to the proof of an exponential lower bound for AC^0-Frege proofs of the PHP principle. We shall also apply the method to some related questions and discuss some open problems.

The non-existence of polynomial AC^0-Frege proofs of PHP was proved by Ajtai [5]. This is, in my view, the single most important paper in proof complexity, not so much because of the particular result (we have stronger bounds and cleaner methods now) or its style (actually it was rather reader-unfriendly) but because it opened a completely new world for proof complexity, freeing it from the straightjacket of combinatorics. Until then proof complexity was perceived as a mere propositional proof theory, but Ajtai's paper demonstrated deep connections with mathematical logic and circuit complexity.

The overall idea of the argument can be explained, in retrospect, rather simply. We do have strong lower bounds for low-depth systems like $R^*(\log)$, so we may attempt to transform – by a suitable restriction – any AC^0-Frege proof into one that looks like an $R^*(\log)$-proof of the same, or similar, formula. This is, viewed from a high level, similar to the strategy we employed in the preceding chapter. But, because we are aiming at a specific formula (the PHP) which already has a small depth, we need to use restrictions that will not collapse it into a constant. This causes a significant shift: the restrictions will collapse AC^0-formulas to simpler ones but these simpler formulas will not be equivalent to the original ones in the semantic sense as Boolean functions. Instead they will be equivalent via short proofs (in a weak system) from the axiom $\neg PHP_n$ (augmented by axioms defining the particular restriction; see Section 15.3). This is why I use the qualification *logical* restrictions in the chapter title.

The method can be also interpreted as defining – for every set Γ of AC^0-formulas that is not too large – notions of Γ-true and Γ-false formulas from Γ that satisfy the usual truth table conditions but make PHP_n Γ-false. Taking for Γ all formulas occurring in a proof of PHP_n that is not too long, we can then use a number of arguments to reach a contradiction and prove the actual lower bound using a soundness argument (all logical axioms are Γ-true and the Γ-truth is preserved by the inferences but the last formula is Γ-false) or an adversary argument as in Section 13.1 (find a Γ-false logical axiom) or a winning strategy for Liar in the Buss–Pudlák game (Section 2.2) or other arguments.

In my view the most faithful interpretation of the argument is the one from the original paper by Ajtai [5]: true is interpreted as being *forced* true in the sense of model-theoretic forcing. In particular, Paris and Wilkie [389] used a simple version of model-theoretic forcing to prove the independence of PHP from the least number principle for $\Sigma_1^b(R)$-formulas (these translate into $\Sigma_1^{S,t}$-formulas), and Ajtai's argument can be seen as using suitable restrictions to reduce any bounded-depth formula into this form and then applying the Paris–Wilkie argument. We shall explain this in Section 20.3 but for now, not to strain the reader's patience too much, we will avoid model theory.

Our argument does not involve any proof theoretic constructions where an exact estimate for the depth would matter (in fact, no non-trivial proof theory is involved). For this reason we shall use Frege systems instead of the much more proof-theoretically precise sequent calculus used in the preceding chapter. Also, to decrease the number of cases to consider in various definitions, we take as our propositional language 0, 1, $\neg$ and $\vee$ ($\wedge$ is defined by a depth-2 formula in this language so constant-depth systems in the DeMorgan language and in the restricted language simulate each other with a proportionate change in the depth).

The definition of the depth of DeMorgan formulas in the context of the sequent calculus LK in Section 3.4 uses unbounded-arity $\bigvee$ and $\bigwedge$ and it did not count negations. This is not suitable for Frege systems, and we need to define the depth for formulas in the ordinary DeMorgan language with binary $\vee$ ($\wedge$ would be treated analogously). The **depth** of a formula in the restricted language is 0 if and only if it is an atom or a constant, it is $\mathrm{dp}(\neg A) = 1 + \mathrm{dp}(A)$ if A does not start with $\neg$ and it is $\mathrm{dp}(A)$ if it does; furthermore

$$\mathrm{dp}(A \vee B) = \begin{cases} \max(\mathrm{dp}(A), \mathrm{dp}(B)) & \text{if both } A \text{ and } B \text{ start with } \vee, \\ 1 + \max(\mathrm{dp}(A), \mathrm{dp}(B)) & \text{if both } A \text{ and } B \text{ start with } \neg, \\ \max(1 + \mathrm{dp}(A), \mathrm{dp}(B)) & \text{if } B \text{ starts with } \vee \text{ and } A \text{ does not,} \\ \max(\mathrm{dp}(A), 1 + \mathrm{dp}(B)) & \text{if } A \text{ starts with } \vee \text{ and } B \text{ does not.} \end{cases}$$

Despite having only binary $\vee$ we shall write formulas (and clauses, in particular) using an unbounded-arity $\bigvee$: this is meant only as an abbreviation for a disjunction formed from the binary $\vee$ by any bracketing.

For the rest of the chapter we shall fix an arbitrary Frege system F in the language $0, 1, \neg, \vee$. For $d \geq 0$, F_d is the subsystem whose proofs are allowed to use only formulas of depth at most d.

15.1 PHP-Trees and *k*-Evaluations

We shall consider the pigeonhole principle in its weaker formulation when talking about bijections between $[n+1]$ and $[n]$, as for the more general *onto*WPHP in

Section 11.4. This will make the lower bound stronger. We shall represent it by a set of clauses formalizing the negation of the principle. That is, $\neg onto\text{PHP}_n$ is the set of clauses formed from the atoms p_{ij}, $i \in [n+1]$ and $j \in [n]$:

1. $\bigvee_j p_{ij}$, one clause for each i;
2. $\neg p_{i_1 j} \vee \neg p_{i_2 j}$, one clause for each triple $i_1 < i_2$ and j;
3. $\neg p_{ij_1} \vee \neg p_{ij_2}$, one clause for each triple i and $j_1 < j_2$;
4. $\bigvee_i p_{ij}$, one clause for each j.

Note that even the clauses in the first two items are unsatisfiable; the third item implies that the principle talks about functions as opposed to multi-functions, and the last item restricts them further to bijections.

Let us also fix $n \geq 1$. The truth value of a formula built from the atoms p_{ij} can be determined by a decision tree (Section 1.4) branching at a node according to the truth values of some atoms p_{ij}. The paths in such a tree determine partial assignments to the atoms. It may happen that a path gathers information about an assignment that will certify that the assignment violates a clauses of $\neg onto\text{PHP}_n$; for example, it may contain information that $p_{i_1 j} = 1$ and also $p_{i_2 j} = 1$ for some $i_1 \neq i_2$. However, to violate a clause in the first or the last item the path would have to have length at least n.

The notion of a PHP-tree stems from these two simple observations and incorporates into its definition conditions that prevent both. We will later consider how PHP-trees change after a restriction and thus we now give a definition that is somewhat more general than the discussion so far motivates.

Let $D \subseteq [n+1]$ and $R \subseteq [n]$. The notion of a PHP-**tree over** D, R is defined by induction as follows.

- A single node, a root, is a PHP-tree over any D, R.
- For every $i \in D$, the following is a PHP-tree over D, R:
  - at the root the tree branches according to all $j \in R$, labeling the corresponding edge p_{ij};
  - at the end-point of the edge labeled by p_{ij} the tree continues as a PHP-tree over $D \setminus \{i\}, R \setminus \{j\}$.
- For every $j \in R$, the following is a PHP-tree over D, R:
  - at the root the tree branches according to all $i \in D$, labeling the corresponding edge p_{ij},
  - at the end-point of the edge labeled by p_{ij} the tree continues as a PHP-tree over $D \setminus \{i\}, R \setminus \{j\}$.
- A PHP-tree over $[n+1], [n]$ is simply called a PHP-tree.

The **height** of a tree T, to be denoted by $h(T)$, is the maximum number of edges on a path through the tree (it is sometimes called also the depth of the tree but we want to avoid any confusion with the depth of formulas). A PHP tree of height $\leq k$ will be called a k-PHP **tree**. For the present we shall not consider trees other than PHP trees and hence until further notice we shall call them just **trees** and k-**trees**.

It is useful to think of a tree as branching according to the queries $f(i) = ?$ and $f^{(-1)}(j) = ?$, where f is the name for a possible bijection between $[n+1]$ and $[n]$. Every path in a tree determines a partial injective map between $[n+1]$ and $[n]$; it is often convenient to identify the path with the partial map and the tree with the set of all such maps corresponding to all maximal paths.

Consider the simplest example: a tree of height 1 branching according to all answers to $f(i) = ?$. The formula $\bigvee_j p_{ij}$, an axiom of $\neg onto\mathrm{PHP}_n$, is intuitively true at every leaf of the tree because at any leaf one p_{ij} is made true. However, if we think of f as being defined everywhere (and, in particular, as $f(i)$ being thus defined) then the tree describes all possibilities. Hence the formula $\bigvee_j p_{ij}$ is true in the sense that it holds for all possibilities described by the tree.

Now take the formula $\neg p_{ij} \vee \neg p_{ik}$, for $j \neq k$, another axiom of $\neg onto\mathrm{PHP}_n$. A suitable tree to use for evaluating the formula is a tree branching first according to $f^{(-1)}(j) = ?$ and then, at a branch corresponding to any $u \in [n+1]$, according to $f^{(-1)}(k) = ?$ with answers from $[n+1] \setminus \{u\}$. At every path through this height 2 tree either $f(i) \neq j$ or $f(i) \neq k$ and hence the formula is satisfied. As before, thinking of f as an injective map that is onto, the branching of the tree describes all possibilities. Hence again the formula is true in the sense of being true in all situations described by the tree.

Our general strategy is thus the following. We assign to all formulas a tree and a subset of (the set of paths in) the tree where the formula is true, thinking of the subset as the truth value of the formula in the Boolean algebra of all subsets of (the set of all paths in) the tree. Boolean algebras satisfy all Frege axioms (all tautologies, in fact) and are sound with respect to all Frege rules, and hence we ought to be able to use this as a model where Frege systems are sound while $onto\mathrm{PHP}_n$ fails.

Of course, this cannot work so simply. All tautologies get the maximum value in every Boolean algebra and so does $onto\mathrm{PHP}_n$. In our case the problem is that there is no bijection f and so no tree can decide the truth of all atoms. This implies that formulas will have to have *different* trees attached to them and acquire truth values in different Boolean algebras. We shall need a way to compare them. Explaining more informally would, I think, rather obfuscate things so we launch into a formal treatment.

Let *Maps* be the set of all partial bijections between $[n+1]$ and $[n]$, including the empty map $\emptyset$. Maps from *Maps* will be denoted by small Greek letters $\alpha, \beta, \ldots$ The size of α is the size of its domain and will be denoted by $|\alpha|$.

If $\alpha \cup \beta \in Maps$ we say that α and β are **compatible maps**; this will be denoted by $\alpha || \beta$. However, if $\alpha \cup \beta \notin Maps$, α and β are **incompatible** (or **contradictory**), denoted by $\alpha \perp \beta$. Note that this means that either for some i in the domains of both maps $\alpha(i) \neq \beta(i)$ or for some j in the range of both maps $\alpha^{(-1)}(j) \neq \beta^{(-1)}(j)$.

- For $H \subseteq Maps$ and T a tree, T **refines** set H, denoted by $H \lhd T$, if and only if whenever an $\alpha \in T$ is compatible with some $\beta \in H$ then it contains some $\gamma \in H$.

- For two trees T and S, $T \times S := \{\alpha \cup \beta \mid \alpha \in T, \beta \in S, \alpha || \beta\}$ is the **common refinement** of T and S.
- For $H \subseteq Maps$ and S a tree, the **projection** of H onto S is the set $S(H) := \{\alpha \in S \mid \exists \gamma \in H, \gamma \subseteq \alpha\}$.

In the next few lemmas we use the letters $H, K, \ldots$ for arbitrary subsets of *Maps* while we reserve the letters S and T for trees, without stressing this explicitly. Trees are also thought of as subsets of *Maps* (a tree determines the set of all paths in it) and hence definitions for sets H apply to them as well (for example, $T(S)$ is well defined).

Lemma 15.1.1 *If $|\delta| + h(S) \leq n$ then $\exists \gamma \in S$, γ and δ are compatible maps, i.e. $\gamma || \delta$.*

Proof Walk through the tree S answering all queries according to δ whenever it applies, and arbitrarily but consistently with δ otherwise. The assumption that $|\delta| + h(S) \leq n$ implies that we do not run into a contradiction before reaching a leaf of S. The map γ is determined by the particular path. □

Lemma 15.1.2 *Assume that $h(S) + h(T) \leq n$ and $H \lhd S \lhd T$. Then also $H \lhd T$.*

Proof Assume that $\delta \in T$ is compatible with some $\alpha \in H$. We want to show that δ contains some element of H.

By Lemma 15.1.1, $\exists \gamma' \in S, \gamma' || \delta$. By this, and by $S \lhd T$, we have $\exists \gamma \in S, \gamma \subseteq \delta$. Such a γ is necessarily compatible with α and hence, by $H \lhd S$, $\exists \alpha' \in H, \alpha' \subseteq \gamma$. Hence $\alpha' \subseteq \delta$ too. □

Lemma 15.1.3 *Assume that $h(S) + h(T) \leq n$. Then $S \times T$ is a tree and $h(S \times T) \leq h(S) + h(T)$ and is such that $S \lhd S \times T$ and also $T \lhd S \times T$.*

Proof To see that $S \times T$ is indeed a tree, place copies of T under each leaf of S and delete inconsistent paths and duplications. The bound to the height of $S \times T$ is obvious. We will prove that $S \lhd S \times T$: the second statement is proved in an identical way.

Assume that $\beta \cup \gamma \in S \times T$, with $\beta \in S$ and $\gamma \in T$, is compatible with some $\alpha \in S$. Then necessarily $\alpha = \beta$, i.e. $\beta \cup \gamma$ contains an element of S. □

The next two lemmas will be very useful when comparing the truth values acquired by different formulas in different trees and also when actually computing truth values.

Lemma 15.1.4 *Assume that $h(S) + h(T) \leq n$ and $H \lhd S \lhd T$. Then*

(i) $T(S(H)) = T(H)$,
(ii) $T(S) = T$,
(iii) $S(H) = S$ *if and only if* $T(H) = T$.

Proof The inclusion $T(S(H)) \subseteq T(H)$ follows from the definition. For the opposite inclusion, assume that $\beta \in T(H)$ because $\beta \supseteq \gamma$ for some $\gamma \in H$. Using

Lemma 15.1.1, $S \lhd T$ implies that $\exists \alpha \in S, \alpha \subseteq \beta$. Such a β is then compatible with γ and hence, as $H \lhd S$, we conclude that $\exists \gamma' \in H, \gamma' \subseteq \alpha$. So we have $\gamma' \subseteq \alpha \subseteq \beta$ and thus $\beta \in T(S(H))$. This proves item (i).

Item (ii) follows from item (i) by taking $H := \{\emptyset\}$. For item (iii), assume first that $S(H) = S$. By items (ii) and (i), $T(S) = T$ and $T(S(H)) = T(H)$. So $T = T(H)$.

Finally, assume that $T(H) = T$. Let $\alpha \in S$. By Lemma 15.1.1 there is a $\beta \in T$ compatible with α. Also, by the assumption, $\beta \in T(H)$ and so $\exists \gamma \in H, \gamma \subseteq \beta$. But such a γ is compatible with α and hence, by $H \lhd S$, it holds that $\exists \gamma' \in H, \gamma' \subseteq \alpha$. So $\alpha \in S(H)$ as we required. □

Lemma 15.1.5

(i) $S(\bigcup_i H_i) = \bigcup_i S(H_i)$.
(ii) If $H_0, H_1 \subseteq T$ *and* $H_0 \cap H_1 = \emptyset$ *then* $T(H_0) \cap T(H_1) = \emptyset$.
(iii) If $S \lhd T$, $h(S) + h(T) \leq n$ *and* $H \subseteq S$ *then* $T(S \setminus H) = T \setminus T(H)$.

Proof The first two items follow directly from the definitions. By Lemma 15.1.4 $T(S) = T$, hence the last item follows from the first two. □

After verifying these preliminary and somewhat tedious facts we are ready to define the key notion of k-evaluations and prove its basic properties. Recall that we have some fixed $n \gg 1$. Let $1 \leq k \leq n$ be a parameter and let Γ be a set of formulas built from the atoms of *onto*PHP$_n$ that is *closed under subformulas*. A **k-evaluation of Γ** is a map

$$(H, S) \colon \varphi \in \Gamma \to H_\varphi \subseteq S_\varphi$$

assigning to a formula $\varphi \in \Gamma$ a k-tree S_φ and its subset H_φ, such that the following four conditions are satisfied.

1. $S_0 := S_1 := \{\emptyset\}$ (i.e. the tree consisting of the root only) and $H_0 := \emptyset$ and $H_1 := S_1$;
2. $S_{p_{ij}}$ is the depth 2 tree that first branches according to $f(i) = ?$ and then according to $f^{(-1)}(j) = ?$; $H_{p_{ij}} := \{(i,j)\}$, the only path in $S_{p_{ij}}$ of length 1;
3. $S_{\neg\varphi} := S_\varphi$ and $H_{\neg\varphi} := S_\varphi \setminus H_\varphi$, whenever $\neg\varphi \in \Gamma$.
4. Assuming that $\varphi = \bigvee_i \phi_i$ is in Γ (where, as mentioned earlier, the large disjunction symbol abbreviates arbitrarily bracketed binary disjunctions),

$$\bigcup_i H_{\phi_i} \lhd S_\varphi \quad \text{and} \quad H_\varphi := S_\varphi(\bigcup_i H_{\phi_i}).$$

If $H_\varphi = S_\varphi$, we say that φ is **true with respect to the k-evaluation**.

Lemma 15.1.6 *Assume that (H, S) is a k-evaluation of all formulas occurring as subformulas in a clause of $\neg$onto*PHP$_n$*, and that $k \leq n - 2$.*

Then the clause is true with respect to the evaluation.

Proof Consider the clause

$$\varphi = \bigvee_j p_{ij}$$

for some fixed $i \in [n+1]$. The subformulas of φ are also the atoms $p_{ij}, j \in [n]$, and we thus assume they are in Γ as well. By the definition $H_{p_{ij}} = \{(i,j)\}$ and S_φ must refine the set $H = \{(i,j) \mid j \in [n]\}$. Note that H itself is a 1-tree and that $H(H) = H$.

Hence $T(H) = T$ holds also for the common refinement T of H and S_φ by Lemma 15.1.4, and again by the same lemma we also have $S_\varphi = S_\varphi(H) = H_\varphi$.

We leave the reader to check the statements for the other clauses of $\neg onto\mathrm{PHP}_n$.

□

The following statement, formalizing that k-evaluations are sound with respect to Frege rules, is essential for the method to work.

Lemma 15.1.7 *There exists a constant $c_F \geq 1$ depending just on the particular system F such that if (H, S) is a k-evaluation of all formulas occurring as subformulas in an instance of an F-rule, $k \leq n/c_F$, and all hypotheses of the instance of the rule are true with respect to the evaluation, then also the conclusion of the rule is true with respect to the evaluation.*

Proof Consider an s-ary F-rule as in Section 2.1 of the form

$$\frac{A_1(q_1, \ldots, q_t), \ \ldots \ , A_s(q_1, \ldots, q_t)}{A_{s+1}(q_1, \ldots, q_t)}.$$

Let r be a number larger than the number of subformulas in the rule.

Assume that $k \leq n/r$ and that (H, S) is a k-evaluation of the set Γ of formulas occurring as subformulas in some instance

$$\frac{A_1(\beta_1, \ldots, \beta_t), \ \ldots \ , A_s(\beta_1, \ldots, \beta_t)}{A_{s+1}(\beta_1, \ldots, \beta_t)}.$$

Assume further that all hypotheses of the inference

$$A_1(\beta_1, \ldots, \beta_t), \ldots, A_s(\beta_1, \ldots, \beta_t)$$

are true with respect to the evaluation:

$$H_{A_i(\beta_1,\ldots,\beta_t)} = S_{A_i(\beta_1,\ldots,\beta_t)} \quad \text{for } 1 \leq i \leq s\,.$$

Let the set $\Gamma_0 \subseteq \Gamma$ consist of formulas γ of the form $A'(\beta_1, \ldots, \beta_t)$, where A' is a subformula of some A_i, $i \leq s+1$. By the choice of r we have $|\Gamma_0| < r$, and so there is a common refinement T of all S_γ for $\gamma \in \Gamma_0$, and $h(T) \leq (r - 1/r)n$ (by Lemma 15.1.3). In particular, $h(T) + h(S_\gamma) \leq n$ for all $\gamma \in \Gamma_0$.

Claim *The map defined by*

$$\gamma \in \Gamma_0 \to T(H_\gamma)$$

is a map of formulas in Γ_0 into the Boolean algebra of subsets of (the set of all paths in) T such that:

(i) the negation corresponds to the complement, $T(H_{\neg\gamma}) = T \setminus T(H_\gamma)$;
(ii) the disjunction corresponds to the union, $T(H_{\gamma\vee\delta}) = T(H_\gamma) \cup T(H_\delta)$;
(iii) all hypotheses $\gamma = A_i(\beta_1, \ldots, \beta_t)$, $i \le s$, of the instance of the rule acquire the maximum value in the Boolean algebra, $T(H_\gamma) = T$.

For item (i), if $\neg\gamma \in \Gamma_0$, $H_{\neg\gamma} = S_\gamma \setminus H_\gamma$, and hence $T(H_{\neg\gamma}) = T \setminus T(H_\gamma)$ by Lemma 15.1.5. For item (ii) let $\gamma \vee \delta \in \Gamma_0$. We need to consider cases distinguished by the form of γ and δ; we shall treat only the hardest case, when both γ and δ are themselves disjunctions. Assume $\gamma = \bigvee_u \gamma_u$ and $\delta = \bigvee_v \delta_v$. By Lemma 15.1.5,

$$H_{\gamma\vee\delta} = S_{\gamma\vee\delta}(\bigcup_u H_{\gamma_u}) \cup S_{\gamma\vee\delta}(\bigcup_v H_{\delta_v});$$

hence, by Lemmas 15.1.4 and 15.1.5,

$$T(H_{\gamma\vee\delta}) = T(S_{\gamma\vee\delta}(\bigcup_u H_{\gamma_u})) \cup T(S_{\gamma\vee\delta}(\bigcup_v H_{\delta_v}))$$

$$= T(\bigcup_u H_{\gamma_u}) \cup T(\bigcup_v H_{\delta_v}) = T(S_\gamma(\bigcup_u H_{\gamma_u})) \cup T(S_\delta(\bigcup_v H_{\delta_v}))$$

$$= T(H_\gamma) \cup T(H_\delta) .$$

Item (iii) follows by Lemma 15.1.4:

$$T(H_{A_i(\overline{\beta})}) = T(S_{A_i(\overline{\beta})}) = T$$

for $i \le s$.

The lemma follows on noting that any Frege rule preserves the maximum truth value in any Boolean algebra, because the implication

$$\bigwedge_{i \le s} A_i(q_1, \ldots, q_t) \ \rightarrow \ A_{s+1}(q_1, \ldots, q_t)$$

is a tautology. □

Our strategy for proving a lower bound for F_d-refutations of $\neg onto\mathrm{PHP}_n$ is now clearer. Having such an alleged refutation π we take a k-evaluation (with small enough k) of the set of all formulas occurring in π as subformulas. This leads to a contradiction, by Lemmas 15.1.6 and 15.1.7. We cannot hope to be able to do this for all π (after all, there is a trivial exponential-size refutation in R). But if we manage to construct a k-evaluation of any *small* set of formulas then we can conclude that no F_d-refutation of $\neg onto\mathrm{PHP}_n$ can be small.

15.2 The Existence of k-Evaluations

The task for this section is to construct k-evaluations of small sets of formulas. The qualification *small* will mean of size at most $2^{n^{\Omega(1)}}$. At first sight this does not look promising as it is quite easy to find small sets which have no k-evaluation with $k < n$.

The strategy is to employ a simplification procedure by a suitable restriction before trying to find a k-evaluation with small k.

We shall think of the set *Maps* as a set of partial bijections between a subset of some domain D and some range R. Initially we have $D = [n + 1]$ and $R = [n]$ as earlier but in the process both D and R will shrink.

For $\alpha, \rho \in Maps$, define the **restriction of** α **by** ρ to be

$$\alpha^\rho = \begin{cases} \alpha \setminus \rho & \text{if } \alpha || \rho, \\ \text{undefined} & \text{if } \alpha \perp \rho. \end{cases}$$

Further define:

1. $H^\rho := \{\alpha^\rho \mid \alpha \in H\}$;
2. $D^\rho := D \setminus dom(\rho)$;
3. $\mathrm{R}^\rho := R \setminus rng(\rho)$;
4. $n_\rho := |R^\rho| (= n - |\rho|)$.

Our strategy for the construction of a k-evaluation of a set Γ will be the following. We shall construct the evaluation in steps. We have defined already what values must be assigned by any evaluation to the atoms and to the constants. At every step we extend the k-evaluation to negations and to disjunctions (which are themselves in Γ) of formulas for which it is already defined (hence the number of steps is bounded by the maximum depth of a formula in Γ), i.e. by some fixed d.

Negations are again determined by the definition of a k-evaluation and only the case of disjunctions will cause us a difficulty. To extend the definition to disjunctions we will need to apply a restriction by some ρ. The following lemma essentially says that the part of the evaluation already constructed in prior steps (i.e. before a restriction is applied) will work after the restriction as well. The lemma is straightforward. We continue using the convention that $S, T, \ldots$ denote trees.

Lemma 15.2.1 *Let $\rho \in Maps$ be arbitrary. Then:*

(i) if $H \lhd S$ then $H^\rho \lhd S^\rho$;
(ii) if $|\rho| + h(S) \leq n$ then S^ρ is a tree over D^ρ and R^ρ;
(iii) if $H \lhd S$ then $S^\rho(H^\rho) = (S(H))^\rho$.

The next lemma is the key technical statement needed in the construction of k-evaluations. Sometimes it is called the PHP **switching lemma**. Its proof is far from straightforward. We shall extended the notation $h(S)$ to any H:

$$h(H) := \max\{|\alpha| \mid \alpha \in H\}.$$

Lemma 15.2.2 *Let $0 < \delta < \epsilon < 1/5$ and $H_i \subseteq Maps$, for $i \leq s$. Assume that $h(H_i) \leq k$ for all $i \leq s$, that $n \gg 1$ and that*

$$k \leq n^\delta \quad \textit{and} \quad s \leq 2^k.$$

Then there exists a $\rho \in Maps$ such that $n_\rho = n^\epsilon$ and such that there exist trees S_i, $i \leq s$, over D^ρ and R^ρ, satisfying:

(i) $H_i^\rho \lhd S_i$;
(ii) $h(S_i) \leq k$.

Proof Assume first that we have just one H; we shall consider the case of s sets H_i at the end; this will require only a minor extension of the argument.

We shall describe a game determined by the set H and played by two players. In the proof it will be actually a game determined by H^ρ, but we consider H first so as not to complicate the notation.

At the beginning player I picks an $h_1 \in H$. Player II replies by some $\delta_1 \in Maps$ such that $dom(h_1) \subseteq dom(\delta_1)$ and $rng(h_1) \subseteq rng(\delta_1)$ and such that no proper submap of δ_1 has this property. It may be that $\delta_1 = h_1$ or that $\delta_1 \supseteq h$, for some $h \in H$: in that case the game ends. Otherwise, necessarily $\delta_1 \perp h_1$ and the play moves to the next round.

Generally, before round $t \geq 2$, the players have constructed two sequences of moves:

$$h_1, \ldots, h_{t-1} \text{ (the moves of } I) \quad \text{and} \quad \delta_1 \subseteq \cdots \subseteq \delta_{t-1} \text{ (the moves of } II) .$$

At the tth round player I picks some $h_t \in H$ compatible with δ_{t-1}; if no such h_t exists then, the play stops. If h_t was chosen, player II extends δ_{t-1} to some $\delta_t \in Maps$ such that $dom(h_t) \subseteq dom(\delta_t)$, $rng(h_t) \subseteq rng(\delta_t)$, and such that no proper submap of δ_t containing δ_{t-1} has this property. If δ_t contains some $h \in H$ then the play stops; otherwise, the players move to the next round.

The use of this game is described in the following claim, which follows immediately from the definition of when a play stops.

Claim 1 For any fixed strategy of player I, consider the set

$$S := \{\delta_t \mid \delta_1 \subseteq \cdots \subseteq \delta_t \text{ is a finished play in some strategy of } II\}.$$

Then the set S is a tree and $H \lhd S$.

To further simplify the situation we shall adopt a particular strategy for I: fix an ordering $h^1, h^2, \ldots$ of H and that player I always picks in his or her move the first h in the ordering that is compatible with the previous move of II. We shall call player I when using this strategy I_{fix}.

Let us call all pairs (i,j) in all $h_\ell \setminus \delta_{\ell-1}$ the *critical pairs* of the play. These are exactly the pairs for which II is required to specify $f(i)$ and $f^{(-1)}(j)$. If the number of critical pairs in all finished plays against I_{fix} is bounded by r then clearly $h(S) \leq 2r$. Hence we would like to show that the number of critical pairs is bounded by $k/2$. However, it is easy to construct a set of small maps from *Maps* such that any finished play must contain $\geq n/2$ critical pairs. To restrict the number of critical pairs, we employ a restriction by a suitable map ρ.

Assume that we fix $\rho \in Maps$ and first restrict H by ρ and play the game on H^ρ (we continue to use for I_{fix} the ordering of H^ρ induced by the original ordering of H). This is the same as if we defined $\delta_0 := \rho$ and required h_1 to be compatible with, and δ_1 to contain, δ_0.

Claim 2 There exists a $\rho \in Maps$, $n_\rho = n^\epsilon$, such that every play (tacitly against I_{fix}) on H^ρ contains at most $k/2$ critical pairs.

We shall prove the claim by contradiction. Assume that there is no such ρ. Hence, for every ρ there is a play, resulting in the moves $\delta_1 \subseteq \cdots \subseteq \delta_t$ of *II*, that contains at least $k/2 + 1$ critical pairs. In fact, we will truncate the play when it reaches the $(k/2 + 1)$th critical pair, so we shall assume that there are exactly $k/2$ critical pairs (this is only to simplify the computation). Fix one such play for each ρ.

Now concentrate on one fixed ρ and the associated fixed play. Note that all the critical pairs are disjoint, and are also disjoint from ρ. Hence the set τ containing ρ and all critical pairs is actually an element of *Maps*, and $|\tau| = |\rho| + k/2$.

Having τ we cannot determine ρ a priori but we can determine the first move h_1^ρ of I_{fix}: it is the first $h^\rho \in H^\rho$ that is compatible with τ, i.e. the first $h \in H$ that is compatible with τ.

Now note that we can actually encode the critical pairs in h_1^ρ and the first move δ_1 of *II* by a small number: if there are k_1 critical pairs in h_1^ρ then their set is one of its $\leq \binom{k}{k_1}$ subsets (here we use that $|h_1| \leq h(H) \leq k$), and the move of *II* is determined by giving a value (resp. an inverse value) of f for every i (resp. j) occurring in the critical pairs in h_1^ρ. There are k_1 such i's and j's, and the values player *II* chooses must be outside the domain (resp. the range) of ρ and $n - |\rho| = n^\epsilon$. Hence there are at most $((n^\epsilon + 1)n^\epsilon)^{k_1}$ possibilities for *II*'s action on the critical pairs. All together, we can encode *II*'s first move δ_1 by a number $\leq \binom{k}{k_1}((n^\epsilon + 1)n^\epsilon)^{k_1} \leq (k^{1/2}n^\epsilon)^{2k_1}$.

Once we know δ_1 we can replace in τ all the critical pairs in h_1^ρ by δ_1, obtaining some τ'. However, we know that can also reconstruct the second move h_2^ρ of I_{fix}: it is the first $h^\rho \in H^\rho$ compatible with τ'. Hence we proceed as before: encode *II*'s second move by a number $\leq (k^{1/2}n^\epsilon)^{2k_2}$ (where k_2 is the number of critical pairs in h_2^ρ), and replace in τ' all the critical pairs in h_2^ρ by δ_2, etc.

The whole (truncated) play is in this way encoded by the map τ together with a number

$$\leq \prod_i (k^{1/2}n^\epsilon)^{2k_i} \leq (k^{1/2}n^\epsilon)^{\sum_i 2k_i} \leq (k^{1/2}n^\epsilon)^k,$$

using that $\sum_i k_i \leq k/2$. Because τ, together with the auxiliary information, determine ρ, the numbers

$$\begin{aligned} a :&= \text{the number of different } \rho\text{'s of size } n - n^\epsilon \\ &= \binom{n+1}{n^\epsilon + 1}\binom{n}{n^\epsilon}(n - n^\epsilon)! \end{aligned}$$

and

$$b := \text{the number of different } \tau\text{'s of size } n - n^{\epsilon} + k/2$$
$$= \binom{n+1}{n^{\epsilon} - k/2 + 1}\binom{n}{n^{\epsilon} - k/2}(n - n^{\epsilon} + k/2)!$$

must satisfy the inequality

$$a \leq b(k^{1/2}n^{\epsilon})^k .$$

This argument applies to one set H. However, if we had s sets then we would just encode by a number $\leq s$ which set is the one in which we have, for a given ρ, a play with at least $k/2$ critical pairs. Hence, if no suitable ρ exists, it would have to hold that

$$a \leq sb(k^{1/2}n^{\epsilon})^k.$$

It is not difficult to compute that this inequality does not hold for $n \gg 1$ if the parameters satisfy the hypotheses of the lemma. □

Now we are going to use a restriction ρ in order to construct a k-evaluation. We will need the notion of a formula restricted by ρ defined, as follows. Put

$$p_{ij}^{\rho} = \begin{cases} 1 & i \in dom(\rho) \wedge \rho(i) = j, \\ 0 & \{(i,j)\} \perp \rho, \\ p_{ij} & \text{otherwise}, \end{cases}$$

and then take for φ^{ρ} the formula φ with all atoms p_{ij} replaced by p_{ij}^{ρ}.

Lemma 15.2.3 *Let $0 < \delta < \epsilon < 5^{-d}$. Then, for sufficiently large $n \geq 1$, every set Γ of formulas of depth $\leq d$, $|\Gamma| \leq 2^{n^{\delta}}$, and Γ closed under subformulas there exists a map ρ, $|\rho| = n - n^{\epsilon}$, and an n^{δ}-evaluation of Γ^{ρ}.*

Proof Let $s = 2^{n^{\delta}}$ and $k = n^{\delta}$. Assume that $|\Gamma| \leq s$. Pick $0 < \epsilon_0 < 1/5$ such that $\epsilon_0^d = \epsilon$. We shall construct the restriction ρ and the k-evaluation of Γ^{ρ} in d steps.

Put $\rho_0 := \emptyset$ and let ν_0 be the canonical 2-evaluation of the depth-0 formulas in Γ, i.e. the constants and the atoms, from Section 15.1. In the tth step, $1 \leq t \leq d$, we assume that we already have restrictions $\rho_0 \subseteq \cdots \subseteq \rho_{t-1}$ with $n_{\rho_\ell} = n^{\epsilon_0^{\ell}}$ and a k-evaluation ν_{t-1} of all formulas of depth $\leq t-1$ in $\Gamma^{\rho_{t-1}}$.

To extend the evaluation ν_{t-1} to depth-t formulas we apply Lemma 15.2.2 with $n := n_{\rho_{t-1}}$ and the parameter ϵ_0 chosen earlier. This will give us a restriction on the universe $([n+1] \setminus dom(\rho_{t-1})) \times ([n] \setminus rng(\rho_{t-1}))$, i.e. a restriction $\rho_t \supseteq \rho_{t-1}$ on the original universe $[n+1] \times [n]$. By Lemma 15.2.1, $\nu_{t-1}^{\rho_t}$ will still work for all formulas of depth $\leq t-1$, and this evaluation is extended to an evaluation ν_t of all formulas as depth $\leq t$ in Γ^{ρ_t} by virtue of Lemma 15.2.2.

The final $\rho := \rho_d$ and $\nu := \nu_d$ satisfy the requirements of the lemma. □

15.3 Ajtai's Theorem

Theorem 15.3.1 (Ajtai's theorem [5], as improved by [324, 400]) *Let $d \geq 2$ and $0 < \delta < 5^{-d}$ be arbitrary. Then for $n \gg 1$ it holds that*

- *in any F_d-refutation of $\neg onto\mathrm{PHP}_n$ at least 2^{n^δ} different formulas must occur as subformulas.*

In particular, any such refutation must have size at least 2^{n^δ}.

Proof Assume for the sake of contradiction that π is an F_d-refutation of $\neg onto\mathrm{PHP}_n$ with fewer than 2^{n^δ} different formulas.

Let Γ be the set of all formulas occurring in π as subformulas and put $k := n^\delta$. Take a restriction ρ and a k-evaluation (H, S) of Γ^ρ guaranteed to exist by Lemma 15.2.3. For $n \gg 1$ it holds that $n^\delta < n/c_F$, where c_F is the constant from Lemma 15.1.7. By Lemmas 15.1.6 and 15.2.1, the axioms of $(\neg onto\mathrm{PHP}_n)^\rho = \neg onto\mathrm{PHP}_{n_\rho}$ are true with respect to the evaluation. By Lemma 15.1.7 all steps in π are true with respect to the evaluation too. But the last formula, the constant 0, is not true with respect to any evaluation. That is a contradiction. □

It is simple and instructive to use k-evaluations for an adversary-type argument when proving Theorem 15.3.1. By increasing the depth by 1 we may assume without loss of generality by Theorem 2.2.1 that π is tree-like and in a balanced form, i.e. the height $h := \mathbf{h}(\pi)$ of the proof tree is proportional to $\log \mathbf{k}(\pi)$.

Build a path $B_1, B_2, \ldots$ through π from the last formula $B_1 = 0$ up to some axiom, always going from B_i to one of the hypotheses B_{i+1} of the inference that yielded B_i in π and such that no B_i is true with respect to the evaluation. Note that this does not mean that $H_{B_i} = \emptyset$, and we need to witness how false B_i is. We build also a sequence of partial bijections $\alpha_1, \alpha_2, \ldots$, all from *Maps*, such that:

- $\alpha_i \in S_{B_i} \setminus H_{B_i}$, for all i;
- $\bigcup_i \alpha_i \in Maps$.

For α_1, take any map from S_{B_1}. Assume that we have the sequence up to B_i and that we also have maps up to α_i. If B_i was derived in π from the formulas $C_1, \ldots, C_s$, take (by Lemma 15.1.3) a common refinement T of all trees $S_{C_1}, \ldots, S_{C_s}$ and S_{B_i}. Lemma 15.1.1 guarantees that there is a $\beta \in T$ compatible with $\alpha_1 \cup \cdots \cup \alpha_i$. Because $\alpha_i \notin H_{B_i}$, β is not in $\bigcap_{j \leq s} T(H_{C_j})$; assume $\beta \notin T(H_{C_j})$. Hence, for some $\gamma \in S_{C_j} \setminus H_{C_j}$, $\gamma \subseteq \beta$ and we put $B_{i+1} := C_j$ and $\alpha_{i+1} := \gamma$.

As we cannot reach any axiom (by Lemma 15.1.6), the length of the path must be too long for us to apply Lemma 15.1.1 in its construction. That is, it must be that $\sum_i |\alpha_i| \leq kh \leq O(n^\delta \log(\mathbf{k}(\pi)))$ is bigger than $n_\rho \geq n^\epsilon$. This yields

$$\mathbf{k}(\pi) \geq 2^{\Omega(n^\epsilon)}.$$

Once we have used tree-like proofs, we can also phrase the argument using the Buss–Pudlák game from Section 2.2. Define a strategy for Liar as follows. Take for

Γ the set of all formulas that Prover may ask for in some play against Liar, in a game restricted to h rounds. Let (H, S) be its k-evaluation (after a restriction ρ).

For each answer to the questions $B_i = ?$, Liar keeps a witness $\alpha_i \in S_{B_i}$, i.e. if the answer was "true" then $\alpha_i \in H_{B_i}$ and if it was "false" then $\alpha_i \notin H_{B_i}$. Further, he maintains the consistency of his answers by making sure that $\alpha_1 \cup \cdots \cup \alpha_i \in Maps$. The rest of the argument is same as above.

Another way to look at the argument is to view it as a reduction, of sorts, of F_d to R(log); this motivated the qualification *logical* in the chapter title. Let π be an F_d-refutation of $\neg onto\text{PHP}_n$. Let Γ stand for all the formulas occurring in it and assume we have a restriction ρ and a k-evaluation (H, S) of Γ^ρ with the terms and the parameters having the same meaning as above.

For an $\alpha \in Maps$, denote by $\hat{\alpha}$ the conjunction $\bigwedge_{i \in dom(\alpha)} p_{i\alpha(i)}$. The idea is that if the refutation π has steps $D_1, \ldots, D_r$ then we replace each D_i by a k-DNF formula

$$\bigvee_{\alpha \in H_{D_i}} \hat{\alpha},$$

which we write as a clause C_i of k-terms

$$\{\hat{\alpha} \mid \alpha \in H_{D_i}\}$$

and fill in more clauses of k-terms to get an R(log)-refutation $\hat{\pi}$ (note that $k = n^\delta$ and $|\pi| \geq 2^k$) of $\neg onto\text{PHP}_{n_\rho} \cup Ax_\rho$, where

$$Ax_\rho := \{\{p_{ij}\} \mid i \in dom(\rho) \wedge j = \rho(i)\}.$$

For this to work, we need first to make one more technical maneuver with π: consider $\neg D$ to mean the formula in negation normal form, with the negations pushed down to atoms and constants, and define the hat translation of $\neg p_{ij}$ to be

$$\bigvee_{\alpha \in H_{\neg p_{ij}}} \hat{\alpha}$$

(and translate a clause by translating all literals in it). Then Lemma 15.1.5 implies that tertium non datur (Section 2.2) is valid and so implies short R(log) refutations of $\widehat{\neg D}, \hat{D}$ and proofs of $\widehat{\neg D} \cup \hat{D}$, as well as of $D \hat{\vee} E$ from $\hat{D}$ and from $\hat{E}$, and of $\hat{D} \cup \hat{E}$ from $D \hat{\vee} E$, all from $\neg onto\text{PHP}_{n_\rho} \cup Ax_\rho$. Note also that $\hat{\pi}$ will be tree-like if π is.

Let us conclude this section by pointing out a problem that escapes solution, it seems, by the method with which we proved Ajtai's theorem. We know by Corollary 11.4.5 that the propositional formulas

$$onto\text{WPHP}_n^{2n} \quad \text{and} \quad onto\text{WPHP}_n^{n^2}$$

have size-$n^{O(\log n)}$ proofs in R(log) and, in fact, that for $onto\text{WPHP}_n^{n^2}$ the upper bound can be improved to $n^{O(\log^{(k)} n)}$ in F_d, for any fixed $k \geq 1$ and a suitable d depending on k (Theorem 11.4.6). The following problem is natural and is likely to force us to come up with new ideas transcending the method described in this chapter.

Problem 15.3.2 (The WPHP problem) *Are there polynomial-size proofs of* WPHP_n^{2n} *or of* $\mathrm{WPHP}_n^{n^2}$ *in* AC^0*-Frege proof systems?*

15.4 Other Combinatorial Principles

Because Ajtai's theorem 15.3.1 speaks about bijections it can be used fairly easily to derive exponential lower bounds for various other combinatorial principles. We give several examples to illustrate the idea of the reductions.

Recall from Section 11.1 the modular counting principle $Count^m(n, R)$ stating that no set $[n]$ with $m|n$ admits a partition R into blocks all but one having size m and the remaining exceptional block having size between 1 and $m-1$ (in Section 11.1 we spoke about an m-partition R with $0 < rem(R)$). The propositional translations are the formulas $Count_n^m$ for any $n \geq m$ (see Section 11.1).

Lemma 15.4.1 *For any fixed* $m \geq 2$*, the theory* $I\Delta_0(R) + \forall x Count^m(x, \Delta_0(R))$ *proves the principle* $onto\mathrm{PHP}(R)$.

Proof Argue contrapositively. Assume that we have a bijection $f\colon [n+1] \rightarrow [n]$ whose graph is R. For $N = m(n+1)$, identify $[N]$ with the disjoint union of m copies of $[n+1]$ and define an m-partitioning S of $[N]$ consisting of the blocks

$$(i, \ldots, i, f(i)) \quad \text{for all } i \in [n+1]$$

plus one size-1 block $\{n+1\}$ whose element $n+1$ is taken from the last copy of $[n+1]$.

Such an S is $\Delta_0(R)$-definable and violates $Count^m(N, S)$. □

Corollary 15.4.2 *For all fixed* $m \geq 2$ *and for* $n \gg 1$ *divisible by* m*, the formulas* $Count_n^m$ *require* AC^0*-Frege proofs of size* $2^{n^{\Omega(1)}}$.

Proof The arithmetical proof guaranteed to exist by the previous lemma translates into a p-size AC^0-Frege proof (via Theorem 8.2.2) of $onto\mathrm{PHP}_n$ from an AC^0-instance of the $Count^m$ principle. Hence a subexponential-size AC^0-proof of the counting principle would yield a subexponential size-proof of the $onto$PHP principle, contradicting Theorem 15.3.1. □

Next we turn to the Ramsey principle and the formulas RAM_n considered in Section 13.1. The formulas RAM_n formalize the true Ramsey relation $n \longrightarrow (k)_2^2$ (saying that every undirected graph on n vertices contains either a clique or an independent set of size k) for $k := \lfloor (\log n)/2 \rfloor$. We shall now consider more general formulas $\mathrm{RAM}(n, k)$, where we pick the parameters n, k rather than setting k in some canonical way as in RAM_n. In particular, let r_k be the minimum n for which the statement holds.

Recall from Section 13.1 that $\mathrm{RAM}(n, k)$ has size $O(n^k)$ and thus the size of $\mathrm{RAM}(r_k, k)$ is at most $O(4^{k^2})$, owing to the bound $r_k \leq 4^k$; see [175].

Assume that $n+1 = r_k$ and hence there is a graph $G = ([n], E)$ which has no homogeneous subgraph of size k. Assume further for the sake of contradiction that $f: [n+1] \to [n]$ is a bijection. We can use f to pull back E from $[n]$ onto the larger set $[n+1]$, defining (by a bounded formula) a graph $H = ([n+1], R)$. Because k is small, $k \leq \log n$, we can count sets of cardinalities up to k both in $I\Delta_0(E)$ and in AC^0-Frege systems. Thus having a proof that H must contain a small homogeneous subgraph will imply that G does as well: f, being a bijection, must preserve cardinalities as long as we can count them. This simple idea yields the following lemma (we shall prove it directly in propositional logic to make clear the idea of counting small sets).

Lemma 15.4.3 *For every $d \geq 2$, there is an $\epsilon > 0$ such that for $k \geq 2$ every depth d Frege proof of* RAM(r_k, k) *must have size at least $2^{r_k^\epsilon}$.*

Proof Let $p_{i,j}$, $i \in [n+1], j \in [n]$ be the $(n+1)n$ atoms of *onto*PHP$_n$. Assume that π is a size s F_d-proof of RAM(r_k, k); we shall denote its atoms by q_e, for two-element subsets e of $[r_k]$.

Use the graph G defined above to define a substitution for the atoms q_e as follows:

$$\sigma(q_e) := \bigvee_{\{i,j\} \in E} p_{u,i} \wedge p_{v,j}$$

if $e = \{u, v\}$.

In the following claim we use the notation *Cli*(A) and *Ind*(A) from Section 13.2.

Claim 1 For any t such that $1 \leq t \leq n$ and any size t subset $A \subseteq [r_k]$, there are AC^0-Frege proofs of size $n^{O(t)}$ of both the formulas

$$\sigma(Cli(A)) \wedge \neg onto\mathrm{PHP}_n \longrightarrow \bigvee_{B \subseteq [n], |B| = t} \bigwedge_{\{i,j\} \subseteq B} E(i,j)$$

and

$$\sigma(Ind(A)) \wedge \neg onto\mathrm{PHP}_n \longrightarrow \bigvee_{B \subseteq [n], |B| = t} \bigwedge_{\{i,j\} \subseteq B} \neg E(i,j).$$

(The $E(i,j)$ are Boolean constants determined by the edge relation of graph G.) The claim is readily verified by induction on t. As G has no homogeneous subset of size k, both the conjunctions

$$\bigwedge_{\{i,j\} \subseteq B} E(i,j) \quad \text{and} \quad \bigwedge_{\{i,j\} \subseteq B} \neg E(i,j)$$

are false for $|B| = k$. This entails the next claim.

Claim 2 For each $A \subseteq [r_k]$ of size k, there are constant-depth Frege proofs of size $n^{O(k)}$ of *onto*PHP$_n$ from both the formulas $\sigma(Cli(A))$ and $\sigma(Ind(A))$.

We can now combine a substitution instance $\sigma(\pi)$ of the original proof (i.e. a proof of $\sigma(\mathrm{RAM}(r_k, k))$) with the proofs from the previous claim: this is an AC^0-Frege proof of *onto*PHP$_n$ of size at most

$$O(sn^2) + 2\binom{r_k}{k} n^{O(k)} \leq O(sn^2) + n^{O(\log(n))}.$$

The lower bound then follows from Theorem 15.3.1. □

Our last example will be of a slightly different form. While it is conceivable that the method of restrictions and k-evaluations could be modified for RAM(r_k, k), it is hard to see how it could be done for algebraic structures, such as fields, that are more rigid than partitions or graphs.

Consider a language with two ternary relations $A(x, y, z)$ and $M(x, y, z)$ and write a first-order sentence *Field* that is satisfied in a structure interpreting the relations if and only if the structure is a field and A and M are graphs for its addition and multiplication. Let $Field_n$ be its $\langle \dots \rangle$ translation for the size parameter n. Clearly $Field_n$ is satisfiable if and only if n is a power of a prime.

Lemma 15.4.4 *For every $d \geq 2$, there is an $\epsilon > 0$ such that, for infinitely many n, every F_d-refutation of $Field_n$ must have size at least 2^{n^ϵ}.*

Proof The proof strategy is analogous to that in the proof of Lemma 15.4.3. Assume that $n \gg 1$ is such that $n + 1$ is a power of a prime. Let **B** be a field structure on $[n + 1]$ interpreting, in particular, A and M in such way that $Field_{n+1}$ is satisfied.

Assume that f violates *onto*PHP$_n$ and use it to define from **B** a field structure on $[n]$ satisfying $Field_n$. Hence a short refutation of $Field_n$ would yield a short refutation of $\neg$*onto*PHP$_n$, which would contradict Theorem 15.3.1. □

15.5 Relations Among Counting Principles

Lemma 15.4.1 derives *onto*PHP from instances of $Count^m$; we can investigate such proof-theoretic reductions by means of either provability in bounded arithmetic or short provability in AC^0-Frege systems. For stating positive results the former view gives stronger results; for negative results – such as the one presented in this section – the latter is better.

There are a number of questions that we can ask about the mutual reducibility of various versions of PHP (onto, non-onto or dual, analogously to the dual WPHP from Section 7.2 etc.) and of the modular counting principles for different moduli (and possibly for different remainders). In this section we shall cover perhaps the most interesting and instructive example and show that $Count^p$ cannot be proved feasibly from instances of $Count^q$, for $p \neq q$ primes. In fact, we shall restrict to $p = 3$ and $q = 2$ as this example reveals the idea that works for all pairs p, q and is notationally simplest. The mutual relations of all conceivable cases of versions of PHP and $Count^m$, even those with composite moduli, are known and Section 15.7 gives the references.

Let us consider the following situation. The parameter $n \gg 1$ is not divisible by 3 and we have a short F_d-proof π of $Count^3_n$ (this formula uses the atoms p_e, for 3-

element subsets e of $[n]$) from some instances of $Count^2_{m_i}$, for certain odd parameters m_i. We can use definition by cases and assume without loss of generality that the proof employs just one such instance of the $Count^2$ principle for one odd parameter m, say

$$Count^2_m(q_{\{i,j\}}/Q_{\{i,j\}}),$$

with the $q_{\{i,j\}}$ atoms of $Count^2_m$ substituted by formulas $Q_{\{i,j\}}$ in the atoms p_e. This simplification may increase the depth of π by a constant and its size polynomially and we shall simply assume that already π satisfies it.

In the previous section we proved a lower bound for $Count^p$ via a reduction to the lower bound for *onto*PHP. We could have equally well developed k-evaluations for $Count^3$ and argued as for Theorem 15.3.1, but the reduction to *onto*PHP is much simpler. Now we shall need the k-evaluation machinery, however.

Let us describe what we need to change in the set-up of PHP trees and k-evaluations from Section 15.1 in order for the method to apply to $Count^3$. The set *Maps* is replaced by the set *Partitions* consisting of partial 3-partitions α of $[n]$. Such an α is a set of disjoint 3-element subsets (called blocks) of $[n]$; their number is the size of α.

The PHP-trees are replaced by **3-trees**.

- At the root the tree branches according to all valid answers e to a query $i \in ?$, for some $i \in [n]$, asking to which block of a partition i belongs. The edges leaving the root are labeled by all possible e's containing i.
- At a node reached from the root by a path whose edges are labeled $e_1, \ldots, e_u$, the tree may ask $j \in ?$ for any $j \in [n] \setminus (e_1 \cup \cdots \cup e_u)$, and it branches according to all f's containing j but disjoint from all $e_1, \ldots, e_u$.

The height of a tree is the maximum length of a path in it. Where in the PHP case we needed assumptions like $h(T) \leq n$ or $h(T) + |\alpha| \leq n$ we now have $3h(T) \leq n$ or $3(h(T) + |\alpha|) \leq n$, because a partition contains three times more elements of $[n]$ than is its size. Hence, if ρ is a partition, the size n_ρ of $[n]$ not covered by ρ is $n_\rho := n - (3|\rho|)$.

The notion of a k-evaluation using 3-trees is defined in analogy with the PHP case, with the base case for the atoms modified as follows:

- S_{p_e} for $e = \{i_1, i_2, i_3\}$ is the tree branching first according to $i_1 \in ?$, then according to $i_2 \in ?$ and finally according to $i_3 \in ?$, consistently with earlier answers;
- H_{p_e} is the unique path of length 1 corresponding to the answer e to the first query.

The lemmas from Section 15.2 leading to the existence statement about k-evaluations (Lemma 15.2.3) are proved analogously. In particular, the following existence lemma holds. Because we are explaining only the changes from the earlier proof but are not actually computing the particular estimates, we shall formulate it without an explicit bound to δ; that can be found in the references given in Section 15.7.

Lemma 15.5.1 *For every $d \geq 2$, there are $0 < \delta < \epsilon$ such that for sufficiently large $n \geq 1$, every set Γ of formulas of depth $\leq d$ that is closed under subformulas and is of size $|\Gamma| \leq 2^{n^\delta}$ there exists $\rho \in$ Partitions, $n_\rho \geq n^\epsilon$, and an n^δ-evaluation of Γ^ρ.*

As before we say that a formula $\varphi \in \Gamma$ is *true with respect to the evaluation* if and only if $S_\varphi = H_\varphi$.

Lemma 15.5.2 *All axiom clauses of $\neg Count_n^3$ are true with respect to any k-evaluation and the qualification true with respect to the evaluation is preserved by all Frege rules if $k \leq o(n)$.*

Apply the last two lemmas to π and to Γ consisting of the formulas occurring in π as subformulas, fix an n^ρ-evaluation of Γ^ρ and conclude, assuming $|\pi| \leq 2^{n^\delta}$, that

- $\neg Count_m^2(Q_{\{i,j\}})$ is true with respect to the evaluation.

For notational simplicity, we omit the reference to the restriction ρ, think of having a k-evaluation (H, S) of Γ itself, and put

$$H_{\{i,j\}} := H_{Q_{\{i,j\}}} \quad \text{and} \quad T_i := T_{\bigvee_j Q_{\{i,j\}}}.$$

Lemma 15.5.3 *For all $i \in [m]$ and all $\alpha, \alpha' \in \bigcup_j H_{\{i,j\}}$, if $\alpha \neq \alpha'$ then $\alpha \perp \alpha'$. Further, $T_i(\bigcup_j H_{\{i,j\}}) = T_i$.*

Proof If α, α', $\alpha \neq \alpha'$, in the first statement are from the same $H_{\{i,j\}}$ then they are incompatible because they belong to the same 3-tree. If they belong to $H_{\{i,j\}}$ and $H_{\{i,j'\}}$ for $j \neq j'$, then the statement follows because the axiom $\neg Q_{\{i,j\}} \vee \neg Q_{\{i,j'\}}$ is true with respect to the evaluation.

The second statement follows from the fact that the axiom $\bigvee_j Q_{\{i,j\}}$ is true with respect to the evaluation. □

At the beginning of Chapter 6 we considered how to transcribe a clause for algebro-geometric proof systems, and we represented it either by a polynomial equation (6.0.1) or by an integer linear inequality (6.0.2). We used that in Section 6.3 to transcribe $\neg \mathrm{PHP}_n$ to a set of integer linear inequalities suitable for the cutting plane proof system CP. Now we shall need the algebraic transcription of $\neg Count_n^3$; we shall use instead of the atoms p_e the variables x_e; to stress the algebraic context.

The algebraic formulation of $\neg Count_n^3$ is given by the following set of polynomial equations:

$$\sum_{e:i \in e} x_e = 1, \quad x_e x_f = 0 \quad \text{and} \quad x_e^2 = x_e$$

for all $i \in [n]$, all $e \perp f$ and all e, respectively. Because we shall talk about proofs in the algebraic proof systems NS and PC we shall write them as equations with the right-hand sides equal to 0:

$$u_i = 0, \quad v_{e,f} = 0 \quad \text{and} \quad w_e = 0, \tag{15.5.1}$$

where

$$u_i := (\sum_{e:i\in e} x_e) - 1, \quad v_{e,f} := x_e x_f \quad \text{and} \quad w_e := x_e^2 - x_e.$$

We will use the notation

$$\text{NS: } \neg Count_n^3 \vdash_h g$$

to denote that g has a Nullstellensatz proof from polynomials (15.5.1) of degree at most h. Recall from Lemma 6.2.4 that this is the same as having a tree-like PC-proof such that the sum of the degrees of the polynomials used in the multiplication rule is at most $h-2$ (in particular, the height of the PC proof is $\leq h-2$ if we multiply only by variables); this is often easier to see.

For $\alpha = \{e_1, \dots, e_t\} \in Partitions$, put $x_\alpha := x_{e_1} \cdot \dots \cdot x_{e_t}$.

Lemma 15.5.4 *Let $A \subseteq Partitions$ and assume that any two different $\alpha, \alpha' \in A$ are incompatible: $\alpha \perp \alpha'$. Further assume that $T(A) = T$ for some 3-tree T such that $|\alpha| + h(T) \leq n/3$ for all $\alpha \in A$.*

Then

$$\text{NS: } \neg Count_n^3 \vdash_{h(T)} (\sum_{\alpha\in A} x_\alpha) - 1.$$

Proof First show by induction on $h(T)$ that

Claim 1

$$\text{NS: } \neg Count_n^3 \vdash_{h(T)} (\sum_{\beta\in T} x_\beta) - 1.$$

The claim is best verified by constructing a suitable tree-like PC proof of

$$(\sum_{\gamma\in S} x_\gamma) - 1$$

(see the remark about Lemma 6.2.4 above), for successively larger subtrees S of T; the structure of the PC proof copies the structure of T.

Now, by the hypothesis, $T(A) = T$, so, on the one hand

1. $\forall \beta \in T \exists \alpha \in A\ \alpha \subseteq \beta$.

On the other hand,

2. for no $\beta \in T$ can there exist two different $\alpha, \alpha' \in A$ such that $\alpha, \alpha' \subseteq \beta$,

as that would imply $\alpha || \alpha'$; contradicting the hypothesis about A of the lemma.

Moreover, for $\alpha \in A$,

3.

$$\text{NS: } \neg Count_n^3 \cup \{x_\alpha\} \vdash_{h(T)} (\sum_{\beta\in T, \beta\supseteq\alpha} x_\beta) - 1$$

because

$$\sum_{\beta \in T, \beta \supseteq \alpha} x_\beta = \sum_{\gamma \in T^\alpha} x_\gamma = 1,$$

by Claim 1.

Now we can count as follows:

$$1 = \sum_{\beta \in T} x_\beta = \sum_{\alpha \in A} \sum_{\beta \in T, \beta \supseteq \alpha} x_\beta = \sum_{\alpha \in A} x_\alpha \left(\sum_{\gamma \in T^\alpha} x_\gamma\right) = \sum_{\alpha \in A} x_\alpha .$$

□

The following theorem holds for all primes $p \neq q$ (in fact, more generally) but we state it just for the primes 2 and 3.

Theorem 15.5.5 (Ajtai [7], as improved by [52, 124]) *For every $d \geq 2$, there is $\delta > 0$ such that for $n \gg 1$ it holds that*

- *in any F_d-proof of $Count_n^3$ from instances of the $Count^2$ formulas, at least 2^{n^δ} different formulas must occur as subformulas.*

In particular, any such proof must have size at least 2^{n^δ}.

Proof Take an $m \times m$ matrix of polynomials f_{ij} over $\mathbf{F}_2$ with the variables of $\neg Count_n^3$, $i, j \in [m]$:

$$f_{ij} := \sum_{\alpha \in H_{\{i,j\}}} x_\alpha .$$

Hence

$$f_{ii} = 0 \text{ and } f_{ij} = f_{ji}. \tag{15.5.2}$$

By Lemma 15.5.4, for each $i \in [m]$,

$$\text{NS: } \neg Count_n^3 \vdash_{n^\rho} \left(\sum_j f_{ij}\right) - 1$$

and hence also

$$\text{NS: } \neg Count_n^3 \vdash_{n^\rho} \sum_i \left(\left(\sum_j f_{ij}\right) - 1\right). \tag{15.5.3}$$

But m is odd and therefore this is the polynomial

$$\left(\sum_{i,j} f_{ij}\right) - 1,$$

in which – by (15.5.2) – the sum is 0 mod 2. Therefore the NS-proof in (15.5.3) is an NS-refutation of $\neg Count_n^3$. The theorem then follows from the following

Key claim *Every* NS*-refutation of $\neg Count_n^3$ over $\mathbf{F}_2$ must have degree $n^{\Omega(1)}$.*

The proof of this key fact is postponed to Section 16.1 (Theorem 16.1.2 there). □

15.6 Modular Counting Gates

In Section 10.1 we defined $AC^0[m]$-Frege systems as the constant-depth subsystems of a Frege system in the DeMorgan language extended by the modular counting connectives $MOD_{m,i}$ from Section 1.1 and augmented by several axiom schemes for handling these connectives. The lower bounds for constant-depth formulas in this language are known (items (ii) and (iii) in Theorem 1.4.3); in particular, $MOD_{p,j}$ cannot be defined in $AC^0[q]$, if $p \neq q$ are primes.

Proof complexity (more precisely, Ajtai [5]) borrowed the idea of using random restrictions and a form of switching lemma (like the lemmas 14.2.1 and 15.2.2 that we used earlier) from circuit complexity (item (i) in Theorem 1.4.3), where it was very successful. Hence, after lower bounds for constant-depth circuits with gates counting modulo a prime were obtained, it looked as though the method used there ought to be transferable to proof complexity to yield lower bounds for $AC^0[p]$-Frege systems. That hope has not materialized, so far, and the following problem has been open since the late 1980s.

Problem 15.6.1 (The $AC^0[p]$-Frege problem) *Establish super-polynomial lower bounds for* $AC^0[p]$*-Frege systems, for p a prime.*

This problem is often described as the hardest and the most interesting of those deemed solvable but it may actually turn out to be the easiest and the least interesting of those not solvable at the present stage of the development of proof complexity. Switching lemmas and related techniques were indeed successful in proof complexity but I cannot help but view them in part as the gifts of Danae. Their great success stimulated researchers in trying to import other combinatorial techniques from circuit complexity, but the development of genuinely new proof complexity techniques was neglected. We should not forget that while combinatorics had some early successes in circuit complexity there has been no significant progress on lower bounds since the late 1980s, and the approach can hardly be branded as a success. Hence, even if we manage to use this method and to solve the problem above, it will not take us very far: there is little we can use after that. As the reader can guess, in the age of scientific brinkmanship I am in a tiny minority of researchers who take this somewhat somber view.

Switching back to the positive mode, I will describe two results about the $AC^0[p]$-Frege systems that, I think, illuminate their properties and also their differences from the AC^0-Frege systems. The first result is a proof-theoretic formulation of the method of approximation; the second is the collapse of $AC^0[p]$-Frege systems to $F_4(MOD_p)$ systems. To simplify the notation and some arguments below, we shall restrict to $p = 2$, i.e. to the $AC^0[2]$-Frege systems, which we shall denote by $F_d(\oplus)$ as earlier.

We start with the former result but first we replace the $AC^0[2]$-Frege systems by other, p-equivalent, proof systems. While developing the method of k-evaluations in Section 15.1 we faced a technical nuisance (e.g. in the proof of Lemma 15.1.7) that the AC^0-Frege systems are defined using binary $\vee$ and $\wedge$ while the depth is measured as if these connectives were of unbounded arity. In addition, if we plan to compare a logic system with an algebraic system we also have to translate the formulas we are proving, and this may not be canonical (and is further obfuscated by the fact that some authors insist that the truth value true ought to be represented in algebraic proof systems by 0).

To avoid both these issues, we shall define the proof system $AC^0[2]$-R(PC) and its variant $AC^0[2]$-R(LIN), using the proof systems R(PC) and R(LIN) from Section 7.1, and we shall confine the whole discussion to refutations of R(PC)-clauses, i.e. clauses formed from polynomials. These are easily defined by depth-3 formulas of $F(\oplus)$ (of the form $\bigvee_i \bigoplus_j \bigwedge_k \ell_{i,j,k}$).

The system $AC^0[2]$-R(PC) extends R(PC) by adding to an initial set $\mathcal{C}$ of R(PC)-clauses a new set $\mathcal{A}$ of clauses forming **extension axioms** and having the following form. For $f_1, \dots, f_m$ any polynomials and z a new variable not occurring in them or in $\mathcal{C}$, the block of clauses denoted by $Ax[f_1, \dots, f_m; z]$ consists of:

- $\{f_i + 1, z\}$, for all $i \in [m]$;
- $\{z + 1, f_1, \dots, f_m\}$.

The variable z is the **extension variable**. If the clauses are satisfied then $\bigvee_i f_i \equiv z$.

In addition, it is required that the set $\mathcal{A}$ is stratified into levels $\mathcal{A}_1 \cup \dots \cup \mathcal{A}_\ell$ such that each block $Ax[f_1, \dots, f_m; z]$ belongs to one level and:

- the variables in the polynomials $f_1, \dots, f_m$ in the axioms $Ax[f_1, \dots, f_m; z]$ in $\mathcal{A}_1$ are among the variables occurring in $\mathcal{C}$ and z does not occur among them or in other extension axioms in $\mathcal{A}_1$;
- the variables in the polynomials $f_1, \dots, f_m$ in the axioms $Ax[f_1, \dots, f_m; z]$ in $\mathcal{A}_{t+1}$, $1 \le t < \ell$, are among the variables occurring in $\mathcal{C} \cup \bigcup_{s \le t} \mathcal{A}_s$ (including the extension variables from these levels) and z does not occur among them or in the other extension axioms in $\mathcal{A}_{t+1}$.

The axioms in $\mathcal{A}_t$ are called the **level-t axioms**. **Level-ℓ** $AC^0[2]$-R(PC) may use extension axioms $\mathcal{A}$ that can be stratified into ℓ levels and $AC^0[2]$-R(PC) is the collection of level $\ell = 1, 2, \dots$ systems (in analogy to how AC^0-Frege systems are defined). The degree of an $AC^0[2]$-R(PC)-refutation is the maximum degree of a polynomial occurring in it. The systems $AC^0[2]$-R(LIN) can use only linear polynomials.

The following lemma is proved analogously to how *ER* is linked with *EF* in Section 5.8.

Lemma 15.6.2 *$AC^0[2]$-Frege systems and $AC^0[2]$-R(PC) p-simulate each other with respect to refutations of sets of R(PC)-clauses. In particular:*

(i) If there is an $F_\ell(\oplus)$-refutation π of $\mathcal{C}$, $|\pi| = s$, and d_0 is the maximum degree of a polynomial in $\mathcal{C}$ then there exists a set $\mathcal{A}$ of extension axioms with $\le 2\ell$ levels and an R(PC)-refutation ρ of $\mathcal{C} \cup \mathcal{A}$ of size $\le s^{O(1)}$ and degree d_0.

Moreover, if π is tree-like and of height h, ρ is also tree-like of height $O(h)$.

(ii) *If $\mathcal{A}$ are extension axioms stratified into ℓ levels and there is an* R(PC)*-refutation ρ of $\mathcal{C} \cup \mathcal{A}$ of size $|\rho| = s$ then there is an $F_{O(\ell)}(\oplus)$-refutation of $\mathcal{C}$ of size $\leq s^{O(1)}$.*

Moreover, if ρ is tree-like and of height h, π is also tree-like and of height $O(h)$.

Both statements are witnessed by polynomial-time algorithms constructing the refutations.

The same holds with $AC^0[2]$-R(LIN) *in place of* $AC^0[2]$-R(PC) *with respect to refutations of* R(LIN)*-clauses.*

The idea invented by Razborov [431] and perfected by Smolensky [474] for proving lower bounds for $AC^0[q]$ circuits is to replace a constant-depth circuit C with MOD_q gates by a polynomial f over $\mathbf{F}_q$ such that $deg(f)$ is small and f equals C on most inputs, if C is small. One then argues that a low-degree polynomial over $\mathbf{F}_q$ cannot compute $MOD_{p,i}$ (or some other function) with a small error and hence C cannot compute the function exactly. This is the so-called *approximation method* (one of two, both due to Razborov, in circuit complexity). We shall need only the simpler part of it, namely how to approximate disjunctions by low-degree polynomials.

Assume that we want to test whether at least one of the Boolean values $y_1, \ldots, y_m$ is equal to 1, i.e. whether the disjunction $\bigvee_{i \in [m]} y_i$ is true. Pick a random subset $I \subseteq [m]$ and compute $f_I := \bigoplus_{i \in I} y_i$. If all $y_i = 0$ then $f_I = 0$ but, if for at least one, $y_i = 1$ then we have the probability $1/2$ that $f_I = 1$. Hence, if we choose $a \geq 1$ random subsets $I_1, \ldots, I_a \subseteq [m]$, independently of each other, then if $\bigvee_{i \in [m]} y_i = 1$ also $\bigvee_{j \leq h} f_{I_j} = 1$, with the probability of failing bounded above by 2^{-a}.

This can be formalized algebraically as follows. Let

$$\mathsf{disj}_{m,a}(\overline{y}, \overline{r}^1, \ldots, \overline{r}^a) := 1 - \prod_{u \leq a} [1 - \sum_{j \in [m]} r_{u,j} y_j]$$

be a polynomial over $\mathbf{F}_2$, where $\overline{y}, \overline{r}^1, \ldots, \overline{r}^a$ are m-tuples of different variables. Think of $\overline{r}^u$ as defining $I_u := \{j \in [m] \mid r_{u,j} = 1\}$. Then, for any given $\overline{w} \in \{0,1\}^m$,

$$Prob_{\overline{r}}[\bigvee_i w_i \equiv \mathsf{disj}_{m,a}(\overline{w}, \overline{r}) = 1] \geq 1 - 2^{-a}.$$

By averaging, this implies that we can choose a particular $\overline{b} \in \{0,1\}^{ma}$ such that

$$Prob_{\overline{y}}[\bigvee_i y_i \equiv \mathsf{disj}_{m,a}(\overline{y}, \overline{b}) = 1] \geq 1 - 2^{-a}.$$

That is, $\mathsf{disj}_{m,a}(\overline{y}, \overline{b}) = 1$ approximates $\bigvee_i y_i$ very well: they agree on all but a fraction of 2^{-a} evaluations $\overline{w}$ of $\overline{y}$.

The way to transport this approximation into proof complexity is to replace *random* bits by *generic* bits and to introduce axioms (in the form of polynomial equations) that will formalize the statement that the approximation works for all m-tuples $\overline{y}$.

Define the **extension polynomial axiom** to be

$$E_{m,a,i}(\overline{y}, \overline{r}^1, \dots, \overline{r}^a) = 0$$

where the **extension polynomial** is

$$E_{m,a,i} := y_i[1 - \mathsf{disj}_{m,a}(\overline{y}, \overline{r}^1, \dots, \overline{r}^a)] .$$

We know that if all the $y_i = 0$, then $\mathsf{disj}_{m,a}(\overline{y}, \overline{r}^1, \dots, \overline{r}^a) = 0$ also, irrespective of the $\overline{r}$. But if some $y_i = 1$, then by $E_{m,a,i} = 0$ it must be that $\mathsf{disj}_{m,a}(\overline{y}, \overline{r}^1, \dots, \overline{r}^a) = 1$. Hence modulo the extension polynomial axioms we can replace a disjunction of low-degree polynomials by one low-degree polynomial and iterate this construction. The variables $r_{u,j}$ in the extension axiom are called the **extension variables** and the parameter a is called the **accuracy**.

Let us state for the record that adding the extension polynomial axioms is sound: it cannot violate the solvability of a polynomial system not involving the extension variables themselves.

Lemma 15.6.3 *For all $\overline{a} \in \{0, 1\}^m$, there are $\overline{b} \in \{0, 1\}^{ma}$ such that all the extension axioms $E_{m,a,i}(\overline{a}, \overline{b}) = 0$, $i \in [m]$, hold.*

Proof Given an assignment $\overline{y} := \overline{a} \in \{0, 1\}^m$, define

$$r_{u,j} := \begin{cases} 0 & \text{if } u > 1, \\ 1 & \text{if } u = 1,\ a_j = 1 \text{ but, for all } j' < j;\ a_{j'} = 0, \\ 0 & \text{otherwise.} \end{cases}$$

□

The proof of Lemma 15.6.3 also shows that having $a > 1$ is not necessary; having a PC-or NS-refutation using some extension polynomials as initial polynomials, we can always use the above substitution and reduce the accuracy to $a = 1$. Allowing $a > 1$ offers another proof of the lemma: choose an assignment $\overline{b} \in \{0, 1\}^{ma}$ for the extension variables at random. If the total number of extension axioms is S and $a > \log S$ then, with positive probability, all extension polynomial axioms will be satisfied by $\overline{a}, \overline{b}$. This argument replaces the *logical* definition of the values for $\overline{r}$ by a definition by cases in a *logic-free* randomized choice. That gives the impression (but nothing more that we can formalize at this point) of freeing our hand to try a new argument.

To simplify matters, from now on we shall use only extension polynomials of accuracy $a = 1$. We shall use the extension polynomial axioms to strengthen the Nullstellensatz proof system NS from Section 6.2 and to link it with the $AC^0[2]$-Frege systems. The **extended Nullstellensatz proof system** (ENS) allows us to add to the initial set $\mathcal{F}$ of polynomials a new set $\mathcal{E}$ of extension polynomials $E_{m,1,i}$ under the following conditions. The set $\mathcal{E}$ can be stratified into levels $\mathcal{E}_1 \cup \cdots \cup \mathcal{E}_\ell$ such that:

- if $E_{m,1,i}(\bar{f}, \bar{r}) \in \mathcal{E}_t$, for some t and i, then all the companion axioms $E_{m,1,j}(\bar{f}, \bar{r})$ are also in $\mathcal{E}_t, j \in [m]$;
- the variables in the polynomials f_i in $E_{m,1,i}(\bar{f}, \bar{r})$ in $\mathcal{E}_1$ are among the variables occurring in $\mathcal{F}$ and no variable from $\bar{r}$ occurs among them or in other extension axioms in $\mathcal{E}_1$ except in the companion axioms;
- the variables in the polynomials f_i in the axioms $E_{m,1,i}(\bar{f}, \bar{r})$ in $\mathcal{E}_{t+1}$, $1 \leq t < \ell$, are among the variables occurring in $\mathcal{F} \cup \bigcup_{s \leq t} \mathcal{E}_s$ (including the extension variables from these levels) and no variable from $\bar{r}$ occurs among them or in the other extension polynomials in $\mathcal{E}_{t+1}$ except the companion axioms.

The polynomials in $\mathcal{E}_t$ will be called **level-t extension polynomials**.

Lemma 15.6.4 *Let $\mathcal{F}$ be a set of polynomials (i.e. singleton* R(PC) *clauses) of degree at most d_0, and put $\mathcal{F}^{+1} := \{1 + f \mid f \in \mathcal{F}\}$.*

Assume ρ is a tree-like $AC^0[2]$-R(PC) *refutation of $\mathcal{F}$, each polynomial in ρ having degree at most d, for some $d \geq d_0$. Let the size and height of ρ be s and $O(\log s)$ (i.e. ρ is balanced), respectively. Assume further that ρ uses a set $\mathcal{A}$ of extension axioms stratified into $\ell \geq 1$ levels.*

Then there is a set $\mathcal{E}$ of extension polynomials (tacitly of accuracy 1*) and an* NS-*refutation σ of $\mathcal{F}^{+1} \cup \mathcal{E}$ such that*

- *$\mathcal{E}$ has ℓ levels,*
- *$|\mathcal{E}| \leq s^{O(1)}$,*
- *the degree of σ is at most $O((d+1)^{2^\ell} \log s)$.*

Proof Let us first see how to treat one axiom $Ax[f_1, \ldots, f_m; z]$. Substitute:

$$z := \mathsf{disj}_{m,1}(f_1, \ldots, f_m).$$

We shall say that NS *proves* an R(PC) clause $\{g_1, \ldots, g_t\}$ from polynomials $\mathcal{G}$ if and only if NS refutes the set $\mathcal{G} \cup \{g_1, \ldots, g_t\}$. This looks as though we have forgotten to add the negations (i.e. $+1$) but here the logic set-up with true being given by 1 and the NS set-up, with axiom polynomials meaning that they are equal to 0, kicks in: the clause $\{g_1, \ldots, g_t\}$ means that $\bigvee_i g_i = 1$, hence its negation is the list of axioms $\{g_i = 0\}_i$ which are represented for NS just by the g_i themselves.

The following claim is easy; note that the degree of $\mathsf{disj}_{m,1}(f_1, \ldots, f_m)$ is bounded by $(1 + \max_i deg(f_i)) \leq (d + 1)$.

Claim 1 There is an NS-proof of each clause of the substituted axiom

$$Ax[f_1, \ldots, f_m; \mathsf{disj}_{m,a}(f_1, \ldots, f_m)]$$

from the extension polynomials $E_{m,1,i}(\bar{f}, \bar{r})$, $i \in [m]$, of degree at most $(d + 1)$.

Now list all the extension axioms in $\mathcal{A}$ in a linear order in which all level-1 axioms come first, followed by all level-2 axioms etc. For the axioms in this list, perform a substitution as in the claim, in the order in which they are listed. Take for $\mathcal{E}$ the extension polynomials that correspond to all these substitutions, and stratify them into ℓ levels shadowing the stratification of $\mathcal{A}$.

When we make all the substitutions for the level-1 axioms, the degrees of the polynomials in higher levels could go up by a multiplicative factor $d+1$ to $(d+1)^2$, because the polynomials in them may contain the extension variables from the first level. Hence the NS proofs for the axioms at level t provided by Claim 1 will have degree $(d+1)^{2^t}$, and so the degree of all these refutations for all levels can be bounded by $(d+1)^{2^\ell}$.

This implies that all initial clauses in ρ have an NS-proof from $\mathcal{F}^{+1} \cup \mathcal{E}$ with the required parameters. Use induction on the number of steps in ρ to simulate its inferences in NS and to construct an eventual refutation σ. This uses

Claim 2 A clause in ρ which is the end-clause of a subproof of height h has an NS-proof from $\mathcal{F}^{+1} \cup \mathcal{E}$ of degree bounded by $O(h(d+1)^{2^\ell})$.

For example, assume we have NS-proofs η_L and η_R for both the left and the right hypothesis in the resolution inference

$$\frac{D \cup \{g\} \quad D' \cup \{1+g\}}{D \cup D'}.$$

First multiply the axiom $1+g$ that the proof η_L uses by g (this causes an increase in the degree), obtaining an NS-proof of g, and then use it as a subproof in η_R, yielding the initial polynomial g. □

Lemma 15.6.5 $AC^0[2]$*-Frege systems simulate ENS with a constant number of levels of extension polynomials, in the following sense.*

Assume that σ is a degree-d NS-refutation of an initial set of polynomials $\mathcal{F}$ in n variables, $|\mathcal{F}| = k$, that uses a set $\mathcal{E}$ of extension polynomials (tacitly of accuracy 1), $|\mathcal{E}| = s$, stratified into ℓ levels. Assume (without loss of generality, for simplicity of estimation) that s also bounds k, n: $s \geq k, n$.

Then there is an $AC^0[2]$-Frege refutation π of $\mathcal{F}^{+1}$ of depth $O(\ell)$ and of size $|\pi| \leq s^{O(d)}$.

Proof From the bottom level of $\mathcal{E}$ up, substitute for the extension variables $r_{1,j}$ corresponding to $f_1, \ldots, f_m$ the $F(\oplus)$-formulas defining their value as in the proof of Lemma 15.6.3:

$$r_{1,j} := f_j \wedge \bigwedge_{j'<j} (\neg f_{j'}),$$

where a polynomial f is represented as the parity of a set of conjunctions of some variables.

It is easy to see that the substitutions reduce all polynomials in $\mathcal{E}$ to formulas easily refutable in an $AC^0[2]$-Frege system in some $O(1)$ depth. Repeating this consecutively for each level of $\mathcal{E}$, the depth of the eventual $AC^0[2]$-Frege proof will be $O(\ell)$.

Now we have to estimate its size. Each polynomial in σ has degree $\leq d$ and there are at most $n+s$ variables (n variables from $\mathcal{F}$ and at most s extension variables), and hence it may contain at most $(n+s)^{O(d)}$ monomials (of degree $\leq d$) and thus

$$d(n+s)^{O(d)}$$

is an estimate of its size as an $AC^0[2]$-formula. The number of polynomials in an extension axiom is also bounded by s (because of our rule that all companion extension polynomials must also be in $\mathcal{E}$. Hence the size of the formula substituted for $r_{1,j}$ at the first level is at most $d(n+s)^{O(d)}$.

In the whole refutation σ there are $(k+s)(n+s)^{O(d)}$ monomials (k axioms from $\mathcal{F}$ and s from $\mathcal{E}$) of total size at most $d(k+s)(n+s)^{O(d)}$, and the substitution may increase this by the above multiplicative factor $d(n+s)^{O(d)}$.

This is performed for all ℓ levels of $\mathcal{E}$, and hence the size of the eventual proof π is bounded above by

$$d(k+s)(n+s)^{O(d)}d^{\ell}(n+s)^{O(d\ell)} = s^{O(d)}.$$

□

Lemmas 15.6.3–15.3.5 imply the following result.

Corollary 15.6.6 ([124]) *Assume that $\mathcal{F}_n$ are sets of at most n constant-degree polynomials over $\mathbf{F}_2$ in $\leq n$ variables. Let $s(n) \geq n$ be any non-decreasing function.*

Then the following two statements are equivalent:

- *there is a constant ℓ such that, for $n \gg 1$, $\mathcal{F}_n$ has an $F_\ell(\oplus)$-refutation of size $2^{(\log s(n))^{O(1)}}$;*
- *there is a constant ℓ' such that, for $n \gg 1$, $\mathcal{F}_n^{+1}$ has a degree $\leq (\log s(n))^{O(1)}$ ENS-refutation using sets $\mathcal{E}$, of size $2^{(\log s(n))^{O(1)}}$, of extension polynomials stratified into ℓ' levels.*

We shall study degree lower bounds for NS in Section 16.1, but the method used there does not apply to ENS. It thus remains to be seen whether this reduction of the $AC^0[p]$-Frege problem 15.6.1 to a lower bound for the degree in ENS refutations will be useful.

Although we do not have a lower bound for these systems at present, there is an interesting result going in a sense in the opposite direction. The original proof proceeds via a formalization of Toda's theorem from computational complexity (implying that all $Q_2\Sigma_0^{1,b}$-formulas collapse to the bottom level) in a suitable bounded arithmetic theory (a fragment of the theory $IQ_2\Sigma_0^{1,b} + \Omega_1$ from Section 10.1). However, the bounded arithmetic background required by that argument exceeds what we are covering in this book.

Theorem 15.6.7 (Buss, Kolodziejczyk and Zdanowski [129]) *An* $AC^0[2]$*-Frege system of any depth can be quasi-polynomially simulated by a depth 4 subsystem (operating with disjunctions of conjunctions of poly-logarithmic-degree polynomials).*

Note that, in particular, the depth-d vs. depth-$(d+1)$ Problem 14.3.2 has a negative answer for the systems $F_d(\oplus)$, although there still can be a super-polynomial separation analogous to that in Theorem 14.5.1.

Theorem 15.6.7 together with Corollary 15.6.6 implies an analogous statement about the ENS proof system.

Corollary 15.6.8 *Assume that $\mathcal{G}_n$ are sets of at most n constant-degree polynomials over $\mathbf{F}_2$ in $\leq n$ variables. Then there is a constant $\ell_0 \geq 1$ such that for any $\ell \geq 1$ the following holds for any parameter $s \geq n$:*

- *if $\mathcal{G}_n$ has for $n \gg 0$ an* ENS *refutation of degree $\leq (\log s)^{O(1)}$ and using sets $\mathcal{E}$ of size $2^{(\log s)^{O(1)}}$ of extension polynomials stratified into ℓ levels then $\mathcal{G}_n$ has an* ENS *refutation also of degree $\leq (\log s)^{O(1)}$ and using sets $\mathcal{E}$ of size $2^{(\log s)^{O(1)}}$ of extension polynomials stratified into ℓ_0 levels.*

We remark that one can choose $\ell_0 \leq 3$. There ought to be an elementary proof of Corollary 15.6.8 but I do not know of one. It would offer, together with Corollary 15.6.6, a new (and elementary) proof of Theorem 15.6.7.

The simplest open instance of the $AC^0[p]$-Frege systems problem 15.6.1 is perhaps that for $p = 2$ and where the subsystems R(LIN) or R(PC_d) have only a small degree d (Section 7.1).

15.7 Bibliographical and Other Remarks

The particular formulation in Theorem 15.3.1 of Ajtai's [5] original lower bound as the non-existence of a polynomial upper bound is a mere artifact of the use of the compactness of first-order logic in the proof. The argument does yield a specific super-polynomial lower bound, as was shown by Bellantoni, Pitassi and Urquhart [63]. The lower bound they extracted was not enough to derive the independence of PHP(R) in the bounded arithmetic $S_2(R)$. For that, a super quasi-polynomial lower bound was needed. The first such lower bound was the exponential lower bound from Theorem 14.3.1 (Krajíček [276]). That lower bound is not for the PHP formulas. The first strong enough lower bound for PHP was the improvement of the bound in Theorem 15.3.1 to an exponential one by Krajíček, Pudlák and Woods [324] and by Pitassi, Beame and Impagliazzo [400], announced jointly in [53]. The joint extended abstract was required by the STOC editors and perhaps it would have been more natural then to write the full paper jointly. I objected to that: I could not swallow the lack of acknowledgements in [400] of the ideas I shared during my visit to Toronto in Spring 1991. May my excuse be that I offered to my coauthors that I would step down as an author, to enable joint publication, but they both refused.

Secs. 15.1, 15.2 and 15.3 are based on [324] as presented in [278], but I used the elegant and natural notion of PHP-trees introduced by [400] instead of the more general but less elegant notion of *complete systems of maps* used in [324]. The adversary argument presented after the proof of Theorem 15.3.1 is closer to [400], and the Buss–Pudlák game formalism was worked out by Ben-Sasson and Harsha [68]. The set-up of the proof of Theorem 15.3.1 was generalized to the notion of *partial Boolean valuations* in [277, 278] and to *local Boolean valuations* by Buss and Pudlák [132]. Both these set-ups prove to be universal: if a lower bound is valid then in principle it can be proved in this way.

Hastad [223] proved an exponential lower bound for the AC^0-Frege systems with a better dependence on the depth; the lower bound has the form 2^{n^δ} with the exponent δ being $\Omega(d^{-1})$ rather than 5^{-d} as in Theorem 15.3.1. Note that if one could arrange that δ is $\omega(d^{-1})$ then Theorem 2.5.6(ii) would yield a super-polynomial lower bound for general Frege systems. However, such an exponential lower bound is not known at present even for AC^0-circuits.

Sec. 15.4 The idea of manipulating the size of the universe in various combinatorial principles, assuming that *onto*PHP (or even WPHP) fails, and getting lower bounds for them is from [288] and was further used in [296], under the name *structured* PHP, and in [303] (Lemma 15.4.3 is from there: its proof follows closely [303, Sec. 2]). Ben-Sasson [67] proved a lower bound also for AC^0-Frege proofs of Tseitin's formulas from Section 13.3.

Sec. 15.5 The original proof of Theorem 15.5.5 by Ajtai [7] gave a non-polynomial lower bound by an argument using the representation theory of the symmetric group; we shall discuss this in Section 16.5 as that work had other implications for algebraic proof systems. Another proof of the non-existence of a polynomial upper bound was given by Beame *et al.* [52]; that paper introduced the reduction to degree lower bounds for the Nullstellensatz proof system and proved (via Ramsey theory) the Key claim in the proof of the theorem as the non-existence of a constant-degree upper bound. This was subsequently improved by another argument by Buss *et al.* [124], giving a lower bound degree of $n^{\Omega(1)}$ (Section 16.1). They also classified all possible mutual reducibilities among the counting principles for composite moduli and for fixed remainders of the partitions involved.

Beame and Pitassi [56] proved an exponential separation between the PHP and the counting principles, improving upon Ajtai [6].

Sec. 15.6 By virtue of Lemma 10.1.4, lower bounds for NS and PC over $\mathbf{F}_p$ are prerequisites for Problem 15.6.1; these lower bounds will be treated in Chapter 16.

There are various other results about Problem 15.6.1 besides the two theorems we explained. In [282] an exponential lower bound was proved for a subsystem of an $AC^0[p]$-Frege system that extends both AC^0-Frege systems and $PC/\mathbf{F}_p$. Maciel and Pitassi [347] showed a quasi-polynomial simulation of $AC^0[p]$-Frege systems by a proof system operating with depth 3 threshold formulas. Impagliazzo and Segerlind [247] proved that AC^0-Frege systems with counting axioms modulo a prime p do not polynomially simulate PC/$\mathbf{F}_p$ (but it is open whether a quasi-polynomial simulation exists). More recently Garlík and Kolodziejczyk [197] refined the lower bounds in [282], considering tree-like and dag-like versions of the systems (named $\mathrm{PK}_d^c(\oplus)$ by them) and linked them with AC^0-Frege systems with modular counting principles as extra axioms (Section 15.5) and with R(LIN) (Section 7.1) (and they also bridged some gaps in the arguments in [247, 464]). Krajíček [309] reduced Problem 15.6.1 to computational complexity lower bounds for search tasks involving search trees branching upon values of maps on the vector space of low-degree polynomials over $\mathbf{F}_p$. There is a model-theoretic approach to the problem in [304, Chapter 22] (Section 20.4).

Beame *et al.* [124] considered also **unstructured** ENS, denoted by UENS: this system allows extension polynomials of the form

$$(g_1 - r_1) \cdot \dots \cdot (g_h - r_h),$$

where the r_i do not occur in any g_j (but may occur in other such unstructured extension polynomials), subject to the condition that the number of these polynomials is less than 2^h. A probabilistic argument analogous to the one showing the soundness of ENS (Section 15.6) yields the soundness of UENS as well. In [124] it was proved that UENS is surprisingly strong: it p-simulates *EF*. It is not known whether *EF* also p-simulates UENS and whether the systems are actually p-equivalent. Jeřábek [256] proved that at least the WPHP-Frege system *WF* (Section 7.2) does p-simulate UENS.

Note that, in analogy to the systems $AC^0[2]$-R(PC), one can consider systems Γ-R or Γ-R(PC) for any of the usual circuit classes; this will play some role in Section 19.3.

Let us conclude the chapter by mentioning the intriguing relations between the WPHP problem 15.3.2, the Δ_0-PHP problem 8.1.3 and the finite axiomatizability problem 8.1.2. Paris, Wilkie and Woods [392] proved that if $I\Delta_0$ does not prove the $\mathrm{WPHP}_n^{n^2}(\Delta_0)$ principle then it is not finitely axiomatizable. If the hypothesis of this implication holds then, in particular, $I\Delta_0(R)$ does not prove $\mathrm{WPHP}_n^{n^2}(R)$. Such an unprovability would follow from the existence of a super-polynomial lower bound for AC^0-Frege proofs of $\mathrm{WPHP}_n^{n^2}$, but that is open (recall that by Theorem 11.4.6 there are almost polynomial-size proofs). Paris and Wilkie [389] gave an earlier model-theoretic variant of a similar result. Another related result is in [278]: if $S_2 + Con_{S_2}$ does not prove $\mathrm{PHP}(\Sigma_\infty^b)$ then it also does not prove that NP $=$ coNP (cf. [278, Sec. 15.3] for details and further references).

16

Algebraic and Geometric Proof Systems

In this chapter we shall study the main lower bounds for the algebraic and geometric proof systems introduced in Chapter 6 that can be proved by a combinatorial argument. We postpone to Chapter 18 those lower bounds obtained via the feasible interpolation method. In particular, we give the degree and size lower bounds for NS and PC (but not a lower bound for the number of steps in PC, under any reasonable but not too restrictive hypothesis about degree or size of the proof) and for the Positivstellensatz, and we give rank lower bounds for LS and LS_+.

16.1 Nullstellensatz

Let $\mathcal{F}$ be a set of polynomials over a field $\mathbf{F}$ in the variables $\overline{x} = x_1, \ldots x_n$. As we are primarily interested in propositional logic, we shall assume that $\mathcal{F}$ contains the Boolean axioms (Section 6.2) $x_i^2 - x_i$ for all variables x_i. We want to prove, for some interesting $\mathcal{F}$, that every NS-refutation of $\mathcal{F}$ must have a large degree, i.e. if

$$\sum_{f \in \mathcal{F}} h_f f = 1$$

in $\mathbf{F}[\overline{x}]$ then $\max_f deg(h_f f)$ is large. This will allow us to prove also a lower bound on the *size* of any refutation (for those specific $\mathcal{F}$).

The degree lower bounds will use the following simple notion. A **degree-t design** for $\mathcal{F}$ is a function

$$D: \mathbf{F}[\overline{x}] \rightarrow \mathbf{F}$$

such that

1. $D(0) = 0$ and $D(1) = 1$,
2. D is an $\mathbf{F}$-linear map,
3. for $f \in \mathcal{F}$ and any $g \in \mathbf{F}[\overline{x}]$ such that $deg(gf) \leq t$, $D(gf) = 0$.

Lemma 16.1.1 *Assume that there exists a degree t design for $\mathcal{F}$. Then every* NS-*refutation of $\mathcal{F}$ must have degree larger than t.*

Proof Assume that $\max_f deg(h_f f) \leq t$ in an NS-refutation

$$\sum_{f \in \mathcal{F}} h_f f = 1$$

of $\mathcal{F}$, and let D be a degree t design for $\mathcal{F}$. Apply D to both sides of the refutation. By the third item in the above list, $D(h_f f) = 0$ for all f and hence by the linearity and the first item the left-hand side evaluates to 0. But again by the first item the right-hand side is 1. Hence no refutation can have degree $\leq t$. □

We shall now apply this idea to get a degree lower bound for the $\neg Count_3$ polynomial system (15.5.1) over $\mathbf{F}_2$ and to prove a lower bound that we have used already in the proof of Theorem 15.5.5 (the key claim there).

Theorem 16.1.2 (Beame *et al.* [52], Buss *et al.* [124]) *For $n \geq 3$ and not divisible by* 3, *every* NS/$\mathbf{F}_2$*-refutation of the $\neg Count_n^3$ system* (15.5.1) *must have degree* $\Omega(n^{1/\log 5})$.

The theorem follows easily from Lemma 16.1.4 below. Before stating and proving that lemma let us first simplify a little the situation regarding the definition of a design for the specific case of the $\neg Count_n^3$ system. Recall the definition of the set *Partitions* from Section 15.5.

Lemma 16.1.3 *Assume that* $\sum_{i \in [n]} h_i u_i + \sum_{e \perp f} h_{e,f} v_{e,f} + \sum_e h_e w_e = 1$ *is an* NS*-refutation of* (15.5.1) *of degree $\leq t$. Then there is another refutation,*

$$\sum_{i \in [n]} h'_i u_i + \sum_{e \perp f} h'_{e,f} v_{e,f} = 1,$$

of the system, of degree $\leq t$, such that for each monomial

$$x_\alpha := x_{e_1} \cdot \dots \cdot x_{e_d} \tag{16.1.1}$$

occurring in h'_i with a non-zero coefficient it holds that $\alpha := \{e_1, \dots, e_d\} \in Partitions$ and $i \notin \bigcup \alpha := e_1 \cup \dots \cup e_d$.

Proof Observe that for $i \in e$, $x_e u_i = w_e + \sum_{f: i \in f, f \perp e} v_{e,f}$. Hence all monomials (16.1.1) from h'_i containing i can be removed at the expense of changing the coefficients of some axioms $v_{e,f}$ and w_e. After this procedure is carried out for all h_i, change all occurrences of x_j^k, for all $j \in [n]$ and $k \geq 2$, to x_j: this can make some monomials equal that were not previously equal but not the other way around; hence the refutation remains a refutation. □

By the linearity of designs it is sufficient to define a suitable design D on monomials only. For notational simplicity, we shall write $D(\alpha)$ for $D(x_\alpha)$, with x_α as in (16.1.1).

If α contains two blocks e_i, e_j that are incompatible, so that $e_i \perp e_j$ in the notation of Section 15.5, the submonomial $x_{e_i} x_{e_j}$ is from $\neg Count_n^3$, and thus x_α must acquire the value 0. By Lemma 16.1.3 we do not need to consider monomials that are not multi-linear, and owing to the stated property of the monomials in the coefficient polynomials h_i' we need only to arrange the following condition for monomials of as high degree d as possible:

- for any $i \in [n]$ and any monomial x_α as in (16.1.1) such that α is a partial partition of $[n] \setminus \{i\}$, it holds that

$$D(\alpha) = \sum_{f: f \cap \bigcup \alpha = \emptyset \wedge i \in f} D(\alpha \cup \{f\}). \tag{16.1.2}$$

For the purpose of the next lemma we shall call any D satisfying this condition a *degree-d design on* $[n]$.

Lemma 16.1.4 *Let $n \geq 3$ and not divisible by* 3. *Assume that there exists a degree d design D on $[n]$. Then there is a degree $2d + 1$ design D' on $[5n]$.*

(The constant 5 appears because $5 = 2 + 3$; for a general pair of different primes p, q it would be $p + q$; see Section 16.5 below.]

Proof Let D be a degree-d design on $[n]$ and assume without loss of generality that $D(\beta) = 0$ if β has more than d blocks.

Consider the set $[5n]$ partitioned (completely) into blocks of size 5. To distinguish this from the usage of the term *block* in 3-partitions we shall call these 5-blocks *orbits*. We shall also fix a permutation *Next* whose cycles are exactly the orbits. Hence, for $v \in [5n]$, $\{v^0, v^1, \dots, v^4\}$ is the orbit containing v, where we set $v^5 := v^0 := v$ and $v^{t+1} := Next(v^t)$.

For $X \subseteq [5n]$, denote by $Orbit(X)$ the set of all points from $[5n]$ that share a common orbit with some element of X. We shall denote by $Partitions_n$ and $Partitions_{5n}$ the sets of partial 3-partitions on $[n]$ and $[5n]$, respectively ($Partitions_n$ was denoted just by *Partitions* in Section 15.5).

Now assume that $V \subseteq [5n]$ is a set that intersects each orbit in exactly one point; in particular, $|V| = n$. We shall call such sets **choice sets** and put $V^t := \{v^t \mid v \in V\}$, i.e. $V^0 = V$. Given a choice set V, define first only a degree d design D^V on $[5n]$ by the following construction.

Define a partial mapping $V^* : Partitions_{5n} \to Partitions_n$ by identifying $[n]$ with the set of orbits and proceeding as follows.

- If all blocks e in $\beta \in Partitions_{5n}$ satisfy
  - either $e \subseteq V^1$ or $e \subseteq V^2$ (such an e will be called a **cross-block**)
  - or $e = \{v^3, v^4, v^5\}$ (such an e will be called an **inner-block**),

  and the set

$$\{Orbit(e) \mid e \in \beta \text{ is a cross-block }\}$$

  is a partial 3-partition, we define $V^*(\beta)$ to be this partition (of orbits).
- We let $V^*(\beta)$ remain undefined in all other cases.

Having defined the partial map V^*, put

$$D^V(\beta) = \begin{cases} D(V^*(\beta)) & \text{if } V^*(\beta) \text{ is defined,} \\ 0 & \text{otherwise.} \end{cases}$$

The definition of D^V implies readily that

Claim 1 D^V is a degree d design on $[5n]$. In addition, for any β and $i \in [5n] \setminus \bigcup \beta$, if

- $|V(\beta)| < d$,
- or $V(\beta)$ is undefined,
- or i is in an orbit intersecting a cross-block of β

then the design condition (16.1.2) holds for i and β.

Now we are ready to define the wanted design D'. Put

$$D'(\beta) := \sum_V D^V(\beta) \pmod 2$$

where V ranges over all possible choice sets $V \subseteq [5n]$.

Claim 2 Assume that there is a cross-block $e \in \beta$ such that no other block in β intersects any of the three orbits that e intersects. Then $D'(\beta) = 0$.

For each choice, set V such that $V^*(\beta)$ is defined there is exactly one other choice set W agreeing with V on the orbits which e does not intersect, and such that $W^*(\beta)$ is also defined. By symmetry $D^W(\beta) = D^V(\beta)$, and hence their sum modulo 2 is 0. This holds for all choice sets V and thus indeed $D'(\beta) = 0$.

The following is a key claim.

Claim 3 D' is a degree $2d + 1$ design on $[5n]$.

The condition for $D'(\emptyset)$ is clearly satisfied. Now consider any partial 3-partition β on $[5n]$ with at most $2d$ blocks and any $i \in [5n] \setminus \bigcup \beta$.

We may assume without loss of generality that $V(\beta)$ is defined for at least one choice set V, and we fix one such V. Claims 1 and 2 imply that the design condition is satisfied in the following situations.

1. The orbit of i intersects some cross-block of β: D' obeys the design condition since it is a linear combination of the designs that satisfy it (Claim 1).
2. $|V(E)| < d$, so the same is true for all choice sets W for which $W(\beta)$ is defined, and the design condition is satisfied by Claim 1 again.
3. The orbit of i contains some inner-block of β: then either there exists a block satisfying the condition of Claim 2 or one of the above two items holds.

   If the second item above fails, β has at least d cross-blocks with non-intersecting orbits, and the orbits of every other block intersect with at most one of the d cross-blocks (since $V(\beta)$ is a partition by its definition). Because

β has at most $2d$ blocks, either the first item holds or we can use Claim 2 to conclude that the design condition holds.

It remains to consider the situation when the block containing i does not intersect any block of β, and $V(\beta)$ has exactly d blocks for each choice set V.

Let W range over all selections of representatives from the orbits not containing i, i.e. these are choice sets outside the orbit of i. Put $W_t := W \cup \{v^t\}$. By the symmetry of the construction, $D^{W_t}(\beta) = D^W(\beta)$ for all t and so

$$D'(\beta) = \sum_W \sum_{t=1}^{5} D^{W_t}(\beta) = 5 \sum_W D^W(\beta).$$

Define for $j = 0, 1, 2$ the blocks

$$e_j := \{v^{-j}, v^{-j+1}, \ldots, v^{-j+2}\}$$

and note that, by our assumption on D, these three blocks are the only blocks containing i for which $D^{W_t}(\beta \cup \{e\})$ might possibly be non-zero; otherwise $|W_t(\beta \cup \{e\})| > d$.

Moreover, $D^{W_t}(\beta \cup \{e_j\}) = D^W(\beta)$ if $i = -1 - j$ (i.e. e_j is an inner-block of W_t) and $D^{W_t}(\beta \cup \{e_j\}) = 0$ otherwise. Putting all this together,

$$\sum_{j=0}^{2} D'(\beta \cup \{e_j\}) = \sum_W \sum_{t=1}^{5} \sum_{j=0}^{2} D^{W_t}(\beta \cup \{e_j\}) = 3 \sum_W D^W(E),$$

which is equal to $D'(\beta)$ modulo 2.

This completes the proof of the claim and of the lemma. □

It is much easier to construct designs over fields of characteristic 0; it suffices to assign to each monomial a suitable weight.

Lemma 16.1.5 *Let $n \geq 3$, n not divisible by 3. Then every Nullstellensatz refutation of $\neg Count^3_n$ over a field of characteristic 0 must have degree larger than $\lfloor n/3 \rfloor$.*

Proof By Lemma 16.1.1 it suffices to define a design of degree $\lfloor n/3 \rfloor$. Put $D(\emptyset) :=$ 1 and for a β with $t + 1$ blocks put $D(\beta) := c_{t+1}$, where we define

$$c_{t+1} := c_t \binom{n - 3t - 1}{2}^{-1}.$$

This works as long as $n - 3t - 1 \geq 2$ (i.e., $n/3 \geq t$), as $\binom{n-3t-1}{2}$ is the number of partitions extending a fixed partition β with t blocks by one new block containing a fixed element $i \in [n] \setminus \bigcup \beta$. □

Note that if we apply the strategy that we used to show that instances of $Count^2$ do not feasibly imply $Count^3_n$ to prove that also instances of *onto*PHP do not imply $Count^3_n$, we would end up trying to show that there is no $(m + 1) \times m$ matrix of polynomials f_{ij} (as in the proof of Theorem 15.5.5) summing to 1 in each row and in each column. But, to get a contradiction, we can count in any field and, in particular, over the rationals. Lemma 16.1.5 then gives a statement analogous to the key claim in the proof of Theorem 15.5.5.

When we use the dense representation of polynomials (Section 6.2), a lower bound d for the degree of an NS-refutation of some $\mathcal{F}$ in n variables also yields a lower bound $n^{\Omega(d)}$ for the number of different monomials and hence for the size: there has to be at least one polynomial of degree $\geq d$ and it has that number of monomials. But if we represent a polynomial simply as a sum of the monomials that occur in it with a non-zero coefficient then the degree may be high even for just one monomial.

We shall see in Theorem 16.2.4 a statement analogous to Theorem 5.4.5 allowing us to estimate the number of monomials from below in terms of a degree lower bound. That will be stated in the context of PC. Although there is an analog of Theorem 5.4.3, which involves R*, for PC* and hence for NS, we shall give instead a simpler argument for the specific case of refutations of $\neg Count_n^3$.

Lemma 16.1.6 *Assume that π is an* NS*-refutation of $\neg Count_n^3$ over any field, $n \gg 1$. Then, for any $0 < \delta < 1/(\log 5)$, at least n^{n^δ} different monomials must occur in π.*

Proof Choose a random subset $D \subseteq [n]$, with $|D|$ divisible by 3 and $n - |D| = n^\epsilon$, for an arbitrary $0 < \epsilon < 1$. Then choose randomly a 3-partition ρ of D.

Given a 3-element subset $e \subseteq [n]$, ρ is incompatible with e with probability at least $1 - O(n^{-2})$. Hence a degree d monomial $x_{e_1} \cdot \cdots \cdot x_{e_d}$ gets killed by ρ with the probability of failing to do so bounded above by $O(n^{-\Omega(d)})$ for $d \ll n$.

By averaging, it follows that if there are fewer than $n^{\Omega(d)}$ monomials of degree $\geq d$, there is a ρ of the above form that kills them all. Further, a restriction of the refutation π by ρ is again an NS-refutation π^ρ, but of $\neg Count_{n^\epsilon}^3$. By Theorem 16.1.2 it follows that the degree of π^ρ must be at least $\Omega(n^{\epsilon/\log 5})$ for $n \gg 0$. □

16.2 Polynomial Calculus

There is a variety of lower bound results for polynomial calculus (PC) and we will refer to some of these in Section 16.5. Here we concentrate on the first strong PC degree lower bound, which has the remarkable feature of describing explicitly the space of all polynomials derivable from the starting-axiom polynomials within a bounded degree. Let $\neg \mathrm{WPHP}_n^m$ be the polynomial system $\neg \mathrm{PHP}_n$ from Section 11.7 but with m in place of $n+1$. That is, for $m > n \geq 1$ the $\neg \mathrm{WPHP}_n^m$ polynomial system has variables x_{ij}, $i \in [m]$ and $j \in [n]$ and consists of the following polynomials:

1. $1 - (\sum_{j \in [n]} x_{ij}) = 0$, for each $i \in [m]$;
2. $x_{ij}x_{ij'} = 0$, for all $i \in [m]$ and $j \neq j' \in [n]$;
3. $x_{ij}x_{i'j} = 0$, for all $i \neq i' \in [m]$ and $j \in [n]$.

The polynomials on the left-hand sides of these equations will be denoted by

$$Q_i, \quad Q_{i;j,j'} \quad \text{and} \quad Q_{i,i';j}$$

respectively.

Theorem 16.2.1 (Razborov [436]) *Let* $\mathbf{F}$ *be an arbitrary field and let* $m > n \geq 1$ *be arbitrary. Then there is no degree* $\leq n/2$ PC/$\mathbf{F}$*-refutation of* $\neg\mathrm{WPHP}^m_n$.

The theorem is an immediate consequence of Lemma 16.2.2 below; see the comment after its statement. The proof of that lemma will have another interesting corollary (Lemma 16.2.3): although in general PC is stronger than NS (Section 16.5) this is not so for proofs from $\neg\mathrm{WPHP}^m_n$:

- *if a polynomial* f *(over any* $\mathbf{F}$ *with variables of* $\neg WPHP^m_n$*) has a degree* $d \leq n/2$ PC/$\mathbf{F}$ *proof from* $\neg\mathrm{WPHP}^m_n$ *then it has also a degree* d NS/$\mathbf{F}$*-proof from* $\neg\mathrm{WPHP}^m_n$.

We need to define first several vector spaces (everything is over $\mathbf{F}$) and their bases. We fix n and omit it from the already baroque notation. The tuple of all variables of $\neg\mathrm{WPHP}^m_n$ will be denoted by $\overline{x}$.

We write $\hat{S}$ for the ring $\mathbf{F}[\overline{x}]$ factored by the ideal generated by all the polynomials

$$x_{ij}^2 - x_{ij}, \quad Q_{i_1,i_2;j} \quad \text{and} \quad Q_{i;j_1,j_2}.$$

For $f \in \mathbf{F}[\overline{x}]$, $\hat{f} \in \hat{S}$ denotes the corresponding element in the quotient space.

Maps will denote, analogously to Section 15.1, the set of partial injective maps α but this time from $[m]$ into $[n]$. As in Section 15.5 each $\alpha \in Maps$ defines a monomial $x_\alpha := \prod_{i \in dom(\alpha)} x_{i\alpha(i)}$. In particular, $\emptyset \in Maps$ and $x_\emptyset = 1$.

For $t \geq 0$, define $Maps_t := \{\alpha \in Maps \mid |\alpha| \leq t\}$ and also $\hat{S}_t$ as the set of elements of $\hat{S}$ of degree at most t. Let $T_t := \{x_\alpha \mid \alpha \in Maps_t\}$ be the canonical vector space basis of $\hat{S}_t$.

The vector space for which we shall want to find some combinatorially transparent basis is the set of polynomials $\hat{f} \in \hat{S}$ such that $f \in \mathbf{F}_2[\overline{x}]$ has a degree $\leq t$ PC-proof from $\neg\mathrm{WPHP}^m_n$. It will be denoted by V_t. The following elegant notion will allow us to define such a basis.

Let $\alpha \in Maps$. A **pigeon dance** of α is the following non-deterministic process. Take the first (i.e. the smallest) pigeon $i_1 \in dom(\alpha)$ and move it to any unoccupied hole j that is *larger* than $\alpha(i_1)$. Then take the second smallest pigeon $i_2 \in dom(\alpha)$ and move it to any currently unoccupied hole larger than $\alpha(i_2)$, etc., as long as this is possible. We say that the pigeon dance is defined on α if this process can be completed for all the pigeons in $dom(\alpha)$.

The set $Maps^*$ consists of all maps α: $[m] \to [n] \cup \{0\}$ that are injective outside $\alpha^{(-1)}(0)$. That is, if $i, i' \in dom(\alpha)$ and $\alpha(i) = \alpha(i') \neq 0$ then $i = i'$. Put $Maps^*_t := \{\alpha \in Maps^* \mid |\alpha| \leq t\}$. Also, for $\alpha \in Maps^*$ denote by $\alpha^- \in Maps$ the map α restricted to the domain $dom(\alpha) \setminus \alpha^{(-1)}(0)$.

We need to extend the notation x_α to $Maps^*$. For $\alpha \in Maps^*$ the symbol x_α will be the element of $\hat{S}$ defined by:

$$x_{\alpha^-} \cdot Q_{i_1} \cdots Q_{i_k}, \quad \text{where } \{i_1, \ldots, i_k\} = \alpha^{(-1)}(0).$$

In particular, it is no longer just a monomial.

The last three definitions of subsets of $\hat{S}$ that we shall need are:

- $B_t \subseteq \hat{S}$, the set of all x_α for $\alpha \in Maps^*$ such that a pigeon dance is defined on α^-; in particular, $1 \in B_0$;
- $C_t := B_t \setminus T_t$ and $\Delta_t := B_t \cap T_t$; in particular, Δ_t consists of the monomials x_α for $\alpha \in Maps$ such that a pigeon dance is defined on α.

Lemma 16.2.2 *For $0 \le t \le n/2$, as vector spaces we have*

$$\hat{S}_t \;=\; V_t \bigoplus \mathbf{F}\Delta_t$$

and

$$V_t \;=\; \mathbf{F}C_t.$$

In particular, $1 \notin V_t$ and Δ_t is a basis of the vector space $\hat{S}_t/V_t$.

(Note that $1 \notin V_t$ implies the degree lower bound t for PC-refutations of $\neg\mathrm{WPHP}^m_n$ and hence the lemma implies Theorem 16.2.1.)

Proof We now describe a rewriting procedure for monomials which will show that $C_t \cup \Delta_t$ spans $\hat{S}_t$ and that Δ_t spans $\hat{S}_t/V_t$. We shall need an additional argument for the linear independence.

For the monomials

$$x_\gamma \;=\; x_{i_1 j_1} \cdots x_{i_k j_k} \quad \text{and} \quad x_\delta \;=\; x_{u_1 v_1} \cdots x_{u_\ell v_\ell}$$

from T_t, define a partial ordering $x_\gamma \preceq x_\delta$ by the condition that

- either $k < \ell$, or $k = \ell$ and for the *largest* w such that $j_w \ne v_w$ it holds that $j_w < v_w$.

We shall use induction on $\preceq$ to prove

Claim 1 The monomial x_γ can be expressed as

$$\sum_{x_\alpha \in X} c_\alpha x_\alpha \;+\; \sum_{x_\beta \in Y} c_\beta x_\beta$$

for some $X \subseteq C_t$ and $Y \subseteq \Delta_t$, and non-zero coefficients c from $\mathbf{F}$.

Further, for all $x_\alpha \in X$ and $x_\beta \in Y$, $dom(\alpha) \cup dom(\beta) \subseteq dom(\gamma)$.

Assume $i_1 < \cdots < i_k$ in x_γ above and also $x_\gamma \notin \Delta_t$ (as otherwise there is nothing to prove) and that the claim holds for all terms $\preceq$-smaller than x_γ.

Use the polynomial axiom Q_{i_1} and rewrite x_γ as

$$x_{i_2 j_2} \cdots x_{i_k j_k} \;+\; \sum_{j_1' < j_1} x_{i_1 j_1'} \cdots x_{i_k j_k} \;+\; \sum_{j_1' > j_1} x_{i_1 j_1'} \cdots x_{i_k j_k},$$

computing in $\hat{S}$, i.e. deleting the terms with $j_1' \in \{j_2, \ldots, j_k\}$. The first term and the terms in the first summation are all $\preceq$-smaller than x_γ and their domain is included in $dom(\gamma)$; thus the statement for them follows by the induction hypothesis. The collection of all terms in the second summation can be interpreted as describing all possible moves of i_1 in the attempted pigeon dance of γ. To simulate other steps in all possible pigeon dances, rewrite each of these terms analogously using Q_{i_2} (i.e. simulating the dance of pigeon i_2), and so on. But we have assumed that $x_\gamma \notin \Delta_t$, i.e. that the pigeon dance cannot be completed. Therefore the rewriting process must eventually produce only terms $\preceq$-smaller than x_γ.

We need to establish the linear independence of B_t. For this we shall eventually define an injective map from B_t into T_t, showing that $|B_t| \leq |T_t|$. Because B_t spans $\hat{S}_t$ and T_t is its basis, B_t must be a basis – and, in particular, linearly independent – as well.

Consider first the *minimal* pigeon dance. Given α, it is defined by the following instructions.

- Put $\alpha_1 := \alpha$.
- For $i = 1, 2, \ldots, m$, if $i \in dom(\alpha_i)$, move it to the smallest free hole and let α_{i+1} be the resulting map. If $i \notin dom(\alpha_i)$, do nothing and put $\alpha_{i+1} := \alpha_i$.
- $D(\alpha)$ is the result of this process applied to α.

Claim 2 For all $t \leq n/2$, D is defined on the whole of B_t.

It is not difficult to see that if a pigeonhole dance is defined on α^-, the minimal dance is defined on α^- too. Now, if $\alpha \in B_t$, D is defined on α^-. If $s_1 = |dom(\alpha)|$ and $s_2 = |\alpha^{(-1)}(0)|$, then either s_2 and $\alpha = \alpha^-$ or $2s_1 + s_2 \leq n$. Hence there are at least s_2 holes not used by D on α^- and thus they can be used for the pigeons mapping to 0.

Claim 3 D is an injective map from B_t into T_t, for $t \leq n/2$.

It suffices to show that each of the m steps defining D is injective. Consider the ith step and assume that $\alpha \neq \alpha'$ are mapped by it to β. The domains of these maps are the same and they all agree on all $i' \neq i$ in the domain. But, if $\alpha(i) < \alpha'(i)$ then $\alpha'(i)$ is in a free hole smaller than $\beta(i)$, which contradicts the definition of the minimal dance. The other possibility, $\alpha'(i) < \alpha(i)$, leads to an analogous contradiction. Hence it must be that $\alpha = \alpha'$.

It remains to show that C_t is a basis of V_t. Clearly $C_t \subseteq V_t$ and $\mathbf{F} \cdot C_t$ contains all the axioms Q_i and is closed under the addition rule of PC. For the closure of the space under the multiplication rule, assume that $x_\alpha \in C_t$, $deg(x_\alpha) < t$. We want to show that $x_{ij}x_\alpha \in \mathbf{F} \cdot C_t$ as well.

Proceed by induction on t. By the definition of B_t we can write $x_\alpha = x_\beta Q_{i'}$ for some $i' \in [m]$. As $deg(x_\beta) < deg(x_\alpha) \leq t - 1$, $x_{ij}x_\beta$ is in $\mathbf{F} \cdot B_{t-1}$. Multiplying the $\mathbf{F}$-linear combination of elements of B_{t-1} turns each term into either an element of C_t or 0. □

Note that the reduction process described in the proof of Lemma 16.2.2 also yields the following statement.

Lemma 16.2.3 *If f has a degree $\leq n/2$ PC-proof from $\neg\mathrm{WPHP}_n^m$ then it also has a degree $\leq n/2$ NS-proof from $\neg\mathrm{WPHP}_n^m$.*

We shall conclude by stating a theorem that gives a lower bound for the minimum number of monomials in a PC-refutation, in terms of the minimum degree. It can be proved in a way analogous to the proof of Theorem 5.4.5 and we shall not prove it here.

First, we need a technical definition tailored specifically for application to the WPHP. For a set $\mathcal{F}$ of polynomials in n variables $x_1, \ldots, x_n$ over a field $\mathbf{F}$ and $k \geq 1$, we say that $\mathcal{F}$ has a **k-division** if and only if there are sets $C_1, \ldots, C_k \subseteq \{x_1, \ldots, x_n\}$ such that

$$x_u x_v \in \mathcal{F} \quad \text{whenever } x_u, x_v \in C_i \quad \text{for some } i \leq k.$$

Theorem 16.2.4 (Impagliazzo, Pudlák and Sgall [246]) *Let $\mathcal{F}$ be a set of polynomials in n variables over a field* $\mathbf{F}$. *Assume that the degree of all polynomials in $\mathcal{F}$ is at most d_0. Then:*

(i) if the minimum degree of a PC*-refutation of $\mathcal{F}$ is $d \geq d_0$ then any* PC*-refutation of $\mathcal{F}$ must contain at least $2^{\Omega(d^2/n)}$ monomials;*
(ii) if the minimum degree of a PC*-refutation of $\mathcal{F}$ is $d \geq d_0$, $k \geq 1$ and $\mathcal{F}$ has a k-division then any* PC*-refutation of $\mathcal{F}$ must contain at least $2^{\Omega(d^2/k)}$ monomials.*

The second part of the theorem yields a lower bound for refutations of $\neg\mathrm{WPHP}_n^m$ as $C_j := \{x_{ij} \mid i \in [m], j \leq n\}$ is an n-division for $\neg\mathrm{WPHP}_n^m$.

16.3 Positivstellensatz

An elegant characterization of the non-existence of degree t NS-refutations is the existence of degree-t designs (see Lemma 16.1.1 and the remark about it below in Section 16.5). For Positivstellensatz refutations of systems of polynomial equations and inequalities over $\mathbf{R}$ as (6.4.2),

$$f = 0, \ \text{ for } f \in \mathcal{F}, \ \text{ and } g \geq 0, \ \text{ for } g \in \mathcal{G},$$

a similar notion also exists. A function:

$$E\colon \mathbf{R}[\overline{x}] \to \mathbf{R}$$

is a **degree** t SOS **design** for $\mathcal{F} \cup \mathcal{G}$ if it satisfies the three conditions from the definition of a degree d design in Section 16.1, that is,

1. $E(0) = 0$ and $E(1) = 1$,
2. E is an $\mathbf{R}$-linear map,
3. for $f \in \mathcal{F}$ and any $h \in \mathbf{R}[\overline{x}]$ such that $deg(hf) \leq t$, we have $E(hf) = 0$

and, in addition, two more conditions,

4. for any $h \in \mathbf{R}[\overline{x}]$, if $deg(h) \leq t/2$ then $E(h^2) \geq 0$,
5. for $g_1, \ldots, g_\ell \in \mathcal{G}$ and $h \in \mathbf{R}[\overline{x}]$, if $deg(g_1 \ldots g_\ell h^2) \leq t$ then $E(g_1 \ldots g_\ell h^2) \geq 0$.

Note that if the system (6.4.2) is actually solvable by some $\overline{a} \in \mathbf{R}^n$ then

$$E: h(\overline{x}) \to h(\overline{a})$$

is an SOS-design of any degree t.

The following is proved analogously to Lemma 16.1.1.

Lemma 16.3.1 *Assume that there exists a degree t SOS-design for $\mathcal{F} \cup \mathcal{G}$. Then every SOS-refutation of $\mathcal{F} \cup \mathcal{G}$ must have degree larger than t.*

In analogy to the relation between NS-refutations and designs, SOS-designs are dual objects to SOS-refutations, and Lemma 16.3.1 has a form of converse; see Section 16.5. We shall not use SOS-designs to prove a degree lower bound for the Positivstellensatz, however. Instead we shall employ the following elegant reduction to the complexity of LEC (linear equational calculus) proofs from Section 7.7.

The LEC-**width** of an LEC-derivation is defined analogously to the width of an R-derivation: it is the maximum number of variables occurring in a proof line.

Lemma 16.3.2 (Schoenebeck [461]) *Let $c \geq 3$ be a constant. Assume that, for $n \gg 1$, $A \cdot \overline{x} = \overline{b}$ is an unsolvable system of linear equations in n variables with at most c variables in each equation.*

If every LEC-refutation of the system must have LEC-width at least $w \geq c$ then every Positivstellensatz proof of the unsolvability of the system must have degree at least $w/2$.

The statement is proved in [461, Lemma 13] in the framework of relaxations rather than proof systems and we shall not repeat it here.

Theorem 16.3.3 (Grigoriev [207])

(i) *There is a $c \geq 3$ such that for all $n \gg 1$ there is a graph G with vertices $[n]$, each vertex having degree at most c, such that, for all $f: [n] \to \{0, 1\}$ with $\sum_{i \in [n]} f(i)$ odd, every Positivstellensatz refutation of Tseitin's contradiction $TSE_{G,f}$ (Lemma 13.3.6) must have degree at least $\Omega(n)$.*
(ii) *For $n \gg 1$ and odd, every Positivstellensatz refutation of $Count_n^2$ (Section 11.1) must have degree at least $\Omega(n)$.*
(iii) *There is a $c \geq 3$ such that for all $n \gg 1$ there is an unsolvable system of linear equations $A \cdot \overline{x} = \overline{b}$ with n variables $\overline{x}$ and at most c variables in each equation such that every Positivstellensatz proof of the unsolvability of the system must have degree at least $\Omega(n)$.*

The constant c in items (i) and (iii) can be as small as 3; G is a graph with certain expansion properties and $A, \overline{b}$ is the system of equations expressing $TSE_{G,f}$.

Proof Let us treat the third statement; the first two can be reduced to it. In fact, we shall prove a weaker lower bound $n^{1-\delta}$, for any $\delta > 0$, because that will allow us to refer to a construction and an argument used already.

Let $\delta > 0$ be arbitrary small. By Lemma 16.3.2 it suffices to demonstrate that, for a suitable linear system $A \cdot \overline{x} = \overline{b}$ with n variables and at most c variables in each equation, every LEC-refutation needs an LEC-width at least $n^{1-\delta}$. A suitable linear system is that provided by Theorem 13.3.1 and the argument yielding a suitable LEC-width lower bound is completely analogous to the argument in the proof of the R-width lower bound in Corollary 13.4.6, which uses Lemmas 13.4.4 and 13.4.5. □

We have stated the above theorem for Positivstellensatz refutations, although the original result was formulated for a slightly stronger Positivstellensatz calculus $\mathrm{PC}_{>}$ (see Section 6.7). The bound does not hold for SAC, though, as the next statement shows.

Theorem 16.3.4 (Atserias [27]) *Let $A \cdot \overline{x} = \overline{b}$ be a system of m linear equations in n unknowns over $\mathbf{F}_2$ that is unsolvable. Assume that every equation in the system contains at most ℓ variables.*

Then the unsolvability of the system has an SAC-proof of constant degree (independent of n) and of size polynomial in n and 2^{ℓ}.

The linear system is represented for SAC by $m2^{\ell-1}$ inequalities that have a solution in $\{0, 1\}$ if and only if the original system has a solution in $\mathbf{F}_2$.

16.4 CP and LS: Rank and Tree-Like Size

In this section we shall study rank and tree-like size lower bounds for CP and LS, postponing the size lower bounds for general proofs to Chapter 18. Recall the definition of the rank of CP- and LS-proofs from Section 6.6.

It is easy to prove a rank lower bound via Grigoriev's theorem 16.3.3.

Lemma 16.4.1 *Assume that $\mathcal{F} \cup \mathcal{G}$ is a system of linear equations and inequalities that has an LS-refutation π of rank r. Then it has a Positivstellensatz refutation ρ of degree $d \leq r$.*

If π is an LS_+-refutation then a Positivstellensatz refutation ρ of degree $d \leq 2r$ exists.

Proof We may assume that π is tree-like and simulate it in Positivstellensatz analogously to the way in which Nullstellensatz simulates tree-like PC in Lemma 6.2.4. The multiplicative factor 2 accounts for the starting quadratic inequalities $Q^2 \geq 0$ in LS_+. □

Corollary 16.4.2 *The three principles in Theorem 16.3.3 (when the CNFs are encoded via linear inequalities) require LS_+-refutations of rank at least $\Omega(n)$.*

Note that Lemma 16.4.1 generalizes straightforwardly to LS^d and LS^d_+ and, in fact, to SAC, analogously to how Lemma 6.2.4 relates PC* (tree-like PC) and NS. Hence, so does Corollary 16.4.2; for example, the rank of any LS^d_+-refutation of the principles must be $\Omega(n/d)$.

The next theorem allows us to transform rank lower bounds to size lower bounds. It is proved by an argument analogous to that in the proofs of Theorems 5.4.5 and 16.2.4 and we shall not repeat it.

Theorem 16.4.3 (Pitassi and Segerlind [403]) *Assume that a system of linear inequalities $\mathcal{G}$ in n variables has a tree-like* LS- *(resp. tree-like* LS_+-*) refutation of size s. Then it has another* LS- *(resp.* LS_+-*) refutation (not necessarily tree-like) of rank at most $O((n \log s)^{1/2})$.*

16.5 Bibliographical and Other Remarks

Ajtai's work [7, 8] on the mutual independence of modular counting principles (Section 15.5) led him to study the solvability of symmetric systems of linear equations over various fields (and $\mathbf{F}_p$, in particular). The term *symmetric* means that the variables are indexed by tuples of elements of $[n]$ and any permutation of $[n]$ keeps the polynomial system unchanged. For example, $\langle \dots \rangle$ translations of first-order sentences result in formulas that are symmetric in this sense, and when a formula (or a set of clauses) is translated into a set of equations this property is preserved. Using the representation theory of the symmetric group he proved a strong theorem that showed that the solvability of such a system depends (in a certain precise technical sense) only on the remainder of n modulo some prime power of the characteristic of the underlying field. Examples of symmetric systems of linear equations are equations for the unknown coefficients of polynomials in an NS-refutation of $\neg Count^3_n$ of any fixed degree (the polynomials forming $\neg Count^3_n$ are symmetric in the sense of [8]). Thus Ajtai's result [8] implied immediately a non-constant degree lower bound for NS-refutations of the modular counting principles, the PHP etc., even before the NS proof system was actually defined. In fact, [286] showed that [8] also implies, with a little extra work, non-constant degree lower bounds for PC. This in turn implies, for example, a non-constant degree lower bound for $PC/\mathbf{F}_p$-refutations of $\neg Count^q_n$, for different primes $p \neq q$, and also lower bounds for a number of polynomials systems (e.g. as $Field_n$ in Section 15.4) not amenable to the direct combinatorial arguments. With rather more work, it was shown [286, 287] that the method can give a general model-theoretic criterion related to Theorem 13.1.2 (see also Section 20.5) that implies $\log n$ (for NS) and $\log \log n$ (for PC) degree lower bounds. Further work on this approach to NS and PC was hampered by the technical difficulty of the underlying characteristic free representation theory of the symmetric group of James [254, 253] (there is too much work for too little gain).

Sec. 16.1 The notion of a design was defined by Pudlák in the email Prague–San Diego seminar that ran in the 1990s (see [52]). Buss [114] showed, using the duality of linear programming, that Lemma 16.1.1 is actually an equivalence. Lemma 16.1.3 is from Beame *et al.* [52] and the key technical lemma 16.1.4 is from Buss *et al.* [124, Lemma 4.3]. Our proof of Lemma 16.1.4 follows closely the original proof: we simplified it by the choice $m = 2$ and $q = 3$ that we made but otherwise kept its structure. Both these results and Theorem 16.1.2 hold for all pairs of m, q, where q is a prime not dividing (a possibly composite) m.

Grigoriev [206] gave an example of a constant-degree polynomial system requiring a linear-degree NS-refutation (in characteristics different from 2), using what he called *Boolean multiplicative Thue systems* and related it to Tseitin formulas.

Sec. 16.2 The proof of Theorem 16.2.1 follows Razborov's [436] original proof, as simplified in places by Impagliazzo, Pudlák and Sgall [246]: the definition of the pigeon dance and the basis Δ_t is due to Razborov [436]; the basis C_t was explicitly defined by Impagliazzo *et al.* [246]. The first part of Lemma 16.2.2 is from Razborov [436] (Claims 3.4 and 3.11 there, showing that Δ_t is a basis of $\hat{S}_t/V_t$), and the proof of that (and the second statement in the lemma) are from Impagliazzo, Pudlák and Sgall [246] (Proposition 3.8 and the proof of Theorem 3.9 there). We here kept the notation close to these two papers. The statement about NS- and PC-proofs from $\neg\mathrm{WPHP}_n^m$ in Lemma 16.2.3 is from Razborov [436, Remark 3.12].

Buss *et al.* [122] extended [206] to PC and gave $\Omega(n)$ degree lower bounds for PC, optimal degree lower bounds for the Tsetin formulas and for the modular counting principles and also a linear separation of the minimum degree of PC-proofs over fields of different prime characteristics. This was further strengthened by Ben-Sasson and Impagliazzo [69] (and subsequently by Alekhnovich *et al.* [12]), who proved PC lower bounds for random CNFs. These works utilized a Fourier basis and the representation of binomials by linear equations, and a translation of the *binomial calculus* (PC restricted to using binomials) into linear equational calculus (cf. Section 7.1). Alekhnovich and Razborov [15] substantially extended the method to the non-binomial case and formulated some general conditions implying degree lower bounds. Clote and Kranakis [145] gave an exposition of part of these developments.

A result separating the minimum degree of NS- and PC-proofs is already to be found in Clegg, Edmonds and Impagliazzo [142]; the following principle (called there the **house-sitting principle** but originally called the **iteration principle** in [138]) has degree-3 PC-proofs but requires degree-$\Omega(n)$ NS-proofs: if f: $[n] \to [n]$ then either $0 < f(0)$ fails or $\forall i \in [n]\ i < f(i) \to f(i) < f(f(i))$ fails.

Nordström and Mikša [355] considered a modification of the resolution width arguments (and expansion graphs) to PC (Section 13.3). For resolution these can be formulated using the following game-like set-up. For a suitable CNF-formula, partition the set of clauses into sets C_i and the set of variables into sets V_j, defining a bipartite graph G (the two parts being the sets C_i and the sets V_j, respectively) having an edge (C_i, V_j) if and only if C_i contains a variable from V_j. A width lower bound is then established if G has a suitable expansion property and, by showing that for any edge (C_i, V_j) in G, given any assignment to all variables (by an adversary), we may flip the assignment on V_j to satisfy the clauses in C_i. This itself does not imply PC degree lower bounds (for example, *onto*PHP_n^{n+1} is a suitable formula for the argument but has small degree PC proofs). Nordström and Mikša [355] proved that PC degree lower bounds do actually follow if we can switch the order of things: we first give a partial assignment to the variables in V_j; then the adversary extends this in any way to a global assignment, but the clauses in C_i should be still satisfied by it.

Sec. 16.3 The functions that we called *degree t SOS-designs* were called, in the base case of $\mathcal{F} = \mathcal{G} = \emptyset$ considered in Barak, Kelner and Steurer [46], **level** t **pseudo-expectation operators**, and their definition is attributed there to Barak *et al.* [45]. If $\mathcal{F} \neq \emptyset$, Barak, Kelner and Steurer talked about a pseudo-expectation operator *solving* $f = 0$ for $f \in \mathcal{F}$ if it satisfies what corresponds to our third condition. The terminology comes from the observation that if $\mathcal{D}$ is any distribution on $\mathbf{R}^n$ then the expectation

$$E\colon h(\overline{x}) \in \mathbf{R}[\overline{x}] \rightarrow \mathbf{E}_{\overline{a}\sim\mathcal{D}}[h(\overline{a})]$$

satisfies the SOS-design conditions 1, 2 and 4 (i.e. the SOS-design conditions for $\mathcal{F}$ and $\mathcal{G}$ empty). Just as the existence of degree-t designs is equivalent to the non-existence of degree-t NS-refutations, Lemma 16.3.1 has inverses. Barak and Steurer [47, Theorem 2.7] formulated it for the case $\mathcal{G} = \emptyset$ and Atserias and Ochremiak [41] did so for any $\mathcal{F}$ and $\mathcal{G}$, but in both cases under an additional technical condition (saying that the boundedness of $\sum_i x_i^2$ as represented by an equation $\sum_i x_i^2 + g = M$ for some sum-of-squares g and some $M \in \mathbf{R}^+$ can be derived in a particular way from the initial equations and inequalities). Barak, Kelner and Steurer [46] gave a number of results showing that pseudo-expectations have properties analogous to ordinary expectations.

Grigoriev and Vorobjov [210] proved an exponential (resp. a linear) degree lower bound for a so-called *telescopic system of equations* in the Positivstellensatz (SOS) proof system and for a system called there Positivstellensatz calculus, respectively (and they also derived size lower bounds). This hard example is not, however, Boolean and thus does not provide a propositional hard example. Grigoriev's theorem 16.3.3 was the first propositional hard example (he stated that $c \leq 6$ but it is known that $c = 3$ suffices; see Alon and Spencer [19]). Grigoriev's original proof built on [206] and on Buss *et al.* [122]. The lower bound was re-proved by Schoenebeck [461] and generalized by Itsykson and Kojevnikov [251]. The complementary theorem 16.3.4 is from Atserias [27], who generalized an earlier result by Grigoriev, Hirsch and Pasechnik [209]; we refer the reader for a proof to [27].

Grigoriev [208] gave a degree lower bound for Positivstellensatz calculus proofs of the knapsack formulas over the reals. The lower bound depends on the coefficients of the instance: sometimes it is a linear lower bound and sometimes there is a constant upper bound, implying that the Positivstellensatz calculus is stronger than

PC/**R** in the sense that the PC degree for all instances of the knapsack is linear; see Impagliazzo, Pudlák and Sgall [246].

Lauria and Nordström [337] constructed 4CNFs in n variables that have degree-$d(n)$ Positivstellensatz refutations but each Positivstellensatz refutation must contain at least $2^{\Omega(d(n))}$ monomials, for a wide range of $d(n)$, i.e. from a large enough constant to $n^{\Omega(1)}$. The method uses the idea of relativization (Section 13.2) to amplify (quite generally) a degree lower bound to a lower bound for the number of monomials.

Sec. 16.4 Grigoriev, Hirsch and Pasechnik [209] gave a survey of all conceivable variants, static as well as dynamic, of systems like CP and LS: LS^d (a degree-d version of CP in which $\mathrm{LS} = \mathrm{LS}^2$), and their extensions LS^d_+ (and other) analogous to LS_+ (Section 6.5). They also showed that extending these systems with a division rule allows one to simulate CP with p-size coefficients, and they proved lower bounds on LS rank and on the size of *static* and *tree-like* LS^d-proofs. Further they proved an exponential lower bound for tree-like LS_+ (static LS_+ in their terminology) proofs of the unsolvability of a system of 0/1 linear *inequalities* but not for a translation of a propositional formula. A number of open problems are mentioned in [209].

Razborov [443] studied the width of proofs in CP and LS and the associated algorithms (Section 6.6), where the *width* is defined as the number of variables involved in cuts specific to the proof system or the algorithm in question, and he proved lower bounds for this measure for $Count^2_n$, for Tseitin formulas and for random k-CNFs.

Regarding only CP, Buresh-Oppenheim *et al.* [105] introduced a game-theoretic method for rank lower bounds for CP and LS, and applied it to random k-CNFs and Tseitin's formulas. Atserias, Bonet and Levy [30] formulated a rank lower bound criterion for CP and the Chvátal rank. We shall treat the size of general CP-proofs in Chapter 18.

There are a number of open problems regarding algebro-geometric proof systems. Some are linked to problems about logical calculi (e.g. the problem of proving a lower bound for the number of steps in PC under various restrictions to degree or size is intriguing), and some have been solved only conditionally (e.g. a size lower bound for LS_+ that we shall discuss via interpolation in Chapter 18).

I think that another key step is to prove a lower bound for any proof system combining logical reasoning with algebraic reasoning when the *top system is logical*, for example, systems like R(LIN), R(PC_d) for small degrees d (even $d = 1$) and R(CP), LK(CP), etc. from Section 7.1. So far we have lower bounds for systems where the top reasoning is algebraic, such as $F^c_d(\mathrm{MOD}_p)$ (Section 7.7).

17

Feasible Interpolation: A Framework

We have seen already three instances of what we shall call the feasible interpolation property: Theorems 3.3.1 and 3.3.2 (for tree-like and dag-like cut-free LK) and the feasible interpolation theorem 5.6.1 (for resolution); more examples will be given in Chapter 18. This chapter is devoted to the presentation of key concepts and constructions underlying the feasible interpolation method. It offers a framework in which to approach lower bounds and provides model statements. But it will be often convenient to tweak it a bit.

I shall use the informal name *feasible interpolation* when describing various instances of the method, as that has become customary over the years. The original name was *effective interpolation*, which is a bit more flexible as the adjective *feasible* traditionally refers to polynomial (maybe randomized or non-uniform) algorithms only. For this reason, I shall reserve the formal term **feasible interpolation** for the specific case when an interpolant is computed by a Boolean circuit. We shall consider a number of generalizations which could be described, to use a paraphrase of a famous quote, as a *continuation of feasible interpolation by other means*: we will consider other computational models X that can be shown to compute an interpolant and this will be distinguished by saying that a proof system P has a **feasible interpolation by** X.

17.1 The Set-Up

The situation in which we shall work is that of Section 13.5. We start with two disjoint NP sets U and V. By Lemma 13.5.1 there are clauses

$$A_1(\overline{p},\overline{q}),\dots,A_m(\overline{p},\overline{q}),B_1(\overline{p},\overline{r}),\dots,B_\ell(\overline{p},\overline{r})$$

containing only literals corresponding to the displayed atoms, the disjoint tuples $\overline{p}$, $\overline{q}$ and $\overline{r}$ where

$$\overline{p}=p_1,\dots,p_n,\quad \overline{q}=q_1,\dots,q_s \quad \text{and} \quad \overline{r}=r_1,\dots,r_t,$$

with

$$m,\ell,s,t \le n^{O(1)}, \tag{17.1.1}$$

which are such that U_n and V_n are defined by CNFs as in (13.5.3) and (13.5.4):

$$U_n = \{\overline{a} \in \{0,1\}^n \mid \bigwedge_{i \le m} A_i(\overline{a}, \overline{q}) \in \text{SAT}\}$$

and

$$V_n := \{\overline{a} \in \{0,1\}^n \mid \bigwedge_{j \le \ell} B_j(\overline{a}, \overline{r}) \in \text{SAT}\}.$$

Using this set-up is more convenient than talking about implications,

$$\bigwedge_{i \le m} A_i \rightarrow \neg \bigwedge_{j \le \ell} B_j, \tag{17.1.2}$$

as most systems to which we shall apply feasible interpolation are naturally refutation systems and can represent clauses in a more or less direct way.

Because we shall always consider one n at a time, we do not include n in the notation for the clauses A_i, B_j or for the parameters m, ℓ, s, t, although they all depend on n. Regarding this dependence only that of the polynomial bound (17.1.1) on the parameters matters. That is, we could allow non-uniformity and consider U and V to be NP/*poly* sets and everything would work in the same way. But there do not seem to be natural examples to take advantage of this.

In the **monotone case** we shall assume that either U is closed upwards or V is closed downwards (often both) and that the atoms from $\overline{p}$ occur only positively in the A_i or only negatively in the B_j, respectively (cf. Lemma 13.5.1).

We shall refer to the situation just described as the **(monotone) feasible interpolation set-up**, abbreviated often to the **(monotone) FI set-up**.

Our primary mental image is that of the pair U, V and not of their definitions via the clauses A_i, B_j. That is, the constructions needed to certify a feasible interpolation in a particular case that we want to develop ought to work for all definitions and not just for some. This will give us room to choose particular definitions when proving the disjointness of some specific pairs, in order to rule out feasible interpolation and automatizability (Section 17.3). The opposite case, choosing definitions for which the disjointness is hard to prove, is not particularly interesting. After all, if we take $C(\overline{p})$, an unsatisfiable CNF that is hard for the proof system at hand, we can redefine any NP set W by adding $C(\overline{p})$ as a disjunct

$$\overline{p} \in W \vee C(\overline{p})$$

and putting it into a CNF.

With these modified definitions, there is no short proof that any pair of NP sets is disjoint in such a proof system as disjointness would imply that C is unsatisfiable, which is supposed to be hard to prove therein.

The key definition is as follows.

Definition 17.1.1 A proof system P has the **feasible interpolation** property (or we say that it **admits feasible interpolation**) if and only if there is a function f:

$\{0,1\}^* \to \{0,1\}^*$ such that whenever π is a P-refutation of the clauses $A_1, \ldots, A_m$, $B_1, \ldots B_\ell$ obeying the FI set-up then:

(i) $f(\pi)$ is an interpolation circuit $I(\overline{p})$ for the implication (17.1.2), i.e. both the formulas

$$\bigwedge_i A_i \to I \quad \text{and} \quad I \to \neg \bigwedge_j B_j \tag{17.1.3}$$

are tautologies;

(ii) f is polynomially bounded: $|f(\pi)| \leq |\pi|^{O(1)}$.

We say that P admits **monotone feasible interpolation** if and only if such an f exists satisfying items (i) and (ii) for clauses A_i, B_j satisfying the monotone FI set-up and, in addition, for these clauses it also holds that

(iii) $f(\pi)$ is a monotone circuit.

The function f will be sometimes referred to as a **feasible interpolation construction**.

Most constructions that we shall study provide an f satisfying more than the definition requires. The following additional properties occur most often:

- if π is tree-like then $f(\pi)$ is a formula;
- f is p-time computable;
- f produces P-proofs of both implications in (17.1.3);
- f has only linear growth: $|f(\pi|)| \leq O(|\pi|)$.

The feasible interpolation construction provided by the proof of the feasible interpolation theorem 5.6.1 (feasible interpolation for R) has all four of these additional properties (because Theorems 3.3.1 and 3.3.2 about LK$^-$, on which it is based, provide such an f).

The point of the feasible interpolation method is that, in principle, it reduces the task of proving a length-of-proof *lower* bound for a proof system P to the task of establishing an *upper* bound on the growth of a feasible interpolation construction function f and a *lower* bound for the circuits separating two NP sets. That is, in a sense it reduces proof complexity to circuit complexity, which appears to be more rudimentary. The following lemma follows immediately from the definition.

Lemma 17.1.2

(i) Assume that U and V are two disjoint NP*-sets such that the pairs U_n and V_n are inseparable by a circuit of size $\leq s(n)$, for all $n \geq 1$. Assume that P admits feasible interpolation.*

Then any P-refutation of $A_1, \ldots, A_m, B_1, \ldots, B_\ell$ must have size at least $s(n)^{\Omega(1)}$.

(ii) Assume that U and V are two disjoint NP*-sets with U closed upwards (or V downwards) such that the pairs U_n and V_n are inseparable by a monotone*

circuit of size $\leq s^+(n)$*, for all* $n \geq 1$*. Assume that P admits monotone feasible interpolation.*

Then any P-refutation of $A_1, \ldots, A_m, B_1, \ldots, B_\ell$ *must have size at least* $s^+(n)^{\Omega(1)}$.

Let us remark that if P does not admit feasible interpolation but there exists a function f satisfying the conditions put on feasible interpolation constructions except that it grows faster than polynomially, say $|f(\pi)| \leq g(|\pi|)$ for some non-decreasing function $g(w) \geq w$ on $\mathbf{N}$, then weaker but perhaps still non-trivial lower bounds $g^{(-1)}(s(n))$ and $g^{(-1)}(s^+(n))$, respectively, follow.

As only very weak lower bounds for general circuits are known, we can obtain at present via Lemma 17.1.2 only conditional lower bounds in the non-monotone case. The situation is much better for monotone circuits (e.g. Theorem 13.5.3) and thus we shall always attempt to establish monotone feasible interpolation, even at the cost of introducing a computational model stronger than monotone Boolean circuits.

For completeness of the discussion let us mention a simple general statement illustrating that conditional lower bounds can be obtained in the non-monotone case.

Lemma 17.1.3 *Assume that* NP $\not\subseteq$ P/*poly. Then no proof system admitting feasible interpolation is polynomially bounded.*

Proof Assume that a proof system P admits feasible interpolation but is also p-bounded. Then NP = coNP (Theorem 1.5.2) and hence NP = NP $\cap$ coNP. The hypothesis of the lemma then implies that NP $\cap$ coNP $\not\subseteq$ P/*poly*.

Take any $U \in$ NP $\cap$ coNP witnessing this non-inclusion and put $V := \{0,1\}^* \setminus U$. By the p-boundedness of P the disjointness of the NP pair U and V has a p-size P-proof and hence, by feasible interpolation, the sets U_n, V_n can be separated by p-size circuits. But that means that the circuits compute U_n exactly, contradicting the choice of U. □

We shall formulate now, as a specific consequence of feasible interpolation, a feasible version generalizing Beth's theorem 3.3.3.

Lemma 17.1.4 *Let U be an* NP *set and assume that clauses* A_i, $i \leq m$*, define* U_n *in the sense of the* FI *set-up,* $n \geq 1$*. Assume that a proof system P admits feasible interpolation.*

If π *is a P-refutation of*

$$A_1(\overline{p},\overline{q}), \ldots, A_m(\overline{p},\overline{q}), A_1(\overline{p},\overline{q}'), \ldots, A_m(\overline{p},\overline{q}'), q_1, \neg q_1'$$

then there is a circuit $C(\overline{p})$ *of size* $|\pi|^{O(1)}$ *computing* q_1*:*

$$\bigwedge_i A_i(\overline{p},\overline{q}) \;\rightarrow\; q_1 \equiv C(\overline{p}).$$

That is, if P gives a short proof that the first bit of a witness $\overline{q}$ *for the membership* $\overline{p} \in U_n$ *is unique (i.e. it is implicitly defined), then it is explicitly defined by a small circuit.*

Proof Just note that any interpolant of

$$(\bigwedge_i A_i(\overline{p},\overline{q}) \wedge q_1) \rightarrow (\neg \bigwedge_i A_i(\overline{p},\overline{q}') \vee q_1')$$

computes q_1. □

Let us note two simple consequences of this lemma. If we take for the clauses A_i the clauses of Def_C defining a circuit C with $\overline{p}$ representing an input and $\overline{q}$ and $\overline{q}'$ representing computations, respectively, then already resolution proves in $O(|C|)$ steps that the computation is unique (Lemma 12.3.1). By the feasible interpolation theorem 5.6.1, R admits feasible interpolation and hence each bit of $\overline{q}$, and the output bit in particular, can be computed by an interpolant. Hence, unless every circuit can be replaced by a formula of a polynomially related size, we cannot strengthen the requirement in the feasible interpolation property and demand that $f(\pi)$ is always a formula.

The second consequence is a variant of Lemma 17.1.3. The *unambiguous non-deterministic polynomial-time* class UP is the subclass of NP consisting of the languages accepted by a non-deterministic p-time machine such that on every accepted input there is exactly one accepting computation. Then it follows from Lemma 17.1.4 that UP $\not\subseteq$ P/*poly* implies that no proof system admitting feasible interpolation is p-bounded. This is weaker than Lemma 17.1.3, but the class UP contains some particular problems that have been extensively studied from the algorithmic point of view but for which no p-time algorithm, even probabilistic, has been found yet. The hypothesis that such a problem has no small circuits seems to have been well tested. The most prominent such problem is perhaps the **prime factorization** problem: given a natural number, find its prime factorization (this is turned into a decision problem by the usual maneuver: ask for the values of particular bits in some canonical form of the factorization). Note that the factorization is also in NP $\cap$ coNP (and in UP $\cap$ coUP). Using this problem we can formulate the implication in a more concrete way.

Corollary 17.1.5 *Assume that prime factorization cannot be computed by p-size circuits. Then no proof system that admits feasible interpolation is p-bounded.*

We shall formulate a more general statement of this sort for future reference.

Corollary 17.1.6 *Let h: $\{0,1\}^* \rightarrow \{0,1\}^*$ be an injective p-time computable function such that $|h(u)| \geq |u|^{\Omega(1)}$ for all $u \in \{0,1\}^*$. Assume that a proof system P admits feasible interpolation.*

Then, if π is a P-proof of the (propositional translation of the) fact that each $z \in \{0,1\}^n$ has at most one pre-image in h,

$$h(x) = z = h(y) \rightarrow x = y$$

(see below for a precise formulation), then the inverse function $h^{(-1)}$ can be computed on $\{0,1\}^n$ by a circuit of size $|\pi|^{O(1)}$.

Proof Let $U_n := rng(h)$ and let A_i be the clauses formalizing the natural definition of U_n,

$$\overline{z} \in U_n \equiv [\exists \overline{x} \in \{0,1\}^{n^{O(1)}} h(\overline{x}) = \overline{z}]$$

(the $n^{O(1)}$ upper bound on the length of $\overline{x}$ follows from our assumption about the growth of h). In the clauses the tuple of atoms $\overline{p}$ represents $\overline{z}$ and the tuple $\overline{q}$ represents the bits, say $q_1, \ldots, q_k$, of $\overline{x}$ and the bits of the computation of $h(\overline{x})$.

The injectivity is then expressed by the unsatisfiability of the set of clauses as in Lemma 17.1.4 but now for all q_i, for $i \leq k$:

$$A_1(\overline{p},\overline{q}), \ldots, A_m(\overline{p},\overline{q}), A_1(\overline{p},\overline{q}'), \ldots, A_m(\overline{p},\overline{q}'), q_i, \neg q_i'.$$

Lemma 17.1.4 implies that each q_i, $i \leq k$, is computed by a circuit of size $|\pi|^{O(1)}$ and so is $h^{(-1)}$ on U_n. □

17.2 Disjoint NP Pairs Hard to Separate

Before we start developing feasible interpolation constructions we shall discuss several examples of disjoint NP pairs that are hard to separate. Without such pairs the whole set-up would be meaningless. In the monotone case their hardness are proven theorems but in the non-monotone case they are merely well established conjectures.

Obviously, it is better to get unconditional lower bounds and thus our primary concern will always be monotone feasible interpolation. In fact, there is a simple lemma showing that it is also stronger than the general case.

Lemma 17.2.1 *Assume P is a proof system that p-simulates R* and admits monotone feasible interpolation. Then it also admits general feasible interpolation.*

Proof Assume that clauses A_i, B_j obey the FI set-up but do not satisfy the monotonicity condition. For each atom p_i introduce a new atom $\tilde{p}_i$ and transform the A-clauses as follows:

- replace each occurrence of the literal $\neg p_i$ by $\tilde{p}_i$;
- add to the list of clauses all clauses $\{\tilde{p}_i, p_i\}$, $i \in [n]$

and let $\mathcal{A}'$ be the set of these clauses.
Transform the B-clauses analogously but now using the negation of the new atom:

- replace each positive occurrence of p_i by $\neg\tilde{p}_i$;
- add to the list of clauses all clauses $\{\neg\tilde{p}_i, \neg p_i\}$, $i \in [n]$

and let $\mathcal{B}'$ be the set of these clauses. Note that after the transformation all the atoms p_i and $\tilde{p}_i$ occur only positively in $\mathcal{A}'$ and only negatively in $\mathcal{B}'$.

The clauses introduced in the second items of the two translations imply together that $\tilde{p}_i \equiv \neg p_i$. Using this we can derive in R* all the original clauses from the new clauses, in a size proportional to the total size of the clauses, and thus turn any P-refutation of the clauses A_i, B_j into a P-refutation of $\mathcal{A}' \cup \mathcal{B}'$.

The assumed monotone feasible interpolation then provides us with an interpolating circuit C with inputs p_i and $\tilde{p}_i$, $i \in [n]$, and, by substituting into it $\tilde{p}_i := \neg p_i$ we get a circuit interpolating the original set of clauses. □

We shall now present three examples of monotone NP pairs. The first is the most prominent and has been used already in Section 13.5. We repeat it here in order to have all examples in one place.

17.2.1 Clique-Coloring

For parameters $n \geq \omega > \xi \geq 3$, the pair is

$$U := \text{Clique}_{n,\omega} \quad \text{and} \quad V := \text{Color}_{n,\xi},$$

where U is the set of graphs on $[n]$ that have a clique of size ω and V is the set of graphs on $[n]$ that are ξ-colorable. As before, graphs are coded by strings of length $\binom{n}{2}$. The set U is closed upwards and the set V downwards, and the sets of clauses defining them and obeying the monotone FI set-up are:

- A-clauses with atoms p_{ij}, $\{i,j\} \in \binom{n}{2}$ and ωn atoms q_{ui}, $u \in [\omega]$ and $i \in [n]$,
  - $\bigvee_{i \in [n]} q_{ui}$, for all $u \leq \omega$,
  - $\neg q_{ui} \vee \neg q_{uj}$, for all $u \in [\omega]$ and $i \neq j \in [n]$,
  - $\neg q_{ui} \vee \neg q_{vi}$, for all $u \neq v \in [\omega]$ and $i \in [n]$,
  - $\neg q_{ui} \vee \neg q_{vj} \vee p_{ij}$, for all $u \neq v \in [\omega]$ and $\{i,j\} \in \binom{n}{2}$;
- B-clauses in the atoms $\overline{p}$ and $n\xi$ atoms r_{ia}, $i \in [n]$ and $a \in [\xi]$,
  - $\bigvee_{a \in [\xi]} r_{ia}$, for all $i \in [n]$,
  - $\neg r_{ia} \vee \neg r_{ib}$, for all $a \neq b \in [\xi]$ and $i \in [n]$,
  - $\neg r_{ia} \vee \neg r_{ja} \vee \neg p_{ij}$, for all $a \in [\xi]$ and $\{i,j\} \in \binom{n}{2}$.

The hardness of this pair is provided by Theorem 13.5.3.

17.2.2 Broken Mosquito Screen (BMS)

This is close to the clique-coloring example but is more symmetric, and some researchers consider this to be an important feature. Let $m = n^2 - 2$ for $n \geq 3$. The set BMS_n^+ consists of graphs on $[m]$ that can be partitioned into n blocks having size n, except for one block which has size $n-2$, such that each block is a clique. Dually, the set BMS_n^- is the set of graphs on $[m]$ that can be also partitioned into n blocks having size n, except for one block which has size $n - 2$, but now we require that each block is an independent set.

The sets BMS_n^+ and BMS_n^- are closed upwards and downwards, respectively, and it is easy to write down clauses defining them and obeying the monotone FI set-up, in a manner analogous to the clique and coloring clauses above. The hardness of this pair is similar to the hardness of the clique-coloring pair:

Theorem 17.2.2 (Haken [217]) *For $3 \leq n$, every monotone Boolean circuit separating BMS_n^+ from BMS_n^- must have size at least $2^{n^{\Omega(1)}}$.*

17.2.3 Bipartite Matching Problem (Hall's Theorem)

This pair consists of the set BMP_n^+ of bipartite graphs on $[n] \times [n]$ that contain a perfect matching i.e. the graph of a bijection between the two copies of $[n]$, and the set BMP_n^- of those who do not. Clearly BMP_n^+ and BMP_n^- are closed upwards and downwards, respectively, and BMP_n^+ is an NP set. Hall's theorem implies that BMP_n^- is an NP set as well: a bipartite graph with edges E is in BMP_n^- if and only if there is a subset $X \subseteq [n]$ that has fewer than $|X|$ E-neighbors (the set of $j \in [n]$ such that $E(i,j)$ for some $i \in X$). We shall formulate this pair using linear inequalities instead of clauses, as we shall use it in connection with geometric proof systems.

Let us denote by I and J the two copies of $[n]$ forming vertex sets, so the graphs G under consideration are given by $E \subseteq I \times J$. As we shall be discussing integer linear inequalities, we shall use variables x, y, z in place of atoms p, q, r.

Let y^I_{ai} and y^J_{aj}, $a \in [n]$, $i \in I, j \in J$, be $2n^2$ variables. Define the following inequalities:

1. $\sum_i y^I_{ai} \geq 1$, for all $a \in [n]$;
2. $1 - y^I_{ai} + 1 - y^I_{a'i} \geq 1$, for all $a \neq a' \in [n]$;
3. $\sum_j y^J_{aj} \geq 1$, for all $a \in [n]$;
4. $1 - y^J_{aj} + 1 - y^J_{a'j} \geq 1$, for all $a \neq a' \in [n]$;
5. $1 - y^I_{ai} + 1 - y^J_{aj} + x_{ij} \geq 1$, for all $a \in [n]$, $i \in I$ and $j \in J$.

The inequalities 1 and 2 imply that y^I_{ai} determines a bijection $f\colon [n] \to I$, and similarly 3 and 4 imply that y^J_{aj} determine a bijection $g\colon [n] \to J$. Conditions 5 imply that the edges $\{(f(a), g(a)) \in I \times J \mid a \in [n]\}$ form a perfect matching in a graph G. Let E_i comprise these $2n + 2\binom{n}{2} + n^3$ inequalities. Clearly the set

$$U := \{\overline{a} \in \{0,1\}^{n^2} \mid \exists \overline{y}^I, \overline{y}^J \bigwedge_i E_i(\overline{a}, \overline{y}^I, \overline{y}^J)\}$$

is the set BPM_n^+.

The set BPM_n^- can be defined as follows. We shall describe the inequalities (there are $O(n^4)$) but not write them down. They use tuples of variables x, z^X, z^I, z^J. The x-variables are those for the unknown E. There are $n-1$ z^X-variables, which determine a subset $X \subseteq I$ that contains n (and hence is non-empty). The inequalities formalize that the z^I- and z^J-variables encode bijections $f\colon X \cap [n] \to I$ and $g\colon X \cap [n-1] \to J$ (or dually $f\colon X \cap [n] \to J$ and $g\colon X \cap [n-1] \to I$), and that all neighbors of vertices in $rng(f)$ are in $rng(g)$. We shall denote the set of all inequalities defining the two sets by $Hall_n$. It has no $0/1$ solution.

The pair of sets BPM_n^+ and BPM_n^- is hard for monotone *formulas* only, and we shall utilize it for tree-like proofs.

Theorem 17.2.3 (Raz and Wigderson [429]) *Every monotone Boolean circuit separating BMP_n^+ from BMP_n^- must have depth at least $\Omega(n)$ and every monotone formula separating the two sets must have size at least $2^{\Omega(n)}$.*

The link between circuit depth and formula size is provided by Spira's lemma (see Lemmas 1.1.4 and 1.1.5).

Let us summarize, following part (ii) in Lemma 17.1.2.

Lemma 17.2.4 *Assume that P admits monotone feasible interpolation. Then P-proofs of the disjointness of the clique-coloring pair and of the broken mosquito pair must have size at least $2^{n^{\Omega(1)}}$.*

Assume that P admits monotone feasible interpolation witnessed by a feasible interpolation construction which produces interpolating formulas (this will often be the case when P allows only tree-like proofs). Then P-proofs of the disjointness of the bipartite matching problem must have size at least $2^{n^{\Omega(1)}}$.

Now we turn to the non-monotone case. Before we resort to computational complexity conjectures let us see that logic provides an example pair as well.

17.2.4 Reflection (a.k.a. Canonical) Pair

We have already proved a result that can be interpreted as the hardness of proving the disjointness of two NP sets. In particular, we showed in Chapter 8 (see Section 8.6) that if P proves the reflection principle $Ref_{Q,2}$ formalizing that any Q-provable DNF is satisfiable then P p-simulates Q (with respect to proofs of DNFs). Recall that the reflection principle as formalized in (8.4.4),

$$Prf_Q^0(x,y,z,P,F,G) \rightarrow Sat^0(x,y,E,F,H),$$

says that if a size-y formula F in x atoms has a size-z proof P (as witnessed by an additional string G) then any assignment E to the x atoms satisfies F (as witnessed by an additional string H). If we denote by $U_Q(n,m)$ the set of formulas of size at most n that have a Q-proof of size $\leq m$ and take $\text{FALSI}_n := \text{FALSI} \cap \{0,1\}^n$, where FALSI is the NP set of falsifiable formulas α, then this is equivalent to

$$U_Q(n,m) \cap \text{FALSI}_n \ = \ \emptyset$$

for $x \leq y := n$ and $z := m$. To replace the collection of $U_Q(n,m)$ by one U_n, $m \geq 1$, we may choose $m := 10n$. This will work for most proof systems Q: if α has a size m proof then $\alpha' := \alpha \vee 0 \vee \cdots \vee 0$, padded to length m, will have – under some mild conditions on Q – a Q-proof of size $\leq 10m$. Thus $U_Q(n,m) \cap \text{FALSI}_n \neq \emptyset$ implies that $U_Q(m,10m) \cap \text{FALSI}_m \neq \emptyset$ for such Q. The set $\bigcup_m U_Q(m,10m)$ is an NP set.

It is, however, more elegant to change the format of the two NP sets and think of the succedent of Ref_Q as also depending on z; we define two NP sets

$$\text{PROV}(Q) \ := \ \{(\alpha, 1^{(m)}) \mid \alpha \ \text{ has a size} \leq m \ Q\text{-proof}\}$$

and

$$\text{FALSI}^* \ := \ \text{FALSI} \times \{1^{(m)} \mid m \geq 1\}.$$

The soundness of Q is then equivalent to the disjointness of PROV(Q) and FALSI*. In particular, refuting the clauses corresponding to $\text{PROV}(Q) \cap \{0,1\}^k$ and $\text{FALSI}^* \cap$

$\{0,1\}^k$ means proving $\langle Ref_Q \rangle_{n,m}$ for $n+m \le k$. Thus, if we take for Q a proof system that cannot be simulated by P, these refutations cannot be of polynomial size, and if the speed-up of Q-proofs over P-proofs is actually exponential then the refutations cannot have a subexponential size.

This is not a satisfactory argument, as one length-of-proofs lower bound is derived from another (the non-existence of a simulation). In Chapter 21 we shall return to these pairs in connections with the optimality problem 1.5.5. Now we derive another statement showing that one can always establish feasible interpolation relative to a presumably not too strong oracle.

Lemma 17.2.5 *Assume that P simulates ER. Let U and V be disjoint* NP *sets. Assume that U_n and V_n are defined by clauses A_i, B_j obeying the* FI *set-up, for $n \ge 1$, and that P admits p-size proofs of the disjointness of U_n and V_n.*

Then there is a p-time function f such that, for all $n \ge 1$ and $x \in \{0,1\}^n$,

$$x \in U_n \Rightarrow f(x) \in \text{PROV}(P) \quad \textit{and} \quad x \in V_n \Rightarrow f(x) \in \text{FALSI}^*.$$

That is, composing f with any algorithm that separates PROV(P) *from* FALSI* *will separate U from V.*

Proof Given two disjoint sets U, V, let g be some canonical p-time reduction of V to FALSI such that S^1_2 proves that g is a p-reduction of V to FALSI, and hence ER admits p-size proofs of these facts for all input lengths (Section 12.4).

It holds that $g(U) \subseteq \text{TAUT}(= \{0,1\}^* \setminus \text{FALSI})$ and we can define a p-time function Q by

$$Q(w) := \begin{cases} g(u) & \text{if } w = \langle u, v \rangle \wedge \bigwedge_i A_i(u,v), \\ 1 & \text{otherwise.} \end{cases}$$

The function Q is a possibly incomplete proof system and the disjointness of U and V is equivalent to its soundness. Thus if P admits p-size proofs of the disjointness of U_n and V_n, it follows that P p-simulates Q. Hence we can define the required function f by

$$x \rightarrow (g(x), 1^{(m)}),$$

where $m = |x|^{O(1)}$ is a parameter stemming from the size increase in the p-simulation of Q by P. □

Let us note that in Lemma 17.2.5 it suffices to assume that P contains R, as some canonical p-reduction of an NP set to FALSI can be defined even in the theory $T^1_1(\alpha)$ (Section 10.5).

The next disjoint NP pair relates to Corollary 17.1.6 and generalizes Corollary 17.1.5. It is based on the notions of one-way functions and permutations that play a prominent role in theoretical cryptography.

A function f: $\{0,1\}^* \rightarrow \{0,1\}^*$ is **one-way** (OWF) if and only if

1. f is computable in p-time,

2. for any probabilistic p-time algorithm $\mathcal{A}$, we have

$$Prob_{w\in\{0,1\}^n}[f(\mathcal{A}(f(w), 1^{(n)})) = w] \leq n^{-\omega(1)},$$

where the distribution on $\{0,1\}^n$ is uniform.

The second item means that any such algorithm $\mathcal{A}$ succeeds in finding at least one pre-image of $f(w)$ with **negligible probability**, i.e. with probability less than n^{-c}, for all $c \geq 1$ and $n \gg 1$. We need to reformulate item 2 as follows:

2′. If $S(n)$ is the minimum $s \geq 1$ such that there is a size $\leq s$ circuit C such that

$$Prob_{w\in\{0,1\}^n}[f(C(w, 1^{(n)})) = w] \geq 1/s$$

then $S(n) = n^{\omega(1)}$, i.e. $S(n) \geq n^c$, for all $c \geq 1$ and $n \gg 1$.

This stronger requirement is also widely used in theoretical cryptography and computational complexity, and functions obeying it are called **strong one-way** functions or, perhaps more descriptively, one-way functions super-polynomially hard with respect to circuits.

A (strong) one-way function f is a (strong) **one-way permutation** (OWP) if and only if it also satisfies:

3. f is a bijection, and
4. it is **length-preserving**: $|f(w)| = |w|$ for all $w \in \{0,1\}^*$.

Such an f permutes each $\{0,1\}^n$. When it is hard to invert a permutation it must be hard to find the ith input bit, with i an additional input for the algorithm $\mathcal{A}$ above. But for some specific i this may be easy (e.g. if we factor an odd number then all prime factors will be odd). One specific hard bit is provided by the following concept.

A function B: $\{0,1\}^* \to \{0,1\}$ is a **hard bit predicate** (or just a hard bit) for a permutation h if and only if, for any algorithm $\mathcal{A}$ computed by a p-size circuit, it holds that

$$Prob_{w\in\{0,1\}^n}[\mathcal{A}(h(w), 1^{(n)}) = B(w)] \leq n^{-\omega(1)}. \tag{17.2.1}$$

By the Goldreich–Levin theorem [204], if a strong OWP exists then there is also a strong OWP that has a hard bit.

17.2.5 Hard Bit of OWP Pair

Let h be a strong OWP with a hard bit predicate B. Define the pair

$$U_{h,B} := \{w \in \{0,1\}^* \mid \exists u \in \{0,1\}^* \; B(u) = 1 \wedge h(u) = w\}$$

and

$$V_{h,B} := \{w \in \{0,1\}^* \mid \exists u \in \{0,1\}^* \; B(u) = 0 \wedge h(u) = w\}.$$

As $V_{h,B} = \{0,1\}^* \setminus U$, separating $U_{h,B}$ from $V_{h,B}$ is the same as computing the predicate $B(h^{(-1)}(w))$, which is assumed to be hard.

Lemma 17.2.6 *Assume that P admits feasible interpolation. Let h be an OWP with a hard bit predicate B. Then the (propositional translation of the) disjointness of the pair* $U_{h,B}, V_{h,B}$ *does not have p-size proofs in P.*

For lower bounds it seems important to have specific examples of supposedly hard formulas. Hence we want a pair based on a specific OWP h.

17.2.6 RSA Pair

The RSA is the well-known public key cryptosystem of Rivest, Shamir and Adleman [450]. We take $N := pq$, where p, q are two sufficiently large primes. We use an upper-case N as we want to reserve lower-case n for $n := \lceil \log N \rceil$, the bit length of N. A number $u < N$ (think of a string $u \in \{0,1\}^n$) is encoded using a public key, i.e. a pair (N, e) with $1 < e < N$, by the number

$$v := u^e \pmod N.$$

The decoding of u from v is done using a secret key, which is a number d such that

$$e \cdot d \equiv 1 \pmod{\varphi(N)},$$

where $\varphi(N)$ is the Euler function (in our case $\varphi(N) = (p-1)(q-1)$), which computes

$$u := v^d \pmod N.$$

We can encode a single bit via the RSA by encoding a random even (resp. odd) number $u < N$ if the bit is 0 (resp. 1). The security of this probabilistic encryption is known to be as good as that of RSA; see Section 17.9. We use this to define the disjoint pair of NP sets

$$\begin{aligned} RSA_i := \{(N, e, v) \mid \exists u, d, r < N\, (u \equiv i \pmod 2)) \wedge (u^e \equiv v \pmod N)) \\ \wedge (v^d \equiv u \pmod N)) \wedge (v^r \equiv 1 \pmod N)) \wedge (e, r) = 1\} \end{aligned}$$

for $i = 0, 1$. It may seem that we have included redundant information in the definition but in fact this information will help us to prove the disjointness of the two sets efficiently in Section 18.7.

The generally accepted hypothesis that the RSA is a secure method of encryption against adversaries formalized by non-uniform p-time functions (i.e. in the class $\mathcal{P}/poly$) implies that this pair cannot be separated by a set in P/*poly*; see Alexi *et al.* [17].

17.3 Win–Win Situation: Non-Automatizability

If we establish feasible interpolation for a proof system, we can rightly think that progress has been made: we have at least a conditional lower bound (Lemma 17.1.3). And if it is monotone feasible interpolation, we can celebrate: we have an unconditional lower bound as in the feasible interpolation theorem 5.6.1 for R. But what if we have established neither? Thanks to an elegant notion of automatizability introduced by Bonet, Pitassi and Raz [97] (the notion is attributed to Impagliazzo there), this has an interesting consequence as well (Corollary 17.3.2).

Recall from Section 1.5 that for a proof system P and a formula A, the symbol $\mathbf{s}_P(A)$ denotes the minimum size of a P-proof of A, if it exists. We shall denote analogously by $\mathbf{s}_P(\mathcal{C})$ the minimum size of a P-refutation of a CNF $\mathcal{C}$. A proof system P is **automatizable** if and only if there exists a deterministic algorithm $\mathcal{A}$ which, when given as an input an unsatisfiable CNF $\mathcal{C}$, finds a P-refutation in a time polynomial in $\mathbf{s}_P(\mathcal{C})$.

The connection to feasible interpolation is simple but important.

Lemma 17.3.1 *Assume that P is a proof system having the following property: whenever a set of clauses $\mathcal{C}$ has a P-refutation of size s then any set of clauses $\mathcal{D}$ resulting from $\mathcal{C}$ by substituting for some atoms in $\mathcal{C}$ any truth values has a P-refutation of size $O(s)$.*

Then, if P is automatizable, it also admits feasible interpolation.

Proof Let A_i, B_j be clauses obeying the FI set-up and let π be a P-refutation of the set of these clauses.

Assume $\mathcal{A}$ is an algorithm witnessing the automatizability of P and assume that it succeeds in finding a refutation of any unsatisfiable $\mathcal{C}$ in time $\leq \mathbf{s}_P(\mathcal{C})^c$, for some fixed $c \geq 1$. We shall define a separating algorithm $\mathcal{B}$ for U and V that will use π and $\mathcal{A}$, as follows:

1. given $\overline{a} \in \{0,1\}^n$, run $\mathcal{A}$ on the set $\mathcal{D}$ of all clauses $A_i(\overline{a}, \overline{q})$, $i \leq m$, for time $|\pi|^{c+1}$;
2. if $\mathcal{A}$ outputs a refutation, we know that $\overline{a} \notin U$ and $\mathcal{B}$ outputs 0;
3. otherwise $\mathcal{B}$ outputs 1.

If it were the case that $\overline{a} \in V$ as witnessed by an assignment $\overline{r} := \overline{c} \in \{0,1\}^*$ then, by the condition about substitutions that we imposed on P, the set of clauses

$$\{A_i(\overline{a}, \overline{q})\}_i \cup \{B_j(\overline{a}, \overline{c})\}_j$$

has a size $O(|\pi|)$ P-refutation, but at the same time this follows from the set of clauses $A_i(\overline{a}, \overline{q})$ (all $B_j(\overline{a}, \overline{c})$ contain 1) in size $O(|\pi|)$. Hence $\mathcal{A}$ would find some refutation of clauses $A_i(\overline{a}, \overline{q})$ in time $(O(|\pi|))^c \leq |\pi|^{c+1}$ for $|\pi| \gg 0$.

This shows that $\mathcal{B}$ is sound. □

Corollary 17.3.2 *Assume that P is a proof system having the following property: whenever a set of clauses $\mathcal{C}$ has a P-refutation of size s then any set of clauses $\mathcal{D}$ resulting from $\mathcal{C}$ by substituting for some atoms in $\mathcal{C}$ any truth values has a P-refutation of size $O(s)$.*

Assume further that h is a one-way permutation and that P proves its injectivity on $\{0,1\}^n$, $n \geq 1$, by proofs of size $n^{O(1)}$. Then P is not automatizable.

Proof Under the hypothesis P does not admit feasible interpolation, by Lemma 17.2.6. Hence it is not automatizable, by Lemma 17.3.1. □

17.4 Communication Protocols

A **multi-function** defined on $U \times V$, $U, V \subseteq \{0,1\}^n$, with values in $I \neq \emptyset$ is a ternary relation $R \subseteq U \times V \times I$ such that

$$\forall \overline{u} \in U, \overline{v} \in V \exists i \in I\, R(\overline{u}, \overline{v}, i) .$$

The prefix *multi* refers to the fact that there does not need to be a unique valid value i. The task of finding, given $\overline{u}$ and $\overline{v}$, at least some valid value i leads to a two-player **Karchmer–Wigderson game**. The two players will be called the **U-player** and the **V-player**; the U-player receives $\overline{u} \in U$ while the V-player receives $\overline{v} \in V$. They attempt to find $i \in I$ such that $\mathrm{R}(\overline{u}, \overline{v}, i)$ and to do so they exchange bits of information. This is done according to a KW-**protocol**, a finite binary tree T such that

- each non-leaf is labeled by U or V and the two edges leaving such a node are labeled by 0 and 1,
- each leaf is labeled by an element of I

and there are strategies S_U and S_V for the players, i.e. functions from $U \times T_0$ (resp. $V \times T_0$) into $\{0,1\}$, where T_0 are the non-leaves of T.

Having the protocol, the players start at the root of T. The label of a non-leaf $x \in T_0$ tells them who should send a bit, and if it is the U-player she sends $S_U(\overline{u}, x)$ (similarly if it is the V-player he sends $S_V(\overline{v}, x)$), determining the edge out of x and hence the next node $y \in T$. If y is non-leaf, this procedure is repeated. If it is a leaf, its label i is the output of the play on the pair $(\overline{u}, \overline{v})$. It is required that the label is always a valid value for the multi-function, i.e. that $R(\overline{u}, \overline{v}, i)$ holds. The **communication complexity** of R, denoted by $CC(R)$, is the minimum height of the protocol tree taken over all protocols (T, S_U, S_V) that compute R. At present we are not interested in the size of the KW-protocol trees but later on we shall consider more general protocols (both tree-like and dag-like), and their size will be important.

A multi-function of special interest for us is the **Karchmer–Wigderson multi-function**. It is defined on any disjoint pair $U, V \subseteq \{0,1\}^n$ with values in $[n]$, and a valid value for $(\overline{u}, \overline{v}) \in U \times V$ is $i \in [n]$ such that $u_i \neq v_i$. This multi-function will be denoted by $KW[U, V]$. It also has a monotone version. For that version we assume that either U is closed upwards or V downwards and that a valid value is any $i \in [n]$ such that $u_i = 1 \wedge v_i = 0$. We shall denote the monotone multi-function by $KW^m[U, V]$.

The communication complexity of these two functions has a clear logical meaning.

Theorem 17.4.1 (Karchmer and Wigderson [264]) *Let $U, V \subseteq \{0,1\}^n$ be two disjoint sets. Then $CC(KW[U,V])$ equals the minimum logical depth of a DeMorgan formula $\varphi(\overline{p})$ in the negation normal form (Section 1.1), where we count the logical depth of a literal as 0; the formula $\varphi(\overline{p})$ is identically 1 on U and 0 on V, respectively (i.e. φ separates U from V).*

If, in addition, either U is closed upwards or V downwards then $CC(KW^m[U,V])$ equals the minimum logical depth of a DeMorgan formula $\varphi(\overline{p})$ without negations that separates U from V.

Proof Assume that a separating formula φ in negation normal form exists. The players will use it as a KW communication protocol as follows. They start at the top connective and walk down to smaller subformulas until they reach a literal, maintaining the property that the current subformula gives the value 1 on $\overline{u} \in U$ and the value 0 on $\overline{v} \in V$.

This is true at the start by the hypothesis that the formula separates U from V. In order to decide which subformula to continue, if the current connective is a disjunction then the U-player indicates by one bit whether the left or the right subformula is true and if it is a conjunction then the V-player indicates which one is false. The literal at which they arrive at is a valid value for $KW[U,V]$, and also for $KW^m[U,V]$ if there is no negation in φ.

For the opposite direction, construct a suitable φ by induction on the communication complexity of $KW[U,V]$ or of $KW^m[U,V]$, respectively. □

We want to construct protocols from proofs and for that we need to generalize the notion of KW-protocols. Let $U, V \subseteq \{0,1\}^n$ be two disjoint sets and, as before, we aim at computing either $KW[U,V]$ or $KW^m[U,V]$. A **protocol** for the Karchmer–Wigderson multi-function is a labeled directed graph G satisfying the following conditions.

1. G is acyclic and has one **source** (the in-degree-0 node) denoted by $\emptyset$ and one or more **leaves** (the out-degree-0 nodes). All other nodes are **inner nodes**.
2. The leaves are labeled by a function lab by one of the following formulas:

$$u_i = 1 \wedge v_i = 0 \quad \text{or} \quad u_i = 0 \wedge v_i = 1$$

   for some $i \in [n]$.
3. There is a function $S(\overline{u}, \overline{v}, x)$ (the **strategy**) such that S assigns to any pair $(\overline{u}, \overline{v}) \in U \times V$ and any inner node x a node $S(\overline{u}, \overline{v}, x)$ that can be reached from x by an edge of G.
4. There is a collection F of sets $F(\overline{u}, \overline{v}) \subseteq G$, for $(\overline{u}, \overline{v}) \times \in U \times V$, called the **consistency condition**, such that the following hold:
   (a) $\emptyset \in F(\overline{u}, \overline{v})$;
   (b) if x is an inner node then $x \in F(\overline{u}, \overline{v}) \rightarrow S(\overline{u}, \overline{v}, x) \in F(\overline{u}, \overline{v})$;
   (c) the label lab(y) of any leaf y from $F(\overline{u}, \overline{v})$ is valid for the pair $(\overline{u}, \overline{v})$.

The protocol (G, lab, S, F) is **monotone** if and only if every leaf in it is labeled by one of the formulas $u_i = 1 \wedge v_i = 0$, $i \in [n]$.

The complexity of the protocol is measured by two parameters: the **size**, which is the number of nodes in G, and its **communication complexity**, which is the minimum number t such that for every inner node $x \in G$, the communication complexities of the predicate $x \in F(\overline{u}, \overline{v})$ and of the multi-function $S(\overline{u}, \overline{v}, x)$ are at most t.

An intuitive but incorrect mental picture of the correctness of a protocol is that when the players start at the root and follow the strategy function, sooner or later they reach a leaf with a valid label. But that is not enough: It is easy to define a linear-size would-be protocol of communication complexity 2 computing $KW[U, V]$: the players first compare their first bits, if they are equal their second bits, etc. The correctness of the proper protocols is guaranteed by the conditions on G, lab, S and F and, in particular, that the players would solve the task if they started their walk at *any* node from $F(\overline{u}, \overline{v})$.

We say that the protocol is **tree-like** if G is a tree. If the adjective *tree-like* is not used it means that we have in mind general, dag-like, protocols. However, tree-like protocols (and proofs, as we shall see later) are not very interesting, owing to the next observation.

Lemma 17.4.2 *Assume that U, V is a disjoint pair of subsets of $\{0, 1\}^n$ and (G, lab, S, F) is a tree-like protocol for $KW[U, V]$ of size s and communication complexity t. Then there is a KW-protocol of height $O(t \log s)$, i.e. $CC(KW[U, V]) \leq O(t \log s)$.*

The same holds in the monotone case.

Proof The players use Spira's lemma 1.1.4 to find an inner node x in G that splits the tree in 2/3–1/3 fashion. Then they decide whether $x \in F(\overline{u}, \overline{v})$ (using $\leq t$ bits). If $x \in F(\overline{u}, \overline{v})$ they concentrate on the subtree of G with root x. If $x \notin F(\overline{u}, \overline{v})$ they concentrate on its complement. It takes $O(\log s)$ rounds, exchanging $\leq t$ bits in each round, to find a leaf that is in $F(\overline{u}, \overline{v})$ and hence is labeled with a valid answer. □

Assume that we have a circuit $C(\overline{p})$ separating U from V. The construction of the KW-protocol in the proof of Theorem 17.4.1 works here as well but it produces a more general protocol: the graph G is the underlying graph of C whose leaves and their labels correspond to the inputs of C, the strategy function is the same as before and the consistency condition $F(\overline{u}, \overline{v})$ contains subcircuits which give different values for $\overline{u}$ and $\overline{v}$ (or give 1 for $\overline{u}$ and 0 for $\overline{v}$ in the monotone case). Hence the size of such a protocol is $|C|$ and its communication complexity is 2.

Crucially, the next statement shows that the opposite construction can also be simulated.

Theorem 17.4.3 (essentially Razborov [435]) *Let $U, V \subseteq \{0, 1\}^n$ be two disjoint sets and let (G, lab, S, F) be a protocol for $KW[U, V]$ which has size s and communication complexity t.*

Then there is a circuit $C(\overline{p})$ of size $s2^{O(t)}$ separating U from V. Moreover, if G is monotone then so is C.

Proof Let (G, lab, S, F) be a protocol satisfying the hypothesis. For a node a and $w \in \{0,1\}^t$, let $R_{a,w}$ be the set of pairs $(\overline{u}, \overline{v}) \in U \times V$ such that the communication of the players deciding $a \in_? F(u, v)$ evolves according to w and ends with the affirmation of the membership. It is easy to see that $R_{a,w}$ is a rectangle, i.e. it is of the form $R_{a,w} = U_{a,w} \times V_{a,w}$ for some $U_{a,w} \subseteq U$ and $V_{a,w} \subseteq V$.

For a node a denote by k_a the number of nodes in G that can be reached from a by a (directed) path. So $k_a = 1$ for a a leaf, while $k_\emptyset = s$ for the source $\emptyset$.

Claim 1 For all $a \in G$ and $w \in \{0,1\}^t$, there is a circuit $C_{a,w}$ separating $U_{a,w}$ from $V_{a,w}$ and of size $\leq k_a 2^{O(t)}$.

(The constant in the power $O(t)$ is independent of a.) This implies the theorem by taking for a the source (which is in all $F(\overline{u}, \overline{v})$).

The claim is proved by induction on k_a. If a is a leaf then the statement is clear. So, assume that a is not a leaf and let $w \in \{0,1\}^t$. For $\overline{u} \in U_{a,w}$ let $\overline{u}^* \in \{0,1\}^{4^t}$ be a vector whose bits $\overline{u}^*_\omega$ are parameterized by $\omega = (\omega_1, \omega_2) \in \{0,1\}^t \times \{0,1\}^t$ and which is such that $\overline{u}^*_\omega = 1$ if and only if there is a $\overline{v} \in V_{a,w}$ for which the communication of the players computing $S(\overline{u}, \overline{v}, a)$ evolves according to ω_1 and the computation of $S(\overline{u}, \overline{v}, a) \in_? F(\overline{u}, \overline{v})$ evolves according to ω_2. Define $\overline{v}^*_\omega \in \{0,1\}^{4^t}$ dually: $\overline{v}^*_\omega = 0$ if and only if there is a $\overline{u} \in U_{a,w}$ such that the communication of the players computing $S(\overline{u}, \overline{v}, a)$ evolves according to ω_1 and the computation of $S(\overline{u}, \overline{v}, a) \in_? F(\overline{u}, \overline{v})$ evolves according to ω_2.

Let $U^*_{a,w}$ and $V^*_{a,w}$ be the sets of all these $\overline{u}^*$ and $\overline{v}^*$ respectively.

Claim 2 There is a monotone formula $\varphi_{a,w}$ (in 4^t atoms) separating $U^*_{a,w}$ from $V^*_{a,w}$ of size $2^{O(t)}$.

Claim 2 follows from Theorem 17.4.1 as there is an obvious way in which the players can find a bit ω for which $\overline{u}^*_\omega = 1$ and $\overline{v}^*_\omega = 0$: they simply compute $S(\overline{u}, \overline{v}, a)$ (this gives them ω_1) and then decide $S(\overline{u}, \overline{v}, a) \in_? F(\overline{u}, \overline{v})$ (this gives them ω_2).

Let us resume the proof of Claim 1. For $\omega_1 \in \{0,1\}^t$ let a_{ω_1} be the node $S(\overline{u}, \overline{v}, a)$ computed for some $\overline{u}, \overline{v}$ with communication ω_1. Define a circuit

$$C_{a,w} := \varphi_{a,w}(\ldots, y_{\omega_1,\omega_2}/C_{a_{\omega_1},\omega_2}, \ldots);$$

that is, we substitute the circuit $C_{a_{\omega_1},\omega_2}$ in the position of the (ω_1, ω_2)th variable in $\varphi_{a,w}$.

As $k_{a_{\omega_1}} < k_a$, the induction hypothesis implies that all $C_{a_{\omega_1},\omega_2}$ work correctly on all $U_{a_{\omega_1},\omega_2} \times V_{a_{\omega_1},\omega_2}$. Then the circuit $C_{a,w}$ works correctly by the definition of the formula $\varphi_{a,w}$.

This concludes the proof of the general case. But the same proof gives also the monotone case (as $\varphi_{a,w}$ is monotone). □

I consider the protocols as defined here, or generalized later, to be the primary objects for feasible interpolation and their possible transformations into circuits of some sort as a technical vehicle when proving lower bounds.

17.5 Semantic Feasible Interpolation

Our proof of a fairly general feasible interpolation theorem (Theorem 17.5.1) does not use any particular information about the inference rules of a proof system. It works with sets of satisfying assignments defined by possible lines in a proof. This means that we can generalize the feasible interpolation concept to a more general situation, which can be grasped from the following notions.

Let $N \geq 1$. The **semantic rule** allows us to infer from two subsets $X, Y \subseteq \{0,1\}^N$ a third,

$$\frac{X \quad Y}{Z},$$

if and only if $Z \supseteq X \cap Y$.

A **semantic derivation** of a set $C \subseteq \{0,1\}^N$ from sets $D_1, \ldots, D_r \subseteq \{0,1\}^N$ is a sequence of sets $E_1, \ldots, E_k \subseteq \{0,1\}^N$ such that $E_k = C$, and such that each E_j is either one of the D_i or derived from two earlier E_{i_1}, E_{i_2}, $i_1, i_2 < j$, by the semantic rule.

For a class $\mathcal{X} \subseteq \mathcal{P}(\{0,1\}^N)$, a semantic derivation $E_1, \ldots, E_k$ is an $\mathcal{X}$**-derivation** if and only if all $E_i \in \mathcal{X}$.

Semantic derivations without a restriction to some class $\mathcal{X}$ would be rather trivial: C is derivable from the D_i if and only if $C \supseteq \bigcap_i D_i$ and if and only if C is in the filter on $\mathcal{P}(\{0,1\}^N)$ (a class closed under intersections and closed upwards) generated by the D_i. But when the family $\mathcal{X}$ is not a filter on $\{0,1\}^N$, the notion of $\mathcal{X}$-derivability becomes non-trivial. An important example is provided by the semantic derivations determined by the family of subsets of $\{0,1\}^N$ definable by clauses. Such semantic derivations are sometimes called *semantic resolution* (see Section 17.6 for more examples).

Note that proofs in any of the usual propositional calculi based on binary (or, more generally, bounded-arity) inference rules translate into semantic derivations: replace a clause (or a sequent, a formula, an equation, etc.) by the set of its satisfying truth assignments. The soundness of the inference rules implies that they translate into instances of the semantic rule.

The following technical definition abstracts a property of sets of truth assignments that we shall use in the proof of the feasible interpolation theorem 5.6.1. We shall fix from now on

$$N := n + s + t,$$

where n, t, s are the parameters from the FI set-up (Section 17.1). In the following we shall talk about strings and strings indexed by strings and, to ease the notation, for a while we shall stop using an over-line to denote strings of bits. There should be no danger of confusion.

Let $A \subseteq \{0,1\}^N$ be an arbitrary set. We shall consider several tasks to be solved by two players, called again the U-player and the V-player, the U-player knowing $u \in \{0,1\}^n$ and $q^u \in \{0,1\}^s$ and the V-player knowing $v \in \{0,1\}^n$ and $r^v \in \{0,1\}^t$. The **communication complexity of** A, to be denoted as for multi-functions by $CC(A)$, is the minimum number of bits that the two players using a KW protocol need to exchange in the worst case, in solving any of the following three tasks.

1. Decide whether $(u, q^u, r^v) \in A$.
2. Decide whether $(v, q^u, r^v) \in A$.
3. If $(u, q^u, r^v) \in A \not\equiv (v, q^u, r^v) \in A$ find $i \leq n$ such that $u_i \neq v_i$.

For U closed upwards, the **monotone communication complexity with respect to** U **of** A, $MCC_U(A)$, is the minimum $t \geq CC(A)$ such that the following task can be solved communicating at most t bits in the worst case.

4. If $(u, q^u, r^v) \in A$ and $(v, q^u, r^v) \notin A$ either find an $i \leq n$ such that

$$u_i = 1 \wedge v_i = 0$$

or learn that there is some u' satisfying

$$u' \geq u \wedge (u', q^u, r^v) \notin A.$$

Recall that $u' \geq u$ means that u' can be obtained from u by changing some 0 bits to 1.

For a set $A \subseteq \{0,1\}^{n+s}$, define a new set $\tilde{A} \subseteq \{0,1\}^N$ by

$$\tilde{A} := \bigcup_{(a,b)\in A} \{(a,b,c) \mid c \in \{0,1\}^t\},$$

where a, b, c range over $\{0,1\}^n$, $\{0,1\}^s$ and $\{0,1\}^t$, respectively, and similarly, for $B \subseteq \{0,1\}^{n+t}$, define

$$\tilde{B} := \bigcup_{(a,c)\in B} \{(a,b,c) \mid b \in \{0,1\}^s\}.$$

Theorem 17.5.1 (Krajíček [280]) *Let $A_1, \ldots, A_m \subseteq \{0,1\}^{n+s}$ and $B_1, \ldots, B_\ell \subseteq \{0,1\}^{n+t}$. Assume that there is a semantic derivation $\pi\colon D_1, \ldots, D_k$ of the empty set $\emptyset = D_k$ from the sets $\tilde{A}_1, \ldots, \tilde{A}_m, \tilde{B}_1, \ldots, \tilde{B}_\ell$.*

If the communication complexity of the D_i, $i \leq k$, satisfies $CC(D_i) \leq t$ then there is a size-$(k+2n)$ protocol, of communication complexity at most t, for computing $KW[U, V]$ for the two sets

$$U = \{u \in \{0,1\}^n \mid \exists q^u \in \{0,1\}^s; (u, q^u) \in \bigcap_{j \leq m} A_j\}$$

and

$$V = \{v \in \{0,1\}^n \mid \exists r^v \in \{0,1\}^t; (v, r^v) \in \bigcap_{j \leq \ell} B_j\}.$$

In particular, U and V can be separated by a circuit of size at most $(k+2n)2^{O(t)}$.

If the sets $A_1, \ldots, A_m$ satisfy a monotonicity condition with respect to U given by

$$(u, q^u) \in \bigcap_{j \le m} A_j \wedge u \le u' \;\rightarrow\; (u', q^u) \in \bigcap_{j \le m} A_j,$$

and $MCC_U(D_i) \le t$ for all $i \le k$, then there is a size $k+n$ protocol of communication complexity at most t computing $KW^m[U, V]$ and thus also a monotone circuit separating U from V of size at most $(k+n)2^{O(t)}$.

Further, if π is tree-like then the protocol is tree-like and the separating circuit is a formula.

Proof Assume that π is a refutation satisfying the hypothesis of the theorem, and let us leave the monotone case aside for a moment.

Define the protocol (G, lab, S, F) as follows.

- The graph G has k inner nodes corresponding to the k steps in π and an additional $2n$ leaves. The leaves are labeled by lab by all $2n$ formulas $u_i = 1 \wedge v_i = 0$ and $u_i = 0 \wedge v_i = 1$, $i = 1, \ldots, n$.

  We shall denote the inner nodes also by $D_1, \ldots, D_k$.
- The consistency condition $F(u, v)$ is formed by those inner nodes D_j that are not satisfied by (v, q^u, r^v), i.e. $(v, q^u, r^v) \notin D_j$, and also by those leaves whose label is valid for the pair u, v.
- The strategy function $S(u, v, x)$ at an inner node $x = D_j$, with D_j derived from earlier steps D_{i_1} and D_{i_2}, is defined as follows.
  1. If $(u, q^u, r^v) \notin D_j$ then

$$S(u, v, D_j) := \begin{cases} D_{i_1} & \text{if } (v, q^u, r^v) \notin D_{i_1}, \\ D_{i_2} & \text{if } (v, q^u, r^v) \in D_{i_1} \text{ (and so } (v, q^u, r^v) \notin D_{i_2}). \end{cases}$$

  2. If $(u, q^u, r^v) \in D_j$ then the players find (item 3 in the definition of CC) an $i \le n$ such that $u_i \neq v_i$, and then, of the two nodes labeled by $u_i = 1 \wedge v_i = 0$ and $u_i = 0 \wedge v_i = 1$, $S(u, v, D_j)$ is the one whose label is valid for the pair u, v.

Because $(u, q^u, r^v) \in \bigcap_{j \le m} A_j$ while $(v, q^u, r^v) \in \bigcap_{i \le \ell} B_i$, the strategy is well defined at all inner nodes. As G has $k+2n$ nodes and communication complexity $\le 2t$, Theorem 17.4.3 yields a circuit separating U from V and having size at most $(k+2n)2^{O(t)}$.

In the monotone case we need to modify the definition of the protocol a little. Assume that the clauses A_i satisfy the monotonicity condition (and hence that U is closed upwards). The graph G still has k inner nodes corresponding to the steps in π but now it has only n leaves, labeled by all n formulas $u_i = 1 \wedge v_i = 0$, $i = 1, \ldots, n$. The consistency condition $F(u, v)$ is defined as before but the strategy function changes somewhat. In the second case in the definition of S above the players use their strategy for item 4 after the definition of MCC_U and either find $i \leq n$ such that

$$u_i = 1 \wedge v_i = 0,$$

in which case the strategy maps from D_j to the corresponding leaf, or learn that there is some u' satisfying

$$u' \geq u \wedge (u', q^u, r^v) \notin D_j,$$

in which case the strategy is defined as before,

$$S(u, v, D_j) := \begin{cases} D_{i_1} & \text{if } (v, q^u, r^v) \notin D_{i_1}, \\ D_{i_2} & \text{if } (v, q^u, r^v) \in D_{i_1}. \end{cases}$$

By the monotonicity condition for every potential u' occurring above it holds that:

$$(u', q^u) \in \bigcap_{j \leq m} A_j.$$

Hence the strategy is well defined. By Theorem 17.4.3 we get a monotone separating circuit of size $(k+n)2^{O(t)}$. □

Given a proof system P whose proofs consist of steps derived using inference rules, we can consider the class $\mathcal{X}_P \subseteq \{0,1\}^N$ consisting of all sets definable by possible lines (clauses, formulas, polynomials, inequalities, etc.) in P-proofs. For example, for the resolution proof system R these are the sets definable by clauses and for the degree d polynomial calculus PC/**F** they are the sets defined by polynomials over **F** of degree at most d. Semantic derivations using only sets from $\mathcal{X}_P$ will be called **semantic *P*-proofs** or, somewhat more loosely, ***P*-like proofs**, and we refer informally to **semantic** P as if it were a proof system.

Using this terminology, Theorems 17.5.1 and 13.5.3 imply the following lower bound criterion.

Corollary 17.5.2 *Let P be a proof system operating with steps derived by inference rules and assume that every set $A \in \mathcal{X}_P$ has the monotone communication complexity $MCC_U(A) \leq t$. Assume that $3 \leq \xi < \omega$ and $\sqrt{\xi}\omega \leq n/(8 \log n)$.*

Then every semantic P-proof of the disjointness of $\text{Clique}_{n,\omega}$ *and* $\text{Color}_{n,\xi}$ *must have at least*

$$2^{\Omega(\sqrt{\xi}) - O(t)} - n$$

steps.

In particular, for $\xi := n^{1/3}$, $\omega := n^{2/3}$ and $t = n^{o(1)}$ the number of steps is at least $2^{\Omega(n^{1/6})}$.

17.6 Simple Examples

In our first example we consider the system LK^-, i.e. cut-free LK; this will illustrate that it is useful to be flexible when applying the feasible interpolation framework. Let clauses A_i, B_j obey the FI set-up and let the cedent (Section 3.1) $\Gamma(\overline{p}, \overline{q})$ consist of formulas $\bigvee A_i$ and the cedent $\Delta(\overline{p}, \overline{r})$ consist of formulas $\bigwedge \neg B_j$ (conjunctions of the negations of literals from the clauses B_j). Assume that π is a negation normal form LK^- proof (Section 3.3) of the sequent

$$\Gamma(\overline{p}, \overline{q}) \longrightarrow \Delta(\overline{p}, \overline{r}).$$

Hence any sequent in π has the form

$$\Pi(\overline{p}, \overline{q}) \longrightarrow \Sigma(\overline{p}, \overline{r}). \tag{17.6.1}$$

Define a protocol (G, lab, S, F) as follows.

- The nodes of G are the sequents in π. The leaves are the initial sequents from (3.3.1) of the form

  $$p_i \longrightarrow p_i \quad \text{and} \quad \neg p_i \longrightarrow \neg p_i$$

  for all $\overline{p}$-atoms p_i. They are labeled by lab by the formulas $u_i = 1 \wedge v_i = 0$ and $u_i = 0 \wedge v_i = 1$, respectively.

  All other sequents in π are the inner nodes of G.
- The consistency condition $F(u, v)$ contains those sequents (17.6.1) in π such that $\Pi(u, q^u)$ is true and $\Sigma(v, r^v)$ is false.
- The strategy function S at a node in F either moves to a hypothesis of an inference that is in F, or otherwise is defined arbitrarily.

Note that S preserves F, as formulas from the antecedents and from the succedents in π never mix, and that the (monotone) communication complexity of the protocol is 2. In addition, if the monotone FI set-up holds then initial sequents of the form $\neg p_i \longrightarrow \neg p_i$ do not occur in π and hence the labels of all leaves are of the form $u_i = 1 \wedge v_i = 0$. Theorem 17.4.3 thus implies a slightly weaker version of Theorem 3.3.2.

Corollary 17.6.1 *Assume that clauses A_i and B_j obey the FI set-up and that the sequent*

$$\bigvee A_1(\overline{p}, \overline{q}), \ldots, \bigvee A_m(\overline{p}, \overline{q}) \longrightarrow \bigwedge \neg B_1(\overline{p}, \overline{r}), \ldots, \bigwedge \neg B_\ell(\overline{p}, \overline{r})$$

has a negation normal form LK^-*-proof with k steps. Then U and V can be separated by a circuit of size $O(k)$.*

If the clauses obey the monotone FI *set-up then the separating circuit can be chosen as monotone too.*

Now we shall consider three examples of subclasses of $\mathcal{P}(\{0,1\}^N)$ that correspond to three proof systems and derive feasible interpolation for their semantic versions from Theorem 17.5.1 and a lower bound from Corollary 17.5.2.

17.6.1 Semantic R: Clauses

Any $A \subseteq \{0,1\}^N$ defined by a clause has $CC(A) \leq \log n$ if $n > 1$: items 1 and 2 in the definition of $CC(A)$ require each player to send one bit (the truth value of the portion of the clause for which they have evaluation) and $\log n$ bits are required for the binary search (item 3). Also, $MCC_U(A) \leq \log n$, as the second option in item 4 can be decided by the U-player only.

In the FI set-up in Section 17.1 we required that all $\overline{p}$ occur only positively in all clauses A_i (and hence U is closed upwards). Note that this condition also implies the requirement

$$[u' \geq u \wedge \bigwedge_i A_i(u, q^u)] \rightarrow \bigwedge_i A_i(u', q^u)$$

in the monotone case in Theorem 17.5.1.

Corollary 17.6.2

(i) *(Theorem 13.5.2 – another derivation)*
Assume that clauses A_i and B_j obey the FI set-up. If they can be refuted in resolution in k steps then there is a circuit separating U from V of size at most $(k + 2n)n^{O(1)}$.
If also the conditions of the monotone FI set-up are obeyed then there is a monotone circuit separating U from V of size at most $(k + n)n^{O(1)}$.

(ii) *Every semantic* R*-proof of the disjointness of* $\text{Clique}_{n,n^{2/3}}$ *and* $\text{Color}_{n,n^{1/3}}$ *must have at least $2^{\Omega(n^{1/6})}$ steps.*

17.6.2 Semantic LEC: Linear Equations

If $A \in \{0,1\}^N$ is defined by a linear equation over $\mathbf{F}_2$ then the same estimates for $CC(A)$ and $MCC_U(A)$ as for clauses apply. Hence Corollary 17.6.2 also holds for the linear equational calculus LEC (Section 7.1).

17.6.3 Semantic CP: Integer Linear Inequalities

Assume that $A \in \{0,1\}^N$ is defined by the inequality

$$\overline{e} \cdot \overline{x} + \overline{f} \cdot \overline{y} + \overline{g} \cdot \overline{z} \geq h,$$

with $\overline{e}, \overline{f}$ and $\overline{g}$ vectors of integers and h an integer. When solving the tasks required in the definitions of $CC(A)$ and $MCC_U(A)$, this time the players will send to each other the values of the parts of the left-hand side that they can evaluate. However, if M is the maximum absolute value of a coefficient occurring in the inequality this may need $\log(Mn^{O(1)}) = O(\log n) + \log M$ bits. Hence both the $CC(A)$ and $MCC_U(A)$ are estimated by $(\log n)(O(\log n) + \log M)$ (the extra $\log n$ factor comes from the binary search in solving tasks 3 or 4, respectively. Hence we get

Corollary 17.6.3

(i) *Assume that clauses A_i and B_j obey the FI set-up and are represented for the* CP *proof system by integer linear inequalities as in* (6.0.2). *If they can be refuted in the semantic* CP *in k steps such that*

(*) *every set occurring in the derivation is defined by an inequality with all coefficients bounded in absolute value by M*

then there is a circuit separating U from V of a size that is at most $(k+2n)(Mn^{O(1)})^{\log n}$.

If also the conditions of the monotone FI *set-up are obeyed, then there is a monotone circuit separating U from V of size at most* $(k+n)(Mn^{O(1)})^{\log n}$.

(ii) *Every semantic* CP*-proof of the disjointness of* $\text{Clique}_{n,n^{2/3}}$ *and* $\text{Color}_{n,n^{1/3}}$ *satisfying (*) must have at least* $2^{\Omega(n^{1/6})}/M^{\log n}$ *steps.*

Note that we could generalize the feasible interpolation (and the lower bounds) from semantic R, LEC and CP to semantic proof systems operating with sets that are defined by Boolean combinations of at most w original sets (in particular, to low-width proofs in the system R(CP) from Section 7.1): this causes an increase in the (monotone) communication complexity of the protocol by a multiplicative factor w and hence in the size of the separating circuits by a multiplicative factor $2^{O(w)}$. Thus we still get lower bounds for $w = n^{o(1)}$.

17.7 Splitting Proofs

The proofs of feasible interpolation for tree-like and dag-like LK^- (cut-free LK) in Section 3.3 gave not only interpolants (a formula or a circuit) but also *proofs* that it works. That is, from a proof of

$$\Gamma(\overline{p},\overline{q}) \longrightarrow \Delta(\overline{p},\overline{r}) \tag{17.7.1}$$

we constructed an interpolating circuit $C(\overline{p})$ (or just a formula if the proof was tree-like) defined by the CNF $\text{Def}_C(\overline{p},\overline{y})$ with output gate y_s and short proofs of the sequents

$$\Gamma(\overline{p},\overline{q}), \text{Def}_C \rightarrow y_s \quad \text{and} \quad \text{Def}_C, y_s \rightarrow \Delta(\overline{p},\overline{r}).$$

If we now substitute $\overline{p} := \overline{a} \in \{0,1\}^n$ and evaluate the subcircuits $y_i := y_i(\overline{a})$ on $\overline{a}$, all clauses in Def_C will contain 1 and hence we get proofs of

$$\Gamma(\overline{a},\overline{q}) \rightarrow y_s(\overline{a}) \quad \text{and} \quad y_s(\overline{a}) \rightarrow \Delta(\overline{a},\overline{r}).$$

If $y_s(\overline{a}) = 0$ then the first proof is a refutation of $\Gamma(\overline{a},\overline{q})$; if $y_s(\overline{a}) = 1$ then the second proof is a proof of $\Delta(\overline{a},\overline{r})$. Moreover, by the constructions in Section 3.3, these proofs are obtained from the original proof by a substitution and some deletions.

Once we have the idea that a proof of (17.7.1) ought to *split*, after any substitution $\overline{p} := \overline{a} \in \{0,1\}^n$, into essentially two separate proofs not mixing $\overline{q}$ and $\overline{r}$, it is easy to do this splitting directly, without reference to the earlier construction. Namely, assume that π is a negation normal form LK$^-$-proof of (17.7.1). For a sequent S in π let π_S be the subproof ending with S. Given an assignment $\overline{a} \in \{0,1\}^n$ we shall mark each sequent in the substituted $\pi(\overline{a})$,

$$\Pi(\overline{a},\overline{q}) \longrightarrow \Sigma(\overline{a},\overline{r}), \tag{17.7.2}$$

either by 0, meaning that

$$\Pi(\overline{a},\overline{q}) \longrightarrow$$

is proved, or by 1, meaning that

$$\longrightarrow \Sigma(\overline{a},\overline{r})$$

is proved, where the proofs of these substituted sequents are obtained from $\pi_S(\overline{a})$, possibly by deleting some sequents.

This is easy for the initial sequents:

- mark $p_i \rightarrow p_i$ by a_i and $\neg p_i \rightarrow \neg p_i$ by $\neg a_i$,
- mark $z, \neg z \rightarrow$ by 0 and $\rightarrow z, \neg z$ by 1, where z is any atom not in $\overline{p}$,

and, by induction on the number of inferences for all other sequents as well,

- mark the conclusion of a unary inference in the same way as its hypothesis,
- if the two hypotheses of a $\vee$:left inference were marked y and z, mark its conclusion by $y \vee z$ and dually for $\wedge$:right – the conclusion is marked $y \wedge z$.

It is easy to check that if a sequent S of the form (17.7.2) is marked 0 (resp. 1), keeping in $\pi_S(\overline{a})$ only sequents marked 0 (resp. 1) and only ancestors of formulas in the respective end-sequent will create a proof of

$$\Pi(\overline{a},\overline{q}) \longrightarrow \quad \text{or} \quad \longrightarrow \Sigma(\overline{a},\overline{r}),$$

respectively. Actually the construction also yields a circuit $C(\overline{p})$ which computes the marking (the initial sequents in the first item in the bullet list above correspond to inputs p_i and $\neg p_i$). Further, note that C is monotone if the atoms p_i occur only positively in $\Gamma(\overline{p},\overline{q})$: there are no initial sequents $\neg p_i \rightarrow \neg p_i$ in that case.

Let us summarize.

Lemma 17.7.1 *There is a p-time algorithm that, upon receiving an* LK$^-$*-proof* π *of* (17.7.1),

(i) builds a circuit $C(\overline{p})$ *whose graph is obtained from the proof graph of* π *by deleting some nodes,*
(ii) and on any input $\overline{a} \in \{0,1\}^n$ *produces an* LK$^-$*-proof* $\sigma(\overline{a})$ *of* $\Gamma(\overline{a},\overline{q}) \rightarrow$ *if* $C(\overline{a}) = 0$ *or of* $\rightarrow \Delta(\overline{a},\overline{r})$ *if* $C(\overline{a}) = 1$.

The proofs $\sigma(\overline{a})$ *are obtained from* π *by substituting* $\overline{p} := \overline{a}$ *and by deleting some sequents and some formulas.*

Moreover, if the atoms $\overline{p}$ occur only positively in $\Gamma(\overline{p},\overline{q})$ then the circuit C can be chosen as monotone.

If the cedents $\Gamma(\overline{p},\overline{q})$ and $\Delta(\overline{p},\overline{r})$ are obtained from sets of clauses A_i and B_j obeying the FI set-up, as at the beginning of Section 17.6, Lemma 17.7.1 together with the derivation of feasible interpolation for R via LK^- in the feasible interpolation theorem 5.6.1 yields a similar statement for R.We shall streamline the proof without reference to LK^- (the construction is analogous, though, to the above construction with LK^-).

We shall need the following ternary propositional connective:

$$sel(x,y,z) := \begin{cases} y & \text{if } x = 0, \\ z & \text{if } x = 1. \end{cases}$$

Theorem 17.7.2 (Pudlák [412]) *There is a p-time algorithm that, upon receiving an* R*-refutation π of clauses A_i, B_j obeying the* FI *set-up,*

(i) it builds a circuit $C(\overline{p})$ whose graph is obtained from the proof graph of π by deleting some nodes and placing at the initial clauses the inputs $p_i, \neg p_i$ and at other clauses the connectives $\vee, \wedge$ or $sel(p_i, y, z)$ or $sel(\neg p_i, y, z)$ for suitable choices of the atoms p_i,
(ii) and on any input $\overline{a} \in \{0,1\}^n$ produces an R*-refutation $\sigma(\overline{a})$ of the clauses $A_i(\overline{a},\overline{q})$ if $C(\overline{a}) = 0$ or of the clauses $B_j(\overline{a},\overline{r})$ if $C(\overline{a}) = 1$.*

The refutations $\sigma(\overline{a})$ are obtained from π by substituting $\overline{p} := \overline{a}$ and by deleting some clauses and some literals.

Moreover, if the atoms $\overline{p}$ occur only positively in all $A_i(\overline{p},\overline{q})$ the circuit C can be chosen as monotone.

Proof We shall describe a two-round procedure that first replaces each clause E in $\pi(\overline{a})$ by its subclause E' and then marks each new clause as 0 or 1. The newly created clauses will be branded as q-clauses or r-clauses; the former implies that no $\overline{r}$-atom occurs in it, the latter that no $\overline{q}$-atom occurs in it. This simulates in resolution refutations the notion of ancestors in cut-free proofs that we used earlier.

The new clauses are created by considering all possible types of inference as follows.

1. The initial clauses are left unchanged; the clauses A_i are branded as q-clauses and B_j as r-clauses.
2. The resolution of a $\overline{p}$-atom,

$$\frac{E, p_i \quad F, \neg p_i}{E \cup F};$$

the hypotheses are replaced by E' and F', respectively; $E \cup F$ is replaced by E' if $a_i = 0$ and by F' if $a_i = 1$, and the clause keeps its brand.

3. The resolution of a $\overline{q}$-atom,

$$\frac{E, q_i \quad F, \neg q_i}{E \cup F};$$

the hypotheses are replaced by E' and F', respectively.

If one of E', F' is branded as an r-clause, it does not contain q_i and we replace $E \cup F$ by this clause. If both E', F' are branded as q-clauses, either we resolve q_i and brand the conclusion as a q-clause, or we replace the conclusion by E' or by F', respectively, if it does not contain q_i or $\neg q_i$, respectively.

4. The resolution of an $\overline{r}$-atom,

$$\frac{E, r_i \quad F, \neg r_i}{E \cup F};$$

the hypotheses are replaced by E' and F', respectively. Proceed dually to item 3.

After this reduction, delete all clauses containing a literal that becomes true after the substitution $\overline{p} := \overline{a}$, and delete all such literals that become false. If the empty end-clause is branded as a q-clause then we delete all r-clauses and get an R-refutation of the clauses $A_i(\overline{a}, \overline{q})$. If it is an r-clause then we dually delete all q-clauses and get an R-refutation of the clauses $B_j(\overline{a}, \overline{r})$.

The required circuit $C(\overline{p})$ computes the branding of clauses as q-clauses (by computing 0) or as r-clauses (by computing 1). To create C do the following:

- place the constant 0 on all initial A_i clauses and the constant 1 on all initial B_j-clauses;
- in the resolution of atom p_i (item 2 above) we compute the value x for the conclusion from the value y of the left-hand hypothesis and the value z of the right-hand hypothesis as

$$x := sel(p_i, y, z);$$

- in the resolution of atom q_i (item 3 above) we put

$$x := y \vee z;$$

- in the resolution of atom r_i (item 4 above) we put

$$x := y \wedge z.$$

To obtain a monotone circuit in the case of a monotone set-up note that, in item 2 above, if E' is a q-clause then it cannot contain $\neg p_i$ and hence we can replace the conclusion of the inference by E' as well. This means that we can replace the *sel* connective by

$$(p_i \vee y) \wedge z,$$

which is monotone. □

The splitting proof approach to feasible interpolation will have some important applications but it has also its limits. For example, it appears very hard to split proofs operating with small Boolean combinations of clauses like these considered at the end of Section 17.6.

17.8 Splitting CNF Formulas

The syntactic format of clauses A_i, B_j obeying the FI set-up and, in particular, the division of the atoms into three disjoint sets seems necessary for any application of feasible interpolation. Hrubeš and Pudlák [237] found a way to circumvent this, however.

Assume that $\mathcal{C}$: $C_1, \ldots, C_m$ are clauses in $2n$ atoms. Split the atoms arbitrarily into two disjoint sets of size n,

$$\overline{p}^0 = p_1^0, \ldots, p_n^0 \quad \text{and} \quad \overline{p}^1 = p_1^1, \ldots, p_n^1,$$

and write each clause C_j as a disjoint union $C_j^0 \cup C_j^1$, where C_j^b contains only literals corresponding to the atoms from $\overline{p}^b$, $b = 0, 1$.

Let $q_1, \ldots, q_m$ be new atoms and define a new set of clauses $\mathcal{D}$ consisting of

$$D_j^0 \ := \ C_j^0 \cup \{\neg q_j\} \quad \text{and} \quad D_j^1 \ := \ C_j^1 \cup \{q_j\}.$$

Lemma 17.8.1 *The set of clauses $\mathcal{C}$ is unsatisfiable if and only if $\mathcal{D}$ is unsatisfiable, and every monotone interpolant $E(\overline{q})$ of unsatisfiable $\mathcal{D}$ satisfies, for every $J \subseteq [m]$,*

- *if $\{C_j^0\}_{j \in [m] \setminus J}$ is satisfiable then $E(\overline{a}_J) = 1$;*
- *if $\{C_j^0\}_{j \in J}$ is satisfiable then $E(\overline{a}_J) = 0$.*

Where $\overline{a}_J$ is the characteristic function of J ($a_j = 1$ if and only if $j \in J$).

Any monotone Boolean function E satisfying the property from Lemma 17.8.1 is called a **monotone unsatisfiability certificate** for the specific splitting of the $2n$ atoms into two sets.

The lemma implies that for every proof system P admitting monotone feasible interpolation (using some computational model), if an unsatisfiable $\mathcal{C}$ has a P-refutation of size s then $\mathcal{C}$ also has a size-$s^{O(1)}$ monotone unsatisfiability certificate for *any* splitting of the set of its atoms. This idea does not extend the class of proof systems for which a lower bound can be established but it allows one to prove lower bounds for *random* CNFs (Section 18.8).

17.9 Bibliographical and Other Remarks

17.9.1 History of the Idea

Mundici [361, 362, 363, 364] pointed out in the early 1980s a relation between the sizes of interpolants of propositional implications *as estimated in terms of the sizes of the implications* and the central problems of computational complexity theory. For example, any set in NP ∩ coNP is trivially the unique set separating two disjoint NP sets. Hence if we could always find an interpolating formula of size polynomial in the size of the implication the (unexpected) inclusion NP ∩ coNP ⊆ NC^1 would follow. For predicate logic, there were earlier lower bounds in terms of recursion theory (Friedman [189] in 1976 or unpublished results of Kreisel [328] from 1961). Gurevich [212] discussed further connections to computer science.

The idea of what we call now *feasible interpolation* was first formulated in my paper [276]; it was published in a journal in 1994 but its preprint circulated in 1991 and was noticed by everybody in the field as it proved the first exponential lower bound for AC^0-Frege systems. The idea is formulated there explicitly, as is the equivalent *feasible Beth's theorem* (Lemma 17.1.4 here), and it was shown that it works for LK^- (cut-free LK). My hope then, as explained in [276] was to extend feasible interpolation (and the implied Beth's property) to AC^0-Frege systems and use it to show the mutual independence of the modular counting principles for different primes, reducing it to $AC^0[p]$ lower bounds for MOD_q. This did not work and the independence was eventually proved differently (Section 15.5).

Then things moved on in 1994. I obtained feasible interpolation for R by reducing it to that for LK^- (the proof of Theorem 5.6.1 in Section 5.6) but I had no particular use for the result. My interest in it was renewed by a remark in a 1994 preprint of Razborov [435] that in the technical developments underlying the unprovability results for bounded arithmetic there are certain interpolation theorems for fragments of second-order bounded arithmetic. I realized that if that is the case then these interpolation theorems and problems ought to be studied at the more rudimentary level of propositional logic and that this must relate to the feasible interpolation idea in [276], and I quickly proved that my feasible interpolation for R implies (together with the known relations between R and bounded arithmetic established in [276]) the unprovability results of Razborov [435]. No propositional proof systems appear in [435] and the feasible interpolation idea is not mentioned there. Nevertheless, this paper was a crucial inspiration for me in several aspects. The interpolation in [435] is present via a special and unusual technical definition of the formal systems studied there (the so-called *split theories*), which are tailored to allow some interpolation. I would not be surprised if the interpolation, and a connection with [435], was also behind the notion of *natural proofs* developed at that time (in a 1994 preprint) by Razborov and Rudich [444]. Then I proved the general feasible interpolation theorem for semantic derivations (Theorem 17.5.1 here), using a modification of the notion of protocols from [435], and applied it to a few proof systems. After hearing a lecture by Bonet at the meeting *Logic and Computational Complexity* (Indianapolis, October 1994) about her work with Pitassi and Raz [95, 96] I realized that their result follows from my general theorem and I replaced the original example of linear calculus, LEC, by CP with small coefficients in [280] (until then it had not occurred to me to consider this version of CP). Originally Bonet *et al.* [95] did not know that their constructions were related to a more general interpolating property (see their remark in [95]).

Razborov also explained to me the remarks on one-way functions that he made (but did not elaborate upon) in [435, Sec. 8] and this resulted in conditional lower bounds for feasible interpolation (for Beth's theorem) in [280]. In 1994 Razborov [433] also produced a preprint about the canonical pairs, but the main result (the completeness of these pairs) was a simple reformulation of the completeness of the reflection principles for proof systems (cf. Section 17.2), as shown in [280], and it was never published by him. Nevertheless the paper stimulated a combinatorial approach to these pairs, and to reflection principles in general, and although that still waits for some significant application it will eventually, I think, be found. Expanding on this we established in 1995 with Pudlák [320, 321] links between the security of RSA and the impossibility of feasible interpolation for *EF* and, at the same time, Pudlák [412] found a beautiful application of feasible interpolation to unrestricted CP. After these developments the method caught on, and many other researchers have contributed to it since then, finding new and unforeseen notions and applications.

The reader can find historical notes also in Bonet, Pitassi and Raz [97], Razborov [433], Pudlák [412, 413] and Beame and Pitassi [58] and in other references from the late 1990s.

Secs. 17.1, 17.4, 17.5 and 17.6 are based on [280]. The definition of the general protocols in Section 17.4 is from [280] and it is a variant of a notion defined by Razborov [435]. The strategy function $S(\overline{u}, \overline{v}, x)$ may seem redundant because the players can test which of the children of $x \in F(\overline{u}, \overline{v})$ belong also to the consistency condition set $F(\overline{u}, \overline{v})$, and choose one. However, this works only for protocols with a bounded (or very small) out-degree (and even for those another strategy is sometimes more natural). The second part of Corollary 17.6.3 is due to Bonet, Pitassi and Raz [95].

Sec. 17.2 The examples of pairs discussed here are from Razborov [430] and Alon and Boppana [18] (the clique-coloring pair), Haken [217] (the broken mosquito screen), Raz and Wigderson [429] and Impagliazzo, Pitassi and Urquhart [244] (the bipartite perfect matching), [280] (the reflection and OWPs) and Krajíček and Pudlák [321] (the RSA pair). The reformatting of the NP pair of sets from the reflection principles into the sets PROV(Q) and FALSI* was done by Razborov [433], who called them the **canonical pair** of the proof system (he denoted our PROV(Q) by *REF*(Q) and FALSI* by SAT*).

Sec. 17.3 exposes an idea of Bonet, Pitassi and Raz [97].

Sec. 17.7 The splitting construction for R is from Pudlák's [412].

Sec. 17.8 is based on Hrubeš and Pudlák [237].

17.9.2 Feasible Disjunction Property

A feasible interpolation construction for P, which, besides an interpolant, also provides a P-proof that it works (as in Section 17.7), implies in particular that if a disjunction of the form

$$D_1(\overline{q}) \vee D_2(\overline{r})$$

in disjoint sets of atoms $\overline{q}, \overline{r}$ has a P-proof of size s then one of the two disjuncts has a P-proof of size $s^{O(1)}$. We shall describe this situation by saying that P has the **feasible disjunction property** (Pudlák [416] studies this property under the name *existential interpolation*). Somewhat more generally, P has the **strong feasible disjunction property** if the same is true also for arbitrarily long disjunctions

$$D_1(\overline{q}^1) \vee \cdots \vee D_m(\overline{q}^m)$$

in disjoint sets of atoms $\overline{q}^i$, $i \leq m$. Note that we do not require that a short proof of one of the disjuncts is found feasibly – that would imply feasible interpolation – but only that such a proof exists.

If P has the feasible disjunction property and proves in p-size the disjointness of two NP sets U, V then there are two *disjoint* coNP sets $\tilde{U}, \tilde{V}$ such that

$$U \subseteq \tilde{U} \quad \text{and} \quad V \subseteq \tilde{V}.$$

Take for $\tilde{U}_n$ the complement of the set of all $\overline{a} \in \{0,1\}^n$ for which P has a short proof of $\overline{a} \notin U_n$, and similarly for $\tilde{V}_n$. The existence of such a disjoint coNP pair for every disjoint NP pair is not known to contradict any established computational complexity conjecture and even strong proof systems may possess the property.

Problem 17.9.1 (The feasible disjunction problem) *Does F or EF admit the (strong) feasible disjunction property?*

It is also unknown whether the property for $m = 2$ implies the strong version.

Note, however, that for the purpose of proving that a proof system is not p-bounded one may *assume* without loss of generality that the system has the (strong) feasible disjunction property: one of the disjuncts must be a tautology and if it does not have a short proof then we already know that the proof system is not p-bounded (these remarks are from [302], where also the handy notation $\dot{\bigvee}_i D_i$ for a disjunction of formulas in disjoint sets of variables was used).

Note an observation linked to model theory in Section 20.1: applying the feasible disjunction property to short P-proofs of Ref_P formulas implies that, for any tautology α, in a short proof P either proves α or proves that it *cannot prove α in a short proof* (this is more memorable than the precise version in terms of parameters).

18

Feasible Interpolation: Applications

This chapter presents several feasible interpolation results, sometimes using new computational models to evaluate an interpolant. We will also prove that strong proof systems above the *ground level* (see Chapter 22, discussing levels of proof systems) do not admit feasible interpolation by Boolean circuits, assuming some plausible conjectures from computational complexity theory. Possible generalizations that may capture even these proof systems are discussed in Section 18.8.

We concentrate mostly on the more general dag-like case of proofs and protocols and consider the tree-like case only when the dag-like case is open. In particular, tree-like protocols are discussed as a proof of the concept of the real game in Section 18.2 (there are no known lower bounds for dag-like protocols in that situation) and in Sections 18.4 and 18.6, where we present lower bounds for the tree-like versions of LS and R(LIN), respectively, as the general cases are open at present.

18.1 CP and Monotone Real Circuits

Recall the definition of the cutting plane proof system CP in Section 6.3. The lines in its proofs are integer linear inequalities and we shall thus use variables $\overline{x}, \overline{y}$ and $\overline{z}$ in place of atoms $\overline{p}, \overline{q}, \overline{r}$ and assume that the set of clauses A_i, B_j obeying the FI set-up is expressed as a set of inequalities like (6.0.2).

To establish feasible interpolation for CP we shall modify the splitting construction from Section 17.7. In the non-monotone case this will be straightforward but for the monotone case we shall need to introduce a new monotone computational model.

Let π be a CP-refutation of a system of integer linear inequalities $A_i(\overline{x}, \overline{y})$ and $B_j(\overline{x}, \overline{z})$ obeying the FI set-up. A step in the proof π has the form

$$\overline{a} \cdot \overline{x} + \overline{b} \cdot \overline{y} + \overline{c} \cdot \overline{z} \geq d, \tag{18.1.1}$$

where $\overline{a}, \overline{b}, \overline{c}$ are tuples of integers and d is an integer. If some of these do not actually occur we treat them as 0.

Given an assignment $\overline{x} := \overline{u}$ we replace each inequality (18.1.1) in π by a pair of inequalities

$$\overline{e} \cdot \overline{y} \geq d_y \quad \text{and} \quad \overline{f} \cdot \overline{z} \geq d_z \tag{18.1.2}$$

and it will always hold that these two relations imply the original one, i.e.,

$$d_y + d_z \geq d - \overline{a} \cdot \overline{u}. \tag{18.1.3}$$

We define the pairs by induction on the number of steps in π. The axioms $x \geq 0$ and $-x \geq -1$ are replaced by $0 \geq 0$ and those for the y- and z-variables are left intact. An initial inequality A_i of the form

$$\overline{a} \cdot \overline{x} + \overline{b} \cdot \overline{y} \geq d$$

is replaced by the pair

$$\overline{b} \cdot \overline{y} \geq d - \overline{a} \cdot \overline{u} \quad \text{and} \quad 0 \geq 0$$

and, analogously, an initial inequality B_j of the form

$$\overline{a} \cdot \overline{x} + \overline{c} \cdot \overline{z} \geq d$$

is replaced by the pair

$$0 \geq 0 \quad \text{and} \quad \overline{c} \cdot \overline{z} \geq d - \overline{a} \cdot \overline{u}.$$

The addition rule is performed simultaneously on both inequalities in (18.1.2) as is the multiplication rule. Clearly the condition (18.1.3) is maintained.

The division rule can be also performed on the equations in the pair because the coefficients of the variables appearing in them are the same as in the original inequality. But we need to check that the condition (18.1.3) will hold too. For this to be the case, it suffices that

$$d_y + d_z \geq d - \overline{a} \cdot \overline{u}$$

implies for any $g > 0$ dividing all coefficients in $\overline{a}$ that

$$\lceil d_y/g \rceil + \lceil d_z/g \rceil \geq \lceil d/g \rceil - \sum_{i \in [n]} \frac{a_i}{g} u_i.$$

But that is obviously true.

Assume that we reach the end step $0 \geq 1$ in π and that it is replaced by two inequalities

$$0 \geq d_y \quad \text{and} \quad 0 \geq d_z,$$

for which

$$d_y + d_z \geq 1 \ .$$

Therefore either $d_y \geq 1$ or $d_z \geq 1$ and thus we have obtained a refutation of the inequalities $A_i(\overline{u}, \overline{y})$ in the former case or of $B_j(\overline{u}, \overline{z})$ in the latter case.

The computation of the pairs and, in particular, of the coefficients in these pairs can be done in a time that is polynomial in the number of steps in π and in the maximum bit size of a coefficient in π. Thus we have proved

Lemma 18.1.1 *There is an algorithm that, upon receiving a* CP*-refutation π of inequalities A_i, B_j obeying the* FI *set-up, with k steps and and maximum absolute value M of a coefficient in π,*

(i) builds, in a time polynomial in $k, n, \log M$, a circuit $C(\overline{p})$,
(ii) and, on any input $\overline{u} \in \{0,1\}^n$, produces a CP*-refutation of the inequalities $A_i(\overline{u}, \overline{y})$ if $C(\overline{u}) = 0$ or of the inequalities $B_j(\overline{u}, \overline{z})$ if $C(\overline{u}) = 1$.*

The occurrence of the bound M in Lemma 18.1.1 spoils it a bit, but fortunately it can be removed via Theorem 6.3.3. By this lemma we may assume that, for some other CP-refutation with possibly $O(k^3 \log n)$ steps, all integers occurring in it have absolute value at most $O(k2^k n^{O(1)})$. As k and n are bounded by the size of a refutation we get

Theorem 18.1.2 (Pudlák [412]) CP *admits feasible interpolation.*

Hence we know, by Lemma 17.1.3, that CP is not p-bounded unless NP $\subseteq$ P/*poly*. But we want an unconditional lower bound and this requires monotone feasible interpolation. For that we need to introduce a stronger model of monotone computation than a monotone Boolean circuit, as follows.

A **monotone real circuit** is a circuit which computes reals and whose gates are computed by unary or binary *non-decreasing* real functions. Such a circuit computes a Boolean function f on $\{0,1\}^n$ if and only if on all inputs from $\overline{a} \in \{0,1\}^n$ it computes the value $f(\overline{a})$ (i.e. we do not care what it computes on other tuples from $\mathbf{R}^n$).

Replacing, in an ordinary Boolean monotone circuit, $\wedge$ by min and $\vee$ by max we get a monotone real circuit computing the same Boolean function. Hence these new monotone circuits are at least as strong as the old ones.

Theorem 18.1.3 (Pudlák [412]) CP *admits monotone feasible interpolation by monotone real circuits. The size of the interpolating circuit is $O(k)$, where k is the number of steps in the original* CP*-refutation, and in particular it does not depend on the bit size of the coefficients occurring in it.*

Proof Assume that A_i, B_j are initial inequalities that obey the monotone FI set-up and in particular, that all the coefficients of all x-variables in all the A_i are non-negative. We shall proceed as in the construction underlying Lemma 18.1.1 but we need to compute only d_y but, for monotonicity reasons, we shall compute $-d_y$ instead. The computation will be done by a monotone circuit as in the case of resolution but the circuit will be a real circuit this time.

The circuit will use various gates that will allow it to compute the initial values of the d_y numbers and then to simulate the inferences. These gates include binary addition and also the addition of a fixed constant, division by a positive number, multiplication by a positive number and unary rounding to the next integer. It also uses a threshold gate that is the identity for numbers w below 1 and identically equal to 1 for $w \geq 1$.

What such gates cannot a priori compute are the initial values when some coefficients of the x-variables are negative. But by our hypothesis this does not happen in any A_i; a negative coefficient occurs only in the axioms $-x_j \geq -1$. For those axioms, we simply put $d_y := 0$ as before. □

To get a lower bound for CP we need a lower bound for monotone real circuits. In fact, it turns out that methods for obtaining such lower bounds for monotone Boolean circuits work here as well.

Theorem 18.1.4

(i) (Pudlák [412]) *Every monotone real circuit separating the clique-coloring pair (Section 17.2) must have size* $2^{n^{\Omega(1)}}$.
(ii) (Cook and Haken [152]) *Every monotone real circuit separating the broken mosquito pair (Section 17.2) must have size* $2^{n^{\Omega(1)}}$.

Corollary 18.1.5 (Pudlák [412]) CP *is not p-bounded. In fact, every* CP*-refutation of the clique-coloring pair must have size at least* $2^{n^{\Omega(1)}}$.

18.2 CP and the Real Game

In this section we generalize the way in which the players communicate in protocols and apply the new protocols to semantic CP refutations. This may seem redundant, especially because it yields unconditional lower bounds for only the tree-like case. But I think that the communication complexity approach is more robust than the splitting-proof argument (see the remark at the end of Section 17.7) and that lower bounds for the new protocols will be proved eventually.

Let U, V be two disjoint subsets of $\{0, 1\}^*$. The **real game** is played by two players as in the KW-game (Section 17.4), one receiving $\overline{u} \in U$ and the other $\overline{v} \in V$. The players do not exchange bits, however, but in every round each player sends one real number (say r_U for the U-player and r_V for the V-player) to a referee. The referee then compares them and announces the value of one bit: it equals 1 if and only if $r_U > r_V$ and 0 if and only if $r_U \leq r_V$. Each player decides on the next move (i.e. what number to send to the referee) based on $\overline{u}$ and $\overline{v}$, respectively, and on the binary string (called a **position**) of the announcements that the referee has made until that point in the play.

In analogy to the definition of the communication complexity in Section 17.4 we can define a notion of complexity using the real game. For a multi-function $R \subseteq U \times V \times I, I \neq \emptyset$, define its **real communication complexity**, denoted by $CC^{\mathbf{R}}(R)$, to be the minimum number h such that there are strategies for the players of the real game on U, V to reach a position where they agree on a valid value of R. That is, there is a function g: $\{0, 1\}^h \to I$ such that, for every $(\overline{u}, \overline{v}) \in U \times V$, if the position in the game after the hth step is w then $R(\overline{u}, \overline{v}, g(w))$.

Analogously to the first part of Theorem 17.4.1 we have (the second part follows via a Spira-type argument) the following:

Lemma 18.2.1 *Let U be closed upwards. The real communication complexity $CC^{\mathbf{R}}(KW^m[U,V])$ is at most the minimum depth of a monotone real circuit C that separates U from V. In fact, it is also at most $\log_{3/2} s$, where s is the minimum size of a monotone real formula separating U and V.*

It is open whether the opposite bound holds too, as it does in the KW-game case.

Working in the FI set-up and using the notation from Section 17.5, we define the **monotone real communication complexity with respect to** U of $A \subseteq \{0,1\}^N$, to be denoted by $MCC_U^{\mathbf{R}}(A)$, to be the minimum t such that each of the tasks 1–4 from Section 17.5 can be solved by the real game in at most t rounds.

It is clear that $MCC_U^{\mathbf{R}}(A)$ is bounded by $\log n$ for sets A defined by integer linear inequalities: the first two tasks are solved in one round and tasks 3 and 4 in $\log n$ rounds via binary search. This is analogous to the case of sets defined by clauses that arises when we are using the ordinary communication complexity $CC(A)$. Considering just the monotone case, entirely in analogy with Theorem 17.5.1 we obtain

Theorem 18.2.2 *Assume that clauses A_i and B_j obey the monotone* FI *set-up and are represented by integer linear inequalities as in* (6.0.2). *If they can be refuted in semantic* CP *in k steps then there is a size $k+n$ protocol computing $KW^m[U,V]$ of monotone real communication complexity at most* $\log n$.

If π is tree-like, the protocol is a tree and, in this case, the real communication complexity of $KW^m[U,V]$ is at most $O((\log n)(\log(k+n)))$.

Proof The protocol is constructed in identical way to that in the proof of Theorem 17.5.1, except that this time its communication complexity is measured by the real game. The estimate of the real communication complexity of $KW^m[U,V]$ in the tree-like case is as in Lemma 17.4.2 for the ordinary case. □

Now we link the real communication complexity of a multi-function R to its probabilistic communication complexity $C_\epsilon^{pub}(R)$ using "public coins" randomness and error $\epsilon > 0$. Let $R_m^{>} \subseteq \{0,1\}^m \times \{0,1\}^m \times \{0,1\}$ be the characteristic function of the lexico-graphic ordering of $\{0,1\}^m$, i.e. the set of all triples $R_m^{>}(\alpha,\beta,1)$ if $\alpha >_{lex} \beta$ and $R_m^{>}(\alpha,\beta,0)$ if $\alpha \leq_{lex} \beta$. The probabilistic communication complexity of this function is known.

Theorem 18.2.3 (Nisan [366]) *For* $0 < \epsilon < 1/2$, $C_\epsilon^{pub}(R_m^{<}) = O(\log m + \log \epsilon^{-1})$.

Using this theorem we can estimate the probabilistic complexity from the real complexity of any multi-function.

Lemma 18.2.4 *Let $R \subseteq U \times V \times I$ be a multi-function. Then, for any* $0 < \epsilon < 1/2$, *we have*

$$C_\epsilon^{pub}(R) \leq CC^{\mathbf{R}}(R)\, O(\log n + \log \epsilon^{-1}).$$

Proof If h bounds the number of rounds in a real game, then at most

$$|U|\,|V|\,2(2^h - 1) < 2^{2n+h+1}$$

different reals will be sent to a referee by the players. If $r_0 < r_1 < \cdots < r_k$, $k < 2^{2n+h+1}$, is their enumeration in increasing order, the players may use the index i in place of the original r_i without affecting what the referee replies.

By Theorem 18.2.3, each step in the resulting game can be simulated by $O(\log m + \log(\epsilon^{-1}h))$, $m = 2n + h + 1$, steps of the probabilistic KW-game with an error ϵh^{-1}. Thus the whole real game can be probabilistically simulated with an error $\epsilon > 0$ in

$$h \cdot O(\log m + \log(\epsilon^{-1}h)) = h\, O(\log n + \log \epsilon^{-1})$$

rounds, as we may bound $h \leq n$. □

Now we shall put these estimates to work. Recall the bipartite matching problem and the two (complementary) sets BPM^+ and BPM^- from Section 17.2.

Lemma 18.2.5 *Let G be a tree-like protocol for $KW^m[BPM^+, BPM^-]$ of size s and real communication complexity at most t. Then*

$$s \geq 2^{\Omega((n/(t\log n))^{1/2})}.$$

Proof By Lemma 18.2.4 (the error is bounded by $s\epsilon^{-1}$):

$$C_\epsilon^{pub}(KW^m[BPM^+, BPM^-]) \leq \log s\, t\, O(\log n + \log \epsilon^{-1} + \log s).$$

By [429, Theorem 4.4],

$$C_0^{pub}(KW^m[BPM^+, BPM^-]) = \Omega(n)$$

while, by [428, Lemma 1.4] for any R,

$$C_0^{pub}(R) \leq (C_\epsilon^{pub}(R) + 2)(\log_{1/\epsilon} n + 1).$$

Taking $\epsilon := n^{-1}$ we get

$$\log^2 s = \Omega(\frac{n}{t \log n})$$

and the lemma follows. □

By combining Theorem 18.2.2 with Lemma 18.2.5 we get

Corollary 18.2.6 (Impagliazzo, Pitassi and Urquhart [244]) *Let π be a tree-like semantic* CP-*refutation of the inequalities $Hall_n$ (Section 17.2) with k steps.*

Then

$$k \geq \exp\left(\Omega(\frac{n}{\log n})^{1/2}\right).$$

18.3 NS, PC and Span Programs

We shall consider the algebraic proof systems NS and PC over a finite field or over the rationals, in order to have a priori an efficient way to represent the field elements and to perform the field operations. In this section the symbol **F** will stand for one of these fields.

Let us first recall a few facts from linear algebra. Let V be a D-dimensional vector space over **F**. For vectors $v_1, \dots, v_a \in V$, $Span(v_1, \dots, v_a)$ is the linear subspace generated by the vectors. The question whether $w \in Span(v_1, \dots, v_a)$, $w \in V$, is equivalent to the question whether a particular system of D linear equations with a unknowns has a solution. This is answered by Gaussian elimination: after bringing the system into a row-reduced form we can immediately see the answer, and if it is affirmative we can read from it the coefficients of a linear combination of the v_i witnessing it. The total number of field operations that this takes is bounded by $O(Da\min(D, a)) \leq O(Da^2)$, if $a \leq D$, and this also bounds the time if **F** is finite. For $\mathbf{F} = \mathbf{Q}$ we need to consider also the bit size of the coefficients that appear in the computation (this is often overlooked). Assume that each coefficient occurring in the vectors w and v_i is expressed explicitly as a fraction c/d of two integers and that b bounds the bit size of all these c, d. The bit size of the input to the problem is now $O(aDb)$. An analysis of the Gaussian process in which we reduce all intermediate fractions to their lowest terms shows that the bit size of each integer appearing in the process can be kept polynomial.

Lemma 18.3.1 *For* **F** *a finite field or* $\mathbf{F} = \mathbf{Q}$*, the problem* $w \in_? Span(v_1, \dots, v_a)$ *can be solved in polynomial time, that is, in a time bounded above by* $(a+D+b)^{O(1)}$*, where* D *is the dimension of the vectors and* b *bounds the bit size of the fractional representation of their coefficients (we may put* $b := 1$ *for finite* **F**).

In the next lemma we talk about bounded-degree NS and PC proof systems although they are not complete. Using Lemma 18.3.1 it is easy to derive

Lemma 18.3.2 *For* **F** *a finite field or* $\mathbf{F} = \mathbf{Q}$ *and any fixed degree* d*, the degree* $\leq d$ *proof systems* NS/**F** *and* PC/**F** *are automatizable.*

In particular, there is a deterministic algorithm that, for any unsolvable system $\mathcal{F}$ *of* m *polynomial equations of degree at most* d_0 *in* n *unknowns, finds in a time* $O(d(m + n^d + b)^{O(1)})$ *an* NS/**F***-refutation of minimum degree* $d \geq d_0$*, where* b *bounds the bit size of the fractional representation of the coefficients in* $\mathcal{F}$ *for* $\mathbf{F} = \mathbf{Q}$ *(and we put* $b := 1$ *for* **F** *finite).*

The same holds also for PC/**F** *with the time bound* $O(d(m + n^d + b)^{O(d)})$.

Proof For $d \geq d_0$, let $\mathcal{G}_d$ be all polynomials of the form

$$m \cdot f,$$

where $f \in \mathcal{F}$, m is a monomial and $deg(m \cdot f) \leq d$. The polynomials $\mathcal{G}_d$ form a subspace of the vector space V_d of all degree $\leq d$ polynomials; its dimension is $D = n^{O(d)}$.

The existence of an NS-refutation of $\mathcal{F}$ of degree $\leq d$ is equivalent to the question whether $1 \in Span(\mathcal{G}_d)$, where 1 is the vector corresponding to the constant-1 polynomial, and any solution of the corresponding linear system contains the coefficients of a refutation. By Lemma 18.3.1 this can be computed in polynomial time.

If d is not given, we try to find a refutation for $d = d_0, d_0+1, \ldots$ until we succeed.

For PC, we proceed similarly but in more rounds. Assume that $d \geq d_0$ is fixed and put $\mathcal{H}_0 := \mathcal{F}$. For rounds $t = 0, 1, \ldots$ do the following:

- for each variable x_i and each $h \in \mathcal{H}_t$ such that $deg(h) < d$, test whether $x_i h \in Span(\mathcal{H}_t)$ and, if not, add it to form $\mathcal{H}_{t+1}$;
- proceed until nothing can be added, so that $\mathcal{H}_t = \mathcal{H}_{t+1}$.

Because the dimension is bounded by $n^{O(d)}$, the process stops after at most $n^{O(d)}$ rounds. Each round involves at most $n|\mathcal{H}_t| \leq nn^{O(d)}$ tests, each solvable – by Lemma 18.3.1 – in polynomial time.

The lemma for PC follows from the next, obvious, claim.

Claim *If $\mathcal{H}_t$ is the least fixed point of the process then, for all $h \in V_d$, $h \in \mathcal{H}_t$ if and only if h has a degree $\leq d$ PC proof from $\mathcal{F}$.*

□

Note that for $\mathbf{F} = \mathbf{Q}$, the size of the interpolating circuit depends also on the maximum bit size of the coefficients in the original refutation. Lemmas 18.3.2 and 17.3.1 yield feasible interpolation.

Corollary 18.3.3 *For any fixed $d \geq 1$ and $\mathbf{F}$ a finite field or $\mathbf{F} = \mathbf{Q}$, the systems* NS/$\mathbf{F}$ *and* PC/$\mathbf{F}$ *restricted to degree $\leq d$ proofs admit feasible interpolation.*

The automatizing algorithms do not seem to yield monotone feasible interpolation but an analysis of the argument for NS in the proof of Lemma 18.3.2 does this. It uses the notion of span programs for computing Boolean functions.

Let $f\colon \{0,1\}^n \to \{0,1\}$ be a Boolean function. A **span program** (tacitly over $\mathbf{F}$) is a list of vectors A: $v_1, \ldots, v_a$ from a D-dimensional vector space over $\mathbf{F}$, each labeled either by x_i or by $\neg x_i$. For $\overline{u} \in \{0,1\}^n$, let $A(\overline{u})$ be the set of those vectors from A that are labeled by x_i if $u_i = 1$ or by $\neg x_i$ if $u_i = 0$, for all i. The **size** of the span program is defined to be the dimension D. The span program computes f if, for all $\overline{u} \in \{0,1\}^n$,

$$f(\overline{u}) = 1 \quad \text{if and only if} \quad 1 \in Span(A(\overline{u})).$$

This program is **monotone** if only positive labels x_i are used. Clearly, a monotone span program computes a monotone Boolean function.

We shall use this notion to establish a form of monotone feasible interpolation for NS. Rather than talking about refutations of sets of clauses represented by polynomials, as in (6.0.1), it is easier to talk directly about refutations of sets of polynomials of a particular form. Let $f_i(\overline{x}, \overline{y})$, $g_j(\overline{x}, \overline{z})$ be degree $\leq d_0$ polynomials over $\mathbf{F}$ in the variables shown. The pair of sets U, V is defined as those $\overline{a} \in \{0,1\}^n$ for which the system $f_i(\overline{a},\overline{y}) = 0$ or the system $g_j(\overline{a},\overline{z}) = 0$, respectively, is solvable.

We say that the polynomial system obeys the **monotone FI set-up with respect to U** if and only if all polynomials f_i have the form

$$\prod_{j \in J_i} (1 - x_j) f'_i(\overline{y}), \tag{18.3.1}$$

where $J_i \subseteq [n]$ (for $J_i = \emptyset$ define the product to be 1): changing some a_i from 0 to 1 can only make more terms $\prod_{j \in J_i}(1 - x_j)$ equal to 0 and hence also preserve $\overline{a} \in U$. Analogously, the system obeys the **monotone FI set-up with respect to V** if and only if all polynomials g_j have the form

$$\prod_{k \in J_j} x_k g'_j(\overline{z}). \tag{18.3.2}$$

For example, the pair of sets $\mathrm{Clique}_{n,\omega}$ and $\mathrm{Color}_{m,\xi}$ can be defined using a system obeying the monotone FI set-up with respect to both sets (in analogy to its definition by clauses in Section 13.5):

- $(\sum_{i \in [n]} y_{ui}) - 1$, for all $u \in [\omega]$,
- $y_{ui}y_{uj}$, for all $u \in [\omega]$ and $i \neq j \in [n]$,
- $y_{ui}y_{vi}$, for all $u \neq v \in [\omega]$ and $i \in [n]$,
- $y_{ui}y_{vj}(1 - x_{ij})$, for all $u \neq v \in [\omega]$ and $\{i,j\} \in \binom{n}{2}$,

and

- $(\sum_{a \in [\xi]} z_{ia}) - 1$, for all $i \in [n]$,
- $z_{ia}z_{ib}$, for all $a \neq b \in [\xi]$ and $i \in [n]$,
- $z_{ia}z_{ja}x_{ij}$, for all $a \in [\xi]$ and $\{i,j\} \in \binom{n}{2}$.

The size of a span program is defined in a way that does not refer to the bit size of the coefficients (i.e. the elements of $\mathbf{F}$ have unit size) and hence we do not need to restrict to finite fields or the rationals.

Theorem 18.3.4 (Pudlák and Sgall [423]) *For any fixed $d \geq 1$ and $\mathbf{F}$ any field,* NS/$\mathbf{F}$ *restricted to degree $\leq d$ proofs admits monotone feasible interpolation by monotone span programs.*

In particular, if $f_i(\overline{x},\overline{y})$, $g_j(\overline{x},\overline{z})$ are degree $\leq d_0$ polynomials over $\mathbf{F}$ that obey the monotone FI *set-up with respect to V and have degree $d \geq d_0$* NS/$\mathbf{F}$ *refutation then there is a monotone span program of size $n^{O(d)}$ that separates U from V.*

Proof First note that we may assume without loss of generality that all the sets J_j in (18.3.2) are either empty or singletons: if $J_j = \{k_1, \ldots, k_r\}$, $r \geq 2$, replace the original polynomial (18.3.2) by

$$x_{k_1} w_2 \ldots w_r g'_j(\overline{z}),$$

together with $r-1$ polynomials

$$x_{k_u} w_u, \quad \text{for } u = 2, \ldots, r,$$

where $w_2, \ldots, w_r$ are new variables, and also include all $w_u^2 - w_u$. It is easy to derive the original polynomials (18.3.2) from this set by a degree-(d_0+1) NS proof. We shall thus assume without loss of generality that the original polynomials g_j already satisfy this restriction.

Let π be the degree $\leq d$ NS-refutation provided by the hypothesis. We want to construct from π a span program A separating U from V. If $\{f_i(\overline{a}, \overline{y}) = 0\}_i$ is solvable, say by $\overline{y} := \overline{b}$, then substituting for $\overline{x}, \overline{y}$ in π the values $\overline{a}, \overline{b}$ yields a degree $\leq d$ NS-refutation of $\{g_j(\overline{a}, \overline{z})\}_j$.

This refutation uses a polynomial $g_j(\overline{a}, \overline{z})$ if no x-variable occurs in g_j, or if x_k occurs in it and $a_k = 1$. In other words, a degree $\leq d$ polynomial in $\overline{z}$ (considered as a vector in a space whose basis consists of all degree $\leq d$ monomials) has a degree $\leq d$ NS proof from $g_j(\overline{a}, \overline{z})$ if and only if it is in the span of the following vectors:

- all degree $\leq d$ polynomials of the form $m(\overline{z}) g_j(\overline{z})$, where m is a monomial and no x-variable occurs in g_j,
- all degree $\leq d-1$ polynomials of the form $m(\overline{z}) g_j(\overline{z})$, where m is a monomial, g_j has the form (18.3.2) and $a_k = 1$.

Hence the span program consisting of all vectors in the first item labeled by 1 and all vectors in the second item labeled by x_k separates U from V. □

Super-polynomial lower bounds for monotone span programs are known (see also Section 18.8). We shall formulate the second part of the following statement rather generically because the pair there is more difficult to define than the previous examples.

Theorem 18.3.5

(i) (Babai *et al.* [43]) *Let* **F** *be any field and let* $f_i(\overline{x}, \overline{y})$, $g_j(\overline{x}, \overline{z})$ *be the polynomials over* **F** *described above that define the pair of sets* $U :=$ Clique$_{n,n/2}$ *and* $V :=$ Color$_{n,n/2-1}$.

Then any monotone span program over **F** *separating* U *from* V *must have size at least* $n^{\Omega(\log n / \log \log n)}$.

(ii) (Robere *et al.* [451]) *There is a set of* $n^{O(1)}$ *polynomials* $f_i(\overline{x}, \overline{y})$, $g_j(\overline{x}, \overline{z})$ *over* **Q** *of* $O(1)$ *degree and in* $n^{O(1)}$ *variables that obeys the monotone* FI *set-up with respect to* V *such that every monotone span program over* **Q** *that separates* U *from* V *must have size at least* $n^{n^{\Omega(1)}}$.

The last two theorems can be used to prove a degree-$n^{\Omega(1)}$ lower bound for NS/**Q** (and NS/**R**) and $\Omega(\log n / \log \log n)$ for any **F**, although this route seems to be more difficult than the original lower bound argument in Section 16.1.

18.4 The Lovász–Schrijver Proof System

Recall from Section 6.5 the proof system LS. We shall extend the splitting proofs approach from Sections 17.7 and 18.1 to establish non-monotone feasible interpolation for the system. In fact, we can allow the union of CP and LS, i.e. we may add the rounding rule of CP to LS.

Let $A_i(\overline{x},\overline{y})$ and $B_j(\overline{x},\overline{z})$ be integer linear inequalities obeying the FI set-up. Assume that π is an LS-refutation of this system. Given an assignment $\overline{x} := \overline{a}$ we want to split π and to get a refutation either of the clauses $A_i(\overline{a},\overline{y})$ or of the clauses $B_j(\overline{a},\overline{z})$, and we want to do this splitting by a polynomial-time algorithm.

Lovász–Schrijver refutations contain in general quadratic inequalities and we cannot split these into two inequalities separating the $\overline{y}$ and the $\overline{z}$ variables, as we did for the linear inequalities in CP refutations in Section 18.1. However, if in a refutation we create some quadratic inequalities then the only thing we can do with them is to form a suitable positive linear combination in which all the degree-2 monomials cancel out. That is, we may look at an LS-refutation as deriving in rounds, from a set of linear inequalities already derived, a set of quadratic inequalities from which a new linear inequality (or more than one) is derived by positive linear combinations. This view of refutations is faithful to the original Lovász–Schrijver lift-and-project method.

From the LS-refutation π construct (in p-time) a sequence of *linear* inequalities

$$\tilde{\pi}: L_1, \ldots, L_k$$

such that each L_t, $t \leq k$, is either one of the initial inequalities A_i or B_j or is obtained from $L_1, \ldots, L_{t-1}$ by taking a positive linear combination of some inequalities of the following types:

1. $L_1, \ldots, L_{t-1}$,
2. u and $1 - u$,
3. $u^2 - u \geq 0$ and $u - u^2 \geq 0$,
4. uv, $u(1 - v)$ and $(1 - u)(1 - v)$,
5. uL_s and $(1 - u)L_s$ for $s < t$,

where u, v are any of the $\overline{x}, \overline{y}, \overline{z}$ variables.

Given an assignment $\overline{x} := \overline{a} \in \{0, 1\}^n$, we substitute it into the refutation to get $\tilde{\pi}(\overline{a})$ and then split $\tilde{\pi}(\overline{a})$ into two proofs, from clauses $A_i(\overline{x},\overline{y})$ and from clauses $B_j(\overline{a},\overline{z})$, respectively, and see which has the end-line $0 \geq 1$.

Assume that we have already created the derivations of two linear inequalities L_s^y, L_s^z in the $\overline{y}$ and the $\overline{z}$ variables, respectively, for $s < t$ such that these two inequalities imply the original L_s (as in Section 18.1). Take the same positive linear combination that gave L_t in $\tilde{\pi}$ but this time with the split L_s^y and L_s^z. Note that all quadratic terms have to cancel out in the two parts of the linear combination that use inequalities of type 3 for u, type 4 for u, v and type 5 for u and L_s^v, where either u is from $\overline{y}$ and v is from $\overline{z}$, or vice versa. That is, in these parts of the linear combination the variables are split already and we need only to see how to split and derive the linear inequality coming from the remaining part of the inference, which is a positive linear combination of polynomials of the form

$$z_e y_f,\quad z_e(1-y_e),\quad y_f(1-z_e),\quad (1-y_f)(1-z_e),$$
$$z_e L_s^y,\quad (1-z_e)L_s^y,\quad y_f L_s^z,\quad (1-y_f)L_s^z.$$

Write this remaining linear inequality as

$$\overline{b}\cdot\overline{y}+\overline{c}\cdot\overline{z}+d\geq 0. \tag{18.4.1}$$

Note that the integrality conditions $u^2-u\geq 0$ and $u-u^2\geq 0$ are not among these inequalities, and hence the inequality (18.4.1) follows from the inequalities L_s^y and L_s^z, $s<t$, over the reals and is thus a positive linear combination of these inequalities. The y- and the z-parts of this combination express d as $d=d_y+d_z$ (not necessarily integers) and hence we derive

$$\overline{b}\cdot\overline{y}+\lfloor d_y\rfloor\geq 0\quad\text{and}\quad\overline{c}\cdot\overline{z}+\lfloor d_z\rfloor\geq 0. \tag{18.4.2}$$

Lemma 18.4.1 *Given π and $\overline{a}$, the splitting of $\tilde{\pi}(\overline{a})$ can be done in polynomial time.*

Proof The process as described is p-time if one can compute the numbers d_y and d_z in (18.4.2) in p-time. But that can be achieved using a p-time algorithm for linear programming: to find $-d_y$ (and analogously $-d_z$) maximize $\overline{b}\cdot\overline{y}$ subject to the inequalities L_s^y, $s<t$, used in the derivation of (18.4.1). □

Lemma 18.4.1 implies the following theorem. Note that the size of the interpolating circuit depends also on the maximum bit size of the coefficients in the original refutation.

Theorem 18.4.2 (Pudlák [414]) LS *with* CP *admits feasible interpolation.*

By Lemma 17.1.3 this implies that LS is not p-bounded unless NP $\subseteq$ P/*poly*. For an unconditional lower bound, we need monotone feasible interpolation by some computational model for which we can establish an unconditional lower bound. That is open at present but Oliveira and Pudlák [376] defined a new model of monotone linear programming circuits by which LS admits monotone interpolation (Section 18.8).

In the tree-like case the situation is better, and we have an unconditional lower bound. If we have a quadratic polynomial and the variables in its monomials happen to be divided between the players in a game (Section 17.4) then the communication complexity (even if real) for finding out whether the polynomial equals 0 is high. Beame, Pitassi and Segerlind [59] came up with the idea of using three-party communication to bypass this problem. A suitable game uses a **number-on-forehead** set-up: each of three players has a (string of) numbers on his or her forehead that is visible to the other two players but not to him- or herself. They announce (according to some agreed on protocol for the communication) bits of information visible to all. Assume that the variables of a quadratic polynomial are split among the three players and that each has on his or her forehead an evaluation of his or her variables. Then the two values needed to evaluate any one quadratic monomial are visible to at least one player. Hence the players can evaluate the polynomial if each announces the value of the portion of the polynomial that he or she can compute. In particular, the players can solve the *CC* (and the *MCC*) tasks for sets appearing in a semantic LS-refutation by communicating three numbers in $O(\log n)$ rounds. In the dag-like case this would lead to a generalization of the communication complexity of protocols and we do not have a lower bound for it, even in the monotone case. But for the tree-like case an adversary argument as in Lemma 17.4.2 produces ordinary tree-like protocols of a bounded height for the *three-party* game. This allows one to deduce size lower bounds for *tree-like* LS (i.e. LS*) from lower bounds for the multi-party communication complexity, as follows.

The communication complexity problem used for this purpose is the **disjointness problem**: the players have on their foreheads subsets $X, Y, Z \subseteq [n]$, respectively, and their task is to find out whether $X \cap Y \cap Z$ is non-empty. The CNF for which the lower bound is deduced is a substitution instance of a Tseitin formula (Section 13.3) for a suitable constant-degree graph.

We state the following result without proof. The difficult part is not the reduction from tree-like refutations to shallow protocols for the multi-party game but the reduction from the CNF to the disjointness problem.

Theorem 18.4.3 (essentially Beame, Pitassi and Segerlind [59]) *There exist* CNF*s* C_n *of total size* $n^{O(1)}$ *such that each tree-like* LS*-refutation of the* CNF *must have size* $2^{n^{\Omega(1)}}$.

Note that the lower bound is for the size and not for the number of steps. The lower bound applies to the semantic version of LS as well, but it still depends on the bit size of the coefficients defining the sets occurring in a derivation.

18.5 The OBDD Proof System

Recall the definition of the system from Section 7.4. It operates with ordered binary decision diagrams (OBDDs), which are branching programs querying variables in an order consistent with one global linear ordering. We defined it in Section 7.4 as a semantic system using only the inference rule

$$\frac{f_P \quad f_Q}{f_R}, \quad \text{if } f_P \wedge f_Q \leq f_R,$$

where f_P, f_Q, f_R are Boolean functions defined by the OBDDs P, Q, R.

We shall work under the FI set-up. Denote by *Var* the set of all atoms in $\overline{p}$, $\overline{q}$ and $\overline{r}$. We shall call a linear ordering π of *Var* **block consistent** with $q < r < p$ if and only if π puts all $\overline{q}$-variables before all $\overline{p}$- and $\overline{r}$-variables, and all $\overline{r}$-variables before all $\overline{p}$-variables.

Lemma 18.5.1 *Assume that π is a linear ordering of Var that is block consistent with $q < r < p$. Let $A \subseteq \{0,1\}^N$ be a set definable by a π-OBDD P of size S. Then both $CC(A)$ and $MCC_U(A)$ are bounded above by $O(\log(S)\ \log(n))$.*

Proof The U-player embarks on a path through P answering q-queries according to q^u. Then he indicates to the V-player using $\log(S)$ bits the last node that he reached. The V-player then takes over and continues in the path answering all r-queries via r^v, and eventually sends to the U-player $\log(S)$ bits indicating the last node that she reached (it is labeled with the first p-query). Call this node Q. From this node onwards they travel on individual paths (answering p-queries according to u or v, respectively), and eventually send each other one bit: the output they have computed.

If these bits differ then they use binary search to find a node whose p-query they answered differently. This will take at most $\log(n)$ rounds, in which they always send each other $\log(S)$ bits naming the node querying the particular p-variable on their paths. Thus $CC(A) \leq O(\log(S)\ \log(n))$.

Under the monotone FI set-up we need to explain how the players solve task 4 from the definition of $MCC_U(A)$ in the previous section. Assume that $(u, q^u, r^v) \in A$ while $(v, q^u, r^v) \notin A$. Let $Q = Q_0, \ldots, Q_w$ (with Q as above) be the path determined by v (the v-path) ending in the leaf Q_w, which is labeled by 0. In particular, $w \leq n$. First the U-player sends one bit indicating whether there is a $u' \geq u$ such that $(u', q^u, r^v) \notin A$. If so, the players have solved task 4 and they stop.

If no such u' exists, the V-player finds the mid-point $Q_{w/2}$ on her path and sends its name ($\log(S)$ bits) to the U-player. The U-player then sends back one bit indicating whether there is a $u' \geq u$ such that the u'-path from Q_0 leads to $Q_{w/2}$. If yes, the players take the sub-path $Q_{w/2}, \ldots, Q_w$. If not, they continue with the sub-path $Q_0, \ldots, Q_{w/2}$.

In this process the U-player does not know the whole v-path from Q but only the endpoints of the current sub-path considered. After r rounds of this process the players will determine a sub-path of length $\leq n/2^r$ (the U-player knowing only its end-points) with the property that there is a $u' \geq u$ for which the u'-path from Q leads to the starting node of the sub-path but no $u' \geq u$ exists which would lead to the end-node of the sub-path. This means that in at most $\log(n)$ rounds, and exchanging in total at most $\log(n)\ (1 + \log(S))$ bits, the players find a node Q_e on the v-path such that for no $u' \geq u$ does the u'-path lead from Q_e to Q_{e+1}. In particular, if p_i is the label of Q_e then it was not queried earlier and it must be that $v_i = 0$ while $u_i = 1$. This solves task 4 and the players stop. □

The next lemma follows from Theorem 17.5.1 and Lemma 18.5.1 (we estimate the number of lines by the size).

Corollary 18.5.2 *Assume an* FI *set-up. Let* π *be a linear ordering of Var that is block consistent with* $q < r < p$. *Assume that there is a semantic* OBDD*-refutation of the sets* $\tilde{A}_1, \ldots, \tilde{A}_m, \tilde{B}_1, \ldots, \tilde{B}_\ell$ *of size S.*

Then the two sets U and V can be separated by a circuit of size at most $S^{O(\log(n))}$.

Moreover, if the monotone FI *set-up is obeyed then there is a monotone circuit separating U from V of size at most* $S^{O(\log(n))}$.

This interpolation works only for orderings that are block consistent with $q < r < p$, not for general orderings. Our strategy now is to introduce an unknown permutation of the atoms and rewrite the clauses using it, and when these modified clauses are refuted under some ordering we substitute a suitable permutation to recover an ordering for which Lemma 18.5.2 works.

We shall construct a new CNF-formula from the set of clauses A_i and B_j. This new formula will be $D(\overline{w},\overline{f})$, where $\overline{w}$ is an N-tuple of atoms w_i and $\overline{f}$ is an N^2-tuple of atoms f_{ij}, with $i,j \in [N]$. First, we consider the following CNF formula, suggestively denoted by $Map(\overline{f})$, which expresses that the atoms f_{ij} define a graph $\{(i,j) \mid f_{ij} = 1\}$ of a permutation on $[N]$:

$$\bigwedge_i \bigvee_j f_{ij} \wedge \bigwedge_j \bigvee_i f_{ij} \wedge \bigwedge_{i_1 \neq i_2, j} (\neg f_{i_1 j} \vee \neg f_{i_2 j}) \wedge \bigwedge_{i, j_1 \neq j_2} (\neg f_{ij_1} \vee \neg f_{ij_2}).$$

Further, define the formulas

- P_j^1 and P_j^0 by

$$P_j^1 := \bigwedge_{i \in [N]} \neg f_{ij} \vee w_i \quad \text{and} \quad P_j^0 := \bigwedge_{i \in [N]} \neg f_{ij} \vee \neg w_i$$

  for $j \in [n]$,
- formulas Q_j^1 and Q_j^0 by

$$Q_j^1 := \bigwedge_{i \in [N]} \neg f_{i(n+j)} \vee w_i \quad \text{and} \quad Q_j^0 := \bigwedge_{i \in [N]} \neg f_{i(n+j)} \vee \neg w_i$$

  for $j \in [s]$
- and formulas R_j^1 and R_j^0 by

$$R_j^1 := \bigwedge_{i \in [N]} \neg f_{i(n+s+j)} \vee w_i \quad \text{and} \quad R_j^0 := \bigwedge_{i \in [N]} \neg f_{i(n+s+j)} \vee \neg w_i$$

  for $j \in [t]$.

Assuming that $Map(\overline{f})$ is true, the formulas P_j^1 and P_j^0, and similarly the Q- and the R-formulas, are complementary.

The CNF formula $D(\overline{w},\overline{f})$ is defined to be the conjunction of the clauses of $Map(\overline{f})$ together with the clauses obtained by the following process. For any 3-clause C, which is one of A_i or B_j, construct a CNF $\tilde{C}$ as follows.

1. Replace each positive literal p_j in C by the formula P^1_j and each negative literal $\neg p_j$ by P^0_j, and similarly for the atoms q and r. The resulting formula is a disjunction of three conjunctions, each being itself a conjunction of N 2-clauses.
2. Use its distributivity to replace this formula by a conjunction of N^3 6-clauses: the formula $\tilde{C}$.
3. Each 6-clause of $\tilde{C}$ is a clause of $D(\overline{w},\overline{f})$.

Lemma 18.5.3 *Assume that A_i, B_j are 3-clauses. Let σ be an arbitrary linear ordering of the variables $\overline{w},\overline{f}$ and let π be an arbitrary linear ordering of the variables $\overline{p},\overline{q},\overline{r}$.*

If $D(\overline{w},\overline{f})$ has a σ-OBDD-refutation of size S then the clauses A_i, B_j have a π-OBDD-refutation of size at most S.

Proof Enumerate the $\overline{w}$ atoms in the order induced by σ:

$$w_{i_1} < w_{i_2} < \cdots < w_{i_N} ,$$

$\{i_1, \ldots, i_N\} = [N]$. Let $F : [N] \to [N]$ be a permutation such that $F(i_a)$, for $a \in [N]$, is defined as

- j, if the ath element of π is p_j,
- $n + j$, if the ath element of π is q_j,
- $n + s + j$, if the ath element of π is r_j.

With this permutation in hand, in the whole refutation substitute $f_{uv} := 1$ if $F(u) = v$, and $f_{uv} := 0$ otherwise. If P_j is the ath element of the ordering π, after the substitution the formula P^1_j reduces to w_{i_a} (the ath variable of $\overline{w}$ in σ) and P^0_j reduces to $\neg w_{i_a}$ (and analogously for the q- and the r-variables). This means that the ordering of these resulting P-, Q- and R-formulas determined by the ordering σ of $\overline{w}$ is identical to the ordering by π of the $\overline{p}$-, $\overline{q}$- and $\overline{r}$-atoms.

The above substitution also satisfies all clauses of *Map*$(\overline{f})$ and so the σ-OBDD refutation with which we started is transformed into a π-OBDD refutation of the set of initial clauses A_i, B_j. □

We shall use the clique-coloring pair as before but rewritten using auxiliary variables so as to replace the large clauses in the Clique and Color formulas by sets of 3-clauses.

Corollary 18.5.4 *Let A_i, B_j be 3-clauses expressing the disjointness of the sets* $\text{Clique}_{n,n^{2/3}}$ *and* $\text{Color}_{n,n^{1/3}}$ *and obeying the monotone* FI *set-up. Let $D(\overline{w},\overline{f})$ be the* CNF *defined from clauses A_i, B_j as above.*

Then every semantic OBDD-*refutation of $D(\overline{w},\overline{f})$ must have size at least $2^{\Omega(n^{1/6}/\log n)}$.*

Proof Take an arbitrary linear ordering π of the p-, q- and r atoms in the clauses A_i, B_j that is block consistent with $q < r < p$. Let σ be any ordering of the atoms $\overline{w}$ and $\overline{f}$ of D and assume that D has a size S σ-OBDD refutation.

Lemma 18.5.3 implies that there is a π-OBDD proof of the disjointness of $\text{Clique}_{n,n^{2/3}}$ and $\text{Color}_{n,n^{1/3}}$ of size at most S. But, by Corollary 18.5.2, any such refutation yields a monotone circuit of size at most $S^{O(\log(n))}$ separating the two sets. Theorem 13.5.3 then implies that $S^{O(\log(n))} \geq 2^{\Omega(n^{1/6})}$, which yields the corollary. □

18.6 R(LIN), CP and Randomized Protocols

Recall from Section 7.1 that R(LIN) is a proof system operating with clauses of linear functions $C := \{f_1, \ldots, f_w\}$. If a set $A \subseteq \{0,1\}^N$ is defined by such a clause then its (monotone) communication complexity $(M)\ CC(A)$ is bounded by $\max(2w, \log n)$; if w is large, say larger than n, then this gives no useful information. However, there is a simple process replacing C with another clause $\tilde{C}$ having a small **linear width** (i.e. cardinality) and defining a set $\tilde{A} \subseteq A$ such that $A \setminus \tilde{A}$ is small. The clause $\tilde{C}$ is constructed by a random process: select randomly and independently $J_1, \ldots, J_\ell \subseteq [w]$ and define

$$\tilde{C} := \{\sum_{j \in J_i} f_j \mid i = 1, \ldots, \ell\}\,.$$

Any $\overline{a} \in \{0,1\}^n$ satisfying $\tilde{C}$ clearly satisfies C as well, and if $\overline{a}$ satisfies C then, with probability $\geq 1 - 2^{-\ell}$ over the choice of the sets J_i it also satisfies $\tilde{C}$. By averaging there is a choice for the sets such that $|A \setminus \tilde{A}| \leq 2^{n-\ell}$. Hence if we have an R(LIN) refutation of clauses A_i, B_j in the FI set-up, we may first replace any C in it that has a large linear width by a random $\tilde{C}$ with a smaller linear width, but approximating C well, and use this modified refutation as the basis for a protocol. The protocol will have some error but not too much.

A similar situation arises with CP. There we have protocols with small real communication complexity and we may use Lemma 18.2.4 to simulate the computation of the strategy and the consistency condition using the ordinary KW-game, again introducing some, but not too much, error. These considerations lead to the notion of a randomized protocol. We will work under the FI set-up.

A **randomized protocol** for a multi-function $R \subseteq U \times V \times I$ with error $\epsilon > 0$ is a random variable $(\mathbf{P_r})_\mathbf{r}$ where each $\mathbf{P_r}$ is a 4-tuple satisfying the conditions 1–3 and 4(a) from Section 17.4 defining protocols but, instead of the conditions 4(b) and 4(c), it satisfies weaker ones:

(4b′) For every $(u, v) \in U \times V$,

$$Prob_\mathbf{r}[\exists x,\ x \in F_\mathbf{r}(u, v) \wedge S_\mathbf{r}(u, v, x) \notin F_\mathbf{r}(u, v)] \leq \epsilon.$$

(4c′) For every $(u, v) \in U \times V$,

$$Prob_\mathbf{r}[\exists \text{leaf}\, x,\ x \in F_\mathbf{r}(u, v) \wedge \mathsf{lab}_\mathbf{r}(x) = i \wedge \neg R(u, v, i)] \leq \epsilon.$$

The **size** of $(\mathbf{P_r})_\mathbf{r}$ is defined to be $\max_\mathbf{r} size(\mathbf{P_r})$ and its **communication complexity** is $\max_\mathbf{r} CC(\mathbf{P_r})$. If each $\mathbf{P_r}$ is tree-like then we call $(\mathbf{P_r})_\mathbf{r}$ also **tree-like**.

One type of error can be eliminated. Introduce for each inner node $w \in G_\mathbf{r}$ a new leaf $\tilde{w}$ and define a new strategy $\tilde{S}_\mathbf{R}$ that first checks whether

$$w \in F_\mathbf{r}(u, v) \to S_\mathbf{r}(u, v, w) \in F_\mathbf{r}(u, v)$$

holds true. If so then it is defined as $S_\mathbf{r}$; if not, it sends w into $\tilde{w}$ and the failure of the condition is the definition of $\tilde{w} \in \tilde{F}_\mathbf{r}(u, v)$. Hence we have

Lemma 18.6.1 *For any randomized protocol* $(\mathbf{P_r})_\mathbf{r}$ *of size S, communication complexity t and error* ϵ *there exists a randomized protocol* $(\tilde{\mathbf{P}}_\mathbf{r})_\mathbf{r}$ *computing the same multi-function such that (4b) never fails and which has size at most* $2S$*, communication complexity at most* $3t$ *and error* ϵ*.*

We shall concentrate on randomized protocols for $KW^m[U, V]$ only, as there are small protocols having a small error that compute the non-monotone $KW[U, V]$ for any disjoint pair (Section 18.8).

The next notion targets semantic derivations. For a set $X \in \{0, 1\}^N$ and a random distribution of sets $\mathcal{Y} = (Y_\mathbf{r})_\mathbf{r}$, which are all subsets of $\{0, 1\}^N$, and $\delta > 0$, we say that $\mathcal{Y}$ is a δ**-approximation of** X if and only if, for all $w \in \{0, 1\}^N$,

$$Prob_\mathbf{r}[w \in X \triangle Y_\mathbf{r}] \leq \delta,$$

where $X \triangle Y$ is the symmetric difference. We define the **(monotone) communication complexity** of $\mathcal{Y}$ to be $\leq t$ if and only if this is true for all $Y_\mathbf{r}$, and the δ**-approximate (monotone) communication complexity** of X to be $\leq t$ if and only if there is a δ-approximation $\mathcal{Y}$ of X with this property.

Theorem 18.6.2 *Assume a monotone* FI *set-up and let* $\pi\colon D_1, \ldots, D_k = \emptyset$ *be a semantic refutation of sets* $\tilde{A}_1, \ldots, \tilde{A}_m, \tilde{B}_1, \ldots, \tilde{B}_\ell$ *such that the* δ*-approximate monotone communication complexity of every* D_i *is at most t.*

Then there is a randomized protocol for $KW^m[U, V]$ *of size at most* $k + n$*, communication complexity* $O(t)$ *and of error at most* $3\delta k$*.*

Moreover, if π *is tree-like, so is* $(\mathbf{P_r})_\mathbf{r}$*.*

Proof The strategy and the consistency condition of the protocol $P = (G, \mathsf{lab}, F, S)$ provided by Theorem 17.5.1 are defined in terms of the sets D_i. In particular, for any $(u, v) \in U \times V$ and an inner node $w \in G$, both the value of $S(u, v, w)$ and the truth value of $w \in F(u, v)$ are defined from at most three truth values of statements of the form $(u, q^u, r^v) \in D_i$ or $(v, q^u, r^v) \in D_i$ for some specific indices $i \leq k$ determined by w.

Assume that $(E^i_\mathbf{s})_\mathbf{s}$ are δ-approximations of the sets D_i, $i \leq k$. We let the sample space of $\mathbf{r}$ for $\mathbf{P_r}$ be the product of the sample spaces of these k δ-approximations and define the strategy $S_\mathbf{r}$ and the consistency condition $F_\mathbf{r}$ as in the original protocol but using the sets $E^i_\mathbf{s}$ (with $\mathbf{s}$ determined by $\mathbf{r}$) in place of the sets D_i. The underlying graph and the labeling lab remain the same.

For any given $(u, v) \in U \times V$ and w a vertex of G, the (truth) value of $S_{\mathbf{r}}$ and $F_{\mathbf{r}}$ differs from the original values with probability bounded above by 3δ. Hence for (u, v) the error in the conditions (4b′) and (4c′) is at most $\epsilon := 3\delta k$. □

Combining this with the discussion at the beginning of the section we get a monotone feasible interpolation for R(LIN) and CP by monotone randomized protocols.

Corollary 18.6.3 *Assume a monotone* FI *set-up.*

(i) Assume that $A_1, \dots, A_m, B_1, \dots, B_\ell$ are defined by R(LIN)*-clauses.*
Let π be a semantic R(LIN) *refutation (i.e. only sets definable by an* R(LIN) *clause are allowed) of these sets with k steps. Let $w \geq 1$ be any parameter.*
Then there is a randomized protocol for $KW^m[U, V]$ of size at most $k + n$, communication complexity $O(w \log n)$ and error at most $3 \cdot 2^{-w}k$.

(ii) Assume that the sets $A_1, \dots, A_m$ and $B_1, \dots, B_\ell$ are defined by integer linear inequalities and that there is a semantic CP*-refutation π of the sets $\tilde{A}_1, \dots, \tilde{A}_m, \tilde{B}_1, \dots, \tilde{B}_\ell$ that has k steps.*
Then for any $\epsilon < 1/k$, there is a randomized protocol for $KW^m[U, V]$ of size $k + n$, communication complexity $O(\log(n/\epsilon))$ and error at most ϵk.

Moreover, in both cases, if the refutation π is tree-like then G is also tree-like.

Proof For the first part we need the following estimate from [312, Theorem 5.1] (we refer the reader to that publication).

Claim *Let D be an* R(LIN)*-clause of linear width w. Then $MCC_U(D) = O(w \log n)$.*

The second part follows from Theorems 18.2.2 and 18.2.3. □

To use the monotone feasible interpolation for length-of-proofs lower bounds we need a lower bound for randomized protocols computing some $KW^m[U, V]$. But in this case we seem to be in a better position than in the cases of the monotone computational models considered in earlier sections: there is a monotone circuit model related to randomized protocols (Section 18.8).

As in the case of LS we have lower bounds for R(LIN) at least in the tree-like case. The next theorem is proved in a completely analogous way to Corollary 18.2.6, using Corollary 18.6.3 (the first lower bound for tree-like R(LIN) was proved by Itsykson and Sokolov [252] differently, for another formula; see Section 18.8). We use the formula $Hall_n$ (Section 17.2) written using clauses instead of linear inequalities.

Theorem 18.6.4 *Any tree-like semantic* R(LIN) *refutation of clauses defining $Hall_n$ must have at least*

$$\exp\left(\Omega(\frac{n}{\log n})^{1/2}\right)$$

steps.

The problem of establishing a lower bound for general R(LIN) can be also approached via protocols where the players compute the consistency conditions and the strategy by the real game (in fact, a weakening of it); see Section 18.8.

18.7 Limits and Beyond

Are there any limits to the feasible interpolation method? If the question is meant in the narrow sense of feasible interpolation by (monotone) Boolean circuits then the answer is yes: an *unconditional* yes for the monotone version and a *conditional* yes for the non-monotone version.

But if we interpret the question broadly and ask whether there are any limits to the feasible interpolation idea then the answer is open and I tend to think that it is no, there are no limits. I shall return to this at the end of the section (and again in Chapter 22) but we start with the narrow (and well-defined) case.

There is an a priori limitation on how strong a proof system can be if it is to admit monotone feasible interpolation in the narrow sense: it must not be able to prove the WPHP. The NP definitions of the sets $\text{Clique}_{n,\omega}$ and $\text{Color}_{n,\xi}$ talk about maps

$$f\colon [\omega] \to [n] \quad \text{and} \quad g\colon [n] \to [\xi],$$

coded by atoms $\overline{q}$ and $\overline{r}$, respectively. If we compose the maps to $h := g \circ f$ then we get a map $h\colon [\omega] \to [\xi]$. Assuming that $\overline{p}$ defines a graph in the intersection $\text{Clique}_{n,\omega} \cap \text{Color}_{n,\xi}$, the map h is injective: the range of f is a clique and any two vertices from it must acquire different colors from g. That is, assuming the FI set-up, from the initial clauses A_i, B_j we can obtain a short proof of $\neg\text{WPHP}^{\omega}_{\xi}(\overline{h})$, where $\overline{h}$ stands for $\omega\xi$ formulas h_{ab}, $a \in [\omega], b \in [\xi]$:

$$h_{ab} := \bigvee_{i \in [n]} q_{ai} \wedge r_{ib}.$$

In particular, this derivation can be performed in R(2) (Section 5.7) within size $n^{O(1)}$. Choosing $\omega := n^{2/3}$ and $\xi := n^{1/3}$ and combining this with Corollary 11.4.5 and Theorem 13.5.3 we get the following statement.

Theorem 18.7.1 *No proof system simulating* R(log) *admits monotone feasible interpolation by monotone Boolean circuits.*

To establish a similar result in the non-monotone case we show that a strong enough proof system gives in a short proof the disjointness of an RSA pair or the hard bit predicate of an OWP pair by proving the injectivity of a suitable OWP; see Lemma 17.2.6. The definition of the RSA pair that follows the lemma uses the exponentiation of numbers modulo some number $N \geq 2$. It is well known that (by repeated squaring modulo N) this function is p-time computable. The algorithm is easily definable in Cook's PV or in S^1_2 and its usual properties, such as $x^{y+z} \equiv$

$x^y x^z \pmod N$ and $x^{yz} \equiv (x^y)^z \pmod N$, are provable there. Note that these proofs need only induction on the *length* of the numbers involved, and that is available in the theories.

Theorem 18.7.2 (Krajíček and Pudlák [321]) *The theory S^1_2 proves that the RSA pair as defined in Section 17.2 is disjoint. Therefore EF does not admit feasible interpolation by Boolean circuits unless RSA is not secure against* P/*poly adversaries.*

Proof It suffices to prove in S^1_2 that for every v there is at most one $u < N$ satisfying the definition of either RSA_0 or RSA_1. Suppose for the sake of contradiction that, for some $u_0, u_1 < N, r_0, d_0, d_1$, it holds that $v^{r_0} \equiv 1 \pmod N$, $(e, r_0) = 1$ and also

$$u_i^e \equiv v \pmod N \wedge v^{d_i} \equiv u_i \pmod N$$

for $i = 0, 1$. Euclid's algorithm is definable in S^1_2 and thus we can prove the existence of an inverse d' to e modulo r_0. Using

$$u_i^{r_0} \equiv (v^{d_i})^{r_0} \equiv (v^{r_0})^{d_i} \equiv 1 \pmod N$$

for $i = 0, 1$, we derive

$$v^{d'} \equiv u_i^{ed'} \equiv u_i \pmod N,$$

and so $u_0 = u_1$ as required.

The statement about *EF* follows from the p-simulation of S^1_2 by *EF* (Theorem 12.4.2). □

The issue one has to confront when trying to establish a similar result for a weaker proof system is to make the arithmetical operations used in the definition of a suitable NP pair definable by formulas of the weaker system and their properties utilized in the computation involved in proving the disjointness of the NP pair provable by a short proof in this proof system.

The addition, and its iteration, and the multiplication of n-bit numbers are definable by TC^0-formulas, and their properties are provable in a corresponding bounded arithmetic theory (in the sense of Theorem 10.2.3) (see e.g. Cook and Nguyen [155, Chapter IX]). It is expected that the iterated multiplication, repeated squaring and exponentiation, all modulo a number, are not in TC^0, but there is no proof of this. Hence, to be able to formalize a computation involving exponentiation modulo a number, one has to find a pair of disjoint sets for which the computations of these terms can be part of the common information but which do not reveal the secret parts. This can be done using the Diffie–Hellman [169] commitment scheme.

Theorem 18.7.3

(i) (Bonet, Pitassi and Raz [97])

The TC^0-Frege systems (and hence also Frege systems) do not admit feasible interpolation unless the Diffie–Hellman scheme is not secure against adversaries in P/*poly.*

(ii) (Bonet *et al.* [93])

There is $d \geq 2$ such that a depth d Frege system (in the DeMorgan language) does not admit feasible interpolation unless the Diffie–Hellman scheme is not secure against non-uniform adversaries working in a subexponential time $2^{n^{o(1)}}$.

The question whether low-depth AC^0-Frege systems including R(log) or even R(2) admit feasible interpolation in the narrow sense is closely linked to the automatizability of the resolution and related proof systems, and we shall discuss this in Section 21.5.

Let us now leave aside the narrow interpretation of feasible interpolation via Boolean circuits or similar computational models and see whether there are variants of the underlying idea that could apply to strong systems.

One option is to use the reflection pair from Section 17.2 and Lemma 17.2.5. This reduces the task to showing (perhaps using some established computational complexity conjecture) that a certain pair of disjoint NP sets (determined by the proof system) is not complete among all such pairs.

Another option is to demand that the proof system proves more than the mere disjointness of two NP sets. An example of this is *chain interpolation*, to be described at the end of Section 18.8. However, these approaches lead to difficult combinatorial problems that are not easily conceptualized for even moderately strong proof systems.

I view the question if and how the feasible interpolation idea can be extended to strong proof systems as a specific instance of the following informal problem:

- *Is proof complexity an essentially different hardness measure from computational complexity?*

To be a little more specific:

- *Given a proof system P, can the proof complexity of a tautology τ be traced back to the computational complexity of a task explicitly associated with τ?*

I think that this problem, although informal at present, is fundamental and underlies the three problems discussed in Section 1.5. In Part IV we shall discuss two theories (proof complexity generators and forcing with random variables) that aim at giving a tentative positive answer (see also Chapter 22).

18.8 Bibliographical and Other Remarks

Sec. 18.1 is largely based on Pudlák [412]. Theorem 18.1.3 and Corollary 18.1.5 hold for the semantic version of CP as well but Filmus, Hrubeš and Lauria [183] showed that, in general, the semantic version is stronger.

Sec. 18.2 The concepts of the real game and the real communication complexity, as well as protocols from semantic CP refutations, are from [283]. Here, we have switched the response of the referee: he announces 1 if $r_U > r_V$ (in [283] it was the other way around). This is in order to make the game itself *monotone* in a sense: if at any point the U-player increases the numbers she sends to the referee (or the V-player decreases them), the verdicts of the referee can only switch from 0 to 1, not the other way around. This could be useful for an eventual lower bound proof for the protocols.

It is clear that a monotone real circuit separating U from V (under the monotone set-up) can be turned into a protocol of about the same size and small real communication complexity. However, it is open whether an opposite transformation also exists, i.e. whether a statement analogous to Theorem 17.4.3 holds. Hrubeš and Pudlák [236] proved such a statement for a subclass of protocols with small real communication complexity.

The lower bound in Corollary 18.2.6 was obtained by Impagliazzo, Pitassi and Urquhart [244] for a slightly different formalization of the Hall theorem.

The real communication complexity is used in Bonet *et al.* [94] to prove a depth lower bound for monotone real circuits and to derive from it exponential separations between (both tree-like and dag-like) R and CP.

The following appears to be a key problem if we want to use protocols with small real communication complexity in feasible interpolation.

Problem 18.8.1 (Real protocols problem) *Prove a super-polynomial (or an exponential) lower bound for the size of protocols having small real communication complexity that compute $KW^m[U, V]$ for two* NP *sets obeying the monotone* FI *set-up.*

The qualification *small* ought to correspond to something like $n^{o(1)}$ or maybe even $n^{1-\Omega(1)}$, as in the applications of such a lower bound for tree-like protocols (Lemma 18.2.5).

Another problem is to extend feasible interpolation by such protocols or similar devices to the systems R(CP) and LK(CP) of Section 7.1. Note that [284] proves feasible interpolation for these systems when the size of the coefficients in the inequalities in a proof is bounded. The system LK(CP) is fairly strong; it proves, for example, all counting principles $Count^m$ from Section 11.1, for any fixed $m \geq 2$, in polynomial size (see [284]).

The systems R(CP) and LK(CP) were further studied by Hirsch and Kojevnikov [231] who proved, among other results, that the cut-free subsystem of LK(CP) p-simulates the *lift-and-project* proof system defined there. They also listed a number of open problems about these systems and their relations to the Lovász–Schrijver system LS.

Sec. 18.3 The automatizability of bounded-degree PC (part of Lemma 18.3.2) was first noted by Clegg, Edmonds and Impagliazzo [142] using the Gröbner basis algorithm (they did not consider the bit size of coefficients and counted only the number of field operations). Another algorithm is in Clote and Kranakis [145].

Span programs were defined by Karchmer and Wigderson [265]. Theorem 18.3.4 (and its proof) is from Pudlák and Sgall [423]. We formulated the monotonicity condition dually (it seems more natural). They also showed that PC has monotone interpolation by monotone **polynomial programs**, a notion that they invented. It is somewhat unsatisfactory as it was defined using the notion of bounded degree derivability in PC; no lower bounds for this model are known. They also considered monotone **dependency programs** as a means to study the strength of span and polynomial programs (these programs provide a weaker model) and proved strong lower bounds for them.

Theorem 18.3.5 was improved by Robere *et al.* [451] to the exponential case for the real field **R** (and hence for the rationals **Q**, in which we are interested).

Sec. 18.4 is based on Pudlák [414]; the construction and Lemma 18.4.1 are from that paper. Pudlák also notes an observation of Sgall that the use of linear programming is inevitable, as solving instances of LP can be reduced to finding interpolants for specific sets of linear inequalities refutable by a short proof in LS.

Dash [165] found an extension of the argument to LS_+. The problem one encounters there is to express a sum of the squares of linear functions $\sum_u L_u(\overline{y}, \overline{z})^2$ as

$$\sum_v M_v(\overline{y})^2 + \sum_w N_w(\overline{z})^2 + j \sum_{v,w} 2M_v(\overline{y}) N_w(\overline{z}),$$

in such a way that the quadratic terms get canceled by the terms resulting from the lifting rule and the integrality conditions. This task leads to a problem that can be solved by semidefinite linear programing in p-time.

Oliveira and Pudlák [376] defined a new model of monotone linear programming circuits with respect to which LS has monotone interpolation. A **monotone linear programming circuit** is defined in a similar way to monotone real circuits (Section 18.1): one can use constants but otherwise only *partial* monotone functions g: $\mathbf{R}^n \to \mathbf{R} \cup \{*\}$ of a specific form. The function g must be defined by a linear program as follows:

$$g(\overline{y}) \ := \ \max\{(\overline{c} + C \cdot \overline{y}) \cdot \overline{x} \mid A \cdot \overline{x} \leq b + B \cdot \overline{y},\ \overline{x} \geq 0\},$$

where the matrices B, C have non-negative entries only. Oliveira and Pudlák proved that their model is super-polynomially stronger than monotone Boolean circuits and even exponentially stronger than monotone span programs (and extends both).

Theorem 18.4.3 was proved by Beame, Pitassi and Segerlind [59] as a conditional statement but in the mean time a lower bound for the multi-party communication complexity of the disjointness problem sufficient to yield an exponential lower bound on the size of tree-like LS was proved by Lee and Shraibman [339]. Allowing polynomials of degree d (i.e. aiming at tree-like LS^d), an analogous game with $d + 1$ players can be used (and the lower bound still holds).

Sec. 18.5 is based on [299]. Segerlind [465, 464] proved a lower bound for the *tree-like* version (the preprints of [299, 465] appeared in the same week). One can combine R with the OBDD proof system into a system R(OBDD); it operates with clauses $\{P_1, \ldots, P_k\}$ formed by π-OBDDs (the same ordering π is used for the whole refutation) and has just one rule:

$$\frac{\Gamma \cup \{P\} \quad \Delta \cup \{Q\}}{\Gamma \cup \Delta \cup \{R\}}, \quad \text{if } f_R \geq f_P \wedge f_Q.$$

The target end-clause is $\{0\}$, where 0 denotes the reduced OBDD representing the constant 0. Mikle-Barát [354] considered a similar but syntactic system. The system R(OBDD) p-simulates systems R(k) and it is an open problem to prove some lower bounds for it. Also, its relation to R(CP) of [284] is open.

Itsykson, Knop, Romashchenko and Sokolov [250] considered proof systems extending the OBDD system by adding more inference rules or allowing some form of a reordering of the OBDDs during the course of a proof. In particular, the disjointness of the clique-coloring pair has a short proof in one of these systems. The proof is similar to the proof in LS^4 (as LS but with degree-4 polynomials) by Grigoriev, Hirsch and Pasechnik [209].

Sec. 18.6 is based on [312]. The remark after Lemma 18.6.1 that it makes sense to consider only randomized protocols in the monotone case (i.e. for $KW^m[U, V]$) stems from Raz and Wigderson [428]: for any disjoint pair U, V and any $\epsilon > 0$ there is a tree-like randomized protocol $(\mathbf{P_r})_\mathbf{r}$ computing $KW[U, V]$ of size $S = (n + \epsilon^{-1})^{O(1)}$, communication complexity $t = O(\log n + \log(\epsilon^{-1}))$ and error ϵ. This is based on checking the parity of random subsets of the coordinates of $\overline{u}, \overline{v}$ and using binary search.

The monotone circuit model mentioned towards the end of the section comprises **monotone circuits with a local oracle** (monotone CLO for short). A monotone CLO separating U from V is determined by an ordinary monotone Boolean circuit $D(x_1, \ldots, x_n, y_1, \ldots, y_e)$ with inputs $\overline{x}$ and $\overline{y}$ and by a set $\mathcal{R}$ of combinatorial rectangles $U_j \times V_j \subseteq U \times V$, for $j \leq e$, called the **oracle rectangles** of the CLO, satisfying the following condition:

- for all monotone Boolean functions f_j: $\{0, 1\}^n \to \{0, 1\}, j \leq e$, such that

$$f_j(U_j) \subseteq \{1\} \quad \text{and} \quad f_j(V_j) \subseteq \{0\},$$

  the function

$$C(\overline{x}) \ := \ D(\overline{x}, f_1(\overline{x}), \ldots, f_e(\overline{x}))$$

  separates U from V, so that

$$C(U) = \{1\} \quad \text{and} \quad C(V) = \{0\}.$$

The **size** of the CLO is the size of D and its **locality** is

$$\frac{|\bigcup_{j \leq e} U_j \times V_j|}{|U \times V|}.$$

It was proved in [312, Lemma 3.1] that a randomized protocol for $KW^m[U, V]$ yields a small monotone CLO with a small locality separating U from V. Krajíček and Oliveira [314] proved a lower bound for monotone CLOs under additional restrictions.

An exponential lower bound for tree-like R(LIN) (Theorem 18.6.4) was proved first by Itsykson and Sokolov [252] for the PHP formula and for a substitution instance of a Tseitin formula for a particular graph.

The problem of establishing a lower bound for R(LIN) can be also approached via the real game protocols. Let f be a linear polynomial. Given a partitioning of its variables between two players, we can write $f = f_1 + f_2$ such that player $i = 1, 2$ owns the variables of f_i and can evaluate it on any assignment to them. Hence if an assignment $\overline{a}$ to all variables is split into two parts corresponding to this partitioning, say $\overline{b}$ and $\overline{c}$, then $f(\overline{a}) = 0$ if and only if $f_1(\overline{b}) = f_2(\overline{c})$. More generally, given w linear polynomials $f_j, j \in [w]$, with variables partitioned between two players, it holds that

$$\bigwedge_j f_j(\overline{a}) = 0 \quad \Leftrightarrow \quad \bigwedge_j f_{j,1}(\overline{b}) = f_{j,2}(\overline{c}) \quad \Leftrightarrow \quad \beta = \gamma,$$

where β, γ are reals defined by

$$\beta := \sum_j f_{j,1}(\overline{b})\, 2^j \quad \text{and} \quad \gamma := \sum_j f_{j,2}(\overline{c})\, 2^j .$$

Hence the real communication complexity of deciding membership in an R(LIN) clause is 2. In fact, as pointed out by Sokolov [478], it may be advantageous in this situation to modify the real game so that the referee announces the equality of the two reals rather than their ordering (then the complexity becomes 1).

Sec. 18.7 The limits for monotone feasible interpolation in the narrow sense were pointed out in [280].

The reader may wonder why one could not use the hardness of prime factorization as a tool to establish that a proof system does not admit feasible interpolation via Corollary 17.1.5, by just giving a short proof in the proof system in question (the propositional translations of the fact) that the prime factorization is unique. However, for this one needs an NP definition $Prime_{NP}(x)$ of primality such that its soundness

$$Prime_{NP}(x) \rightarrow Prime(x)$$

(where $Prime(x)$ is the usual coNP definition) has feasible proofs. Krajíček and Pudlák [321, Sec. 2] showed that the uniqueness of the prime factorization follows in S^1_2 from such a soundness statement and further that (over S^1_2) the soundness of the well-known NP definition by Pratt [405] implies the implicit definability of the discrete logarithm. Jeřábek (unpublished) showed that the soundness of the Pratt definition is provable in the extension of S^1_2 by instances of the WPHP for p-time functions.

We refer the reader to Bonet, Pitassi and Raz [97] for details of the formalization involved in the proof of the first part of Theorem 18.7.3. Note that the second part follows from the first by Theorem 2.5.6.

18.8.1 Further Generalizations

Hrubeš [233, 234, 235] extended the feasible interpolation method to some formal systems for modal and intuitionistic propositional logic (Section 7.7).

In Section 18.7 we saw that having a proof of the disjointness of two NP sets in a proof system containing a small-depth Frege system probably does not imply the existence of a feasible separating set. The idea proposed in [301] is that we require the proof system to prove a stronger property than just the disjointness and show that any proof of the property must yield a feasible interpolant (or some other non-trivial computational information). The specific situation considered in [301], termed *chain interpolation*, is the following.

For a first-order language L, take two Σ_1^1 L-sentences Φ and Ψ that cannot be satisfied simultaneously in any finite L-structure (by Fagin's theorem 1.2.2 this is the same as talking about two disjoint NP sets). Then the following principle $Chain_{L,\Phi,\Psi}(n,m)$ holds.

- *For any chain of finite L-structures $C_1, \dots, C_m$, all with the universe $[n]$, one of the following conditions must fail*:
  1. $C_1 \models \Phi$;
  2. $C_i \cong C_{i+1}$, for $i = 1, \dots, m-1$;
  3. $C_m \models \Psi$.

The principle translates (for fixed L) into a DNF-formula of size polynomial in n, m. In [301] it was proved that if L contains only constants and unary predicates then there is a constant c_L such that if an AC^0-Frege system admits a size s proof of (the translation of) $Chain_{L,\Phi,\Psi}(n,m)$ then the class of finite L-structures with universe $[n]$ satisfying Φ can be separated from the class of those L-structures with the same universe satisfying Ψ by a depth 3 formula of size $2^{\log(s)^{O(1)}}$ and with bottom fan-in $\log(s)^{O(1)}$. It is open to extend this to any relational language L or to some stronger proof systems (some limits are discussed in [301]).

Part IV

Beyond Bounds

Part IV presents results and ideas (and in a few cases emerging theories) that attempt to say something relevant about the fundamental problems from Section 1.5. We aim at statements that are not specific to some weak proof system but are possibly valid for *all* proof systems.

Some researchers believe that by proving lower bounds for stronger and stronger proof systems we will accumulate knowledge that will lead eventually to solutions of these fundamental problems. I think that this is unlikely to happen without trying to formulate statements (and methods to prove them) that apply – at least in principle – to all proof systems. Given such general statements it would then be reassuring, of course, to establish them for weak systems. Likewise it would be good to know that a method aimed at all proof systems actually works in some particular cases. But I doubt that ideas that are specific to a weak system can be of use in a general situation. For this reason I present, in this part, results that are perhaps weaker in the technical sense but are formulated and proved in the spirit of what we believe to be true for all proof systems.

The gradual approach also has, I think, the negative psychological effect that we feel we ought first to solve all problems about weak systems before moving on to strong ones. But weak systems are mathematically rich objects and there will always be interesting open problems about them.

By a **strong proof system** I shall informally mean any proof system that simulates *EF*. In Chapter 19 we will discuss which tautologies form plausible candidates for being hard for strong proof systems. This is complemented in Chapter 20, which looks at possible methods to prove their hardness. Both these issues seem of key importance in attacking the main problem 1.5.3, the NP vs. coNP problem. The choice of material may seem subjective and it is indeed close to my own research. I wish I could present alternative approaches but I am not aware of any other systematic proposals.

Chapter 21 will present results about the optimality problem 1.5.5. There is a fairly rich theory relating the problem to topics in proof theory, structural complexity theory and finite model theory. The proof search problem 1.5.6 is discussed in Section 21.5 and we focus there mostly on what is the right formulation of the problem.

When investigating strong proof systems, i.e. *EF* and beyond, it is often more convenient to use single-sort bounded arithmetic theories and the $\|\dots\|$ translation (as discussed in Chapter 12); this will be the case especially in Chapter 21. For *EF* itself the $\langle\dots\rangle$ translation is also helpful because of the particular theory V_1^1 to which it corresponds (Section 10.3); we shall take advantage of this in Chapter 19.

When dealing with strong proof systems we may consider without loss of generality tautologies in DNF, as theories corresponding to *EF* prove the NP-completeness of the set of satisfiable CNFs. In fact, we do not need to be too specific about precisely how formulas look when describing them as formalizing this or that principle. While we shall formulate some statements for proof systems containing *EF* such statements will often hold for systems containing R only (although to see this may require careful formalization of the various concepts involved).

19
Hard Tautologies

It seems that a good starting point with the main problem 1.5.3 is to define candidate hard tautologies (i.e. tautologies that are hard to prove) and to formulate a method of approaching a potential lower bound proof. Two classes of formulas are often mentioned as possible hard examples: random DNFs and formulas expressing one of the combinatorial principles that has been proved to be hard for Peano arithmetic PA or even for set theory ZFC. The former suggestion appears to be void while the latter is incorrect.

Most researchers in proof complexity would agree, I suppose, with the folkloristic informal conjecture that for any proof system most formulas are hard. It is more customary in this connection to talk about refuting random CNFs, and a statement that random CNFs with a specific distribution are hard to prove is just a more formal version of the conjecture. Lower bound arguments invented for some specific formulas as in Part III are sometimes recognized as using only those properties of the formulas that are shared by random formulas, and the lower bound is extended to random CNFs. But I am not aware of any example where the first non-trivial lower bound for a proof system has been proved for random formulas (see also the discussion at the beginning of Section 13.4).

The latter suggestion, to use a finite combinatorial principle unprovable in very strong theories, unfortunately does not work. These principles have the logical Π^0_2 form $\forall x \exists y \ldots$, and their unprovability is deduced from the enormous growth rate that any function finding a witness for y in terms of x must have. Consider the most famous (though not the logically strongest) example, the **Paris–Harrington principle**. This principle says that for any $k, e, r \geq 1$ there is an M so large that, for any partition

$$P\colon \binom{M}{e} \to [r]$$

of the e-element subsets of $[M]$ into r parts there is a homogeneous set $H \subseteq [M]$ (all the e-element subsets of H are in one part) such that

$$|H| \geq k \ \wedge\ |H| \geq \min H. \tag{19.0.1}$$

If we had only the first conjunct in (19.0.1) then the statement would be the well-known finite Ramsey theorem. It is the second conjunct that makes the principle very strong. In particular, Paris and Harrington [384] proved that the principle is true but not provable in PA. To be able to formulate it in propositional logic we need to take M as the main parameter, in order to turn it into a $\Pi_1^{1,b}$-statement and use the $\langle \dots \rangle$ translation. In particular, if f: $\langle k, e, r\rangle \to M$ is a non-decreasing witnessing function for the principle, the propositional translation with parameter M will express that

- *for any* $\langle k, e, r\rangle \leq f^{(-1)}(M)$, *for any partition* P *of* $\binom{M}{e}$ *into* r *parts there is a homogeneous set* H *satisfying* (19.0.1).

The existence of H is expressed using a disjunction of size $\binom{M}{k} \ll M^{\log M}$ over all possibilities. Unfortunately, the unprovability of the principle in PA stems from the fact that every witnessing function f grows so fast that it is not provably total in PA. But once we may assume that there is a number at least as large as the minimum M witnessing the statement (for parameters k, e, r) the principle is provable in bounded arithmetic $T_2(P)$. In particular, propositional translation admits polynomial-size AC^0-Frege proofs.

There are also Π_1^0 statements unprovable in ZFC provided by Gödel's theorem: they express the consistency of the theory or some related property. These statements can be reformulated to have a more combinatorial or number-theoretic flavor. For example, by the negative solution to Hilbert's 10th problem (Matiyasevich's theorem) the consistency of ZFC can be expressed as a statement that a particular Diophantine equation has no integer solution. But its meaning is more about encoding the syntax of first-order logic into the equation than about number theory.

19.1 Levels of Uniformity

Recall the function $\mathbf{s}_P$ from Section 1.5. A set $H \subseteq \mathrm{TAUT}$ is a **hard set** for a proof system P if and only if

- *for any constant* $c \geq 1$, *infinitely many* $\phi \in H$ *require a P-proof of size larger than* $|\phi|^c$: $\mathbf{s}_P(\phi) > |\phi|^c$.

It follows that such an H is infinite and that there is a sequence $\{\alpha_k\}_k \subseteq H$, $|\alpha_k| \geq k$, such that, for all $c \geq 1$, for $k \gg 1$, $\mathbf{s}_P(\alpha_k) > |\alpha_k|^c$. By filling the sequence with some tautologies (not necessarily from H) this implies that there is a sequence $\{\beta_k\}_k \subseteq$ TAUT satisfying

1. $\beta_k \in \mathrm{TAUT} \wedge k \leq |\beta_k| \leq k^{O(1)}$,
2. for all $c \geq 1$, for $k \gg 1$, $\mathbf{s}_P(\beta_k) > k^c$.

We shall call any such sequence of tautologies a **hard sequence** for P. By strengthening the lower bound requirement we could analogously define quasi-polynomially or exponentially hard sets and sequences, respectively.

Sets of tautologies that we hope to be hard for all proof systems cannot be too simple.

Lemma 19.1.1 *No* NP *set* $H \subseteq$ TAUT *is hard for all proof systems.*

Proof Assume that H is a set of tautologies with an NP definition of the form

$$\phi \in H \quad \Leftrightarrow \quad \exists u(|u| \leq |\phi|^c) R(\phi, u),$$

where R is p-time and c is a constant. Define a proof system P by stipulating that a string w is a P-proof of ϕ if and only if it is either an EF-proof of ϕ or it is a witness for $\phi \in H$, i.e. $|w| \leq |\phi|^c \wedge R(\phi, w)$. Clearly H is not hard for P. □

This statement means that on the one hand we can target a specific proof system P and define a set of tautologies hard for P only, and such a set could be computationally simple. After all, the sequences of tautologies hard for specific proof systems that we studied in Part III were all p-time constructible. This is true for every particular proof system as the next lemma shows under a condition.

On the other hand, we may aim at all proof systems and try to define a set that is hard for all of them but then it must be computationally unfeasible to certify the membership of it. In fact, we can go from the former situation to the latter. Let P_i, $i \geq 1$, be an enumeration of all proof systems and assume that $\{\beta_k^i\}_k$, $i \geq 1$, are hard sequences for P_i. Then $\{\bigwedge_{j \leq i} \beta_i^j \mid i \geq 1\}$ is a set hard for all proof systems but it is not computationally simple even if all the sequences $\{\beta_k^i\}_k$ are p-time constructible.

We shall use results about the correspondence between theories and proof systems from Chapter 12.

Lemma 19.1.2 *Assume that a proof system P is not optimal. Then there is $\forall\Pi_1^b$ a true L_{BA}-sentence $\forall x A(x)$ such that the sequence $\{\|A\|^n\}_n$ is hard for P. The sequence is computed by a p-time algorithm.*

Proof The hypothesis implies that there is a proof system Q such that P does not simulate Q. Take for $\forall x A(x)$ the reflection principle RFN_Q for Q defined in Section 12.4. The lemma then follows from Corollary 12.4.3 (and the p-constructibility follows from Lemma 12.3.2). □

We know from Chapter 8 that proving lower bounds for a proof system P is equivalent to constructing models of the corresponding theory T with particular properties (Ajtai's argument, Sections 8.5 and 8.6). We shall discuss this in detail in Chapter 20. For EF this means constructing models of the theory S_2^1 in the first-order set-up or of the theory V_1^1 in the second-order set-up (or suitable sub-theories of them, as we shall see later) and that is difficult. So we may ask whether if we could not simplify the matter and replace the model-theoretic criterion by mere unprovability in S_2^1. Unfortunately this is not possible.

Lemma 19.1.3 *There is $\forall\Pi_1^b$ an L_{BA}-sentence $\forall x A(x)$ such that*

(i) $S_2^1 \not\vdash \forall x A(x)$*;*

(ii) *there are polynomial-size EF-proofs π_n of $\|A\|^n$; in fact, π_n can be constructed by a p-time function from $1^{(n)}$.*

Proof Let $B(y)$ be a Π_1^b L_{BA}-formula such that $\forall y B(y)$ is true but unprovable in the theory $S_2^1 + Exp$, where Exp is the sentence

$$Exp\colon \forall u \exists v\; u = |v|.$$

Gödel's theorem provides such a formula. Then put

$$A(x) \;:=\; B(|x|)\;.$$

It follows that S_2^1 does not prove $\forall x A(x)$, but to prove $\|A\|^n$ one need only consider n assignments $\overline{a} \in \{0,1\}^{\log n}$ to the atoms representing $|x|$, take for each the unique assignment to the auxiliary variables in the formula making it true and evaluate it to the value 1. The overall size of the proof is $n\,(\log n)^{O(1)} \leq n^2$, for $n \geq 1$ large enough. □

We have discussed so far three levels of uniformity of the formulas β_k and their propositional proofs π_k; informally,

1. *β_k is a translation of an arithmetic formula and π_k is a translation of an arithmetic proof,*
2. *β_k and π_k are computed by a p-time algorithm from $1^{(k)}$,*
3. *β_k and π_k are bounded in size by $k^{O(1)}$ but are not feasibly constructible.*

In fact, there is a fourth variant lying between 1 and 2:

1.5. *β_k and π_k are computed bit by bit, by circuits of size $O(\log k)$, that is, as strings β_k and π_k are the truth tables of functions computed by circuits of size $O(\log k)$, and the circuits are computable by a p-time algorithm from $1^{(\log k)}$.*

We have considered this already when discussing implicit proof systems in Section 7.3, and will use it again in Sections 19.3 and 21.4.

19.2 Reflection and Consistency

The bounded formulas Ref_Q from Section 8.4 expressing the reflection principle for a proof system Q are of the form

$$Prf_Q(x,y,z,P,F) \rightarrow Sat_2(x,y,E,F)$$

when we are talking about size z Q-proofs P of a size y DNF formula F in x variables, or, more generally, as in (8.4.4),

$$Prf_Q^0(x,y,z,P,F,G_1,\ldots,G_t) \;\rightarrow\; Sat^0(x,y,E,F,H_1,\ldots,H_r),$$

where the extra sets G_u and H_v help to express the provability predicate for Q or the satisfiability relation for general formulas or circuits (see Section 8.4 for the details).

A special case of particular interest is for $F := 0$; F is unsatisfiable and the principle reduces to the **consistency formula** Con_Q:

$$\neg Prf_Q(0, 1, z, P, 0).$$

Here we denote the set representing the formula 0 also as 0, and we think of the possible witnesses G_u as being part of the proof P. That is, the formula expresses that there is no size z Q-proof by contradiction (as represented by 0). The reader who might have skipped the treatment of propositional translations in Part II may appreciate an informal description of propositional (translations of the) consistency statements. The formula $\langle Con_Q \rangle_n$ has $n^{O(1)}$ atoms representing a size n Q-proof of a formula and the computation of the provability predicate. The conditions posed on the provability predicate are in a 3CNF form. The formula $\langle Con_Q \rangle_n$ says that either one condition fails or the proven formula is not 0.

The consistency of Q is, however, a weaker statement than its soundness, and Con_Q implies Ref_Q (say over EF), only under some conditions. These conditions are met for all the proof systems that we have encountered but may not hold for a general proof system.

Lemma 19.2.1 *Let Q be a proof system and assume that the theory V^1_1 proves the following two conditions:*

- *if a formula $F(p_1, \ldots, p_n)$ is Q-provable and $\overline{a} \in \{0,1\}^n$ then also its substitution instance $F(a_1, \ldots, a_n)$ is Q-provable, and*
- *for any formula $F(p_1, \ldots, p_n)$ and any $\overline{a} \in \{0,1\}^n$, F can be evaluated on $\overline{a}$ and if b is the truth-value of $F(a_1, \ldots, a_n)$ then b has a Q-proof from $F(a_1, \ldots, a_n)$.*

Then V^1_1 proves the implication $Con_Q \to Ref_Q$.

Proof We shall show that V^1_1 proves that $\neg Ref_Q$ implies $\neg Con_Q$. Work in V^1_1 and assume that G is a Q-proof of $F(\overline{p})$ but $F(\overline{a}) = 0$ for some $\overline{a} \in \{0,1\}^n$. Use the first condition to get a Q-proof $G(\overline{a})$ of $F(\overline{a})$ and then the second condition to extend it to a Q-proof G' of 0. □

For many proof systems Q that we have encountered (R or ER, Frege systems and its subsystems, G and its subsystems, algebro-geometric systems, etc.), substituting constants a_i for the atoms p_i in the whole Q-proof is allowed and proves the first condition (and the theory corresponding to R formalizes this). The second condition is slightly more subtle because, in the general form of the reflection above the succedent

$$Sat^0(x, y, E, F, H_1, \ldots, H_r)$$

with witnesses H_v allows even very weak systems to talk about an evaluation of F on an assignment E without being able to prove that the witnesses (representing the evaluation) exist. For example, in $I\Delta_0(\alpha)$ we can prove their existence only for AC^0-formulas F. However, we concentrate in Part IV on strong proof systems, ER and above, and the theory V_1^1 corresponding to *ER* proves that any circuit can be evaluated on any input (*ER* simulates such an argument using the extension variables).

For *ER* we shall often use the formula Con_{ER}, as the conditions in Lemma 19.2.1 are valid for *ER*. But for a general proof system we shall use the formula Ref_Q, although Con_Q is simpler, because we do not want to condition various results on the hypotheses of Lemma 19.2.1. The following statement elucidates the role of the $\langle Ref_Q \rangle_n$ formulas as hard formulas. For an implicationally complete proof system P and any proof system Q, denote by

$$P + Ref_Q$$

the proof system extending P by allowing any substitution instance of any formula $\langle Ref_Q \rangle_n$, $n \geq 1$, as an axiom. In the format of refutation systems, as is the case for *ER*, we denote by $\neg \langle Ref_Q \rangle_n$ the set of clauses expressing the negation of the translations of Ref_Q, and a refutation of a set of clauses $\mathcal{D}$ in the refutation system

$$ER + Ref_Q$$

is the collection of *ER*-derivations from $\mathcal{D}$ of all clauses of a substitution instance (which could involve extension variables) of $\neg \langle Ref_Q \rangle_n$, for some $n \geq 1$.

In the following lemma we use the refutation system $ER + Ref_Q$ in order to be able to talk about simulations of proofs of all formulas or circuits (see the discussion above). If we were interested only in simulations of proofs of DNFs (or of AC^0-formulas) we could use R (or AC^0-Frege systems) instead of *ER*.

Lemma 19.2.2 *Let Q be any proof system. Then:*

(i) $ER + Ref_Q \geq_p Q$*;*
(ii) if $H \subseteq$ TAUT *is a hard set for* $ER + Ref_Q$ *then it is also hard for* Q*;*
(iii) if $P \geq ER$ *does not simulate* Q *then*

$$\langle Ref_Q \rangle_n, \quad n \geq 1,$$

is a hard sequence for P.

Proof The p-simulation in the first statement is constructed in analogy to the special case of AC^0-Frege systems in Theorem 8.4.3 and consequence (2) in Section 8.6. The remaining two statements follow from the first. □

Lemma 19.2.2 (and the underlying argument) is reminiscent of a well-known fact of mathematical logic that all Π_1^0-consequences of a theory T in the language L_{PA} containing Robinson's arithmetic Q follow (over a fixed weak theory) from the Π_1^0-sentence Con_T expressing the consistency of T. The next lemma breaks this parallel: a statement analogous to Gödel's second incompleteness theorem fails.

Lemma 19.2.3 *The system ER p-simulates* $ER + Ref_{ER}$. *In fact, let* $P \geq_p R$ *and assume that there is a theory T corresponding to P (Section 8.6). Then P p-simulates* $P + Ref_P$ *with respect to the proofs of all* DNF*s.*

If $P \geq_p ER$ then P p-simulates $P + Ref_P$ with respect to proofs of all formulas (including circuits).

Proof By the hypothesis (condition (C) of the correspondence in Section 8.6), T proves in particular the soundness of P with respect to proofs of depth 2 formulas (the formulas $Ref_{P,2}$). As P p-simulates T, $P \geq_p P + Ref_P$ follows.

If $P \geq_p ER$, T may be assumed to contain the theory V_1^1 (corresponding to ER), which proves that any circuit can be evaluated on any assignment (condition (D) in the correspondence). □

It follows from the above discussion that if there is no optimal proof system then for each proof system P we can find p-time constructible hard sequences of the form $\langle Ref_Q \rangle_n$. However, we do not have an a priori definition of a suitable Q from P, and Lemma 19.2.3 shows that a Gödel-like choice $Q := P$ does not work. We shall return to these formulas, and to issues related to how to find Q given P, in Chapter 21.

19.3 Combinatorics of Consistency

The propositional consistency statements appear to be plausibly hard but we proved their hardness for a proof system P (Lemma 19.2.2) using only a hypothesis about the non-optimality of P. That hypothesis looks, in fact, stronger than the hypothesis that P is not p-bounded. In this section we shall give an example of a combinatorial investigation of the consistency statements (and offer more references in Section 19.6) that could possibly lead to an argument not using other length-of-proof conditions.

The ambient area in which our investigation lives are NP-search problems, although we shall formulate the main result without explicit reference to them in order to give a precise quantitative version. These problems play a crucial role in bounded arithmetic in the so-called witnessing theorems, which characterize the classes of provably total functions in theories (see Section 19.6 for examples and references).

An NP **search problem** is a binary relation $R(x, y)$ such that:

- *there is a constant $c \geq 1$ such that $R(x, y) \rightarrow |y| \leq |x|^c$;*
- *the relation R is p-time decidable;*
- *the relation R is a graph of a multi-function, $\forall x \exists y R(x, y)$.*

The class of NP search problems is also called **total** NP **functions** and denoted TFNP (we shall discuss the background further in Section 19.6). From the logical point of view it is convenient to consider **relativized** NP-search problems: R is p-time relative to an oracle. For example, an oracle may be a unary function α: $\{0, 1\}^n \rightarrow \{0, 1\}^n$ satisfying $\alpha(\alpha(z)) = z$ for all $z \in \{0, 1\}^n$ and the relation $R(x, y)$ may be defined as:

$$\alpha(x) \neq x \ \vee \ (x \neq y \wedge \alpha(y) = y).$$

Owing to the condition on α, the pairs $\{\{z, \alpha(z)\} \mid z \in \{0,1\}^n \wedge z \neq \alpha(z)\}$ form a 2-partition of the set $\{0,1\}^n \setminus \{\, z \in \{0,1\}^n \mid z = \alpha(z)\}$. Hence, if $x = \alpha(x)$ then there must be at least one other y for which $y = \alpha(y)$. This is a reformulation of the counting modulo-2 principle $Count^2$ from Section 11.1.

Given two NP search problems $R(x,y)$ and $S(u,v)$, we say that S can be **p-reduced** to R if and only if there are two p-time functions $f(u)$ and $g(u,y)$ such that for all u, y it holds that

$$R(f(u), y) \;\rightarrow\; S(u, g(u,y)). \tag{19.3.1}$$

That is, from an R-solution y for $x := f(u)$ we can compute an S-solution $v := g(u,y)$ for u. The pair of functions is called a **p-reduction**.

When the search problems are relativized and β is the oracle for S, both f and g may use β; f poses an instance of R with an oracle α that is Turing p-time reducible to β. For example, if S is defined as R in the $Count^2$ example above but with an oracle giving a partial 2-partition β as a set of unordered pairs, we may reduce S to R by taking $f(u) := u$, $g(u,y) := y$ and $\alpha(z)$ defined as the other element in the same β-pair as z, or as z if z is not covered by β.

The consistency statement about a proof system Q formalizes, in particular, the fact that no potential string Π is a Q-refutation of 1. In particular, for any Π there is a place in Π where the correct format of Q-proofs is broken. We shall interpret this as an oracle NP search problem: Π is determined by an oracle, an input specifies the size parameter of the consistency statement and a valid solution is the index (bounded above by the size parameter) identifying the place where the break occurs.

The fact that the consistency statement allows us to p-simulate Q over a weak proof system (Lemma 19.2.2) can be restated as follows. An unsatisfiable set of clauses $\mathcal{D}$ determines the search problem Search($\mathcal{D}$) from Section 5.2: find a clause unsatisfied by a particular truth assignment. A Q-refutation of a set of clauses $\mathcal{D}$ determines a form of reduction of Search($\mathcal{D}$) to a particular search problem concerning the proper format of Q-proofs. This is analogous to an extent to the fact that if the implication

$$\forall x \exists y R(x,y) \;\rightarrow\; \forall u \exists v S(u,v)$$

is provable in PV or S^1_2 then S can be reduced to R (in the Turing sense in the latter case – see Section 19.6 for an explanation).

We shall proceed formally now in order to be able to formulate later on a precise statement about the relation between the reductions and the sizes of ER refutations. Let C be a size s circuit with n Boolean inputs $\overline{x} = x_1, \ldots, x_n$ over the DeMorgan basis. The s instructions of C about how to compute the Boolean values $\overline{y} = y_1, \ldots, y_s$ are collected in the 3CNF $\mathrm{Def}_C(\overline{x}, \overline{y})$ from Section 1.4, and this time we shall indicate in the notation the parameters n and s:

$$\mathrm{Def}_C^{n,s}(\overline{x}, \overline{y}).$$

The formula has at most $3s$ 3-clauses.

It is easy to prove in R that the computation of C is unique, meaning that $y_i \equiv z_i$ has a short R proof from $\mathrm{Def}_C^{n,s}(\overline{x},\overline{y}) \cup \mathrm{Def}_C^{n,s}(\overline{x},\overline{z})$ (Lemma 12.3.1). Given an assignment $\overline{x} := \overline{a}$, in R we can analogously derive from $\mathrm{Def}_C^{n,s}(\overline{a},\overline{y})$ for each i the clause $\{y_i^b\}$, where $b := y_i(\overline{a})$ is the value of the subcircuit y_i for the input $\overline{a}$. However, how can we prove that there is no R derivation of the complementary value $\{y_i^{1-b}\}$? The impossibility of such an alternative derivation turns out to be equivalent to the consistency of ER.

For technical reasons we shall work with the system R_w defined in Section 5.4. Recall that it allows the constant 1 in clauses and that any clause containing 1 is an allowed initial clause (called a **1-axiom**). Further, both a variable and its negation may occur in a clause, and there is also the the weakening rule

$$\frac{C}{D}, \quad \text{if } C \subseteq D.$$

The key property is that after substituting some constants for variables in an R_w-derivation it remains an R_w-derivation (resolution inferences on the variables substituted for are replaced by weakenings; see Section 5.4). Also, instead of considering instances of $\mathrm{Def}_C^{n,s}$ for $\overline{x} := \overline{a} \in \{0,1\}^n$ it is simpler to stipulate $n := 0$, i.e. to consider circuits with no input variables. Such circuits are straight line programs for computing Boolean values.

We shall define an unsatisfiable set $\neg Con_{ER}(s,k)$ of narrow clauses such that a satisfying assignment for $\neg Con_{ER}(s,k)$ would be precisely a k-step R_w-refutation of a set $\mathrm{Def}_C^{0,s}(\overline{y})$. We shall show that the sets $\neg Con_{ER}(s,k)$ express the consistency of ER and hence any proof system that refutes these sets by polynomial size proofs simulates ER. The simulation yields, in fact, a reduction of unsatisfiable CNFs $\mathcal{D}$ to $\neg Con_{ER}(s,k)$ (where k depends on the size of an ER-refutation of $\mathcal{D}$) in the sense discussed earlier.

Let us fix $s,k \geq 1$. The clauses of the set $\neg Con_{ER}(s,k)$ concern k-step R_w-refutations of $\mathrm{Def}_C^{0,s}$ for an unspecified C coded by some atoms of $\neg Con_{ER}(s,k)$. The atoms from which the formula is built are $\overline{q}$ and $\overline{p}$. The former will encode the sequence of clauses forming a refutation and the latter will encode the inference structure of the refutation.

In the following discussion we call the clauses in the refutation $P_1,\dots,P_k$. The clauses P_i may contain the constant 1 or literals corresponding to the $\overline{y}$-variables, i.e. all together up to $1+2s$ different objects. The intended meaning of the atoms $\overline{q}$ used by the formula $\neg Con_{ER}(s,k)$,

$$q_i^u, \quad \text{with } u = 1,\dots,k \text{ and } i \in \{-s,\dots,-1,0,1,\dots,s\},$$

is:

- $q_0^u = 1$ if and only if $1 \in P_u$;
- $q_j^u = 1$ for $j = 1,\dots,s$ if and only if $y_j \in D^u$; and
- $q_j^u = 1$ for $j = -1,\dots,-s$ if and only if $y_j^0 \in D^u$.

The atoms $\overline{p}$ are $p_{u,v}$, $u = 1, \ldots, k$, and $v = 1, \ldots, t$ (with the parameter t specified below). Their role is to encode the inference structure of the refutation: an assignment $\overline{a}^u \in \{0, 1\}^t$ to the atoms $\overline{p}_u = p_{u,1}, \ldots, p_{u,t}$ gives complete information about how the clause P_u was derived and, if $P_u \in \mathrm{Def}_C^{0,s}$, it contains additional information guaranteeing that the clauses of $\mathrm{Def}_C^{0,s}$ have the expected form.

We shall assume that $k \geq 3s > 0$ and that the clauses of $\mathrm{Def}_C^{0,s}$ are listed as the first $3s$ clauses $P_1, \ldots, P_{3s}$. Namely, we assume that $P_{3r-2}, P_{3r-1}, P_{3r}$ define the instruction for y_r (one or two clauses are possibly $\{1\}$ if not all three are needed to represent the instruction).

The number of instructions for computing y_r is bounded above by

$$2 + (r - 1) + 2(r - 1)^2 \leq O(k^2),$$

and the string $\overline{a}^u$ has to specify this uniquely for all $u = 1, \ldots, 3s$. For $u = 3s + 1, \ldots, k$, the string $\overline{a}^u$ has to determine the rule and the hypotheses of the inference yielding P_u; the number of different possibilities is bounded above by

$$2 + (u - 1) + (2 + s)(u - 1)^2 \leq k^3.$$

Hence, for $t := 3 \log k$ the string $\{0, 1\}^t$ has enough bits to encode all possible situations.

We shall depict the clauses of $\neg Con_{ER}(s, k)$ as sequents

$$\ell_1, \ldots, \ell_e \rightarrow \ell_{e+1}, \ldots, \ell_f$$

representing the clause

$$\ell_1^0, \ldots, \ell_e^0, \ell_{e+1}, \ldots, \ell_f.$$

Given $\overline{a} \in \{0, 1\}^t$, we denote by $\overline{p}_u(\overline{a})$ the set of the literals for which $\overline{a}$ is the unique truth assignment satisfying all of them:

$$\overline{p}_u(\overline{a}) := (p_{u,1})^{a_1}, \ldots, (p_{u,t})^{a_t}.$$

Now we are ready to define the set $\neg Con_{ER}(s, k)$. It consists of the following five groups of clauses.

G1. For $r = 1, \ldots, s$ and $u \in \{3r - 2, 3r - 1, 3r\}$, if $\overline{a} \in \{0, 1\}^t$ does not specify a valid instruction for computing y_r then $\neg Con_{ER}(s, k)$ contains the clause

$$\overline{p}_u(\overline{a}) \rightarrow .$$

G2. For $r = 1, \ldots, s$ and $u \in \{3r - 2, 3r - 1, 3r\}$, if $\overline{a} \in \{0, 1\}^t$ does specify a valid instruction for computing y_r then we know, for the constant 1 and for every $\overline{y}$-atom, whether it occurs positively or negatively in P_u. We include in $\neg Con_{ER}(s, k)$ for every $\overline{q}$-variable q_i^u exactly one of the clauses

$$\overline{p}_u(\overline{a}) \rightarrow q_i^u \text{ or } \overline{p}_u(\overline{a}) \rightarrow \neg q_i^u,$$

as specified by $\overline{a}$.

G3. For $u = 3s+1, \dots, k$, if $\overline{a} \in \{0,1\}^t$ does not specify a valid inference for P_u then $\neg Con_{ER}(s,k)$ contains the clause

$$\overline{p}_u(\overline{a}) \to .$$

G4. For $u = 3s+1, \dots, k$, if $\overline{a} \in \{0,1\}^t$ does specify a valid inference for P_u then we consider three separate subcases:

(a) P_u was derived from P_v, P_w by resolving the literal ℓ, where $\ell \in P_v$, $\ell^0 \in P_w$ and ℓ is a $\overline{y}$-literal.
For $i \in \{-s, \dots, -1, 1, \dots, s\}$ corresponding to ℓ and $-i$ to ℓ^0,

- $\neg Con_{ER}(s,k)$ contains the clauses

$$\overline{p}_u(\overline{a}) \to q_i^v, \qquad \overline{p}_u(\overline{a}) \to q_{-i}^w, \qquad \overline{p}_u(\overline{a}) \to \neg q_i^u, \qquad \overline{p}_u(\overline{a}) \to \neg q_{-i}^u$$

(they enforce that ℓ and ℓ^0 appear in P_u, P_v, P_w as prescribed by the resolution rule),

- for $j \neq i, -i$, $j \in \{(-s, \dots, -1, 1, \dots, s\}$, $\neg Con_{ER}(s,k)$ contains the clauses

$$\overline{p}_u(\overline{a}), q_j^v \to q_j^u, \qquad \overline{p}_u(\overline{a}), q_j^w \to q_j^u, \qquad \overline{p}_u(\overline{a}), q_j^u \to q_j^v, q_j^w$$

(they enforce that other literals are passed from P_v, P_w to P_u and that no other literals occur there);

(b) P_u was derived by the weakening rule from P_v, $v < u$. Then $\neg Con_{ER}(s,k)$ contains all the clauses

$$\overline{p}_u(\overline{a}), q_i^v \to q_i^u \ ;$$

(c) P_u is a 1-axiom; then $\neg Con_{ER}(s,k)$ contains the clause

$$\overline{p}_u(\overline{a}) \to q_0^u.$$

G5. The set $\neg Con_{ER}(s,k)$ contains all the clauses

$$\to \neg q_i^k, \quad \text{for all } i$$

(they enforce that $P_k = \emptyset$).

Lemma 19.3.1 *For all $k \geq 3s > 0$, the set $\neg Con_{ER}(s,k)$ is not satisfiable and it contains $O(k^5)$ clauses of width at most* $3 + 3\log k$.

The formula $\neg Con_{ER}(s,k)$ expresses conditions posed on a non-existent R_w-refutation of $\mathrm{Def}_C^{0,s}$, and its negation $Con_{ER}(s,k)$ is the consistency statement for ER. A refutation of a set of clauses $\mathcal{D}$ corresponds to proving its negation $\neg \bigwedge \mathcal{D}$. The use of the consistency statements as the strongest formulas provable in a proof system means that we look for R_w-derivations of $\neg \bigwedge \mathcal{D}$ from a substitution instance of $Con_{ER}(s,k)$. In the refutation format this means that we want R_w-derivations from $\mathcal{D}$ of all clauses of some substitution instance of $\neg Con_{ER}(s,k)$. This motivates the following definition.

Let X, Y be two disjoint sets of propositional variables. A map σ assigning to Y-variables either constants $0, 1$ or disjunctions of X-literals is a **clause substitution**, and the maximum size of a disjunction that σ assigns is the **width** of σ.

Let $\mathcal{C}$ be a set of clauses in Y-literals and $\mathcal{D}$ a set of clauses in X-literals. A clause substitution σ is a **clause reduction** of $\mathcal{D}$ to $\mathcal{C}$ if and only if σ substitutes for Y-variables clauses of X-literals such that, for each clause $C \in \mathcal{C}$, one of the following cases occurs.

(a) The clause reduction $\sigma(C)$ is a 1-axiom
 (when $\sigma(C)$ is written in a sequent format then being a 1-axiom may also mean that $\neg 1$ is in its antecedent).

(b) The clause reduction $\sigma(C)$ has the form

$$U, \bigvee E \rightarrow \bigvee F, V, \tag{19.3.2}$$

where $E \subseteq F$ are sets of literals.

(c) The clause reduction $\sigma(C)$ is a superset of a clause from $\mathcal{D}$.

Note that in the cases (a) and (b) $\sigma(C)$ is logically valid.

We shall use the following notation. For z a variable and $\overline{a}, \overline{b} \in \{0, 1\}^t$, define

$$sel(z, \overline{a}, \overline{b})$$

to be the t-tuple from $\{0, 1, z, \neg z\}^t$ whose ith coordinate is

$$sel(z, a_i, b_i) := (a_i \wedge z) \vee (b_i \wedge \neg z) \vee (a_i \wedge b_i).$$

That is, it defines the following function:

$$sel(z, a_i, b_i) := \begin{cases} 0 & \text{if } a_i = b_i = 0, \\ 1 & \text{if } a_i = b_i = 1, \\ z & \text{if } a_i = 1 \wedge b_i = 0, \\ \neg z & \text{if } a_i = 0 \wedge b_i = 1. \end{cases}$$

Theorem 19.3.2 *Let $\mathcal{D}$ be a set of clauses in n variables and of width at most w. Assume that $\mathcal{D}$ has an ER-refutation π with $k(\pi)$ clauses.*

Then for some $k = O(nk(\pi))$ and $s \leq k/3$, there is a clause reduction σ of $\mathcal{D}$ to $\neg Con_{ER}(s, k)$ of width at most $\max(w, 3)$.

Proof Let $\overline{x}$ be the n variables of $\mathcal{D}$. By Lemma 5.8.4 we may assume that there is another refutation π' of $\mathcal{D}$ with $k = O(nk(\pi))$ steps and that the width of π is at most $\max(w, 3)$. Let $\overline{y}$ be the s extension atoms used in this refutation and think of them as defining a circuit C. Further, without loss of generality we may assume that the clauses defining the extension variables $\overline{y}$ are the first $3s$ clauses and are followed by all $|\mathcal{D}|$ clauses from $\mathcal{D}$. Let S_u, $1 \leq u \leq k$, be the steps in the refutation.

We shall define a clause substitution σ such that every clause $\sigma(A)$ for $A \in \neg Con_{ER}(s, k)$ is either logically valid or could be derived by a weakening of a clause of $\mathcal{D}$. In fact, all clauses $A \in \neg Con_{ER}(s, k)$ yield logically valid $\sigma(A)$ (by virtue of items (a) and (b) in the definition of a clause reduction) except when A is from group $G4$(c): in that case either $\sigma(A)$ will be a 1-axiom or it will follow by the weakening of a clauses of $\mathcal{D}$.

The variables of $\neg Con_{ER}(s,k)$ are $\overline{p}$ and $\overline{q}$ and we define the substitution σ as follows.

1. For q_i^u with $i \neq 0$, substitute 0 or 1, depending on whether the $\overline{y}$-literal corresponding to i occurs in S_u or not.
2. For q_0^u, substitute $\bigvee E_u$, where E_u is the set of $\overline{x}$-literals occurring in S_u together with 1, if $1 \in S_u$. (Note that $|E_u| \leq w$.)
3. For $\overline{p}_u$ with $u = 3r-2, 3r-1, 3r$ and $r \leq s$, proceed as follows:
   (a) if S_u is one of the three clauses corresponding to an instruction of the form $y_r := x_j$, put
   $$\sigma(\overline{p}_u) \ := \ sel(x_j, \overline{a}, \overline{b}),$$
   where $\overline{a}$ and $\overline{b} \in \{0,1\}^t$ define the instructions $y_r := 1$ and $y_r := 0$, respectively;
   (b) otherwise, substitute for $\overline{p}_u$ the string $\overline{a}^u \in \{0,1\}^t$ defining the particular instruction of $\mathrm{Def}_C^{n,s}$ in π'.
4. For $\overline{p}_u$ with $u = 3s+1, \ldots, 3s+|\mathcal{D}|$, substitute $\overline{a} \in \{0,1\}^t$ defining the clause S_u as being a 1-axiom.
5. For $u = 3s+|\mathcal{D}|+1, \ldots, k$, we consider several subcases defining the substitution of $\overline{p}_u$:
   (a) S_u is a 1-axiom: substitute for $\overline{p}_u$ as in item 4;
   (b) S_u was derived by weakening from S_v: substitute for $\overline{p}_u$ the $\overline{a}$ specifying this information;
   (c) S_u was derived by resolution from S_e, S_f resolving the variable y_i: substitute for $\overline{p}_u$ the $\overline{a}$ specifying this information;
   (d) as in (c) but for a resolved variable x_i; assume that $x_i \in S_e$ and $\neg x_i \in S_f$; substitute for $\overline{p}_u$ the term
   $$sel(x_i, \overline{a}, \overline{b})$$
   where $\overline{a}, \overline{b} \in \{0,1\}^t$ specify that S_u was derived by weakening from S_f or S_e, respectively.

We need to verify that, for every clause $A \in \neg Con_{ER}(s,k)$, the clause substitution $\sigma(A)$ falls under one of the three cases (a), (b) or (c) in the definition of clause reductions. Consider the groups $G1$–$G5$ forming $\neg Con_{ER}(s,k)$ separately.

If $A = \overline{p}_u(\overline{a}) \rightarrow$ belongs to the groups $G1$ or $G3$, $\sigma(\overline{p}_u(\overline{a}))$ contains a false literal and so $\sigma(A)$ is a 1-axiom.

For an A from group $G2$, the substitution instance $\sigma(A)$ is clearly a 1-axiom by the definition of $\sigma(q_i^u)$ for all instructions for y_r falling under 3(b) above, i.e. when it does not have the form $y_r := x_j$. In the latter case the instruction is represented by the clauses

$$\{y_r, \neg x_j\}, \quad \{\neg y_r, x_j\}, \quad \{1\}$$

and A describes one of them. The definition of σ in 3(a) above using the selection term on x_j yields a $\sigma(A)$ which either contains 0 in the antecedent (and hence $\sigma(A)$ is a 1-axiom) or in which one of the literals x_j, x_j^0 occurs in both the antecedent and the succedent of $\sigma(A)$, and hence σ falls under case (b) of the definition of reductions.

If A is from group $G4$(a) then, by item 5(c) of the definition of σ, $\sigma(A)$ is a 1-axiom. If A is from group $G4$(b) then $\sigma(A)$ is defined by case (b) of the definition of clause reductions; in particular, for $i = 0$, $\sigma(q_i^v)$ is contained in $\sigma(q_i^u)$ (the E and F in that definition). If A is from group $G4$(c) then $\sigma(A)$ is either a 1-axiom as $\sigma(q_0^u)$ contains the constant 1 if S_u was a 1-axiom, or it falls under case (c) of the definition of clause reductions as $\sigma(q_0^u)$ is $S_u \in \mathcal{D}$ (item 4 of the definition of σ).

Finally, for any A from group $G5$, the substitution instance $\sigma(A)$ is trivially a 1-axiom. □

We interpret Theorem 19.3.2 as reducing the task of showing the hardness of $\mathcal{D}$ for *ER* to the task of showing that $\mathcal{D}$ cannot be clause-reduced (with particular parameters) to $\neg Con_{ER}(s,k)$. Given an *ER*-refutation π of $\mathcal{D}$ in n variables $\overline{x}$ with $k(\pi)$ steps, we have $\neg Con_{ER}(s,k)$ for specific s, k bounded by $O(nk(\pi))$ and a clause reduction σ of small width such that it holds that:

- *For any assignment α to variables $\overline{x}$ of $\mathcal{D}$, if we can find a clause A of $\neg Con_{ER}(s,k)$ that is false under the assignment $\alpha \circ \sigma$ to its variables, we obtain a clause of $\mathcal{D}$ that is false under α: $\alpha \circ \sigma(A)$ can only fail if it falls under item (b) of the definition of reductions and hence contains a clause of $\mathcal{D}$ that is false under α.*

The width bound on σ means that, for a fixed π and for $\neg Con_{ER}(s,k)$ with parameters determined by π, evaluating σ requires at most w calls to the assignment oracle α. Hence if w is a constant or at least bounded by $\log(n|\mathcal{D}|)$ then the reduction is polynomial time in the sense of [50].

19.4 Proof Complexity Generators

Let $1 \leq n < m$ and let g_n: $\{0,1\}^n \rightarrow \{0,1\}^m$ be a function computed by a size-s circuit $C_n(\overline{x})$ with m outputs. We shall denote the instructions of the circuit as $y_1, \ldots, y_{s-m}, z_1, \ldots, z_m$, the last m values being the output values of C_n. Let $\mathrm{Def}_{C_n}(\overline{x}, \overline{y}, \overline{z})$ be the 3CNF defining the circuit (Section 1.4).

The set $\{0,1\}^m \setminus rng(g_n)$ is non-empty and for any $\overline{b} \in \{0,1\}^m$ we define the propositional **τ-formula** $\tau(C)_{\overline{b}}$ to be the 3DNF

$$\mathrm{Def}_{C_n}(\overline{x}, \overline{y}, \overline{z}) \rightarrow \bigvee_{i \leq m} b_i \neq z_i.$$

The size of the formula is $O(s)$, and it is a tautology expressing that $\overline{b}$ is not in the range of g_n.

A **proof complexity generator** is an informal name for a function $g = \{g_n\}_n$ for which the set of all τ-formulas is hard. In this and in the next section (but see Section 19.6) we shall discuss mostly uniform functions g computed by a deterministic algorithm. We shall think of such a g as being explicitly associated with a particular algorithm computing it and hence with an explicit sequence of circuits C_n computing $g_n := g \upharpoonright \{0,1\}^n$. Note that in this case, using the language L_{PV} of PV, the formula $\tau(g)_{\overline{b}}$ is simply the translation

$$\|\forall x(|x| \leq |y|)g(x) \neq y\|^m(\overline{y}/\overline{b}).$$

We shall assume that g satisfies the following properties.

- There is an injective function $m\colon \mathbf{N}^+ \to \mathbf{N}$ such that $n < m(n)$ and $|g(\overline{x})| = m(n)$ for any $\overline{x} \in \{0,1\}^n$, for all $n \geq 1$.
- The algorithm computing g runs on inputs from $\{0,1\}^n$ in a time that is polynomial in $m(n)$, and hence $|C_n| \leq m^{O(1)}$.

We shall call such functions **stretching**. In this case we shall denote $\tau(C_n)_{\overline{b}}$ simply by $\tau(g)_{\overline{b}}$ (the length m of $\overline{b}$ determines n and thus also C_n). Note that the length of the τ-formula is polynomial in m. If g is p-time then $m = n^{O(1)}$ and so is $|\tau(g)_{\overline{b}}|$ for $\overline{b} \in \{0,1\}^n$.

Define a stretching function g to be **hard** for a proof system P if and only if for all $c \geq 1$, for all but finitely many $\overline{b} \in \{0,1\}^* \setminus rng(g)$, we have

$$\mathbf{s}_P(\tau(g)_{\overline{b}}) > |\tau(g)_{\overline{b}}|^c. \tag{19.4.1}$$

When (19.4.1) is satisfied by a bound of the form $2^{|\tau(g)_{\overline{b}}|^{\Omega(1)}}$ we call g **exponentially hard** for P. Let us note a simple observation.

Lemma 19.4.1 *A stretching function g is hard for all proof systems if and only if $rng(g)$ intersects all infinite* NP *sets.*

Proof Assume that g is not hard for P, i.e. the condition (19.4.1) fails for some $c \geq 1$ for infinitely many $\overline{b}$. Then

$$\{\overline{b} \in \{0,1\}^* \mid \mathbf{s}_P(|\tau(g)_{\overline{b}}|) \leq |\tau(g)_{\overline{b}}|^c\}$$

is an infinite NP set disjoint with $rng(g)$.

Assume on the other hand that an infinite NP set U with the definition

$$u \in U \quad \Leftrightarrow \quad \exists v(|v| \leq |u|^d)R(u,v),$$

with R a p-time relation, is disjoint with $rng(g)$. Then g is not hard for a proof-system extending R by, say, accepting as a proof of $\tau(g)_{\overline{b}}$ also any string v such that

$$|v| \leq |\overline{b}|^d \wedge R(\overline{b}, v).$$

□

It may seem that we are too strict in demanding that (19.4.1) holds for all but finitely many strings outside $rng(g)$: to obtain formulas hard for P it would suffice if this held for a fraction of such strings or even for just one in $\{0,1\}^{m(n)} \setminus rng(g_n)$ for infinitely many n. It is this strict property, however, that allows one to develop the theory underlying these formulas. With a relaxed definition this would be just a version of hitting-set generators; see Section 19.6 below.

The original motivation for the definition of the formulas related to cryptographic primitives (one-way functions and pseudo-random generators) and their role in bounded arithmetic and proof complexity (we shall discuss this in Section 19.6). This has shifted somewhat now but nevertheless let us present two results about pseudo-random number generators, one positive and one negative. First we need to recall their definition.

For a stretching function $g = \{g_n\}_n$, g_n: $\{0,1\}^n \to \{0,1\}^{m(n)}$, define its **hardness** $H(g)$ to be the function assigning to $n \geq 1$ the minimum S such that there is a circuit $C(\overline{y})$ with $m(n)$ inputs and of size $\leq S$ such that

$$|Prob_{\overline{x}\in\{0,1\}^n}[C(g(\overline{x})) = 1] \ - \ Prob_{\overline{y}\in\{0,1\}^m}[C(\overline{y}) = 1]| \ \geq \ \frac{1}{S}\,.$$

A **pseudo-random number generator** (PRNG for short) is a stretching function g computed by a deterministic algorithm running in time polynomial in n and of super-polynomial hardness $H(g) \geq n^{\omega(1)}$. The pseudo-random number generator is **strong** (SPRNG) if actually $H(g) \geq \exp(n^{\Omega(1)})$.

Lemma 19.4.2 *Let g be a* PRNG *with $m(n) \geq 2n+1$ and define g^*: $\{0,1\}^{2n} \to \{0,1\}^m$ by*

$$g^*(\overline{u},\overline{v}) \ := \ g(\overline{u}) \bigoplus g(\overline{v}),$$

where $\overline{u},\overline{v} \in \{0,1\}^n$ and $\oplus$ denotes the bit-wise sum modulo 2.

Then g^* is a proof complexity generator hard for all proof systems simulating R *and admitting feasible interpolation (Chapter 17).*

Proof Let P be a proof system admitting feasible interpolation and assume that $\tau(g^*)_{\overline{b}}$ has a size s P-proof. Then (the $\|\ldots\|$ translation of)

$$g(\overline{u}) \neq \overline{y} \ \vee \ g(\overline{v}) \neq \overline{y} \bigoplus \overline{b}$$

has a size $s + m^{O(1)} \ = \ s^{O(1)}$ P-proof. Feasible interpolation then provides a size $s^{O(1)}$ circuit I with m inputs $\overline{y}$ defining a set W separating $rng(g)$ from $\overline{b} \oplus rng(g)$:

$$rng(g) \subseteq W \quad \text{and} \quad W \cap \overline{b} \bigoplus rng(g) = \emptyset\,.$$

If W contains at most half of $\{0,1\}^m$ then $\neg I$ defines a subset of measure $\geq \frac{1}{2}$ in the complement of $rng(g)$, and hence $H(g) \leq |I| \leq s^{O(1)}$. Otherwise $I \oplus \overline{b}$ defines such a subset. It follows that $s \geq H(g)^{\Omega(1)}$. □

Recall from Section 17.2 the notions of one-way functions and permutations (OWF and OWP) and of a hard bit predicate. If h is an OWP and B is a hard bit predicate for h then

$$g : \overline{x} \to (h(\overline{x}), B(\overline{x}))$$

is a PRNG (see Yao [509]).

Lemma 19.4.3 *Assume that h is an OWP, B is a hard bit predicate for h and g is the* PRNG *defined above. Let P be any proof system simulating* R *and admitting p-size proofs of (the $\| \dots \|$ translation of) the injectivity of h, i.e. of the formulas*

$$h(u) = h(v) \to u = v.$$

Then g is not a hard proof complexity generator for P.

In particular, if g is constructed in this way from the RSA then g is not hard for EF.

Proof Assume that $\overline{b} = (b_1, \dots, b_n, b_{n+1}) \notin rng(g)$. As h is OWP, its range is the whole of $\{0,1\}^n$ and, for some $\overline{a} \in \{0,1\}^n$,

$$h(\overline{a}) = (b_1, \dots, b_n) \quad \text{and} \quad B(\overline{a}) \neq b_{n+1}. \tag{19.4.2}$$

A proof of $\tau(g_{\overline{b}})$ may look as follows: take $\overline{a}$, verify (19.4.2) and deduce $\overline{b} \notin rng(g)$ using the injectivity of h,

$$h(\overline{x}) = (b_1, \dots, b_n) \to \overline{x} = \overline{a}.$$

Hence if the injectivity has short P-proofs, so does $\tau(g)_{\overline{b}}$.

The statement about *EF* follows, as *EF* has p-size proofs of the injectivity of the RSA – this is what the proof of Theorem 18.7.2 establishes. □

It follows from the lemma that PRNGs are not, a priori, hard proof complexity generators. But it does not rule out that some other specific constructions of PRNGs may produce hard proof complexity generators. A prominent example is the Nisan–Wigderson construction, to be discussed in Section 19.6.

We have seen actually a proof complexity generator already. The following statement is a reformulation of the second part of Corollary 13.4.6.

Lemma 19.4.4 *For every $\delta > 0$ there is an $\ell \geq 1$ such that for all sufficiently large n there exists an ℓ-sparse $n^2 \times n$ matrix A such that the linear map from $\{0,1\}^n$ into $\{0,1\}^{n^2}$ defined by A is an exponentially hard proof complexity generator for* R.

Problem 19.4.5 (Linear generators for AC^0-Frege) *Is the function from Lemma 19.4.4 also (exponentially) hard for* AC^0*-Frege systems?*

Note that by the discussion at the end of Section 13.4, an affirmative answer would imply that random CNFs of some constant width are hard for AC^0-Frege systems.

We shall now describe one particular construction of proof complexity generators, the so-called gadget generators. There is such a generator computed by 2DNFs that is exponentially hard for AC^0-Frege systems and another computed by degree 2 polynomials that is exponentially hard for PC (Lemma 19.4.7). In the class of these generators there is a universal one, so we may speak of the gadget generator (Lemma 19.4.6). We shall present one more example of a proof complexity generator (the truth table function) in Section 19.5.

Given an arbitrary p-time function

$$f\colon \{0,1\}^\ell \times \{0,1\}^k \to \{0,1\}^{k+1},$$

we define another p-time function

$$Gad_f\colon \{0,1\}^n \to \{0,1\}^m,$$

with

$$n := \ell + k(\ell+1) \quad \text{and} \quad m := n+1,$$

as follows.

1. Interpret an input $\overline{x} \in \{0,1\}^n$ as $\ell + 2$ strings
$$v, u^1, \ldots, u^{\ell+1},$$
where $v \in \{0,1\}^\ell$, all the $u^i \in \{0,1\}^k$ and
2. define the output $\overline{y} = Gad_f(\overline{x})$ to be the concatenation of $\ell + 1$ strings $w^s \in \{0,1\}^{k+1}$, where
$$w^s := f(v, u^s).$$

Note that w^s has one more bit than u^s and hence the first ℓ strings w^s have as many bits as v together with the first ℓ strings u^s, and the last $w^{\ell+1}$ produces one extra bit over $u^{\ell+1}$.

The functions Gad_f of this form will be called **gadget generators** and the function f is a **gadget**. We may fix the function f to be the **circuit value function**

$$CV_{\ell,k}(v,u),$$

which, from a description v of a circuit with k inputs and $k+1$ outputs and from $u \in \{0,1\}^k$, computes the value in $\{0,1\}^{k+1}$. The only unknown in this set-up is the value ℓ, which tells us how large compared with k are the circuits that we need to evaluate. In fact, it turns out that we can restrict without loss of generality to size $k^{1+\epsilon}$, for any fixed $\epsilon > 0$. The proof of the universality of such a generator needs somewhat stronger notion of hardness than we have defined above, i.e. **iterability**. We shall introduce this in Section 19.5 but state now the fact about the universal gadget generator without a proof. It is derived by a self-reducibility argument, cf. [300].

Recall that a size-s circuit in $k \leq s$ inputs can be encoded by $10s \log s$ bits.

Lemma 19.4.6 *Assume that, for any $s = s(k) \geq k$, a gadget generator using as the gadget $CV_{\ell,k}$ with $\ell = 10s \log s$ is (exponentially) iterable for a proof system P containing* R.

Then the gadget generator using $CV_{k^2,k}$ is (exponentially) iterable for P as well.

It thus makes good sense to call the generator Gad_f with specific $f = CV_{k^2,k}$ the **gadget generator** and use the term **gadget** for the string v itself rather than for f.

Lemma 19.4.7 *There is a gadget generator Gad_f using f defined by 2DNFs (i.e. by degree 2 polynomials) that is exponentially hard for AC^0-Frege systems (for PC).*

Proof Take $\ell := k(k+1)$ and consider $\overline{v} = (v_{ji})_{ji}$, with $j \in [k]$ and $i \in [k+1]$. Define the ith bit of $f(\overline{v}, \overline{u})$ to be

$$\bigvee_j (u_j \wedge v_{ji}), \quad \text{for } i \in [k+1].$$

Because AC^0-Frege systems cannot rule out by subexponential-size proofs that $\overline{v}$ is the graph of a bijection between $[k]$ and $[k+1]$ (Ajtai's theorem 15.3.1), they also cannot rule out that, for some fixed $\overline{v}$, the function $\overline{u} \to f(\overline{v}, \overline{u})$ is onto and so is Gad_f.

For PC, define the ith bit of f to be

$$\sum_j u_j v_{ji} \, .$$

The lower bound follows by the same argument from Theorems 16.2.1 and 16.2.4. □

The gadget in the previous argument relied on lower bounds that we have proved already. A possible gadget generator aiming at strong proof systems is defined as follows. Given two parameters $1 \leq k$ and $1 \leq c \leq \log k$ we consider gadgets v with $(k+1)^2$ bits and use them to define a map $nw_{k,c} : \{0,1\}^n \to \{0,1\}^{n+2}$, where

$$n := k(k+2) + k((k+1)^2 + 1).$$

An input $\overline{x} \in \{0,1\}^n$ is interpreted as a tuple

$$x = (A, f, u^1, \ldots, u^t),$$

where $u^i \in \{0,1\}^k$ for $i = 1, \ldots, t = (k+1)^2 + 1$, and where $v = (A, f)$ is read from the gadget v as follows:

- the first $(k+1)k$ bits determine a $(k+1) \times k$ 0/1 matrix A, and
- the subsequent $2^c \leq k$ bits in x define the truth table of a function f: $\{0,1\}^c \to \{0,1\}$.

If an A thus determined has exactly c ones per row, we define the Nisan–Wigderson generator $\overline{y} := nw_{k,c}(\overline{x})$ to be $\overline{y} = (v^1, \ldots, v^t)$ with

$$v_i^j := f(u^j(J_i)), \quad \text{for } j \in [t] \text{ and } i \in [k+1],$$

where $\overline{x}(J_i)$ is the c-tuple $x_{j_1}, \ldots, x_{j_c}$ and $J_i = \{j_1 < \cdots < j_c\}$ is the set of ones in the ith row of A. Otherwise, we define $\overline{y} := \overline{0} \in \{0,1\}^{n+1}$.

Note that $nw_{k,c}$ is computed by a size $O(n^{2+c})$ constant-depth formula: assuming the condition posed on A above (definable by a depth 3 formula), the ith bit in v^j is expressible as

$$\bigvee_{J=\{j_1<\cdots<j_c\}\subseteq[n]} [\, (\bigwedge_{r\in J} A_{ir}) \wedge (\bigwedge_{\overline{w}\in\{0,1\}^c} ((\bigwedge_{r\in J} u^j_r \equiv w_r) \to f_{\overline{w}}) \,) \,].$$

It is consistent with present knowledge that the $nw_{k,c}$ are hard for all proof systems, even those with a small constant c. The gadget contains as a special case the PHP-gadget from Lemma 19.4.7 and hence we have at least:

Lemma 19.4.8 *The gadget generator $nw_{k,c}$ with any $c \geq 1$ is exponentially hard for* AC^0*-Frege systems and for* PC.

19.5 Circuit Lower Bounds

A size s circuit with $k \leq s$ inputs can be encoded by $O(s \log s)$ bits; let us say $10s \log s$ for definiteness. For $k \geq 1$ and $k \leq s \leq 2^{k/2}$, the **truth table function**

$$\mathbf{tt}_{s,k}\colon \{0,1\}^n \to \{0,1\}^m$$

computes, from $n := 10s \log s$ bits describing a circuit C with k inputs and of size $\leq s$, the $m := 2^k$ bits of the truth table of the Boolean function computed by C. We define $\mathbf{tt}_{s,k}$ to be $\overline{0}$ on inputs not encoding a circuit with the prescribed parameters.

The formula $\tau(\mathbf{tt}_{s,k})_{\overline{b}}$ formalizes a lower bound s for the size of circuits computing the Boolean function whose truth table is $\overline{b}$. It has therefore an intuitive appeal as circuit lower bounds are the holy grail of complexity theory. Let us make two observations, one on the positive and one on the negative side.

Lemma 19.5.1 *Let $1 \leq k \leq s = s(k) \leq 2^{k/2}$ (for the following complexity classes see Section 1.3).*

(i) Assume that NE $\cap$ coNE $\not\subseteq$ *Size(s). Then $\mathbf{tt}_{s,k}$ is not hard for some proof system.*
(ii) Assume that there is a proof system P such that
  (a) for some $s(k) \geq 2^{\Omega(k)}$ the function $\mathbf{tt}_{s,k}$ is not hard for P. Then BPP $\subseteq_{i.o.}$ NP.
  (b) for some $s(k) \geq k^{\omega(1)}$ the function $\mathbf{tt}_{s,k}$ is not hard for P. Then NEXP $\not\subseteq$ P/*poly*.

Proof To prove the first statement note that for a specific function $g \in$ NE $\cap$ coNE the set of truth tables of $g_k := g \restriction \{0,1\}^k$, $k \geq 1$, is in NP. Hence we may define a proof system whose proofs of a formula φ are either, say, Frege proofs or the witnesses that $\varphi = \tau(\mathbf{tt}_{s,k})_{\overline{b}}$ for $\overline{b}$ the truth table of some g_k.

For part (a) of the second statement, using the hypothesis we can modify the construction of Nisan and Wigderson [367] and Impagliazzo and Wigderson [249] de-randomizing BPP: guess a pair $(\overline{b}, \pi)$, where $\overline{b}$ is the truth table of a function with an exponential circuit complexity $\geq s(k)$, i.e. outside the range of $\mathbf{tt}_{s,k}$, and π is P-proof of $\tau(\mathbf{tt}_{s,k})_{\overline{b}}$ of size $m^{O(1)} = 2^{O(k)}$. Then use $\overline{b}$ as in [367, 249].

Part (b) holds because, by a result of Impagliazzo, Kabanets and Wigderson [241], if one could certify by p-size strings the super-polynomial circuit complexity of a function then NEXP $\not\subseteq$ P/*poly*. □

Hence we should not expect that the truth table function is a proof complexity generator that is hard for all proof systems. However, to find a proof system for which we could demonstrate that it is not hard (unconditionally) appears difficult. In particular, the usual simple counting argument showing that most functions require exponential-size circuits does not lead to any example within NE $\cap$ coNE.

But we have some hardness results nevertheless. The following statement is proved by an argument analogous to the proof of Lemma 19.4.2 combined with the natural proofs of Razborov and Rudich [444] (also recall the strong PRNGs from Section 19.4).

Lemma 19.5.2 *Assume that there exists a strong PRNG. Let $s(k) \geq k^{\omega(1)}$. Then the truth table function $\mathbf{tt}_{s,k}$ is hard for any proof system that simulates* R *and admits feasible interpolation.*

The τ-formulas based on the truth table function not only have an intuitive appeal but are in a particular sense the hardest. Assume that g is a P/*poly* map with g_n: $\{0,1\}^n \to \{0,1\}^{n+1}$, i.e. it stretches the input by one bit. Note that the $\tau(g_n)$-formulas have size polynomial in n. Define g_n**-circuits** to be circuits which have n inputs and whose instructions can use the inputs, the two constants 0 and 1 and g_n as the only connective (they are n-ary with $n+1$ outputs).

Let $s(n) \geq 1$ be a function. A function g as above is $s(n)$**-iterable** for a proof system P if and only if, for any g_n-circuit D computing m outputs, all the formulas $\tau(D)_{\overline{b}}$, $\overline{b} \in \{0,1\}^m$, require P-proofs of size at least $s(n)$.

Note that $s(n)$-iterability implies, in particular, an $s(n)$ lower bound for P-proofs of the $\tau(g_n)$-formulas. The next statement shows why this concept is of relevance.

Theorem 19.5.3 ([291]) *Assume that a proof system P simulates* R. *Then the following two statements hold.*

(i) There exists a P/*poly map g stretching the input by one bit which is (exponentially) iterable for P if and only if, for any $0 < \delta < 1$, the truth table function $\mathbf{tt}_{s,k}$ with $s = 2^{\delta k}$ is (exponentially) iterable for P.*

(ii) There exists a P/*poly map g stretching the input by one bit which is exponentially iterable for P if and only if there is a $c \geq 1$ such that, for $s = k^c$, the truth table function $\mathbf{tt}_{s,k}$ is exponentially iterable for P.*

Informally, if we replace hardness by iterability then the truth table function is a canonical hard function. The chief use of iterability is to establish the hardness of the truth table function by showing that there is some g that is iterable for P: by the theorem $\mathbf{tt}_{s,k}$ is then iterable too and hence it is also hard. At present we have iterable generators only for R and R(f) with f on the order of log log (Section 19.6). However, the next statement gives an indication that the truth-table function may be hard for *EF*.

Lemma 19.5.4 *Assume that EF does not simulate the* WPHP-*Frege system WF (the system from Section 7.2). Then, for any* $\delta > 0$*, the truth table function* $\mathbf{tt}_{s,k}$ *with* $s = 2^{\delta k}$ *is iterable and hence hard for EF.*

In particular, there exists a stretching g computed in P/*poly that is hard for EF.*

Proof The part about $\mathbf{tt}_{s,k}$ is proved via the first statement in Theorem 19.5.3 by showing that an iterable function exists. Here we shall prove only the existence of a suitable hard g.

We shall prove that *WF* is a *CF* that can use in a proof π extra axioms of the form

$$C(D_1, \ldots, D_n) \not\equiv (r_1^C, \ldots, r_m^C) \tag{19.5.1}$$

where C is a circuit with n inputs and m outputs, D_i are arbitrary circuits and the tuple of r-variables satisfies certain restrictions on their occurrence in the proof π (Section 7.2). By padding the output of C with $|C|$ zeros we may assume without a loss of generality that $|C| = O(m)$ (to take into account our earlier assumption that the size of C is polynomial in m).

Assume that for some $\overline{b} \in \{0, 1\}^m$ the formula $\tau(C)_{\overline{b}}$ has a size-$m^{O(1)}$ proof in *EF* and hence also in *CF*. Replace the tuple $(r_1^C, \ldots, r_m^C)$ in π everywhere by $\overline{b}$ and derive the substitution instance (19.5.1) by a substitution instance of the *CF* proof of $\tau(C)_{\overline{b}}$ (it will have size $m^{O(1)} \sum_{i \leq n} |D_i|$).

Removing in this way all the extra axioms (19.5.1) will yield a *CF* proof which can be p-simulated by an *EF* proof, by Lemma 7.2.2. □

19.6 Bibliographical and Other Remarks

More on the Paris–Harrington principle and its unprovability can be found in Hájek and Pudlák [214]. The statement that most formulas are hard for any proof system has been a folkloristic conjecture in proof complexity at least since the Chvátal and Szemerédi [141] 1988 paper about random CNFs in R. It has been proved for several proof systems besides R; for example, for PC by Ben-Sasson and Impagliazzo [69] and for CP by Hrubeš and Pudlák [237] and Fleming *et al.* [187]. The statement is open for AC^0-Frege systems.

A stand-alone tautology suggested by Cook [158] as hard for Frege systems expresses that matrix inverses commute:

$$A \cdot B = I_n \rightarrow B \cdot A = I_n,$$

where the matrices A, B are $0/1$ and $n \times n$ and I_n is the identity matrix. The statement can be expressed by a formula, **Cook's formula**, of size polynomial in n. However, linear algebra proofs of this fact use concepts that are not known to be definable by polynomial-size formulas (but only by NC^2 circuits). If a proof system can express concepts in NC^2 then it also proves in p-size Cook's formula; see Hrubeš and Tzameret [238] and Cook and Tzameret [160]. There does not seem to be a general argument that the hardness of defining proof concepts ought to imply proof complexity hardness. For example, all versions of PHP are proved in elementary combinatorics by counting, but you do not need it to construct short proofs of WPHP (Section 11.4). The definite advantage of Cook's formula is its transparency.

Sec. 19.1 Lemma 19.1.2 relates to the optimality problem 1.5.5 and is a weak version of a lemma from Krajíček and Pudlák [317]; this topic will be studied extensively in Chapter 21. Lemma 19.1.3 was pointed out to me by Cook. Gödel's theorem yields a $B(y)$ formalizing the consistency of $S^1_2 + Exp$ but a much stronger result was proved by Paris and Wilkie [391]: one can take a B formalizing the consistency of Robinson's arithmetic Q. This topic is also related to the formal system $S^1_2 + 1 - Exp$ discussed in Section 12.9.

Sec. 19.2 The role of propositional consistency statements was first pointed out by Cook [149] and further developed in connection with the optimality problem 1.5.5 in Krajíček and Pudlák [317]. Consistency statements also play a significant role in bounded arithmetic and were studied for a number of provability notions weaker than the standard one. See for example Paris and Wilkie [391] or Krajíček and Takeuti [327].

Sec. 19.3 If we allow the relation R in the definition of NP search problems to be itself NP then nothing really changes, and this relaxation seems more natural and is more useful for bounded arithmetic. The area of NP search problems is closely related to bounded arithmetic and to proof complexity; NP search problems can be used to characterize provably total multi-functions in bounded arithmetic theories. In the second-order set-up they relate to relativized NP search problems and, typically, to theories up to V^1_1 and proof systems up to EF. In the first-order set-up they relate to non-relativized search problems, theories beyond S^1_2 and strong proof systems. For example, the well-known class PLS (polynomial local search) characterizes NP search problems provably total in the theory T^1_2 of Section 10.5; see Buss and Krajíček [130].

Having two NP search problems R and S as in (19.3.1), we can obtain a reduction of S to R by proving over a weak theory that S is well defined, assuming that R is well defined:

$$\forall x \exists y R(x, y) \rightarrow \forall u \exists v S(u, v).$$

For example, in the first-order set-up, if the implication is provable in PV then a p-reduction querying R finitely many times exists, and if it is provable in S^1_2 then a p-time **Turing reduction** (repeated calls are made to R) exists; see Hanika [218, 219]. In addition, R can be replaced by a p-equivalent $\tilde{R}$ such that these reductions become p-reductions as defined earlier. In this way one can establish many reductions to familiar NP search problems. However, showing that an implication is not provable typically means proving a lower bound for the associated propositional formula (an inherently non-uniform task), and that yields various relativized separations. For an example see Beame *et al.* [50], who transcribed Nullstellensatz lower bounds for counting principles (Sections 11.1 and 16.1) to proofs of non-reducibility among the various basic (relativized) NP search classes of Papadimitriou [378]. The investigation of separation problems in bounded arithmetic has led also to definitions of new classes of NP search problems; see Chiari and Krajíček [138, 139], Pudlák [417], Thapen [488] or Beckmann and Buss [60] (further references are given in these papers).

In particular, Chiari and Krajíček [139] introduced the search problem RAM: given an undirected simple graph on $\{0, 1\}^n$ (determined by an oracle or by a uniform or non-uniform p-time algorithm), find a homogeneous subgraph of size $n/2$ (such a subgraph exists by Ramsey's theorem; see Section 13.1). Using the so-called structured PHP approach [296] proved that RSA can be broken relative to RAM, and finding a collision in any candidate collision-free hash family can be reduced to RAM as well. A number of other NP search problems and their properties in bounded arithmetic are treated in Cook and Nguyen [155], where more comprehensive references for the topic can be found.

Arguably the link with bounded arithmetic is the most significant and developed theory around NP search problems and has been largely known for over 20 years. I do not understand why researchers in complexity theory prefer to ignore this and to rediscover specific known facts and immediate consequences of the theory.

The construction in Section 19.3 is based on [310], and our argument for Theorem 19.3.2 follows closely the proof of [310, Theorem 2.1]. Theorem 19.3.2 may be also interpreted as a proof-theoretic reduction: each clause of $\sigma(\neg Con_{ER}(s, k))$ can be derived from $\mathcal{D}$ very easily in a weak proof system (e.g. tree-like $R^*(\log)$); see [310]. Theorem 19.3.2 also implies that a relativized NP search problem that is provably total in the bounded arithmetic theory V^1_1 polynomially reduces to $\neg Con_{ER}(s, k)$ (see [310]). This can be generalized further (using the idea of implicit proofs from Section 7.3) to define non-relativized NP search problems (denoted $i\Gamma$ in [310]) that are complete among all provably total NP search problems in the theory V^1_2 of Buss [106] from Section 9.3. Other NP search problems that are complete for this very strong theory were defined by Kolodziejczyk, Nguyen and Thapen [271] and by Beckmann and Buss [61]. These NP search problems can be used to define combinatorial-looking formulations of the consistency of strong proof systems, including *ER*. Avigad [42] proposed earlier another simple-looking combinatorial formulation of the consistency of *ER* (i.e., of his system p-equivalent to it).

Note that it is possible to scale down the construction behind Theorem 19.3.2 from *ER* to weaker proof systems (e.g. the AC^0- , $AC^0[m]$- or TC^0-Frege systems) by restricting the circuits C that can be used in $\mathrm{Def}^{n,s}_C$ to a suitable class of circuits.

In fact Krajíček, Skelley and Thapen [325] defined earlier a problem related to $\neg Con_{ER}(s, k)$ but for R instead of *ER* (problem $1 - Ref(Res)$ in [325]). This problem allows a clean combinatorial formulation as a p-equivalent **colored PLS** problem of [325] or as a **generalized local search** (GLS), defined earlier by Chiari and Krajíček [138].

Let us point out one interesting property of the construction related to random formulas. For a given $s \geq 1$ and $k \geq 3s$, define a set of clauses, to be denoted $\neg Con_{ER}(s,k)(C_{\mathbf{r}})$, by the following random process yielding a set of clauses ($\mathbf{r}$ signifies random bits).

1. Pick s instructions for computing s subcircuits $\overline{y}$ defining a circuit $C_{\mathbf{r}}$ with no input variables (the ith instruction for y_i is picked uniformly at random from all legal instructions).
2. Substitute in $\neg Con_{ER}(s,k)$ for all variables $\overline{p}_u$ and q_i^u with $u \leq 3s$ the bits defining the clauses of $\mathrm{Def}_C^{0,s}$ corresponding to the circuit $C_{\mathbf{r}}$ defined in the first step.

Is it possible that (with a high probability) the set $\neg Con_{ER}(s,k)(C_{\mathbf{r}})$ is hard in any proof system not simulating *ER*? If so, this set would be a random source of hard formulas that are *always* unsatisfiable (this is not true for many other proposed constructions which provide only sets that are unsatisfiable with a high probability). See the last section in [310] for more discussion.

Sec. 19.4 Lemma 19.4.1 is from [291] and appears to be related to a conjecture of Rudich [454] about the existence of a p-time function for which it is hard to define, by a small non-deterministic circuit, a large set disjoint from its range. Lemma 19.4.2 is from Alekhnovich *et al.* [12], Lemma 19.4.3 from [289] and Lemma 19.4.4 is from [291]. Problem 19.4.5 can be viewed as asking for generalizations of AC^0-Frege lower bounds for the Tseitin formulas (Section 13.3) by Ben-Sasson [67] and Hastad [223]: these formulas are the τ-formulas of the linear map associated with the underlying graph, but the map shrinks the input rather than stretching it.

The gadget generators were defined in [300], where Lemmas 19.4.7 and 19.4.8 were proved. The generator $nw_{k,c}$ was defined in [304].

The τ-formulas were defined independently in [288] and in Alekhnovich *et al.* [12]. My original motivation was an investigation of the role of cryptographic primitives (OWFs, OWPs or PRNGs) in bounded arithmetic. In particular, I recognized as fundamental in this context a theory that I called BT (basic theory), i.e. PV extended by all instances of dual WPHP (dWPHP) for p-time functions. This theory is now called APC_1; see Section 12.6. I tried but failed to prove, using some plausible cryptographic assumption, that BT is stronger than PV; the problem is still open. The dWPHP principle has one attractive model-theoretic feature not shared by the original WPHP. Namely, if we have a p-time function f: $\{0,1\}^{n+1} \rightarrow \{0,1\}^n$ in a model of PV that satisfies WPHP (i.e. there are two different $u, v \in \{0,1\}^{n+1}$ such that $f(u) = f(v)$) then it will satisfy it in all extensions of the model (the pair u, v will still be there). But if g: $\{0,1\}^n \rightarrow \{0,1\}^{n+1}$ satisfies dWPHP, say, some $b \in \{0,1\}^{n+1}$ is not in the range of g, you can, in principle, change that: just add $a \in \{0,1\}^n$ to the model with the property that $g(a) = b$. A less detailed but essentially equivalent formulation of the feature is that WPHP is a $\forall\Sigma_1^b$-sentence while dWPHP is $\forall\Sigma_2^b$.

But (as we shall see in Chapter 20) you can adjoin such an a if and only if there is no EF proof of $\tau(g)_b$ in the ground model.

The motivation in Alekhnovich *et al.* [12] was to understand the role of PRNGs in proof complexity and to investigate from a unified perspective various known lower bounds and how they could possibly be generalized to stronger proof systems (see [12] for details of their view). Lemma 19.4.3 shows that PRNGs are not a priori hard proof complexity generators as well, but it does not rule out that a specific construction of PRNGs could yield such generators. In particular, [12] investigated the **Nisan–Wigderson generator**.

The Nisan–Wigderson generator is constructed from an $m \times n$ 0/1 matrix A (we shall assume that $A_{ij} = 1$ if and only if $j \in J_i \subseteq [n]$) and a Boolean function f in the same way as some of the maps we have seen already. Let us state this precisely. There are two additional parameters $d, \ell \geq 1$ and A is required to be a **(d, ℓ)-design**: $|J_i| = \ell$ for all i and $|J_u \cap J_v| \leq d$ for all different u, v. The function f is defined on $\{0,1\}^\ell$, and the ith bit of the output of the Nisan–Wigderson generator $NW_{A,f}(x)$ determined by A and f is

$$f(x(J_i)), \quad \text{where } x(J_i) = x_{j_1}, \ldots, x_{j_\ell} \text{ if } J_i = \{j_1 < \cdots < j_\ell\}.$$

In [12] the hardness of $NW_{A,f}$ was proved for R, PC and their combination PCR (Section 7.5) under various encodings of the maps in the τ-formulas, assuming some expansion properties of A (as in Section 13.3) and some combinatorial properties of f. Razborov [439, 440] subsequently extended these lower bounds to some DNF-resolution systems R(h) (Section 5.7) up to an h that is proportional to log log in [440] (the paper speaks about R($\epsilon \log n$) but in our definition this is R(Ω(log log)); see the discussion in Section 5.9).

We shall not present these lower bounds in detail but instead will discuss a conjecture that Razborov [440] made about the hardness of $NW_{A,f}$ as a proof complexity generator. First note that if we allow g to be in NP$\cap$coNP, meaning that *the predicate if the ith bit of* $g(x)$ *is* 1 is in NP $\cap$ coNP, then we can still express $b \notin rng(g)$ by a propositional formula. Razborov's conjecture ([440, Conjecture 2]) says that

- *If $g = NW_{A,f}$ is based on a matrix A which is a combinatorial design with the same parameters as in Nisan and Wigderson [367] and on any function f in* NP $\cap$ coNP *that is hard on average for* P/*poly then g is hard for EF.*

The parameters mentioned are likely to be those used by Nisan and Wigderson [367, Lemma 2.5]:

$$d = \log(m), \quad \log(m) \leq \ell \leq m, \quad n = O(\ell^2)$$

(there are other sets of parameters in [367] but they lead to easy g). The **hardness on average** is measured by the minimum S for which there is a circuit C of size $\leq S$ such that

$$Prob_{u \in \{0,1\}^\ell}[f(u) = C(u)] \geq \frac{1}{2} + \frac{1}{S}$$

and [367] requires that it is $2^{\Omega(\ell)}$ and also $\geq m^2$. All these constraints together allow the following set of parameters, for any $\epsilon > 0$:

$$m = 2^{\epsilon n^{1/2}}, \quad d = \epsilon n^{1/2}, \quad \ell = n^{1/2}.$$

Note that this means that the $\tau(g)$-formulas are already provable in quasi-polynomial size $m^{(\log m)^{O(1)}}$ in R, by going through all possible arguments and all possible NP witnesses in the definition of f.

Pich [395, 396] proved that the conjecture holds for all proof systems with feasible interpolation in place of EF (in fact, weaker assumptions on both A and f suffice). The conjecture was also proved in [302] to be consistent with the true universal theory in the language L_{BA}(PV) (it contains PV). It is also consistent with the theory that $rng(NW_{A,f})$ intersects every infinite NP set (Lemma 19.4.1). The conjecture is very sensitive to the choice of parameters. The size of $\tau(g)$ will be polynomial in m even if we allow f to be in NTime($m^{O(1)}$) $\cap$ coNTime($m^{O(1)}$). But in [292] it was proved that then the validity of the conjecture implies *unconditionally* that EF is not p-bounded. It is not clear (or discussed in [440]) why the time complexity of f should matter.

There is also something opaque in using NP$\cap$coNP functions in this context. First, we cannot express propositionally that f is a total function. Having only a partial f could only make the τ-formula harder to prove (there are fewer constraints on the output). But what if we cannot prove that f has unique values? Then we do not need to bother with the τ-formulas, as we already have a lower bound. Moreover, in this case the τ-formulas could be hard irrespective of how hard it is to compute f. In particular, if U, V are two disjoint NP sets whose disjointness is hard for EF then EF cannot prove feasibly that $f(u) \neq b$, for $b \in \{0, 1\}$, for any f separating U from V. In particular, we could take $J_1 = \cdots = J_m$ (i.e. no design condition is needed) and still get a hard g. Recall that such a pair U, V exists if EF is not optimal (Lemmas 17.2.5 and 19.2.2). I think that for these reasons it may be good to study the conjecture for some specific f avoiding pathologies, e.g. a hard bit of the RSA (Theorem 18.7.2).

The advantage of the generator from Razborov's conjecture seems to be that there are non-deterministic witnesses (feasibly unreachable from b) to play with. The advantage of the gadget generator is that it reduces the task of proving the hardness of a specific function (the generator) to the task of proving that the existence of a hard function (the gadget) is *consistent* with the proof system (i.e. it is not refutable in a short proof). Maybe it could be combined, similarly as the function $\mathsf{nw}_{k,c}$ in Section 19.4.

A theory around proof complexity generators is slowly growing [289, 439, 291, 440, 292, 296, 298, 300, 302, 306] and [304, Chapters 29 and 30] offers an overview (do not panic: you do not need to read the first 28 chapters in order to understand it).

Sec. 19.5 The truth table function was considered as a proof complexity generator in [291]. Lemma 19.5.1 is [304, Theorem 29.2.2] and Lemma 19.5.2 is [304, Theorem 29.2.3] and its proof can be found in [304]. The definition of iterability and

Theorem 19.5.3 are from [291]. Iterability is a special case of a stronger property, pseudosurjectivity, that has a model-theoretic meaning. **Pseudosurjective** maps g satisfy the same condition as iterable maps for any circuit D using, besides the gate g_n, also the usual DeMorgan connectives.

Iterability and pseudosurjectivity can be interpreted via the student–teacher game from Section 12.2 (see [291, Sec. 3] for details).

It is possible to avoid iterability when proving the hardness of the truth table function if a very weak WPHP with exponentially many pigeons is not shortly provably in the proof system under consideration. Any function on $\{0, 1\}^k$ can be defined by a size $O(k2^k)$ DNF and, when WPHP fails badly, the terms in the DNF can be enumerated by numbers up to $k^{O(1)}$, thus defining a circuit of p-size computing the function. This relates to a construction of Razborov [435]. An analogous construction works if one has a hard generator g with m exponential in n and such that each output bit is computable in a time polynomial in n (Razborov [440]).

At present we have iterable generators only for R and R(f) with f on the order of log log: it is the function from Lemma 19.4.4. It is a bit premature now, but if Problem 19.4.5 has an affirmative solution then it would be natural to ask whether the function is actually also iterable for AC^0-Frege systems. The following problem seems less premature.

Problem 19.6.1 (Iterability for AC^0-Frege) *Is the linear map from Problem 19.4.5 or the gadget generator from Lemma 19.4.7 (exponentially) iterable for* AC^0*-Frege systems?*

If so then the truth table function is hard for AC^0-Frege systems and, in particular, the theory V^0_1 does not prove super-polynomial circuit lower bounds. Such an unprovability result could be expected to hold even for lower bounds for AC^0-circuits but it is unknown at present (see [291, Theorem 4.3]).

Lemma 19.5.4 is a variant of [291, Theorem 5.2]. The original argument was model-theoretic and showed that under the hypothesis, for any $0 < \delta < 1$, the truth table function $\mathbf{tt}_{s,k}$ with $s = 2^{\delta k}$ is iterable and hence hard for EF.

There is an interesting parallel between how PRNGs are used and how proof complexity generators may be used. The former are used to reduce significantly the number of random bits an algorithm needs to use, and possibly to derandomize it completely. In proof complexity we can consider the following situation. Let $\varphi(y_1, \dots, y_m)$ be an unsatisfiable CNF formula with m variables. We say that a subset $E \subseteq \{y_1, \dots, y_m\}$ is a set of **essential variables** (it is not necessarily unique) if knowing the values of the variables in E allows us to quickly deduce – assuming φ – the values of all the other variables. We formalize this as follows. Let $y_1, \dots, y_m = \overline{u}, \overline{v}$. Then the set of u-variables is essential if and only if the formulas $v_i \equiv w_i$ for all possible i have p-size R-derivations from

$$\varphi(\overline{u}, \overline{v}) \wedge \varphi(\overline{u}, \overline{w}).$$

It was proved in [298], using the generator exponentially iterable for R mentioned above, that there is a substitution h: $\{0, 1\}^\ell \to \{0, 1\}^m$ with $\ell = (\log m)^{O(1)}$ such that $\varphi(h(\overline{x}))$ has a set of ℓ essential variables $\overline{x}$ and any R-refutation of size $\leq m^{(\log m)^\mu}$ for a fixed $\mu > 0$ implies that the *original* formula $\varphi(y_1, \ldots, y_m)$ is unsatisfiable.

There are other ways to represent Boolean functions and to formulate circuit lower bounds in bounded arithmetic; see [278]. An overview may be found in Müller and Pich [358]; we shall return to this topic in Chapter 22.

20

Model Theory and Lower Bounds

This chapter presents a model-theoretic framework for thinking about lower bounds for strong proof systems and, ultimately, about the NP $=_?$ coNP problem 1.5.3. The approach relies essentially on the completeness and the compactness theorems for first-order logic. The unprovability of a sentence A in a theory T is equivalent to the existence of a model of T in which A fails. This equivalence replaces the task of showing that something does not exist (a proof) by the task of constructing something else (a model). We would like to proceed in a similar way with proof systems: to replace the task of showing that a tautology φ does not have a short proof in a proof system P by the task of constructing some *model*. Models in propositional logic are truth assignments and it seems that we cannot construct any further new ones. Here the compactness theorem helps. Rather than thinking about a single formula φ we think about sequences of formulas $\{\psi_k\}_k$ and take φ to be ψ_n for a *non-standard* n (Section 8.3); it represents the whole sequence and its proof complexity reflects the asymptotic proof complexity of the ψ_k. We shall see that for this φ it is possible to add new truth assignments.

A simple statement (with remarkable consequences) underlying this informal description will be given in Section 20.1. Section 20.2 gives a general formulation of model-theoretic tasks equivalent to lower bounds (in proof complexity as well as in circuit complexity), and Sections 20.3 and 20.4 present two forcing set-ups.

20.1 Short Proofs and Model Extensions

This section presents a simple model-theoretic construction which has, nevertheless, remarkable consequences (Corollary 20.1.4). It may be useful to refresh the basic model-theoretic example from Chapters 8 and 9. We shall take as our main example *EF* and its corresponding theory PV_1, and will formulate most statements for this pair only. We will work therefore in the first-order set-up and use the translation $\| \dots \|$ (Chapter 12). The arguments work, however, for all pairs of a proof system P and a theory T satisfying the correspondence conditions. It may be useful if the reader refreshes what the correspondence means (Section 8.6). In particular, it ought to hold that:

1. $P \geq_p EF$ and T is a theory in the language $L_{BA}(\mathrm{PV})$, $T \supseteq \mathrm{PV}_1$;
2. T proves the soundness of P, RFN_P, and hence also its consistency, Con_P;
3. if $A(x)$ is an open $L_{BA}(\mathrm{PV})$-formula and T proves $\forall x A(x)$ then there are polynomial-size P-proofs of the formulas $\|A\|^n$; in fact, such proofs can be constructed by a p-time algorithm f from $1^{(n)}$ and T proves that f constructs such proofs.

The first condition about the strength of P and T is added so that arguments for EF and PV_1 can be smoothly generalized to P and T. The second and third conditions were established for EF and PV_1 in Theorem 12.4.2. The part about the provability in PV_1 that f constructs proofs of the formulas was not stated in Theorem 12.4.2 (it was not needed then) but it obviously follows from the construction there.

We shall use the following $L_{BA}(\mathrm{PV})$ formulas:

- $Sat_0(x, y)$; an open formula formalizing that y is a satisfying assignment for a formula x;
- $Sat(x) := \exists y \leq x\, Sat_0(x, y)$;
- $Taut(x) := \forall y \leq x\, Sat_0(x, y)$;
- $Prf_P(x, z)$, an open formula formalizing that z is a P-proof of x;
- $Pr_P(x) := \exists z Prf_P(x, z)$.

We shall often write φ for x and π for z. We shall also write, in $L_{BA}(\mathrm{PV})$-formulas, $\neg\varphi$ as a shorthand for the $L_{BA}(\mathrm{PV})$-function $f_\neg(\varphi)$ that assigns to the formula code φ the code of its negation.

The completeness of EF can be formalized by the sentence

$$\forall \varphi\ (Taut(\varphi) \rightarrow Pr_{EF}(\varphi)). \tag{20.1.1}$$

This sentence is true but it is not provable in PV_1 unless $\mathrm{P} = \mathrm{NP}$. To see this, unwind the definitions of $Taut$ and Pr_{EF}:

$$\forall \varphi \exists y, z\ (Sat_0(\neg\varphi, y) \vee Prf_{EF}(\varphi, z)).$$

If this sentence was provable in PV_1, by Herbrand's theorem 12.2.1 and its Corollary 12.2.3 there would be finitely many $L_{BA}(\mathrm{PV})$-terms $t_j(x)$ such that, for each φ, one such term $t_j(\varphi)$ computes either a falsifying assignment y or an EF-proof z. We can combine them into one $L_{BA}(\mathrm{PV})$-function (p-time, in particular) that would solve SAT.

This observation implies that in a model $\mathbf{M}$ of PV_1 there may exist a tautology φ, $\mathbf{M} \models Taut(\varphi)$, without an EF-proof. Our first lemma characterizes such models.

Lemma 20.1.1 *Let* $\mathbf{M}$ *be a model of* PV_1 *and assume that* $\varphi \in \mathbf{M}$ *is a propositional formula. Then the following two statements are equivalent.*

(i) There is no EF-proof of φ in **M**:

$$\mathbf{M} \models \forall z \neg Prf_{EF}(\varphi, z).$$

(ii) There is an extension $\mathbf{M}' \supseteq \mathbf{M}$ *in which φ is falsified:*

$$\mathbf{M}' \models \exists y \, Sat_0(\neg\varphi, y).$$

Proof Assume that part (ii) holds. Then **M** cannot contain an *EF*-proof π of φ because such a proof would be also in $\mathbf{M}'$ and that would violate the soundness of *EF* (provable in PV_1).

Assume now that (ii) fails, i.e. in all extensions $\mathbf{M}'$, φ remains a tautology. This means that a theory S in the language $L_{BA}(\mathrm{PV})$ augmented by constants for all elements of **M** and by a new constant c, and consisting of the axioms

- PV_1,
- the diagram of **M**, all quantifier-free sentences true in **M**,
- $Sat_0(\neg\varphi, c)$

is inconsistent. Assume that σ is a proof of the inconsistency; we may look at it as a proof in PV_1 of

$$\bigwedge_{i \le t} D_i \rightarrow \neg Sat_0(\neg\varphi, c), \tag{20.1.2}$$

where the D_i are some sentences from the diagram. If we think of all constants in the implication that are not in $L_{BA}(\mathrm{PV})$ as variables, the p-simulation of PV_1 by *EF* implies that each

$$\| \bigwedge_{i \le t} D_i \rightarrow \neg Sat_0(\neg\varphi, c) \|^m$$

has an *EF*-proof π_m constructed – provably in PV_1 – by a p-time f from $1^{(m)}$.

In particular, taking n to be a length in **M** majorizing the lengths of the constants in (20.1.2), in **M** there is an *EF*-proof $\pi_n := f(1^{(n)})$ of

$$\| \bigwedge_{i \le t} D_i \rightarrow \neg Sat_0(\neg\varphi, c) \|^n . \tag{20.1.3}$$

Now, working in **M**, substitute, for all atoms corresponding to the variables standing for constants, for elements of **M** the actual bits that these elements determine. This turns each D_i into a true propositional sentence which has a simple proof in **M** (by evaluation of the sentence). Hence we have a proof of

$$\| \neg Sat_0(\neg\varphi, c) \|^n$$

from which φ can be deduced as in the claim showing (8.4.8) in the proof of Theorem 8.4.3. □

An extension $\mathbf{M}'$ of **M** is **cofinal** if every element of $\mathbf{M}'$ is bounded above by some element of **M**.

Corollary 20.1.2 *Let* **M** *be a countable model of* PV_1. *Then it has a cofinal extension* $\mathbf{M}' \supseteq_{cf} \mathbf{M}$ *such that every tautology in* $\mathbf{M}'$ *has an EF-proof in* $\mathbf{M}'$.

Proof Enumerate in $\varphi_0, \varphi_1, \ldots$ all the formulas in **M** as well as all the formulas that may appear in the following countable chain construction. Put $\mathbf{M}_0 := \mathbf{M}$ and for $i = 0, 1, \ldots$ proceed as follows.

- If $\mathbf{M}_i \models Taut(\varphi_i) \wedge \neg Pr_{EF}(\varphi_i)$, set $\mathbf{M}_{i+1}$ as an extension of $\mathbf{M}_i$ provided by Lemma 20.1.1 (and possibly shortened to a cut so that $\mathbf{M}_i$ is cofinal in it) in which φ_i is falsified.

Then define $\mathbf{M}' := \bigcup_i \mathbf{M}_i$. □

Corollary 20.1.3 *There is a model* $\mathbf{M}^*$ *of* PV_1 *such that any tautology in* $\mathbf{M}^*$ *has an EF-proof and there is a non-standard element* $a \in \mathbf{M}$ *with length* $n := |a|$ *such that for any element* $b \in \mathbf{M}$, $|b| \leq n^k$ *for some standard* k.

Proof Let **M** be any countable model of PV_1 in which the standard powers n^k of some non-standard length n eventually majorize any length. Such a model can be found as a cut (Section 8.3) in an arbitrary non-standard model of PV_1. Then apply to it Corollary 20.1.2. □

This has a bearing on the provability of length-of-proofs lower bounds to *EF*-proofs in PV_1. How can we express a particular super-polynomial lower bound for *EF*, say a quasi-polynomial lower bound $n^{\log n}$? The theory PV_1 does not prove that for any x with length $n := |x|$ there is a string y of length $n^{\log n}$ (see Parikh's theorem 8.1.1) as it can prove only the totality of functions of polynomial growth in length. However, we do not need such a function to express the lower bound. We may write

$$\forall x \exists \varphi \geq x \, [Taut(\varphi) \wedge \forall \pi (|\pi| \leq |\varphi|^{\log |\varphi|}) \neg Prf_{EF}(\varphi, \pi)]. \tag{20.1.4}$$

The next statement follows immediately from the model in Corollary 20.1.3.

Corollary 20.1.4 *The lower bound* (20.1.4) *is not provable in* PV_1.

Considering the fact that PV_1 proves a considerable amount of complexity theory, this unprovability results seems interesting. We shall return to it in Chapter 22.

We now turn our attention to particular models which are relevant, as we shall see, to lower bounds for *EF*-proofs of particular formulas. A **small canonical model M** of PV_1 is any non-standard model which is a cut in a model of true arithmetic in the language $L_{BA}(\mathrm{PV})$ and in which standard powers of some non-standard length eventually majorize any other length. In other words, a small canonical model **M** is obtained from a model of true arithmetic in the language $L_{BA}(\mathrm{PV})$ by picking a non-standard element a and taking for **M** the smallest cut in the ambient model generated by a. These models are, in fact, models of not only PV_1 but of all true universal (Π^0_1, in fact) $L_{BA}(\mathrm{PV})$ sentences.

Lemma 20.1.5 *Let $\{\psi_k\}_k$ be a p-time constructible sequence of tautologies, $|\psi_k| \geq k$, and let f be a p-time function computing ψ_k from $1^{(k)}$. Then the following two statements are equivalent.*

(i) The sequence $\{\psi_k\}_k$ is a hard sequence for EF (Section 19.1).
(ii) For any small canonical model **M** *and any non-standard length n in it there is an extension $\mathbf{M}' \supseteq \mathbf{M}$ to a model of* PV_1 *in which $\psi_n := f(1^{(n)})$ is falsified.*

Proof Assume first that the sequence is hard for *EF*. That is, for any $c \geq 1$ there is a $k_c \geq 1$ such that for all $k \geq k_c$, the formula ψ_k has no *EF*-proof of size $\leq |\psi_k|^c$. The statement

$$\forall a, n\ (|a| = n \wedge n \geq \underline{k_c} \rightarrow \forall z(|z| \leq n^{\underline{c}}) \neg Prf_{EF}(f(1^{(n)}), z)$$

(the constants c and k_c are represented by dyadic underlined numerals from Section 7.4) is true in this case and hence it is also true in any non-standard model of true arithmetic and thus also in any small canonical model. Taking all these statements together for all $c \geq 1$ (and for corresponding k_c) shows that, in any small canonical model, no ψ_n with a non-standard n has an *EF*-proof. The existence of $\mathbf{M}'$ then follows by Lemma 20.1.1.

For the opposite direction, assume that the sequence is not hard for *EF*, i.e. there is a fixed $c \geq 1$ such that infinitely many formulas ψ_k have an *EF*-proof of size $\leq |\psi_k|^c$. Let $\mathbf{M}_0$ be a model of true arithmetic and b an arbitrary non-standard element.

Claim *In $\mathbf{M}_0$, there is a non-standard $a \leq b$ such that for $n := |a|$ the formula ψ_n has an EF-proof of size $\leq |\psi_n|^c$.*

If not, we could define in $\mathbf{M}_0$ the standard cut **N** by the condition that $x \in \mathbf{N}$ if and only if

$$\exists k \leq |b|\ x \leq k \wedge (\exists z(|z| \leq |\psi_k|^c) Prf_{EF}(\psi_k, z)).$$

That would contradict induction in $\mathbf{M}_0$.

It follows that in no small canonical model does a cut in $\mathbf{M}_0$ containing a admit an extension falsifying ψ_n. □

All the statements above can be proved by identical arguments for any pair of a theory T and a corresponding proof system P satisfying the conditions stated at the beginning of the section. While for some proof system Q we may not find a corresponding theory (if Q does not admit p-size proofs of its own consistency), we can construct $P \geq_p Q$ for which a suitable T exists. In particular, we can take $P := EF + Ref_Q$ (as in Section 19.2) and $T := PV_1 + Ref_Q$. Note that this construction applies also to those incredibly strong proof systems coming from PA or ZFC (Section 7.4).

To conclude this section, let us consider a model-theoretic situation that applies to all proof systems at the same time, i.e. it relates to the NP =? coNP problem 1.5.3. Denote by $\forall_1 Th_{PV}(\mathbf{N})$ the **true universal L_{BA}(PV) theory**: the set of all universal

L_{BA}(PV)-sentences that are true in the standard model **N**. This theory is not very transparent; as a consequence of Trakhtenbrot's theorem [490] it is not even recursively enumerable (see the introduction to Section 12.1). It is axiomatized over PV by the set of all RFN_P, for all proof systems P. This allows to formulate statements similar to (and, in fact, stronger than) the following one. We shall omit its proof, which uses properties of non-standard models (overspill, in particular) that we have not discussed.

Lemma 20.1.6 *The following two statements are equivalent:*

- NP $\neq$ coNP.
- *There exists a small canonical model* **M** *containing a tautology* φ *that can be falsified in some extension* $\mathbf{M}' \supseteq \mathbf{M}$ *that satisfies* $\forall_1 Th_{\mathrm{PV}}(\mathbf{N})$.

20.2 Pseudofinite Structures

Models of bounded arithmetic and their expansions can be replaced by a seemingly more elementary, and thus maybe more accessible, model-theoretic framework working with pseudofinite structures. It is essentially equivalent but it can be formulated without the need first to introduce bounded arithmetic theories. The idea is that what really matters in the models **M** and $\mathbf{M}'$ in Section 20.1 is only the initial segment $[0, n]$, where n bounds the length of all the objects under consideration (a formula, a proof, an assignment, etc.), together with the objects themselves, represented by relations on $[0, n]$ coding them.

Let L_{ps} be a finite first-order language. A **pseudofinite L_{ps}-structure** is an infinite structure satisfying the L_{ps}-theory of all finite L_{ps}-structures. There are several equivalent definitions of this notion and the one most useful for us is the following:

- *a pseudofinite L_{ps}-structure is an L_{ps}-structure that is elementary equivalent to an L_{ps}-structure definable in a non-standard model of true arithmetic* **M** *whose universe is a non-standard initial interval of* **M**.

So, we cannot escape non-standard models. Pseudofinite structures encode, in a sense, a sequence of larger and larger finite structures and their properties reflect asymptotic properties of the sequence. Many computational and proof complexity questions concern the asymptotic behavior of sequences of circuits, or propositional proofs or similar objects, all easily encodable by finite structures (as we have seen in Lemma 1.4.4 or Section 8.4).

The general problem of expansions of pseudofinite structures can be formulated as follows. Let **A** be a pseudofinite L_{ps}-structure definable in a non-standard model **M** of true arithmetic such that the universe of **A** is $[0, N]$, with N non-standard. We shall assume that L_{ps} contains symbols for the constants 0, 1 and the binary relation symbols $suc(x, y)$ and $\leq$ that are interpreted in **A** by the elements 0, 1 of **M**, by the successor relation $x < N \wedge x + 1 = y$ and by the ordering $\leq$ of **M**, respectively.

We may add, if needed, ternary relations interpreted by the graphs of addition and multiplication in **M**. Note that if the arithmetic structure is present on $[0, N]$ then it uniquely determines the minimum cut in **M** containing N and closed under addition and multiplication; the cut is a model of $I\Delta_0(L_{ps})$ (see Claim 2 in the proof of Theorem 8.1.1 in Section 8.3).

We represent the successor function (and addition and multiplication) by their graphs because we want relational languages; that is a helpful restriction in finite model theory. Let $L \supseteq L_{ps}$ be a larger finite relational language. The general task is to expand **A** to an L-structure **B** such that **B** satisfies some property. For example, it might model an L-theory T. The model-theoretic term **expands** means that **B** has the same universe as **A**, interprets L_{ps} identically and only adds on top of **A** an interpretation of the symbols from $L \setminus L_{ps}$.

Let us consider a specific example, a reformulation of Ajtai's argument (Section 8.5) for Theorem 8.5.1. Let R be a binary relation symbol not in L_{ps}. The task is to interpret the relation symbol R by a relation R^* on the universe $[0, N]$ of **A** such that the expanded structure satisfies the following:

(a1) the least number principle (LNP)

$$\exists x B(x) \rightarrow (\exists x \forall y < x\, B(x) \wedge \neg B(y))$$

for all $L_{ps}(R)$-formulas,

(a2) R is the graph of an injective function from $[0, N]$ into $[0, N-1]$.

Lemma 20.2.1 *Assume that, for any finite language L_{ps}, any pseudofinite L_{ps}-structure **A** has an expansion to $\mathbf{B} = (\mathbf{A}, R^*)$ satisfying (a1) and (a2).*

Then the PHP*-formulas* PHP_k *do not have p-size* AC^0*-Frege proofs.*

Proof Assume that there are depth d, size k^c Frege proofs of PHP_k, for all $k \geq 1$, for some constants $c, d \geq 1$. If **A** sits inside a non-standard model of true arithmetic **M** then such proofs exist also for all k from the model and, in particular, for $k := N$. A depth d proof of PHP_N of size $\leq N^c$ can be encoded by a relation Π on $[0, N]$ (as in Section 8.4); let L_{ps} have the name S for the relation and let **A** interpret S by Π.

The only reason for talking about expansions is the following claim.

Claim *If **B** is an expansion of **A** then Π is a depth d proof of* PHP_N *in **B** as well.*

This is simple but key: the property that S is a proof does not mention R at all and the two structures **A** and **B** are identical as L_{ps}-structures.

However, the two conditions rule out the existence of such a proof, as in the proof of Theorem 8.5.1. □

Note that it is easy to incorporate into this set-up stronger lower bounds. For example, for $n \in [0, N]$ an element such that in **M** we have $n = (\log N)^{1/100}$, modify (a2) so that it says that R is the graph of an injection from $[0, n]$ into $[0, n-1]$. Then **A** is large enough to encode proofs of size $N^{O(1)} = 2^{O(n^{1/100})}$ and we deduce from the existence of the expansion such a lower bound.

Earlier, Ajtai [4] used a similar strategy for his lower bound for AC^0-circuits computing parity (Theorem 1.4.3). He reported in [4] that his first proof of the lower bound was model-theoretic via non-standard models and only later did he choose to present it in a finitary way. The former argument may run as follows.

Binary strings w of length $n+1$ are in a correspondence with subsets W of $[0,n]$. By Lemma 1.4.4 a property of binary strings is in the class AC^0 if and only if there is a finite relational language L, an $L(X)$-formula $\Phi(X)$, with X a new unary predicate, and a sequence of L-structures $\mathbf{A}_k$ with universe $[0,k]$, $k \geq 1$, such that a string w of length $k+1$ has the property if and only if the corresponding set W satisfies $\mathbf{A}_k \models \Phi(W)$.

Assume now, for the sake of contradiction, that parity is in AC^0 for infinitely many lengths. Let L be a finite relational language and let $\mathbf{A}_k$ be L-structures with universe $[0,k]$ for infinitely many $k \geq 1$ such that the parity of $X \subseteq [0,k]$ has an odd number of elements if and only if $(\mathbf{A}_k, X) \models \Phi(X)$.

Hence, for any non-standard model of true arithmetic $\mathbf{M}$, there will be a non-standard N such that, for all $W \in \mathbf{M}$, $W \subseteq [0,N]$,

(b1) $|W|$ is odd if and only if $(\mathbf{A}_N, W) \models \Phi(W)$.

Here the parity and the satisfiability relation $\models$ are defined in $\mathbf{M}$.

Given a pseudofinite structure $\mathbf{A}_N$, Ajtai [4] constructed two *pseudofinite* structures, both expanding $\mathbf{A}_N$:

$$(\mathbf{A}_N, U, R) \quad \text{and} \quad (\mathbf{A}_N, V, S),$$

with U, V subsets of the universe and R, S binary relations, such that

(b2) in $(\mathbf{A}_N, U, R)$, R is a 2-partition of U,
(b3) in $(\mathbf{A}_N, V, S)$, S is a 2-partition of V with a remainder block of size 1,
(b4) $(\mathbf{A}_N, U)$ is isomorphic to $(\mathbf{A}_N, V)$

(see Section 11.1 for 2-partitions and their remainder blocks). The following is then easy:

Lemma 20.2.2 *There cannot be three pseudofinite structures* $\mathbf{A}_N$, $(\mathbf{A}_N, U, R)$ *and* $(\mathbf{A}_N, V, S)$ *satisfying the four conditions (b1)–(b4).*

We remark that the existence of the two expanded pseudofinite structures satisfying (b1)–(b4) is actually *equivalent* to the statement that parity is not in AC^0.

The condition that the expansions ought to be themselves pseudofinite is less elementary than in the first example. But it can be replaced by the requirement that:

- Y is a 2-partition of X, and
- the remainder of Y is 1 if and only if $\Phi(X)$,

where X (resp. Y) is a unary (resp. binary) relation symbol interpreted by U, V (resp. by R, S).

Both our examples concern AC^0, i.e. a weak proof system and a weak circuit class. We can generalize the expansion task to any proof system P by considering, instead of the LNP-axiom, a Π^1_1-axiom stating that P is sound, as in the formula Ref_P in (8.4.4).

A task related to the NP $=?$ coNP problem can be formulated analogously to Lemma 20.1.6. The role of the theory $\forall_1 Th_{PV}(\mathbf{N})$ is taken over by the requirement that the expansion is pseudofinite and hence automatically satisfies all true Π^1_1-sentences.

Lemma 20.2.3 *The following two statements are equivalent:*

(i) NP $\neq$ coNP.

(ii) Every pseudofinite $\mathbf{A}$ *has a pseudofinite expansion of the form* $(\mathbf{A}, F, E)$ *such that it holds that*

- *in* $(\mathbf{A}, F)$ *F is a tautology, but*
- *in* $(\mathbf{A}, F, E)$ *E is an assignment falsifying F.*

20.3 Model-Theoretic Forcing

Forcing is often used with two extreme meanings. It is either understood in a narrow sense as set-theoretic forcing or in a wide sense as a framework in which many model-theoretic constructions can be formulated. The former is attractive but quite different from what we encounter in arithmetic; just note that PA is essentially equivalent to set theory with the axiom of infinity replaced by its negation. Various notions key to set-theoretic forcing then become undefinable, an essential obstacle to making it work. The latter sense is too general to be helpful. In this section we shall settle in the middle ground and present a specific forcing going under the name model-theoretic forcing or Robinson's (finite) forcing.

Assume that we have a countable pseudofinite $\mathbf{A}$ with universe $[0, N]$ and we want to add to it a binary relation $R \subseteq [0, N]^2$ with some special properties. We shall build R from larger and larger finite chunks. Let $\mathcal{P}$ be a set of some (not necessarily all) relations r on the universe partially ordered by inclusion. The elements of $\mathcal{P}$ are called **forcing conditions**. Consider the following **forcing game** played by two players, Abelard and Eloise. Abelard starts and picks any condition r_0; Eloise replies with $r_1 \supseteq r_0$. In general Abelard picks conditions $r_{2i} \supseteq r_{2i-1}$ with even indices while Eloise those $r_{2i+1} \supseteq r_{2i}$ with odd ones. The game is played for countably many rounds creating a particular play, the infinite sequence

$$r_0 \subseteq r_1 \subseteq r_2 \subseteq \cdots .$$

The play determines the relation $R := \bigcup_i r_i$.

We say that a property Θ of the expanded structure $(\mathbf{A}, R)$ is **enforceable** if and only if Eloise has a strategy that guarantees that the compiled structure will satisfy Θ no matter how Abelard plays.

Let us give an example. Take for $\mathcal{P}_{php}$ the set of all finite (in the absolute sense) binary relations on the universe of **A** that are graphs of partial injective functions from $[0,N]$ into $[0,N-1]$.

Lemma 20.3.1 *It is enforceable that the compiled structure* $(\mathbf{A},R)$ *satisfies the following:*

- *R is the graph of a total injective function from* $[0,N]$ *into* $[0,N-1]$ *(i.e. it violates the* PHP*);*
- *the least number principle* LNP *(Section 9.4) for any existential $L_{ps}(R)$-formula $B(x)$:*

$$\exists x B(x) \rightarrow (\exists u \forall v\, B(u) \wedge (v < u \rightarrow \neg B(v))).$$

Proof The relation R cannot violate the injectivity of the function. To arrange that it is everywhere defined, Eloise enumerates (that is why we assumed above that **A** is countable) all elements of $[0,N]$ as $u_0, u_1, \ldots$ and when she is creating r_{4i+1} she chooses some whose elements domain contains u_i.

To enforce the LNP is more complicated. Let $B(x) = \exists y_1, \ldots, y_k\, B'(x,\overline{y})$ be an existential formula, with B' open. To simplify things assume that L_{ps} is relational. How much of R do we need to know in order to assert for some u that $B(u)$ is true? We need k witnesses $v_1, \ldots, v_k$ for the y-variables and then we need to fix the truth value of some atomic sentences of the form $R(i,j)$, for i,j either one of $u, v_1, \ldots, v_k$ or a constant from L_{ps}. The number of these atomic sentences is bounded by some constant $c \geq 1$ depending on B but not on u.

To enforce that $R(i,j)$ is true, Eloise simply chooses a condition where this is true; to arrange that it is false she chooses a condition in which $r(i,j')$ holds for some $j' \neq j$.

To enforce the LNP she enumerates all the existential formulas as $B_0, B_1, \ldots$ and when she is choosing r_{4i+3} she proceeds as follows. Let $c \geq 1$ be the constant bounding the number of atomic sentences in B_i. She looks *inside* **M** for the minimum $u \in [0,N]$ such that

- there exists a partial relation r satisfying

$$r \supseteq r_{4i+2} \wedge |r \setminus r_{4i+2}| \leq c \tag{20.3.1}$$

  and enforcing that $B_i(u)$ is true.

The key thing is that while $\mathcal{P}$ is *not definable* in **M**, in the requirement (20.3.1) we quantify over its subset, which *is definable*. Hence we can apply the LNP in **M** and the minimum u exists. □

The reader may note the similarity of this forcing notion to the adversary arguments in Chapter 13. The interesting thing is that to improve the result and to enforce, in the expanded structure, the LNP for all formulas needs just a change in the set of forcing conditions.

Lemma 20.3.2 *Let $\mathcal{P}^+$ be the set of graphs of partial injective functions f from $[0,N]$ into $[0,N-1]$ such that, for some standard $\epsilon > 0$, $|dom(f)| \leq N - N^{\epsilon}$.*

Then it is enforceable that R violates the PHP *and that* $(\mathbf{A}, R)$ *satisfies the* LNP *for all $L_{ps}(R)$-formulas.*

The general idea is that the partial restrictions as in Section 15.2 reduce any L_{ps}-sentence on $(\mathbf{A}, R)$ to equivalent (in the sense of what can be enforced) propositional N^{δ}-DNF formulas built with atoms $r(i,j)$, for some fixed $\delta > 0$. We shall not pursue the details here (see Section 20.5 for references).

The challenge is to find forcing conditions that would create a model for a stronger theory than LNP. If aiming at *EF* we could try to force a model for V^1_1. This would mean expanding **A** not by one set or relation but by a family (containing the original structure) of those: this is then the domain for the set-sort variables. This is difficult as such a family is a priori very rich. A combinatorially easier approach may be to use characterizations of bounded first-order consequences of V^1_1, for example, using the results described in Section 19.3. Assuming that $\mathcal{D}$ is an unsatisfiable set of narrow clauses and σ is a clause reduction of $\mathcal{D}$ to $\neg Con_{ER}(s,k)$, as in Theorem 19.3.2, both encoded in **A**, we may try to add a satisfying assignment E for $\mathcal{D}$ while arranging that $\sigma(E)$ violates $\neg Con_{ER}(s,k)$. This would prove that no clause reduction σ with the required properties exists and hence neither does a short refutation of $\mathcal{D}$.

20.4 Forcing with Random Variables

We now turn to another forcing approach that can be used to construct suitable expansions of pseudofinite structures. We give a brief and informal outline of this approach; the interested reader should consult [304], where it was defined and developed. We use the same or similar notation to that in [304] in order to allow an easy comparison between the two texts.

Let L_{all} be a language with a name for every relation and function on the standard model **N**, for convenience of the general development; at any time we need only a finite part of the language L_{all}. Let **M** be a non-standard $\aleph_1$-saturated model of true arithmetic in L_{all}. The saturation is a technical condition imposed on the model, and we state it for correctness but we shall not explain it here.

Let $n \in \mathbf{M}$ be a fixed non-standard element. The cut $\mathbf{M}_n$ in **M** consists of all elements m such that $m < n^k$ for some standard k. This could be denoted informally $m < n^{O(1)}$. This cut plays the role of a small canonical model as defined in Section 20.1.

Let L_n be the language consisting of all relations in L_{all} and all functions f in L_{all} that map the cut into itself. For example, if f is bounded by a polynomial then $f \in L_n$.

Let Ω be an infinite set such that $\Omega \in \mathbf{M}$; it is therefore finite inside **M** and in **M** we can count its cardinality as well as the cardinalities of all its subsets if they are definable in **M**. The set Ω is called the **sample space** and its elements $\omega \in \Omega$ are

samples. Let $F \subseteq \mathbf{M}$ be any family of partial functions

$$\alpha :\subseteq \Omega \to \mathbf{M}_n$$

that are elements of $\mathbf{M}$, i.e. $\alpha \in \mathbf{M}$, such that

$$\frac{|\Omega \setminus dom(\alpha)|}{|\Omega|}$$

is infinitesimal, i.e. smaller than $1/k$ for all standard k (which is equivalent to being smaller than $1/t$ for some non-standard t). We shall call the set $\Omega \setminus dom(\alpha)$ the **region of undefinability** of α. As α is an element of $\mathbf{M}$, the function is definable there, but it is important that we do not require that the family F itself is definable.

The family F is the universe of a Boolean-valued L-structure $K(F)$ for some $L \subseteq L_n$. The symbols of L are interpreted by composing them with elements from F. For example, for a unary $f \in L$, $f(\alpha)(\omega) = f(\alpha(\omega))$. It is required that this function $f(\alpha)$ is also in F, i.e. that F is L-**closed** in the terminology of [304].

Every atomic $L(F)$-sentence A is naturally assigned a subset $\langle A \rangle \subseteq \Omega$ consisting of those samples $\omega \in \Omega$ for which all elements α of F occurring in A are defined and A is true in $\mathbf{M}$. It was proved in [304, Sec. 1.2] that if we factor the Boolean algebra of $\mathbf{M}$-definable subsets of Ω by the ideal of sets of an infinitesimal counting measure we get a *complete* Boolean algebra $\mathcal{B}$. The image of $\langle\langle A \rangle\rangle$ in $\mathcal{B}$ in this quotient is denoted $[\![A]\!]$. This determines the truth value $[\![A]\!] \in \mathcal{B}$ for any $L(F)$-sentence A: $[\![\ldots]\!]$ commutes with all Boolean connectives and

$$[\![\exists x A(x)]\!] := \bigvee_{\alpha \in F} [\![A(\alpha)]\!] \quad \text{and} \quad [\![\forall x A(x)]\!] := \bigwedge_{\alpha \in F} [\![A(\alpha)]\!]$$

(these values are well defined because $\mathcal{B}$ is complete). This defines a Boolean-valued structure $K(F)$. We say that an $L_n(F)$-sentence is **valid** in $K(F)$ if and only if $[\![A]\!] = 1_{\mathcal{B}}$. All logically valid sentences are valid in $K(F)$ and, more generally, if B follows in the predicate calculus from $A_1, \ldots, A_k$ then

$$\bigwedge_{i \leq k} [\![A_i]\!] \leq [\![B]\!]$$

and, in particular, also all axioms of equality are valid in $K(F)$ (see [304, Chapter 1]).

In the expansion tasks we studied in the previous sections we needed to add a relation or a function to a structure. Let us describe what that means here, considering the addition of a unary function as an example. A unary function on $K(F)$ is a function Θ: $F \to F$ such that for all $\alpha, \beta \in F$,

$$[\![\alpha = \beta]\!] \leq [\![\Theta(\alpha) = \Theta(\beta)]\!]. \tag{20.4.1}$$

This requirement arranges that equality axioms remain valid in the expansion. A particular way to arrange (20.4.1) is explained in [304, Chapter 5]. In particular, we may add a family of such functions (and a set represented by their characteristic functions) and aim at a model of V^1_1.

But where in all this is an expansion of a pseudofinite structure? This needs to use some mathematical logic, which we shall omit here. What one gets from a suitable $K(F)$ is, given a countable **A**, its elementary extension **A**$'$, which is then expanded to **B**. As we noted in the proof of Lemma 20.2.1 (in the claim there), **B** is an expansion of **A** that is only useful because it preserves the L_{ps}-theory. But that is preserved in **A**$'$ as well.

20.5 Bibliographical and Other Remarks

Sec. 20.1 This section was based on Krajíček and Pudlák [319], although some statements were generalized a bit. That paper also formulates several open questions. The terminology *small canonical model* is from [304]; [285] gives several constructions of models of PV.

Sec. 20.2 The pseudofinite set-up and the two examples go back to Ajtai's papers [4, 5]. Maté [351] earlier considered a reformulation of NP $\neq$ coNP in this vein, talking about expansions adding a 3-coloring for a previously not-3-colorable graph. Tautologies expressing 3-colorability and graphs for which they are universal in a similar way to reflection principles were defined in [277, 278].

The paper [311] offers an exposition of this topic; the parity example is treated differently there. Vaananen [499] gives an exposition of pseudofinite structures from an elementary perspective.

The various expansion tasks we discussed (and many others) can be put under one umbrella, comprising *expanded end-extensions*: you first end-extend **A** and then expand it. The general question of when does a pseudofinite structure have an expanded end-extension satisfying a given theory was answered by Ajtai [9, 10], who characterized the situation in a similar way to the completeness theorem but using a notion of proofs definable in **A**. In the case when T is a bounded arithmetic theory, it is related to propositional translations. His characterization works for any T (subject to a few technical conditions), however. Garlík [195] gave a simpler alternative construction.

Sec. 20.3 More on a general model-theoretic set-up of forcing can be found in Hodges [232]. The names of the players, Abelard and Eloise, are taken from there and they are supposed to evoke the $\forall$ and $\exists$ quantifiers. Lemma 20.3.1 is due to Paris and Wilkie [389]; Riis [446] generalized it to Σ_1^b-LNP, in the language of L_{BA}. This leads to the criterion in Theorem 13.1.2.

The details of forcing behind Lemma 20.3.2 can be found in [277] or [278, Sec. 12.7], albeit it is treated without games.

Atserias and Müller [39] offered a general framework for (and a comprehensive survey of) forcing in this area.

Sec. 20.4 Forcing with random variables and structures $K(F)$ was defined in [304]. The consistency of Razborov's conjecture mentioned in Section 19.6 was proved first using this method (see [304, Chapter 31] with a correction in [306]).

20.5.1 Further Model Theory

There are other relations between model theory and proof systems, not based on the compactness theorem. Let **D** be any first-order structure for a language containing some constant c. Think of the variables x_{ij} as being represented by the pair (i,j), and depth $\leq d$ monomials by $\leq d$-tuples of such pairs. Then the equations of the polynomial system $\neg$PHP from Section 11.7 can be defined in **D**, where we think of $[n+1]$ being represented by the universe of **D** and $[n]$ by the universe minus c. For example, to define which pairs (i,j) appear in the equation $1 - \sum_{j \in [n]} x_{ij}$ we write $j \neq c$, or to define the monomial $x_{i_1 j} x_{i_2 k}$ from the system we write $i_1 \neq i_2 \wedge j \neq c$. Such a system is infinite if **D** is and a polynomial in it may have infinitely many monomials. But it can still be evaluated on a definable assignment as long as the structure admits a counting function for definable sets (an abstract Euler characteristic). It was proved in [287] that the solvability of such a definable system in an *Euler structure* implies degree lower bounds for refutations of the system for finite n in both Nullstellensatz and PC.

An intriguing link between finite model theory and subsystems of semi-algebraic calculus was shown by Atserias and Maneva [38] and by Atserias and Ochremiak [41]. They studied the indistinguishability of finite structures by sentences with a bounded number of variables and Sherali–Adams degree relaxations. In particular, for two finite graphs we can write down a system of equations over **R** that has a solution if and only if the two graphs are isomorphic (it is analogous to the PHP_n-formula). If the graphs are not isomorphic then the system is unsatisfiable and has a SAC-refutation. In [41] it was shown that the minimum Sherali–Adams degree of such a refutation is equal to the minimum number of variables needed in a sentence of first-order logic with quantifiers expressing that *there are at least k elements x such that* ... in order to distinguish the two graphs.

Another relation of proof systems to first-order structures was considered in [294], defining what was called there the *combinatorics of a structure* and a *covering relation* between a structure and a proof system. These remarks are an attempt to inspire the reader not to be a priori bounded by some framework but to try to invent his or her own.

21
Optimality

This chapter is devoted to the optimality problem 1.5.5. We shall present several equivalent problems from complexity theory and from mathematical logic, as well as some closely related (though probably not equivalent) problems. A solution to this problem, especially in the slightly more likely negative direction, has many implications and, in particular, a negative solution implies that NP $\neq$ coNP. This makes the problem a very attractive avenue that could lead to an insight beyond proof complexity hardness. We shall concentrate mostly on optimality rather than on p-optimality; it yields simpler and more natural statements (and their modifications for p-optimality, where available, are straightforward).

21.1 Hard Sequences

We learned in Sections 19.1 and 19.2 that the tautologies formalizing reflection principles, or just the consistency statements for many strong proof systems, play a significant role in the existence of simulations among proof systems. Recall the definition of the system $P+Ref_Q$ from Section 19.2 and the main fact (Lemmas 19.2.2 and 19.2.3):

- $Q \leq_p ER + Ref_Q$ *(and for those* $Q \geq_p ER$ *having a corresponding theory we even have* $Q \equiv_p Q + Ref_Q$*).*

In this section, we look at various consequences of this fact for the existence of optimal proof systems. We first recall Lemma 19.2.2 (part (ii)), in order to gather the relevant statements together.

Lemma 21.1.1 *Assume that P contains R and that* $\mathbf{s}_P(\tau') \leq \mathbf{s}_P(\tau)^{O(1)}$*, for any* τ' *obtained from* τ *by substituting constants for some atoms (the conditions on Q in Lemma 19.2.1).*

The proof system P is not optimal if and only if there is a proof system Q such that

$$\langle Ref_Q \rangle_n, \quad n \geq 1,$$

is a hard sequence for P.

This has several corollaries. A set $H \subseteq \{0,1\}^*$ is **sparse** if and only if the intersection $H \cap \{0,1\}^n$ has at most $n^{O(1)}$ elements, for all $n \geq 1$.

Corollary 21.1.2 *The proof system P is not optimal if and only if there is an* NP *set* $H \subseteq$ TAUT *that is hard for P if and only if there is a p-time decidable sparse set* $H \subseteq$ TAUT *that is hard for P.*

Now we shall use the fact that the sequence $\|Ref_Q\|^n$ is very uniform. If no optimal proof system exists then, of course, no proof system can be p-bounded and NP $\neq$ coNP. But one can improve this implication. Recall the classes NE (and thus coNE) from Section 1.3.

Corollary 21.1.3 *Assume that there is no optimal proof system. Then* NE $\neq$ coNE. *In particular, the spectra of first-order sentences are not closed under complementation.*

If there is no p-optimal proof system then E $\neq$ NE.

Proof We shall derive from the assumption NE $=$ coNE that there is a proof system for which no sequence $\|Ref_Q\|^n$ for any proof system Q is hard.

To construct the formula $\|Ref_Q\|^n$ for a fixed Q, an algorithm needs to know only the parameter n. This parameter can be given as an input in binary of length proportional to $\ell := \log n$, and then it runs in time $n^{O(1)} = 2^{O(\ell)}$. However, if we want an algorithm constructing the formula for a variable Q (presented as a string) we run into the problem that the exponent $O(1)$ depends on Q. But note that, by padding proofs with a dummy symbol to lengthen them suitably, we get:

Claim 1 Any proof system Q is p-equivalent to a proof system Q' whose provability predicate is decidable in linear time.

All formulas $\|Ref_Q\|^n$ for *linear-time* proof systems Q and all $n \geq 1$ are then constructible in $2^{O(|Q|+\log n)}$ time from Q, n. Let H be the set of these formulas.

The second issue is that an algorithm constructing formulas $\|Ref_Q\|^n$ in H would not know whether Q is indeed a proof system and $\|Ref_Q\|^n$ is a tautology. This is where we use the power of coNE, which allows us to universally quantify over all Q-proofs of size n. We take a set $H' \supseteq H$ consisting of all formulas $\|Ref_Q\|^n$ for linear-time Turing machines Q and all $n \geq 1$ such that $\|Ref_Q\|^n$ is a tautology.

Claim 2 The set H' is in the class coNE.

Using the hypothesis NE $=$ coNE there is a non-deterministic acceptor of all pairs (Q, n), yielding formulas in H' from Claim 2 (and hence, in particular, those yielding the reflection formulas from H), whose time is bounded by $2^{O(\ell)} = n^{O(1)}$. This is the optimal proof system we want.

The conclusion about spectra follows from the discussion about the Spectrum problem in the Introduction at the start of the book. For the version with p-optimality see Section 21.6 for a reference. □

The next statement is for p-optimality. We shall call a deterministic Turing machine M computing the characteristic function of TAUT a **deterministic acceptor for TAUT**.

Lemma 21.1.4 *There is a p-optimal proof system if and only if there is a deterministic acceptor N for* TAUT *which is optimal on the instances from* TAUT*: any deterministic acceptor M for* TAUT *has at most a polynomial speed-up over N on the instances from* TAUT.

Proof For the if-direction, assume that N is an optimal deterministic acceptor for TAUT. We shall construct a proof system P that p-simulates all *linear-time* proof systems Q and hence all proof systems, as in the proof of Corollary 21.1.3. Let H' be the set of all tautologies $\|Ref_Q\|^n$ where Q is a linear-time clocked Turing machine and $n \geq 1$; by Claim 2 in that proof it is in P. Because N has the optimality property it accepts all formulas from H' in a fixed polynomial time.

Let P_N be the proof system in which the unique proof of φ is a computation of N on φ that ends with the output 1. Hence we can construct in p-time (using N) P_N-proofs of $\varphi \in H'$ and so P_N is p-optimal.

For the only-if direction, assume that Q is a p-optimal proof system. We shall use it to define a deterministic acceptor N for TAUT with the optimality property. For any deterministic acceptor M for TAUT let

- P_M be the proof system defined above,
- M^* be a Turing machine that on an input u runs in a time proportional to the time M used on u and produces M's computation on u,
- f_M be some p-simulation of P_M by Q.

The idea behind how a suitable machine N is defined is that it ought to simulate any M by computing $f_M(M^*(\varphi))$ for $\varphi \in$ TAUT. To define N formally we shall proceed analogously (to an extent) with how the set H' was defined in the proof of Corollary 21.1.3. But it is a little more technical this time.

Let $M_1, M_2, \ldots$ be an enumeration of all deterministic Turing machines such that the property of a triple (w, i, u), where

- *w is the computation of M_i on input u,*

is p-time decidable. On the input u, $n := |u|$, the machine N will work in rounds $t = 1, 2, \ldots$ as follows:

1. it simulates one more computational step of $M_1, \ldots, M_n$ on u;
2. it simulates one more computational step of $M_1, \ldots, M_n$ on all $1, \ldots, t$;
3. it checks whether
   (a) for some $i \leq n$ and $j \leq t$, machine M_i accepted u in j steps, and
   (b) for some $k \leq n$ and $\ell \leq t$, machine M_k produced on input ℓ a Q-proof of the (propositional translation of the) statement that computations of M_i on inputs of the length $\leq n$ and running in $\leq t$ steps are sound.

Item 3(b) is analogous to (but more complicated version of) the coNE check that formulas from H' are tautologies in the proof of Corollary 21.1.3. □

21.2 Disjoint NP Pairs

For any two pairs of disjoint sets (A, B) and (U, V), we say that (A, B) is **p-reducible** to (U, V) and write $(A, B) \leq_p (U, V)$ if and only if there is a p-time computable f: $\{0, 1\}^* \to \{0, 1\}^*$ such that

$$f(A) \subseteq U \quad \text{and} \quad f(B) \subseteq V.$$

We considered pairs of disjoint NP sets in Section 17.2 and, in particular, we attached to a proof system P such a pair (the canonical pair),

$$(\text{PROV}(P), \text{FALSI}^*).$$

Lemma 21.2.1 *For any proof systems P, Q:*

$$P \leq Q \to (\text{PROV}(P), \text{FALSI}^*) \leq_p (\text{PROV}(Q), \text{FALSI}^*).$$

Proof Assume that $\mathbf{s}_Q(\tau) \leq \mathbf{s}_P(\tau)^c$, for all $\tau \in \text{TAUT}$. Then

$$(\varphi, 1^{(m)}) \to (\varphi, 1^{(m^c)})$$

is the required p-reduction. □

Let (A, B) be a disjoint NP pair. In Lemma 17.2.5 we proved that if the (propositional translation of the) disjointness of $A_n := A \cap \{0, 1\}^n$ and $B_n := B \cap \{0, 1\}^n$ has p-size proofs in a proof system P then $(A, B) \leq_p (\text{PROV}(P), \text{FALSI}^*)$. This implies the following statement, because we can add the formulas $A_n \cap B_n = \emptyset$ as axioms (to any proof system).

Lemma 21.2.2 *For any disjoint* NP *pair (A, B) there is a proof system Q such that* $(A, B) \leq_p (\text{PROV}(Q), \text{FALSI}^*)$.

A disjoint NP pair is **complete** if and only if all other disjoint NP pairs are p-reducible to it.

Corollary 21.2.3 *Assume that there is no complete disjoint* NP *pair. Then there is no optimal proof system.*

The p-reducibility between the canonical pairs of two proof systems is linked to the notion of the automatizability of proof systems (Section 21.5). We say that a disjoint pair (A, B) is **p-separable** if and only if there is a p-time function g that outputs 1 on A and 0 on B (this is a uniform version of separability, which we considered in Chapter 17). The following statement flows directly from the definitions.

Lemma 21.2.4 *(i) Let P, Q be two proof systems and assume that*

$$(\text{PROV}(Q), \text{FALSI}^*) \leq_p (\text{PROV}(P), \text{FALSI}^*).$$

Then if (PROV(P), FALSI*) *is p-separable, so is* (PROV(Q), FALSI*).

(ii) *If P is automatizable then* (PROV(P), FALSI*) *is p-separable.*

(iii) *If* (PROV(P), FALSI*) *is p-separable then there is a proof system $Q \geq_p P$ that is automatizable.*

A proof system P for which there is no automatizable $Q \geq_p P$ is called **essentially non-automatizable**, in reference to the logic notion of *essentially undecidable* theories. We shall return to this notion in Section 21.5.

21.3 Quantitative Gödel's Theorem

Gödel's incompleteness theorems say that a sufficiently strong and consistent theory whose set of axioms is algorithmically recognizable is incomplete (the first theorem) and, in fact, does not prove the true sentence expressing its own consistency (the second theorem). In this section we shall look at a quantitative version of the second theorem which takes into account the *lengths of proofs* involved.

Recall from Sections 9.2 and 9.3 the language $L_{BA}(\#)$ and the theory S^1_2 in this language, the classes of $L_{BA}(\#)$-formulas Σ^b_1, $s\Sigma^b_1$, Π^b_1 and Δ^b_1 and Lemma 9.3.2 about the coding of finite sequences of numbers. Also recall that a finite binary word w is represented in the theory by the dyadic numeral $\ulcorner w \urcorner$, whose length is proportional to $|w|$ (Section 7.4). For a first-order formula φ (or a term, a proof, etc.) the numeral $\ulcorner \varphi \urcorner$ is often called **Gödel's number** for φ.

We may formalize the syntax of first-order logic in S^1_2 in a natural way and talk within the theory about provability from some theory T definable in S^1_2. Let us stress that a theory is simply a set of sentences in a given language and we do not assume that it is deductively closed. In particular, different theories may have the same consequences. There is an added ambiguity that the same T may have different definitions in S^1_2 that are not necessarily provably equivalent. All this constitutes a standard mathematical logic set-up and is straightforward to deal with. However, to avoid all technicalities we shall concentrate on *finite* theories T, in a language containing $L_{BA}(\#)$, that themselves contain S^1_2. Hence S^1_2 serves as our basic bottom theory here. It is finitely axiomatizable. The canonical formal definition of T inside S^1_2 is then just a list of finitely many axioms, say in alphabetical order for definiteness.

Analogously to how the provability and the consistency of propositional logic were formalized in Section 12.4, there is an $L_{BA}(\#)$-formula $Prf_T(x, z)$ formalizing that z is a T-proof of formula y, and this formula is – provably in S^1_2 – Δ^b_1. The provability predicate itself is p-time decidable. Using the definition we may define more formulas,

$$Pr_T(x) := \exists z Prf_T(x, z) \quad \text{and} \quad Con_T := \neg Pr_T(\ulcorner 0 \neq 0 \urcorner).$$

Gödel's Second Incompleteness Theorem (a special case) then says that a finite consistent $T \supseteq S^1_2$ does not prove Con_T. The theorem is usually proved in textbooks by establishing four properties of the formula Pr_T. The first three are called **Löb's conditions**:

1. if $T \vdash \varphi$ then $S^1_2 \vdash Pr_T(\ulcorner\varphi\urcorner)$,
2. $S^1_2 \vdash Pr_T(\ulcorner\varphi\urcorner) \rightarrow Pr_T(\ulcorner Pr_T(\ulcorner\varphi\urcorner)\urcorner)$,
3. $S^1_2 \vdash (Pr_T(\ulcorner\alpha\urcorner) \wedge Pr_T(\ulcorner\alpha \rightarrow \beta\urcorner)) \rightarrow Pr_T(\ulcorner\beta\urcorner)$,

where α, β, φ are any formulas. The fourth condition is (an instance of) **Gödel's diagonal lemma**:

4. *There is a sentence δ such that $S^1_2 \vdash \delta \equiv \neg Pr_T(\ulcorner\delta\urcorner)$.*

In the quantitative version of the incompleteness theorem we modify the provability formula Pr_T to

$$Pr^y_T(x) \ := \ \exists z(|z| \leq y) Prf_T(x, z).$$

Note that this is *not* a bounded formula because we are not bounding z but only its length.

Using this formula one can reformulate Löb's conditions in the following way. We use the symbol $S \vdash_m \varphi$ expressing that φ has an S-proof size at most m. The revised conditions 1–3 are:

1′. if $T \vdash_m \varphi$ then $S^1_2 \vdash_\ell Pr^{\underline{m}}_T(\ulcorner\varphi\urcorner)$, where $\ell = m^{O(1)}$;
2′. $S^1_2 \vdash_{\ell^{O(1)}} Pr^{\underline{m}}_T(\ulcorner\varphi\urcorner) \rightarrow Pr^{\underline{\ell}}_T(\ulcorner Pr^{\underline{m}}_T(\ulcorner\varphi\urcorner)\urcorner)$, where $\ell = m^{O(1)}$;
3′. $S^1_2 \vdash_{\ell^{O(1)}} (Pr^{\underline{m}}_T(\ulcorner\alpha\urcorner) \wedge Pr^{\underline{m}}_T(\ulcorner\alpha \rightarrow \beta\urcorner)) \rightarrow Pr^{\underline{\ell}}_T(\ulcorner\beta\urcorner)$, where $\ell = O(m)$

(the underlined quantities are dyadic numerals; see Section 7.4). The diagonal formula condition becomes

4′. *There is a formula $\delta(x)$ such, that for all $m \geq 1$,*

$$S^1_2 \vdash_\ell \delta(\underline{m}) \equiv \neg Pr^{\underline{m}}_T(\ulcorner\delta(\underline{m})\urcorner),$$

where $\ell = m^{O(1)}$ (in fact, it can be $\ell = O(\log m)$).

Define the formula

$$Con_T(y) \ := \ \neg Pr^y_T(\ulcorner 0 \neq 0\urcorner)$$

expressing that there is no T-proof of contradiction having size at most y. Using the modified conditions the usual proof of the second theorem goes through and leads to the first part of the following theorem.

Theorem 21.3.1 *Let $T \supseteq S^1_2$ be a finite, consistent, theory. Then it holds that*

(i) (Friedman [190] and Pudlák [409]) *There is an $\epsilon > 0$ such that, for all $m \geq 1$, $T \nvdash_{m^\epsilon} Con_T(\underline{m})$.*
(ii) (Pudlák [409]) *For all $m \geq 1$, $T \vdash_{m^{O(1)}} Con_T(\underline{m})$.*

Note that both the lower and the upper bound are non-trivial. The size of the formula $Con_T(\underline{m})$ is $O(\log m)$ and hence the lower bound is exponential. However, a proof of the formula by an exhaustive search ruling out all potential proofs of size m would itself have size larger than 2^m. Also note that the lower bound implies Gödel's theorem itself: if T proves $\forall y Con_T(y)$ then each instance $Con_T(\underline{m})$ has a T-proof of size $O(\log m)$.

The upper bound in the theorem is much more subtle than the lower bound. It rests upon the possibility of defining efficiently a partial truth predicate for formulas of a bounded size, and of proving its properties by polynomial-size proofs. In particular, one needs to construct short proofs of the formal statement "*A is true*", where A is any axiom of T. Such a proof crucially utilizes the axiom A itself (A implies that *A is true*) and it is not clear at all whether, and if so how, this could be done for a general $A \notin T$. The following is an open problem.

Problem 21.3.2 (Finitistic consistency problem, Pudlák [409]) *Is there a finite consistent theory* $S \supseteq S^1_2$ *such that for any finite consistent theory* $T \supseteq S^1_2$ *there is a* $c \geq 1$ *such that, for all* $m \geq 1$,

$$S \vdash_{m^c} Con_T(\underline{m})?$$

Pudlák [409, after Problem 1] conjectured that no such theory exists and, in fact, that S fails to prove in size $m^{O(1)}$ the instances $Con_{S+Con_S}(\underline{m})$.

It turned out that this problem was equivalent to the optimality problem 1.5.5.

Theorem 21.3.3 (Krajíček and Pudlák [317]) *An optimal proof system exists if and only if Problem 21.3.2 has an affirmative answer.*

Proof Let P be an optimal proof system. Define the theory $S_P := S^1_2 + Ref_P$. Assume now that T is any theory as in Problem 21.3.2. Then the sequence of formulas

$$\|Con_T(y)\|^m \tag{21.3.1}$$

is p-time constructible and hence the formulas all have P-proofs π_m of size m^k, for some fixed $k \geq 1$. An S_P-proof of $Con_T(\underline{m})$ is constructed as follows:

1. S_P verifies that π_m is a P-proof,
2. it uses the axiom Ref_P to conclude that $\|\neg Pr^y_T(\ulcorner 0 \neq 0 \urcorner)\|^m$ is a tautology,
3. it deduces from this $Con_T(\underline{m})$.

For the first step, one needs the following

Claim *Let* $E(x)$ *be a* Σ^b_1*-formula in* $L_{BA}(\#)$*. Then there is an* $e \geq 1$ *such that, for all* $n \geq 1$,

$$\textit{if } \mathbf{N} \models E(m) \textit{ then } S^1_2 \vdash_{m^e} E(\underline{m}).$$

This is established by induction on the logical complexity of E.

For the opposite direction, assume that S is a theory witnessing an affirmative answer to Problem 21.3.2. Define a proof system P_S whose proofs of a propositional formula φ are exactly S-proofs of the arithmetical $Taut(\ulcorner \varphi \urcorner)$. By Lemma 21.1.1 it suffices to show that for any proof system Q, P_S has p-size proofs of the formulas $\|Ref_Q\|^n$.

If Q is any proof system, take $T := S_Q$. By the hypothesis there are size-m^c S-proofs σ_m of $Con_T(\underline{m})$. Hence P_S has size $m^{O(1)}$ proofs of the formulas (21.3.1), from which the formulas $\|Ref_Q\|^m$ follow by p-size P_S proofs. □

21.4 Diagonalization

It is a widespread opinion that diagonalization cannot help to solve the P =?NP or the NP =?coNP problems because *diagonalization relativizes* and there are oracles for which the classes both are and are not equal (see Baker, Gill and Solovay [44]). This concerns the diagonalization of Turing machines. However, the diagonalization of arguments in logic is less straightforward to relativize and there are more options. In Part II we considered various theories T that are formulated in several arithmetic languages such as L_{BA} or L_{BA}(PV). We also studied their versions $T(\alpha)$ that add to the language one or more relation or function symbols and are allowed these symbols to appear in axiom schemes (e.g. IND) defining T. This is often considered to be a *relativization* of T. An example of a property that all the theories T we considered have, while none of the theories $T(\alpha)$ does, is the completeness statement

$$\text{if } \mathbf{N} \models A(n) \text{ then } T \vdash A(\underline{n})$$

for Σ^0_1-formulas A, or the formalized completeness statement

$$T \vdash B(\underline{n}) \rightarrow Pr_T(\ulcorner B(\underline{n}) \urcorner)$$

for Σ^b_1-formulas B. If we allow α (a unary predicate, say) then, taking for A or B the atomic formula $\alpha(x)$, gives a counter-example to both properties for $T(\alpha)$. To save the property theory, $T(\alpha)$ would have to have as axioms all true basic sentences $\alpha(\underline{n})$ or $\neg\alpha(\underline{n})$. Another example is offered by propositional translations. Although $T(\alpha)$ involves an oracle α, the propositional logic related to $T(\alpha)$ via the $\langle \dots \rangle$ translation does not.

In this section we combine the diagonalization in Section 21.3 with the idea of representing long proofs and formulas by small circuits (used for implicit proof systems in Section 7.3) to prove a theorem about p-optimality. We shall continue to use the formalism from Section 21.3. First we restate the diagonal lemma in the form we need it. The formula in it has a free variable and the notation $\dot{x}$ is a formalization of the dyadic numerals in S^1_2.

Lemma 21.4.1 *Let $T \supseteq S^1_2$ be any finite consistent theory. There is an $L_{BA}(\#)$-formula $D(x)$ satisfying*

$$S^1_2 \vdash D(x) \equiv \neg Pr^x_T(\ulcorner D(\dot{x}) \urcorner)$$

and, for any $n \geq 1$, the sentence $D(\underline{n})$ is true and provable in T but any T-proof of $D(\underline{n})$ must have size larger than n.

Let $C(x_1, \dots, x_k)$ be a circuit whose truth table is a 3DNF formula φ_C of size 2^k. We may define an $L_{BA}(\#)$-formula $BigTaut(x)$ such that $BigTaut(\ulcorner C \urcorner)$ formalizes the fact that $\varphi_C \in$ TAUT. It is not a bounded formula, just as $Con_T(y)$ before was not. The formula says that for any truth assignment y (whose length is bounded by a suitable term in C) there is a term in φ_C satisfied by y. The existential quantification over terms in φ_C is a quantification over inputs of C and we do not need to have the whole big formula φ_C at hand.

The formula $D(x)$ in the rest of the section is the diagonal formula from Lemma 21.4.1.

Lemma 21.4.2 *There is a p-time function $g(x)$ such that for any $n \geq 1$ the value $g(n)$ is a size $(\log n)^{O(1)}$ circuit representing the propositional formula $\|D(x)\|^n$. The implication*

$$BigTaut(g(x)) \longrightarrow D(x) \tag{21.4.1}$$

is provable in the theory S^1_2.

Proof The existence of such a function g follows from the uniformity of the $\| \dots \|$ translation (as discussed at the end of Section 19.1). The implication (21.4.1) is clearly valid. Its proof in S^1_2 follows the following argument.

Assume that w witnesses the failure of $D(n)$:

$$|w| \leq n \wedge Prf_T(\ulcorner D(\underline{n}) \urcorner, w). \tag{21.4.2}$$

Then there is a truth assignment u to the atoms of $\|D\|^n$ that violates the formula, and u is computed from n and w by a p-time function f. The theory S^1_2 proves that (21.4.2) implies that $f(n, w)$ does not satisfy any term in φ_C represented by $C := g(n)$, and thus $BigTaut(g(n))$ fails. □

We shall need one more formula, this time propositional. Let P be a proof system, C a circuit in k inputs as above defining φ_C and E a circuit in $O(k)$ variables that defines a p-time verification of a purported P-proof of φ_C. As for the formula $Correct_\beta$ in Section 7.3, there is a propositional formula $\pi^P_{C,E}$ of size polynomial in $|C| + |E|$ which expresses that the verification defined by E is a valid computation of the provability predicate of P on φ_C and the proof encoded in the computation.

The proof of the following theorem uses several times (in steps 2, 9 and 11) the fact that a true Σ^b_1-sentence has a polynomial-size proof in S^1_2 (cf. the claim in the proof of Theorem 21.3.3).

Theorem 21.4.3 (Krajíček [292]) *At least one of the following three statements is true:*

(i) there is a function f: $\{0, 1\}^* \rightarrow \{0, 1\}$ computable in time $2^{O(n)}$ that has circuit complexity $2^{\Omega(n)}$;
(ii) NP $\neq$ coNP;
(iii) there is no p-optimal propositional proof system.

Proof Assume for the sake of contradiction that all three statements are false. We shall derive a contradiction via Lemma 21.4.1 by constructing a proof of $D(\underline{n})$ that is too short. The construction is split into several steps.

1. Assume that P is a p-bounded and p-optimal proof system (i.e. it witnesses the failure of the second and the third statement in the theorem). In particular, $P \geq_p EF$. Let Prf_P be a Σ^b_1-formula formalizing its provability predicate.
2. Define a theory T by a modification of the way in which the theory S_P was defined in the proof of Theorem 21.3.3. The theory T is S^1_2-augmented by the following form of the reflection principle as an extra axiom:
$$\forall x, y, u,\ Prf_P(\ulcorner \pi^P_{x,y} \urcorner, u) \longrightarrow BigTaut(x).$$
The formula $\pi^P_{x,y}$ is p-time constructible from x, y and so the antecedent of the implication is a Σ^b_1-formula. Hence all true instances of the antecedent have p-size proofs in S^1_2.
3. The propositional formula $\|D\|^n$ is a tautology as $D(n)$ is true, and its size is $n^{O(1)}$. By Lemma 21.4.2 there is a circuit $C := g(n)$ of size $(\log n)^{O(1)}$ defining the formula.
4. The set of all formulas $\|D\|^n$, $n \geq 1$, is polynomial time decidable and hence by Corollary 21.1.2 there is a (deterministic) polynomial time algorithm M computing its P-proof from $\|D\|^n$.
5. We can use M to compute in deterministic p-time a particular accepting computation of the provability predicate of P (i.e. an $n^{O(1)} \times n^{O(1)}$ matrix W encoding the computation). The bits $W_{i,j}$ as a function of i, j are computed by a function in time $2^{O(n)}$.
6. Assuming that the first statement in the theorem also fails, there exists a circuit E in $O(\log n)$ variables and of size $2^{\delta \log n} = n^{\delta}$ that represents W, for an arbitrary small $\delta > 0$. We shall choose a particular δ in the last step of the construction.
7. Take an instance of the reflection principle axiom of T by substituting for x and y the codes of C and E respectively:
$$\forall u,\ Prf_P(\ulcorner \pi^P_{C,E} \urcorner, u) \rightarrow BigTaut(C).$$
8. The size of $\pi^P_{C,E}$ is polynomial in $|C| + |E|$, i.e. it is $n^{O(\delta)}$. Because P is p-bounded, there is a P-proof σ of $\pi^P_{C,E}$ of size $n^{O(\delta)}$. Note that the constants implicit in the O-notation are independent of δ. Substituting σ for u in the formula in step 7 we get
$$Prf_P(\ulcorner \pi^P_{C,E} \urcorner, \sigma) \rightarrow BigTaut(C).$$
9. The antecedent of the formula in step 8 is a true Σ^b_1-sentence of size $n^{O(\delta)}$ and thus it has a proof in S^1_2 (and so in T too) of size $n^{O(\delta)}$.
10. From the formulas in steps 8 and 9 we derive
$$BigTaut(C)$$

and the total size of the proof is $n^{O(\delta)}$.

11. The implication

$$BigTaut(C) \rightarrow D(\underline{n}) \tag{21.4.3}$$

can be obtained as an instance of a universal implication provable in S^1_2 (Lemma 21.4.2):

$$BigTaut(g(x)) \rightarrow D(x).$$

Just substitute for x the numeral $\underline{n}$ and prove $C = g(\underline{n})$ (i.e. that it is a true Σ^b_1-sentence of size $(\log n)^{O(1)}$).

Hence the implication (21.4.3) has an S^1_2-proof of size $(\log n)^{O(1)}$.

12. Combining the proofs in steps 10 and 11 we get a size-$s := n^{O(\delta)}$ T-proof of $D(\underline{n})$. If $\delta > 0$ is small enough that $s < n$ we get a contradiction by Lemma 21.4.1.

□

21.5 The Proof Search Problem

We have considered the proof search problem 1.5.6 already but formulated it only informally. In this section we shall propose a formal statement of the problem.

Proof search refers informally to the general task of finding proofs of statements algorithmically. We have seen a notion that is related to that: the automatizability of a proof system (Section 17.3). I like that notion because it gives an interesting meaning to the failure of feasible interpolation: if feasible interpolation fails for a proof system then the proof system is not automatizable (Lemma 17.3.1). However, the notion on its own does not seem to be particularly useful because there are essentially no known automatizable proof systems. The qualification *essentially* leaves room for examples such as the truth table system: the unique proof of a formula is its evaluation on all assignments. But not a single (complete) proof system that we have considered in this book is known to be automatizable.

Automatizable but incomplete systems can be obtained by severely restricting the space of all proofs. For example, if we restrict the allowed width of clauses by a constant w then such a resolution refutation can be found in time $n^{O(w)}$, if it exists. Just close the set of initial clauses under resolution inferences yielding a clause of width at most w, as long as this is possible. Analogously, restricting the degree of NS or PC refutations by a constant yields automatizable systems (Lemma 18.3.2). This is also often claimed to hold for some semi-algebraic systems and, in particular, for SOS, but that appears to be open at present (Section 21.6). Regarding the negative results, we know that proof systems simulating resolution are not automatizable, assuming various plausible computational hypotheses. This is often proved by showing the failure of feasible interpolation; see Section 18.7 for results about AC^0-Frege systems and above and Section 21.6 for R-like proof systems.

A more general notion is that of weak automatizability, introduced by Atserias and Bonet [29]: a proof system P is **weakly automatizable** if and only if there exists a $Q \geq P$ that is automatizable. The following lemma follows from the definition and from Lemma 21.2.4.

Lemma 21.5.1 *The following three conditions are equivalent:*

(i) P is weakly automatizable;
(ii) there exists a proof system Q and a deterministic algorithm $\mathcal{A}$ which, when given as an input a tautology τ, finds its Q-proof in time polynomial in $\mathbf{s}_P(\tau)$;
(iii) the canonical pair (PROV(P), FALSI*) *is p-separable.*

Perhaps the most interesting open problem is whether resolution is weakly automatizable. The following statement clarifies the situation somewhat. Its heart is the construction of a proof of the reflection principle for R in R(2) (that feasible interpolation for R(2) then implies the p-separability of a canonical pair for R).

Theorem 21.5.2 (Atserias and Bonet [29]) *For any constant $k \geq 2$, the following three statements are equivalent:*

(i) R *is weakly automatizable;*
(ii) R(k) *is weakly automatizable;*
(iii) R(k) *admits feasible interpolation.*

We have no non-trivial examples of weakly automatizable systems, and hence the dual notion of essential non-automatizability is perhaps more useful: P is **essentially non-automatizable** if and only if it is not weakly automatizable (the terminology relates to the classical logic notion of essentially undecidable theories). It may be that any proof system simulating R is essentially non-automatizable.

The notion of (weak) automatizability describes the best possible situation (one can find proofs feasibly) and it is in this respect similar to the notion of p-bounded proof systems (Section 1.5): again, we do not expect to find any. Hence a notion comparing the relative efficiency of proof search algorithms may be more useful. We propose the following one.

By a **proof search algorithm** we shall mean any pair (A, P) where P is a proof system and A is a deterministic algorithm that, given a tautology τ, constructs some P-proof of τ. It is natural to compare two proof search algorithms by the time they use. However, if Q is stronger than P and the formulas (and their Q-proofs) separating Q from P are easy to construct by some p-time function f (which was always the case in the separations that we demonstrated in Part III) it would automatically imply that no (A, P) can be as good as (B, Q), for an algorithm B incorporating f. But the ability to cope with a few tricky formulas should not be enough to claim superiority. This leads to the following definition.

Using the notation from Section 1.3 we define that (A, P) **is as good as** (B, Q) if and only if there are a constant $c \geq 1$ and a P/*poly* set $E \subseteq$ TAUT such that, for all $\alpha \in$ TAUT $\setminus E$,

$$time_A(\alpha) \le time_B(\alpha)^c.$$

We shall use the notation $(A, P) \succeq (B, Q)$ for the situation where this holds. Note that $\succeq$ is a quasi-ordering. If we were interested only in proof search algorithms (A, P) for a fixed P then it would also make good sense to consider the same definition but without the set E.

Now we may formulate the proof search problem 1.5.6 formally as follows.

Problem 21.5.3 (The proof search problem – formal statement) *Given a proof system P, is there a $\succeq$-maximum element among all (A, P)? Is there a pair (A, P) that is a $\succeq$-maximum element among all (B, Q)?*

A suitable name for any maximum element (if it exists) could be an **optimal proof search algorithm** (relative to P in the first case).

A somewhat more complicated proof search problem **Find**, which allows the consideration of proof search algorithms without specifying the proof system was considered in [308].

21.6 Bibliographical and Other Remarks

Although our interest lies in TAUT and propositional proof systems, the optimality problem 1.5.5 has a useful meaning for other languages as well, including those in NP. Obviously, any NP language has an optimal proof system and any language in P has a p-optimal proof system. But does SAT have a p-optimal proof system? In particular, is the canonical definition of SAT given by a p-optimal proof system for SAT? This would mean that we can compute a satisfying assignment from *any* certificate of satisfiability in any proof system. Beyersdorff, Köbler and Messner [78] offered an overview of this issue.

Sec. 21.1 Lemma 21.1.1 and the fact recalled before it go back to Cook [149] and Krajíček and Pudlák [317]. They also imply that it suffices to compare any two strong proof systems on p-time constructible sequences of tautologies. Corollary 21.1.3 is from [317] and it has a version for p-simulations: if there is no p-optimal proof system then E $\neq$ NE. This was improved by Messner and Torán [353] to show that the hypothesis implies that not all sparse sets in NEE (doubly exponential time) are in coNEE, and then by Köbler and Messner [269] for unary languages instead of just sparse languages. This was further generalized by Ben-David and Gringauze [64] to some function classes. The details of the proof of Lemma 21.1.4 can be found in [317, see the proof that (6) implies (9) on p. 1071] and it was modified for SAT by Sadowski [455] and for other languages by Messner [352].

Sec. 21.2 The implication opposite to Lemma 21.2.1 does not hold in general; Pudlák [416] gives an example. Lemma 21.2.2 actually holds in a sharper version: any disjoint NP pair is p-equivalent to the canonical pair of some proof system. It is open whether Corollary 21.2.3 holds as an equivalence.

The p-reductions between the canonical pairs of two proof systems P, Q as in Lemma 21.2.1 were called by Pitassi and Santhanam [402] *strong effectively-p simulations* of P by Q. They discussed various examples and pointed out that for quantified propositional logic the notion becomes void. That there is a map sending a (quantified propositional) formula φ to the implication $(\forall \overline{x}\ \|Ref_P\|^m) \to \varphi$ can be proved in a short proof for any proof system that can handle Σ^q_∞-formulas, where $\overline{x}$ stands for all the variables in $\|Ref_P\|^m$ and m is arbitrary. Hence such a proof system effectively-p simulates *all* proof systems with respect to quantified propositional logic.

In Sections 19.3 and 19.6 we discussed total NP search problems and p-reductions among them. It is an open problem whether there is a complete problem with respect to these reductions. It is also open whether the existence or the non-existence of a complete total NP search problem relates in some way to the Optimality Problem 1.5.5. Pudlák [422] showed that by a construction analogous to how one associates the canonical pair with a proof system it is possible to attach a pair of disjoint coNP sets to a total NP search problem. Using this he proved that the non-existence of a complete pair among these sets implies the non-existence of a complete total NP search problem. A number of related conjectures are discussed in [422] under the name *feasible incompleteness thesis*.

It is worth mentioning (and we shall return to this in Chapter 22) that *relative to a theory* T (subject to some technical conditions) there are optimal proof systems, complete disjoint NP pairs or complete NP search problems. This restriction to T means that we take only proof systems for which T proves their soundness, disjoint NP pairs whose disjointness is T-provable or NP search problems whose totality is T-provable (for some definitions). For proof systems, this is a consequence of the correspondence between a theory and a proof system. Note that in general the insistence that the promise in the definition of a particular promise class of problems is provable in T turns the class into a syntactic one.

The existence of an optimal problem in promise classes was studied from a broader perspective by Köbler, Messner and Torán [270]. Beyersdorff and Sadowski [82, 83] studied links of the optimality problem to p-representations of promise classes (and the existence of complete sets therein); see also Sadowski [456].

Sec. 21.3 Gödel [200] encoded finite sequences of numbers differently from how it was done here. His method, based on the Chinese remainder theorem, is more elementary and can be easily defined just in the language L_{PA}. But the codes it provides do not have a feasible size and hence it is not used in bounded arithmetic. A presentation of Löb's conditions and their use in deriving the incompleteness theorems can be found in Smorynski [475].

Pudlák [410] was able to improve both the lower and the upper bounds in Theorem 21.3.1 to $\Omega(m/(\log m)^2)$ and $O(m)$, respectively. Such good bounds depend on the underlying predicate calculus and its efficiency in writing proofs. Roughly, better upper bounds in Löb's conditions translate into a better lower bound in the theorem.

The particular predicate calculus that he used has an inference rule allowing the introduction of constants witnessing existential quantifiers: from $\exists x \phi(x)$ derive $\phi(c)$ for some new constant c. This shortens proofs because one can talk about an element satisfying ϕ without the need to mention the formula.

The theorem can be also proved under weaker assumptions on T. In particular, one does not need to assume that T contains S^1_2 or is even in $L_{BA}(\#)$. It suffices to assume that T is *sequential*, which roughly means that T allows the development of a rudimentary theory of finite sequences. A sufficient condition I noted when trying (unsuccessfully) to simplify and improve Pudlák's argument is that T interprets a very weak set theory, having just extensionality, the axiom that the empty set exists and an axiom saying that $x \cup \{y\}$ exists for all sets x, y.

Given a theory S as in Problem 21.3.2 there are some natural candidates for a theory T such that the formulas $Con_T(\underline{m})$ ought not to have size-$m^{O(1)}$ S-proofs. Take, for example, $T := S + Con_S$ ([410]) or a T that has a truth predicate for $L_{BA}(\#)$ formulas and an axiom saying that all axioms of S are true. The latter theory was proposed by Buss and termed a *jump of S*.

I think that in order to answer the problem 21.3.2 negatively we shall need a genuinely new proof of Gödel's second incompleteness theorem. It is known that proving the consistency of one theory, T, in another, S, implies the existence of an *interpretation* of T in S (in fact, it implies more; see e.g. [408]). The following problem has been in my head since Pudlák's work in the 1980s. Can one prove that Robinson's arithmetic Q cannot interpret the Gödel–Bernays set theory GB (to keep it finite) *without* using Gödel's theorem? For weak theories (algebro-geometric) the non-interpretability can sometimes be shown by demonstrating that models of the stronger theory are in some sense much more complex than those of the weaker theory. But for Q and GB they are both very complex.

Sec. 21.4 This section is based on [292, Theorem 2.1]; some variants and related statements are presented there.

Sec. 21.5 Beame and Pitassi [57] showed that R* is automatizable in *quasi-polynomial* time. Theorem 5.4.5 implies that if a size $n^{O(1)}$ R-refutation exists then one can find some refutation in subexponential time, as the width can be bounded by $O(\sqrt{n \log n})$.

There are several results linking the weak automatizability of R with the existence of p-time algorithms deciding which of the players has a positional winning strategy in various two-player games on graphs. These include Atserias and Maneva [37] (the mean-payoff game), Huang and Pitassi [239] (the simple stochastic game) and Beckmann, Pudlák and Thapen [62] (the parity game). In [62] also a game was defined for which the decision problem is equivalent to the separability of the canonical pair in R. Atserias [26] offered a comprehensive survey of (weak) automatizability under width or degree restrictions.

Alekhnovich and Razborov [16] proved that a hypothesis from the area of parameterized complexity implies that R is not automatizable (but it leaves open possible automatizability in quasi-polynomial time). Note that a similar situation holds for AC^0-Frege systems: while Theorems 18.7.2 and 18.7.3(ii) yield the non-automatizability of *EF*, Frege and TC^0-Frege systems in subexponential time (if the hypothesis about the RSA of the Diffie–Hellman scheme is formulated in terms of adversaries computing by means of subexponential-size circuits), this is not true for AC^0-Frege systems 18.7.3 (part 2).

Atserias and Bonet [29] defined weak automatizability by condition (ii) in Lemma 21.5.1; the lemma is also immediate from Pudlák [416]. Note also that the notion of strong effectively-p simulations defined earlier in this section preserves weak automatizability downwards and hence essential non-automatizability upwards.

The claimed automatizability of SOS was questioned by O'Donnell [375], who pointed out that such claims forgot to take into account the bit size of the coefficients occurring in the constructed SOS proof. He gave an example where the coefficients have exponential size. Raghavendra and Weitz [425] found such an example with Boolean axioms included but they also showed that in a number of common situations where the SOS algorithm is used the bit size is polynomial. The mere fact of a large bit size of the coefficients does not disprove the automatizability, because an automatizing algorithm has at its disposal a time polynomial in the minimum size of a proof (hence a larger minimum bit size also means more time). But clearly the automatizability of SOS needs clarification.

Somewhat related to proof search are situations where short proofs do exist for a non-trivial class of formulas. As an example consider dense random 3CNFs. We know by Theorem 13.4.2 that random 3CNFs require exponentially long R-refutations and that the lower bound is actually valid up to density $n^{3/2-\Omega(1)}$ (the number of clauses when there are n variables); see Ben-Sasson [66]. Interestingly, Müller and Tzameret [360] proved, building on ideas of Feige, Kim and Ofek [179], that random 3CNFs with $\Omega(n^{1.4})$ clauses do have, with probability tending to 1, polynomial-size refutations in a TC^0-Frege system.

21.6.1 Further Topics

There are links of the optimality problem to various other parts of complexity theory. Chen and Flum [136] linked it with descriptive complexity theory by proving that a p-optimal propositional proof system exists if and only if a particular logic related to least-fixed-point logic captures polynomial-time predicates on finite structures. To find such a logic is a well-known open problem in that area. In [135] Chen and Flum also established a link with parameterized complexity (for a parameterized halting problem).

Cook and Krajíček [153] proved that in the class of proof systems with advice (Section 7.6) there is an optimal one. In particular, there is a proof system with advice of length 1 (i.e. one bit) that is optimal among all proof systems using at most $\log n$ advice bits. Beyersdorff, Köbler and Müller [79] offered a comprehensive presentation of this area.

22

The Nature of Proof Complexity

We shall step back a little in this chapter in order to gain a larger perspective of the field and to contemplate the nature of proof complexity. We have seen a number of proof systems based on a variety of ideas and using diverse formalisms. They are quasi-ordered by (p-)simulations and this is often taken as the basis for describing the realm of proof complexity. It seems to me that taking into account the mathematical objects underlying proof systems (algorithms or theories) allows one to paint a somewhat different and more structured picture. Proof systems are – in this view of proof complexity – divided into four levels, which are fairly clearly distinguished from each other. The separation is not absolutely crisp, however, and there are grey zones between the adjacent levels. We shall call the four levels

Algorithmic, *Combinatorial*, *Logical* and *Mathematical*,

and will discuss them in separate sections of the chapter. While describing them we shall draw on concepts and results presented in all parts of the book without explicitly citing them. The relevant places are easy to find. In addition, Section 22.5 lists for each topic the relevant sections or chapters and also gives references to facts that we shall mention but that were not covered in the book.

22.1 Algorithmic Level

The **algorithmic level** (A-level) contains mostly proof systems that are closely related to algorithms of some sort (SAT, optimization or approximation algorithms). The qualification *closely* means literally that: a proof system in this level is just one aspect of the associated algorithm. A proof in the system certifies that the algorithm does not find a solution and is essentially its failing run.

The A-level contains: subsystems of the resolution system such as R^* and $R^*(\log)$ related to the DPLL algorithm and its extensions; the cutting planes system CP and the Lovász–Schrijver system LS related to integer linear programming algorithms; (subsystems of the) semi-algebraic calculus SAC, and SOS in particular, related to semi-algebraic relaxation methods; polynomial calculus PC and Nullstellensatz NS (and the combined system PCR), which are related to ideal membership algorithms and to Gröbner's algorithm, in particular. We also include in the A-level *regular* R and the OBDD system because they are so close to computational devices (read-once branching programs and OBDDs) and very weak systems like truth tables (the unique proof of a formula is some canonical computation of its truth value on all assignments) and LK^- (cut-free LK) for their simplicity of definition. The resolution system R sits in the grey zone between the A-level and the next, combinatorial, level.

The A-level enjoys three widely applicable lower bound methods: adversary arguments using width or degree restrictions; the (random) restriction method; and feasible interpolation. In addition there are some ad hoc arguments for NS and PC describing exactly the space of derivable polynomials of bounded degree.

Upper bounds for the systems in this level are derived by analyzing the time complexity of some specific algorithm solving the search problem $Search(\mathcal{C})$ associated with an unsatisfiable set of clauses $\mathcal{C}$. Resolution R is related to a simply defined first-order theory but in reality it is rarely used to get an upper bound.

Researchers in the area perceive proving an *unconditional* lower bound for LS as the most pressing open problem. Connections of the proof systems in the A-level to mathematics and to computer science are the richest of all the four levels and there will always be interesting open problems (as long as there are interesting algorithms). Research in the A-level forms, I think, the rudiments from which proof complexity can grow.

22.2 Combinatorial Level

As mentioned above, the resolution system R lives in the grey zone between the A-level and the **combinatorial level** (C-level). Most lower bound methods from the A-level apply to R as well, but there is no exact correspondence with an algorithm but rather with a first-order theory behind R.

The bottom of the C-level is occupied by proof systems adding a small amount of logic to algebraic and geometric proof systems: R(CP), R(LIN), R(PC) and R(OBDD). Slightly above R is R(log), an enigmatic and seemingly simple proof system which resists deeper analysis (Problem 13.7.1). Above the system R(CP) sits LK(CP), which is related to the geometry of discretely ordered rings and to some lattice algorithms.

At the top of the C-level are the Frege systems; in fact, they are perhaps in the grey zone between the C-level and the next logical level. The space below them is inhabited by subsystems of Frege systems defined by bounded-depth circuit classes in various languages, the AC^0-, $AC^0[p]$- and TC^0-Frege systems, in particular.

The proof systems at this level do not correspond to algorithms but to first-order theories. In the two-sort set-up these theories allow quantification only over first-order objects (bit positions). The theories are based on induction axioms but are often augmented by some *counting* facilities, i.e. counting quantifiers or counting principles of some type. Proofs in these theories translate into propositional proofs in various languages but of a bounded depth. Often the best way to establish a length-of-proof upper bound for (the translation of) a principle is to translate a proof of the principle in the corresponding theory.

The restriction method applies to AC^0-Frege systems (and their extensions by some counting principles) and essentially collapses them to systems at the A-level. Theorem 2.5.6(ii) offers a way to prove, in principle, super-polynomial lower bounds for Frege systems by proving very strong lower bounds for AC^0-Frege systems. Müller and Pich [358, Proposition 4.14] proposed a reduction of the super-polynomial lower bound for Frege proofs of specific formulas (expressing circuit lower bounds) to very weak lower bounds for AC^0-Frege systems; the qualification *weak* means specifically *polynomial*.

In my view the most pressing open problem is to extend some lower bound methods from the A-level to systems like R(CP) or R(LIN). Feasible interpolation seems to be best positioned for that. In addition, I think it would be quite interesting to extend the correspondence of A-level systems with algorithms to a correspondence of some of these combined systems with algorithms allowing forms of deduction.

There are several other well-known open problems perceived by researchers as important for further developments, including Problems 14.3.2 (the depth-d vs. depth-$(d+1)$ problem), 15.3.2 (the WPHP problem) and 15.6.1 (the $AC^0[p]$-Frege problem). I find also interesting Problems 19.4.5 and 19.6.1 (linear generators and iterability for AC^0-Frege systems).

22.3 Logical Level

The grey zone between the C-level and the **logical level** (L-level) contains Frege systems, but extended Frege systems sit firmly at the bottom of the L-level. The system *EF* is a pivotal proof system, separating in a sense the bottom two levels (A and C) from the top two levels (L and M). It has a number of different characterizations in terms of p-equivalence (*SF*, *CF*, *ER* or G_1^*) and, even better, the *EF*-size measure $\mathbf{s}_{EF}$ can be equivalently characterized by the number-of-steps measure $\mathbf{k}_P$ for several proof systems including, prominently, Frege systems. This is important because many lower bound methods prove lower bounds for $\mathbf{k}_P$ instead of $\mathbf{s}_P$, and $\mathbf{k}_P$ is proof-theoretically more natural than size.

Slightly above *EF* are G_1 and *WF*, and above them $G_2, G_3, \ldots$, all the way up to G. Still at this level but above *G* is *iEF*. The iterates of the implicit construction i_kEF go to the top of the L-level; however, the system $i_\infty EF$ (defined in [293] but see the discussion about internal iteration of the implicit construction in our Section 7.7) is in the grey zone between the L-level and the next, mathematical, level. There are some ad hoc systems (UENS or IPS) floating near the bottom of the L-level.

The theories corresponding to proof systems at this level are second-order or higher-order in the two-sort set-up; they quantify over strings, sets of strings, etc.

In the more suitable (for this level) one-sort set-up the theories start with PV and S^1_2 and go to T^1_2, APC_1 and APC_2, to the theories T^i_2 and then to U^1_2, V^1_2, U^i_2, V^i_2 and the systems $S^1_2 + k - Exp$. A common feature of these theories is that they are all interpretable in Robinson's arithmetic and hence also in the very rudimentary set theory discussed in Section 21.6. They are mostly defined in a purely logical manner, based on induction axioms accepted for larger and large classes of formulas. Their key feature is that they do formalize a large part of finitary mathematics. Computational complexity theory can be formalized in theories at the bottom of this hierarchy (PV, S^1_2 and APC_1, in particular). The formalization does not concern only simple concepts but also concepts that are rather more delicate to formalize, for example, those involving randomness.

The bottom theories define a number of important p-time and probabilistic p-time algorithms and prove their properties and many key results in computational complexity. These include: Cook's theorem and other NP-completeness results; various circuit lower bounds (for AC^0 and parity, for $AC^0[p]$ and MOD_q and for monotone circuits and the clique function); the natural proofs barrier; the PCP theorem; the Goldreich–Levin theorem and the soundness of construction of a pseudo-random generator from a one-way function; derandomization via the Nisan–Wigderson generator; the construction of sorting networks, an expander construction or Toda's theorem; many others (see Section 22.5 for references).

Of particular interest are formalizations of fundamental conjectures such as the $P \neq NP$ conjecture. Assume that A is a p-time algorithm solving SAT. This can be expressed in the language of PV (i.e., it has a name for A) as a universal statement:

$$\forall \varphi, \overline{a},\ Sat(\varphi, \overline{a}) \rightarrow Sat(\varphi, A(\varphi)). \tag{22.3.1}$$

Proof complexity has a bearing on the provability of such statements: by consequence (4) of the correspondence between a proof system P and a theory T in Section 8.6, a super-polynomial lower bound for P-proofs of *any* sequence of tautologies implies that for all p-time algorithms A it is consistent with T that (22.3.1) is false. In other words, $P \neq NP$ is consistent with T (consequence (4) is, in fact, stronger as it speaks about the consistency of $NP \neq coNP$).

What if we are interested in how hard it is to disprove that A is a SAT algorithm? At a first sight it seems that in order to express that A fails for, say, any large enough length we need to use an existential quantifier,

$$\forall 1^{(n)} (n \geq n_0) \exists \varphi, \overline{a}, \quad |\varphi| \geq n \wedge Sat(\varphi, \overline{a}) \wedge \neg Sat(\varphi, A(\varphi)), \tag{22.3.2}$$

and thus this issue will escape propositional logic. But that is not so; we can use A itself to find its own error. In fact, turning to circuits in order to avoid talking about specific algorithms, consider the following formalization of the n^c lower bound for circuits solving SAT:

$$\forall 1^{(n)} (n \geq n_0) \forall C(|C| \leq n^c) \exists \varphi, \overline{a}, \quad |\varphi| = n \wedge Sat(\varphi, \overline{a}) \wedge \neg Sat(\varphi, C(\varphi)). \tag{22.3.3}$$

It is not difficult to construct (Section 22.5) a p-time algorithm F that finds some witnesses $(\varphi, \overline{a})$ to the existential quantifiers when given C, as long as SAT has no size-n^c circuits for large enough n. In particular, (22.3.3) can be then written as a universal statement (here we leave out the universal quantifiers):

$$[|C| \leq n^c \wedge (\varphi, \overline{a}) = F(C)] \rightarrow [|\varphi| = n \wedge Sat(\varphi, \overline{a}) \wedge \neg Sat(\varphi, C(\varphi))]. \tag{22.3.4}$$

It follows from the above discussion that having a super-polynomial lower bound for *EF*, *WF* or even a stronger system has a meta-mathematical significance for complexity theory. Informally: *any* super-polynomial lower bound for these proof systems implies that P $\neq$ NP is consistent with a bulk of present day complexity theory, and a super-polynomial lower bound for P-proofs of the propositional translations of (22.3.4) implies that an n^c lower bound for circuits solving SAT is not provable in a theory T corresponding to P.

In general, if $\mathcal{H}$ is a hypothesis in complexity theory that we conjecture to be true, it seems more interesting to know that it is consistent with a theory like PV or APC_1 rather than to know that it is not provable. This consistency counts towards the validity of $\mathcal{H}$: it is true in a model of the theory, a structure very close to the standard model from the point of view of complexity theory. However, the unprovability says that the methods available in the respective theory are insufficient to prove $\mathcal{H}$. I suppose that to be aware of this can be useful.

Unfortunately we have no lower bounds for any system in the L-level. Perhaps the meta-mathematical connections give a hint why proving them seems so hard.

We know quite a lot about mutual (p-)simulations of the systems at L-level, and about their surprising strength in the sense of what can feasibly be proved in them. To me the most interesting research direction is to attempt to extend the *feasible interpolation idea* (Section 18.7) from the A-level through the C-level up to the L-level. Proof complexity generators (Section 19.4) can be seen as a generalization of formulas expressing the disjointness of an NP pair (Lemma 19.4.1), and forcing with random variables (Section 20.4) reduces the properties of a model $K(F)$ (in particular, of a proof complexity relevant one) to properties of a family of random variables F.

My **working hypothesis** is that these two strands will come together and yield conditional lower bounds for all proof systems in the C-level and some proof systems in the L-level.

22.4 Mathematical Level

The proof systems that we included in the L-level are defined in a combinatorially transparent (though not necessarily elementary) way, using either some logical calculus or some bottom-up combinatorial construction. The proof systems that I have left for the **mathematical level** (M-level) do not seem to allow such definition and are defined simply as first-order theories. That is, a proof system at this level is described by defining a first-order theory T and interpreting a T-proof of the formal statement "*α is a tautology*" (the formula $Taut(\ulcorner\alpha\urcorner)$ in Section 7.4) as a proof of α in the system P_T (cf. the system P_{PA} in Section 7.4). Note that the proof system P_T defined in this way does not correspond to T in the sense of Section 8.6 (see also the notes at the end of Chapter 12). The theory T can be any one that contains Robinson's arithmetic (or PV, for simplicity of formalization), has a p-time decidable set of axioms and is sound with respect to the standard model $\mathbf{N}$. The axioms of T may involve any mathematics available; in particular, they do not reduce to logic.

The grey zone between the L-level and the M-level contains the theory $I\Delta_0 + Exp$ corresponding to the union of the systems i_kEF, $k \geq 1$. It is not interpretable in Robinson's arithmetic. The system $i_\infty EF$ already sits inside the M-level, I think.

If $\forall x A(x)$ is a true universal sentence in the language of PV, say, then proving a super-polynomial lower bound for P_T-proofs of the propositional translations $\|A\|^n$ implies that T does not prove the sentence. The only method that mathematics knows for showing that some universal statements are not provable in a strong theory is diagonalization. It applies to statements talking directly or indirectly about consistency. Propositional consistency statements do have short proofs, however (Section 19.2). Hence we should either try to find a modification of diagonalization for proof complexity or look for non-uniform sets of hard formulas to which the existing diagonalization methods apply (Chapter 19). Perhaps we should try to do both at the same time, in fact. Note that diagonalization is behind the very few results reaching to the M-level that we know, including Theorems 21.3.3 and 21.4.3.

I think that the big question behind the length-of-proof problems at the M-level and, in a sense, behind most of proof complexity, is this:

What is the nature of proof complexity?

That is, what are the intrinsic reasons why some formulas are hard to prove? Can the proof complexity of some formulas be traced to the computational complexity of associated computational tasks? In particular, does NP $\neq$ coNP follow from a plausible computational hypothesis about feasible computations? Or is the *hardness of proving* in fact genuinely different from the *hardness of computing*?

22.5 Bibliographical and Other Remarks

Sec. 22.1 The proof systems included in the A-level (and their links to algorithms) are discussed in Sections 5.2 and 5.3 (R-like systems), in Chapter 6 (algebraic and geometric systems) and in Sections 3.2 (cut-free LK) and 7.4 (OBDD). The lower bound methods are covered in Chapters 13, 16, 17 and 18. The search problem Search($\mathcal{C}$) was defined in Section 5.2.

Sec. 22.2 The theory corresponding to R was discussed in Section 10.5. The combined systems were introduced in Section 7.1 and the system R(log) in Section 5.7. The link of LK(CP) to a lattice algorithm was mentioned in Section 5.7 and more details are given in [284]. Frege systems are discussed in Chapter 2, the various bounded-depth subsystems were introduced in Section 5.7 and modular counting principles in Section 11.1.

The correspondence of these proof systems to various theories and their applications to length-of-proof upper bounds were discussed in Chapters 10 and 11. The restriction lower bound method is presented in Chapters 14 and 15.

The question whether R(CP) is related to some extension of the Chvátal–Gomory cuts is discussed in [284]. Investigating the same issue for LS or SOS could be, I think, quite fruitful.

The AC^0-Frege systems and their extensions by counting principles are linked with SAT algorithms in an indirect way. Given a proof system P, in [305] $Alg(P)$ is defined to be the class of SAT algorithms whose soundness has polynomial-size P-proofs (the soundness of an algorithm is formalized by reflection formulas, as for proof systems). The relation between the provability of reflection principles and the existence of a simulation (Section 8.6) implies that P simulates all algorithms A from $Alg(P)$, thinking of A as of a proof system. In particular, a length-of-proof lower bound for P translates into a time lower bound for A. It is not difficult to verify that most SAT algorithms proposed by researchers belong to the class $Alg(P)$ for some P for which we have a super-polynomial lower bound. Also, for a restriction A' of an algorithm $A \in Alg(P)$, a way to show that it is weaker than A is to show that it belongs to $Alg(P')$ for a P' that is demonstrably weaker than P.

Sec. 22.3 The systems included in this level were defined in Chapter 4 and Sections 2.4, 7.2, 7.3 and 7.5, and the corresponding theories in Section 9.3 and Chapter 12.

The systems $S^1_2 + k - Exp$ are defined analogously to how $S^1_2 + 1 - Exp$ was defined in Section 12.9. The interpretability of these systems (and the theories) in Robinson's arithmetic goes back to Paris and Wilkie [391], utilizing a trick due to Solovay (the shortening of cuts).

The formalization of complexity-theoretic concepts, constructions and results started with Paris and Wilkie [387, 386, 388, 389, 390, 391] and Buss [106] and has been treated in [278], by Hájek and Pudlák in [214] and by Cook and Nguyen in [155]. The mentioned circuit lower bounds were formalized in a particular way by Razborov [434, 435] (see also the discussion in Sections 19.5 and 19.6) and in an intrinsic formalization in the theory APC_1 in [278, 15.2] (for a parity lower bound; WPHP(f) there is what we have called dWPHP(f)) and by Müller and Pich [358]. They also formalized the natural proofs of Razborov and Rudich [444] and gave an overview of various formalizations and their relations. The PCP theorem was formalized in PV by Pich [397, 399]; the article [399] also offers an overview of the formalization of various other results.

Probabilistic classes and computations, and approximate counting, were formalized in the theories APC_1 and APC_2 by Jeřábek [256, 255, 257, 259, 260], who also formalized derandomization via the Nisan–Wigderson generator and constructions using a sorting network and an expander. Based on this work, Le [338] formalized the Goldreich–Levin theorem. Another construction of an expander was formalized by Buss *et al.* [126] in theory VNC^1. Toda's theorem was proved by Buss, Kolodziejczyk and Zdanowski [129] in an extension of APC_2.

Witnessing the failure of a purported SAT algorithm A and replacing (22.3.2) by a universal statement is simple: given $n \gg 1$, apply A itself to a formula ψ with two n-tuples of atoms $\overline{x}, \overline{y}$ expressing that $\overline{x}$ is a formula satisfied by $\overline{y}$ but not by $A(\overline{x})$. This formula is satisfiable, and hence A either finds a satisfying assignment (and hence a counterexample to itself) or does not, and then ψ is a counterexample (for some length $n^{O(1)}$). This is not quite what we want, because in the latter case we do not have a satisfying assignment certifying that $\psi \in$ SAT but we can still use ψ to find a local inconsistency in how A decides satisfiability and certify its failure by that. An analogous idea was generalized to randomized A by Gutfreund, Shaltiel and Ta-Shma [213].

The formula (22.3.3) is denoted in Pich [398] as $SCE(\mathrm{SAT}, n^c)$. Its witnessing by a p-time function is not so simple as for a uniform algorithm but can be done, assuming the existence of a strong one-way permutation (it yields instances of SAT that are hard to solve) and the existence of a language in E that is exponentially hard on average for circuits (this then allows the construction of a small pseudorandom set of instances, and an exhaustive search through it finds an instance hard for the given circuit). See [398, Proposition 4.2] for details and for other ways in which one can witness the failure of circuits for SAT.

An important open problem is, I think, whether it is also possible to feasibly witness the failure of a purported p-bounded proof system. That is, how computationally difficult is it to find hard formulas for a given proof system? In the uniform set-up this relates to the existence of an optimal proof system (Section 21.1). For the non-uniform set-up (in which proof systems are non-uniform sequences of non-deterministic p-size circuits) there are, actually, conditional negative results: assuming the existence of an exponentially hard one-way permutation, no time-$2^{O(n)}$ algorithm can find hard instances (see the problem $Cert(c)$ in [308]).

There are some results establishing the consistency of a plausible conjecture, or its unprovability, with a theory from the L-level. For example, it is consistent with PV (or with S^1_2) that NP $\not\subseteq$ P/*poly* if the polynomial-time hierarchy does not collapse to the Boolean hierarchy (or to the class $\mathrm{P}^{\mathrm{NP}}[\log n]$, respectively) (Cook and Krajíček [153]) and if for any $k \geq 1$ there is a language L in P that has no size-n^k circuits (this is an unconditional result of Krajíček and Oliveira [313]). An example of an unprovable plausible conjecture is Pich's [398] result that the theory T_{NC^1} (analogous to PV but for uniform NC^1-algorithms) does not prove a super-polynomial circuit lower bound for SAT unless all the functions in P have subexponential hardness on average with respect to *formulas* (see [398] for an exact statement).

Problems concerning the L-level include the finite axiomatizability problem 8.1.2, and this can serve as an example that conditional solutions are possible: Krajíček, Pudlák and Takeuti [323] solved it in the negative, assuming that the polynomial-time hierarchy does not collapse. Other problems in the L-level waiting for at least conditional solutions are the Δ_0-PHP problem 8.1.3 or the problem PV $=?$ APC_1 (posed in [289]).

The computational relevance of forcing with random variables is discussed in [304, p. 3].

Sec. 22.4 The proof of the non-interpretability of $I\Delta_0 + Exp$ in Robinson's arithmetic is due to Paris and Wilkie [391].

Pudlák [421] advocated the thesis (called the *feasible incompleteness thesis* by him) that the unprovability of universal statements in strong theories can be traced to their computational complexity.

Another statement similar in form to Theorem 21.4.3 and reaching to the M-level (and based on diagonalization) can be derived from a result of Williams [506]; namely, [506, Theorem 1.1] implies that either there is no subexponentially bounded proof system (i.e. $2^{n^{o(1)}}$-bounded) or NEXP $\not\subseteq$ P/*poly*.

The possibility of reducing proof hardness to computational hardness is also discussed in [295].

Bibliography

The page number(s) on which a publication is cited is or are given at the end of the details of the publication.

[1] K. Aehlig and A. Beckmann, A remark on the induction needed to prove the Ramsey principle, unpublished manuscript (2006). p. 292

[2] J. Aisenberg, M. L. Bonet, S. R. Buss, A. Craciun and G. Istrate, Short proofs of the Kneser–Lovász coloring principle, in: *Proc. 42nd International Colloq. on Automata, Languages and Programming* (ICALP), Springer Lecture Notes in Computer Science **9135** (2015), 44–55. p. 230

[3] J. Aisenberg, M. L. Bonet and S. R. Buss, Quasipolynomial size Frege proofs of Frankl's theorem on the trace of sets, *J. Symbolic Logic* **81(2)** (2016), 1–24. p. 231

[4] M. Ajtai, Σ_1^1-formulas on finite structures, *Ann. Pure and Applied Logic* **24** (1983), 1–48. pp. 29, 183, 449, 454

[5] M. Ajtai, The complexity of the pigeonhole principle, in: *Proc. IEEE 29th Annual Symp. on Foundations of Computer Science* (1988), 346–355. pp. 3, 163, 165, 179, 183, 306, 307, 318, 327, 334, 454

[6] M. Ajtai, Parity and the pigeonhole principle, in: *Feasible Mathematics*, eds. S. R. Buss and P. J. Scott, Birkhauser (1990), 1–24. pp. 230, 335

[7] M. Ajtai, The independence of the modulo p counting principles, in: *Proc. 26th Annual ACM Symposium on Theory of Computing*, ACM Press (1994), 402–411. pp. 230, 326, 335, 349

[8] M. Ajtai, Symmetric systems of linear equations modulo p, in: *Proc. Electronic Colloquium on Computational Complexity* (ECCC), TR94-015 (1994). p. 349

[9] M. Ajtai, Generalizations of the compactness theorem and Gödel's completeness theorem for nonstandard finite structures, in: *Proc. 4th International Conf. on Theory and Applications of Models of Computation* (2007), 13–33. p. 454

[10] M. Ajtai, A generalization of Gödel's completeness theorem for nonstandard finite structures, unpublished manuscript (2011). p. 454

[11] M. Alekhnovich, E. Ben-Sasson, A. A. Razborov, and A. Wigderson, Space complexity in propositional calculus, *SIAM J. Computing*, **31(4)** (2002), 1184–1211. p. 113

[12] M. Alekhnovich, E. Ben-Sasson, A. A. Razborov, and A. Wigderson, Pseudorandom generators in propositional proof complexity, *SIAM J. Computing*, **34(1)** (2004), 67–88. pp. 157, 278, 292, 350, 437, 438

[13] M. Alekhnovich, E. A. Hirsch and D. Itsykson, Exponential lower bounds for the running time of DPLL algorithms on satisfiable formulas, *J. Automated Reasoning*, **35(1–3)** (2005), 51–72. p. 295

[14] M. Alekhnovich, J. Johannsen, T. Pitassi and A. Urquhart, An exponential separation between regular and general resolution, in: *Proc. 34th Annual ACM Symp on Theory of Computing* (STOC) (2002), 448–456. p. 295

[15] M. Alekhnovich and A. A. Razborov, Lower bound for polynomial calculus: non-binomial case, *Proc. the Steklov Institute of Mathematics*, **242** (2003), 18–35. p. 350

[16] M. Alekhnovich and A. A. Razborov, Resolution is not automatizable unless W[P] is tractable, *SIAM J. Computing*, **38(4)** (2008), 1347–1363. p. 470

[17] W. B. Alexi, B. Chor, O. Goldreich and C. P. Schnorr, RSA and Rabin functions: certain parts are as hard as the whole, *SIAM J. Computing*, **17** (1988), 194–209. p. 364

[18] N. Alon and R. Boppana, The monotone circuit complexity of Boolean functions, *Combinatorica*, **7(1)** (1987), 1–22. pp. 38, 288, 382

[19] N. Alon and J. H. Spencer, *The Probabilistic Method*, 3rd edn, John Wiley and Sons, (2011). p. 351

[20] A. E. Andreev, On a method for obtaining lower bounds for the complexity of individual monotone functions, *Sov. Math. Dokl.*, **31** (1985), 530–534. p. 38

[21] N. Arai, Relative efficiency of propositional proof systems: Resolution and cut-free LK, *Ann. Pure and Applied Logic*, **104** (2000), 3–16. p. 112

[22] N. Arai, T. Pitassi and A. Urquhart, The complexity of analytic tableaux, in: *Proc. ACM Symp. on the Theory of Computing* (STOC) (2001), 356–363. p. 112

[23] T. Arai, A bounded arithmetic AID for Frege systems, *Ann. Pure and Applied Logic*, **103** (2000), 155–199. p. 209

[24] E. Artin, Uber die zerlegung definiter Funktionen in Quadrate, in: *Abhandlungen aus dem mathematischen Seminar der Universitat Hamburg*, **5** (1927), 100–115. Springer. p. 132

[25] G. Asser, Das Reprasentenproblem in Pradikatenkalkul der ersten Stufe mit Identitat, *Zeitschrift für Mathematische Logik und Grundlagen der Mathematik*, **1** (1955), 252–263. p. 2

[26] A. Atserias, The proof-search problem between bounded-width resolution and bounded-degree semi-algebraic proofs, in: *Proc. Conf. on Theory and Applications of Satisfiability Testing* (SAT 2013), eds. M. Jarvisalo and A. Van Gelder, **7962**, Lecture Notes in Computer Science (2013), 1–17. p. 470

[27] A. Atserias, A note on semi-algebraic proofs and gaussian elimination over prime fields, unpublished preprint (2015). pp. 348, 351

[28] A. Atserias, I. Bonacina, S. F. de Rezende, M. Lauria, J. Nordström and A. A. Razborov, Clique is hard on average for regular resolution, to appear in: *Proc. 50th ACM Symp. on Theory of Computing* (STOC) (2018). p. 293

[29] A. Atserias and M. L. Bonet, On the automatizability of resolution and related propositional proof systems, *Information and Computation*, **189(2)** (2004), 182–201. pp. 467, 471

[30] A. Atserias, M. L. Bonet and J. Levy, On Chvátal Rank and Cutting Planes Proofs, in: *Proc. Electronic Colloq. on Computational Complexity* (ECCC), TR03-041 (2003). p. 352

[31] A. Atserias and V. Dalmau, A combinatorial characterization of resolution width, *J. Computer and System Sciences*, **74(3)** (2008), 323–334. pp. 105, 113

[32] A. Atserias, J. K. Fichte, and M. Thurley, Clause-learning algorithms with many restarts and bounded-width resolution, *J. Artificial Intelligence Research*, **40** (2011), 353–373. p. 295

[33] A. Atserias, N. Galesi and R. Gavaldá, Monotone proofs of the pigeon hole principle, *Mathematical Logic Quarterly*, **47(4)** (2001), 461–474. p. 158

[34] A. Atserias, N. Galesi and P. Pudlák, Monotone simulations of nonmonotone proofs, *J. Computer and System Sciences*, **65** (2002), 626–638. pp. 146, 158

[35] A. Atserias, P. Kolaitis and M. Y. Vardi, Constraint propagation as a proof system, in: *Proc. 10th International Conf. on Principles and Practice of Constraint Programming* (CP), **3258**, Lecture Notes in Computer Science (2004), 77–91. pp. 147, 159

[36] A. Atserias, M. Lauria and J. Nordström, Narrow proofs may be maximally long, *ACM Trans. Computational Logic*, **17** (2016), 19:1–19:30. pp. 112, 294

[37] A. Atserias and E. Maneva, Mean-payoff games and propositional proofs, *Information and Computation*, **209(4)** (2011), 664–691. p. 470

[38] A. Atserias and E. Maneva, Sherali–Adams relaxations and indistinguishability in counting logics, *SIAM J. Computing*, **42(1)** (2013), 112–137. pp. 133, 455

[39] A. Atserias and M. Müller, Partially definable forcing and bounded arithmetic, *Archive for Mathematical Logic*, **54(1–2)** (2015), 1–33. p. 454

[40] A. Atserias, M. Müller and S. Oliva, Lower bounds for DNF-refutations of a relativized weak pigeonhole principle, *J. Symbolic Logic*, **80(2)** (2015), 450–476. p. 292

[41] A. Atserias and J. Ochremiak, Definable ellipsoid method, sums-of-squares proofs, and the isomorphism problem, to appear in: *Proc. 33rd Annual ACM/IEEE Symposium on Logic in Computer Science* (LICS), July 2018. pp. 351, 455

[42] J. Avigad, Plausibly hard combinatorial tautologies, in: *Proof Complexity and Feasible Arithmetics*, eds S. R. Buss and P. Beame, American Mathematical Society (1997), 1–12. pp. 63, 436

[43] L. Babai, A. Gál, J. Kollár, L. Rónyai, T. Szabó and A. Wigderson, Extremal bipartite graphs and superpolynomial lower bounds for monotone span programs, in: *Proc. 28th Annual ACM Symp. on Theory of Computing* (STOC) (1996), 603–611. p. 393

[44] T. Baker, J. Gill and R. Solovay, Relativizations of the $\mathcal{P} =? \mathcal{NP}$ question, *SIAM J. Comput.*, **4** (1975), 431–442. p. 463

[45] B. Barak, F. G. S. L. Brandao, A. W. Harrow, J. A. Kelner, D. Steurer and Y. Zhou, Hypercontractivity, sum-of-squares proofs, and their applications, in: *Proc. 28th Annual ACM Symp. on Theory of Computing* (STOC) (1996), 307–326. p. 351

[46] B. Barak, J. A. Kelner, D. Steurer, Rounding sum of squares relaxations, in: *Proc. 46th Annual ACM Symp. on Theory of Computing* (STOC) (2014), 31–40. p. 351

[47] B. Barak and D. Steurer, Sum-of-squares proofs and the quest toward optimal algorithms, in: *Proc. ICM Conf.* (Seul) (2014). pp. 133, 351

[48] P. Beame, Proof complexity, in: *Computational Complexity Theory*, eds. S. Rudich and A. Wigderson, AMS, IAS/Park City Math. Ser., **10** (2004), 199–246. p. 6

[49] P. Beame, C. Beck and R. Impagliazzo, Time–space tradeoffs in resolution: superpolynomial lower bounds for superlinear space, in: *Proc. 44th Annual ACM Symp. on Theory of Computing* (STOC) (2012), 212–232. p. 294

[50] P. Beame, S. A. Cook, J. Edmonds, R. Impagliazzo and T. Pitassi, The relative complexity of NP search problems, *J. Computer Systems Sciences*, **57** (1998), 3–19. pp. 426, 436

[51] P. Beame, J. Culberson, D. Mitchell and C. Moore, The resolution complexity of random graph *k*-colorability, *Discrete Applied Mathematics*, **153** (2005), 25–47. p. 293

[52] P. Beame, R. Impagliazzo, J. Krajíček, T. Pitassi and P. Pudlák, Lower bounds on Hilbert's Nullstellensatz and propositional proofs, *Proc. London Mathematical Society*, **73(3)** (1996), 1–26. pp. 131, 230, 326, 335, 338, 349, 350

[53] P. Beame, R. Impagliazzo, J. Krajíček, T. Pitassi, P. Pudlák and A. Woods, Exponential lower bounds for the pigeonhole principle, in: *Proc. Annual ACM Symp. on Theory of Computing* (STOC) (1992), 200–220. p. 334

[54] P. Beame, R. Impagliazzo, T. Pitassi and N. Segerlind, Formula Caching in DPLL, *ACM Trans. on Computation Theory*, **1(3)** (2010), 9:1–933. p. 295

[55] P. Beame, H. Kautz and A. Sabharwal, On the power of clause learning, in: *Proc. 18th International Joint Conf. on Artificial Intelligence* (IJCAI) (2003), 94–99. p. 295

[56] P. Beame, and T. Pitassi, Exponential separation between the matching principles and the pigeonhole principle, in: *Proc. 8th Annual IEEE Symposium on Logic in Computer Science* (LICS) (1993), 308–319. p. 335

[57] P. Beame and T. Pitassi, Simplified and improved resolution lower bounds, in: *Proc. 37th IEEE Symp. on Foundations of Computer Science* (FOCS) (1996), 274–282. pp. 112, 470

[58] P. Beame and T. Pitassi, Propositional proof complexity: past, present, and future, in: *Current Trends in Theoretical Computer Science: Entering the 21st Century*, eds. G. Paun, G. Rozenberg and A. Salomaa, World Scientific (2001), 42–70. pp. 6, 382

[59] P. Beame, T. Pitassi and N. Segerlind, Lower bounds for Lovász–Schrijver systems and beyond follow from multiparty communication complexity, in: *Automata, Languages, and Programming: Proc. 32nd International Colloq.* (2005), 1176–1188. pp. 395, 396, 407

[60] A. Beckmann and S. R. Buss, Improved witnessing and local improvement principles for second-order bounded arithmetic, *ACM Trans. Computational Logic*, **15(1)** (2014), article 2. p. 436

[61] A. Beckmann and S. R. Buss, The NP search problems of Frege and Extended Frege proofs, *ACM Trans. on Computational Logic*, **18(2)** (2017), Article 11. p. 436

[62] A. Beckmann, P. Pudlák and N. Thapen, Parity games and propositional proofs, *ACM Trans. Computational Logic*, **15(2)** (2014), article 17. pp. 210, 470

[63] S. Bellantoni, T. Pitassi and A. Urquhart, Approximation and small depth Frege proofs, *SIAM J. Computing*, **21(6)** (1992), 1161–1179. p. 334

[64] S. Ben-David and A. Gringauze, On the existence of propositional proof systems and oracle-relativized propositional logic, in: *Proc. Electronic Colloq. on Computational Complexity* (ECCC), TR98-021 (1998). pp. 91, 468

[65] J. H. Bennett, On spectra, Ph.D. thesis, Princeton University (1962). pp. 2, 182, 183, 196, 257

[66] E. Ben-Sasson, Expansion in proof complexity, Ph.D. thesis, Hebrew University, Jerusalem (2001). pp. 292, 471

[67] E. Ben-Sasson, Hard examples for the bounded depth Frege proof system, *Computational Complexity*, **11(3-4)** (2002), 109–136. pp. 335, 437

[68] E. Ben-Sasson and P. Harsha, Lower bounds for bounded depth Frege proofs via Buss–Pudlák games, *ACM Trans. on Computational Logic*, **11(3)** (2010), 1–17. p. 335

[69] E. Ben-Sasson and R. Impagliazzo, Random CNF's are hard for the polynomial calculus, *Computational Complexity*, **19(4)** (2010), 501–519. pp. 131, 350, 434

[70] E. Ben-Sasson, R. Impagliazzo and A. Wigderson, Near-optimal separation of general and tree-like resolution, *Combinatorica*, **24(4)** (2004), 585–604. p. 291

[71] E. Ben-Sasson and J. Nordström, Short proofs may be spacious: an optimal separation of space and length in resolution, in: *Proc. 49th Annual IEEE Symp. on Foundations of Computer Science* (FOCS) (2008), 709–718. p. 294

[72] E. Ben-Sasson and J. Nordström, Understanding space in proof complexity: separations and trade-offs via substitutions, in: *Proc. 2nd Symp. on Innovations in Computer Science* (ICS '11) (2011), 401–416. p. 290

[73] E. Ben-Sasson and A. Wigderson, Short proofs are narrow – resolution made simple, in: *Proc. 31st ACM Symp. on Theory of Computation* (STOC) (1999), 517–526. pp. 100, 101, 112, 277, 281, 282, 292, 293

[74] E. W. Beth, On Padoa's method in the theory of definition, *Indag. Math.*, **15** (1953), 330–339. p. 80
[75] E. W. Beth, *The Foundations of Mathematics*. North-Holland (1959). p. 80
[76] O. Beyersdorff, I. Bonacina and L. Chew, Lower bounds: from circuits to QBF proof systems, in: *Proc. Conf. on Innovations in Theoretical Computer Science* (ITCS) (2016), 249–260. p. 92
[77] O. Beyersdorff, N. Galesi and M. Lauria, A characterization of tree-like resolution size, *Information Processing Letters*, **113(18)** (2013), 666–671. p. 291
[78] O. Beyersdorff, J. Köbler and J. Messner, Nondeterministic functions and the existence of optimal proof systems, *Theoretical Computer Science*, **410(38–40)** (2009), 3839–3855. p. 468
[79] O. Beyersdorff, J. Köbler and S. Müller, Proof systems that take advice, *Information and Computation*, **209(3)** (2011), 320–332. p. 471
[80] O. Beyersdorff and O. Kullmann, Unified characterisations of resolution hardness measures, in: *Proc. 17th International Conf. on Theory and Applications of Satisfiability Testing* (SAT), Lecture Notes in Computer Science, **8561** (2014), 170–187. pp. 112, 114, 293
[81] O. Beyersdorff and J. Pich, Understanding Gentzen and Frege systems for QBF, in: *Proc. Conf. on Logic in Computer Science* (LICS) (2016), 146–155. p. 92
[82] O. Beyersdorff and Z. Sadowski, Characterizing the existence of optimal proof systems and complete sets for promise classes, in: *Proc. Conf on Computer Science – Theory and Applications* (CSR), Lecture Notes in Computer Science, **5675** (2009), 47–58. p. 469
[83] O. Beyersdorff and Z. Sadowski, Do there exist complete sets for promise classes? *Mathematical Logic Quarterly*, **57(6)** (2011), 535–550. p. 469
[84] J.-C. Birget, Reductions and functors from problems to word problems, *Theoretical Computer Science*, **237** (2000), 81–104. p. 160
[85] J.-C. Birget, A. Yu. Ol'shanskii, E. Rips and M. V. Sapir, Isoperimetric functions of groups and computational complexity of the word problem, *Ann. Mathematics*, **156(2)** (2002), 467–518. p. 153
[86] A. Blake, Canonical expressions in boolean algebra, Ph.D. thesis, University of Chicago (1937). p. 112
[87] J. Bochnak, M. Coste and M.-F. Roy, *Real Algebraic Geometry*, Springer (1999). p. 132
[88] I. Bonacina, Total space in resolution is at least width squared, in: *Proc. 43rd International Colloq. on Automata, Languages, and Programming* (ICALP) **55** (2016), 56:1–56:13. pp. 113, 294
[89] I. Bonacina, N. Galesi and N. Thapen, Total space in resolution, *SIAM J. on Computing*, **45(5)** (2016), 1894–1909. p. 294
[90] I. Bonacina and N. Talebanfard, Strong ETH and resolution via games and the multiplicity of strategies, *Algorithmica* (2016), 1–13. p. 295
[91] M. L. Bonet and S. R. Buss, On the deduction rule and the number of proof lines, in: *Proc. 6th Annual IEEE Symp. on Logic in Computer Science* (LICS 91), IEEE Computer Society Press (1991), 286–297. p. 62
[92] M. L. Bonet and S. R. Buss, The deduction rule and linear and near-linear proof simulations, *J. Symbolic Logic*, **58** (1993), 688–709. p. 62
[93] M. L. Bonet, C. Domingo, R. Gavaldá, A. Maciel and T. Pitassi, Non-automatizability of bounded-depth Frege proofs, *Computational Complexity*, **13** (2004), 47–68. p. 405
[94] M. L. Bonet, J. L. Esteban, N. Galesi and J. Johannsen, On the relative complexity of resolution refinements and cutting planes proof systems, *SIAM J. Computing*, **30(5)** (2000), 1462–1484. p. 406

[95] M. L. Bonet, T. Pitassi and R. Raz, Lower bounds for cutting planes proofs with small coefficients, in: *Proc. 27th Annual ACM Symp. on the Theory of Computing* (STOC) (1995), 575–584. pp. 157, 381, 382

[96] M. L. Bonet, T. Pitassi and R. Raz, Lower bounds for cutting planes proofs with small coefficients, *J. Symbolic Logic*, **62** (1997), 708–728. pp. 157, 381

[97] M. L. Bonet, T. Pitassi, and R. Raz, On interpolation and automatization for Frege proof systems, *SIAM J. Computing*, **29(6)** (2000), 1939–1967. pp. 365, 382, 404, 410

[98] W. Boone, The word problem, *Proc. Nat. Acad. Sci. USA*, **44** (1958), 265–269. p. 153

[99] W. Boone, The word problem, *Ann. Mathematics*, **70** (1959), 207–265. p. 153

[100] G. Boole, *The Mathematical Analysis of Logic*, Barclay and Macmillan (1847). p. 116

[101] R. Boppana and M. Sipser, The complexity of finite functions, in: *Handbook of Theoretical Computer Science*, Elsevier Science Publishers (1991), 759–804. p. 38

[102] M. Bridson, The geometry of the word problem, in: *Invitations to Geometry and Topology*, Oxford University Press (2002). p. 160

[103] R. E. Bryant, Graph-based algorithms for Boolean function manipulation, *IEEE Trans. on Computing*, **C,35** (1986), 677–691. p. 146

[104] R. E. Bryant, Syntactic Boolean manipulation with ordered binary decision diagrams, *ACM Computing Surveys*, **2493** (1992), 293–318. p. 146

[105] J. Buresh-Oppenheim, N. Galesi, S. Hoory, A. Magen and T. Pitassi, Rank bounds and integrality gaps for cutting planes procedures, *Theory of Computing*, **2(1)** (2006), 65–90. p. 352

[106] S. R. Buss, *Bounded Arithmetic*. Bibliopolis (1986). pp. 79, 164, 166, 183, 192, 193, 196, 236, 436, 478

[107] S. R. Buss, The Boolean formula value problem is in ALOGTIME, in: *Proc. 19th Annual ACM Symp. on Theory of Computing* (STOC) (1987), 123–131. p. 209

[108] S. R. Buss, Polynomial size proofs of the propositional pigeonhole principle, *J. Symbolic Logic*, **52** (1987), 916–927. pp. 53, 183, 213, 215, 230

[109] S. R. Buss, Axiomatizations and conservation results for fragments of bounded arithmetic, in: *Logic and Computation*, Contemporary Mathematics, **106** (1990), 57–84. p. 193

[110] S. R. Buss, Relating the bounded arithmetic and polynomial-time hierarchies, *Ann. Pure and Applied Logic*, **75** (1995), 67–77. p. 183

[111] S. R. Buss, On Gödel's theorems on lengths of proofs II: lower bounds for recognizing *k* symbol provability, in: *Feasible Mathematics II*, eds. P. Clote and J. Remmel, Birkhauser (1995), 57–90. p. 2

[112] S. R. Buss, Some remarks on the lengths of propositional proofs, *Archive for Mathematical Logic*, **34** (1995) 377–394. p. 62

[113] S. R. Buss, Bounded arithmetic and propositional proof complexity, in: *Logic of Computation*, ed. H. Schwichtenberg, Springer (1997), 67–122. pp. 183, 196

[114] S. R. Buss, Lower bounds on Nullstellensatz proofs via designs, in: *Proof Complexity and Feasible Arithmetics*, eds. S. Buss and P. Beame, American Mathematical Society (1998), 59–71. p. 349

[115] S. R. Buss, An Introduction to Proof Theory, in: *Handbook of Proof Theory*, ed. S. R. Buss, Elsevier (1998), 1–78. p. 6

[116] S. R. Buss, First-order proof theory of arithmetic, in: *Handbook of Proof Theory*, ed. S. R. Buss, Elsevier (1998), 79–147. pp. 6, 79, 196

[117] S. R. Buss, Propositional proof complexity: an introduction, in: *Computational Logic*, eds. U. Berger and H. Schwichtenberg, Springer (1999), 127–178. p. 6

[118] S. R. Buss, Bounded arithmetic, proof complexity and two papers of Parikh, *Ann. Pure and Applied Logic*, **96** (1999), 43–55. p. 182

[119] S. R. Buss, Bounded arithmetic and constant depth Frege proofs, in: *Complexity of Computations and Proofs*, ed. J. Krajíček, *Quaderni di Matematica*, **13** (2004), 153–174. pp. 183, 196

[120] S. R. Buss, Towards NP-P via proof complexity and search, *Ann. Pure and Applied Logic*, **163(7)** (2012), 906–917. p. 6

[121] S. R. Buss and P. Clote, Cutting planes, connectivity, and threshold logic, *Archive for Mathematical Logic*, **3591** (1996), 33–62. pp. 124, 132, 209, 231

[122] S. R. Buss, D. Grigoriev, R. Impagliazzo and T. Pitassi, Linear gaps between degrees for the polynomial calculus modulo distinct primes, in: *Proc. 31st Annual ACM Symp. on Theory of Computing* (STOC) (1999), 547–556. pp. 131, 350, 351

[123] S. R. Buss, J. Hoffmann and J. Johannsen, Resolution trees with lemmas – resolution refinements that characterize DLL-algorithms with clause learning, *Logical Methods in Computer Science*, **4** (2008), 4:13. p. 295

[124] S. R. Buss, R. Impagliazzo, J. Krajíček, P. Pudlák, A. A. Razborov and J. Sgall, Proof complexity in algebraic systems and bounded depth Frege systems with modular counting, *Computational Complexity*, **6(3)** (1996/1997), 256–298. pp. 131, 159, 230, 326, 333, 335, 336, 338, 350

[125] S. R. Buss and J. Johannsen, On linear resolution, *J. Satisfiability, Boolean Modeling and Computation*, **10** (2016), 23–35. p. 291

[126] S. R. Buss, V. Kabanets, A. Kolokolova and M. Koucký, Expanders in VNC^1, in: *Proc. Conf. Innovations in Theoretical Computer Science* (ITCS 2017), Leibniz International Proceedings in Informatics (LIPIcs), **67** (2017). pp. 146, 158, 479

[127] S. R. Buss and L. A. Kolodziejczyk, Small stone in pool, *Logical Methods in Computer Science*, **10(2)** (2014), paper 16. p. 295

[128] S. R. Buss, L. Kolodziejczyk and N. Thapen, Fragments of approximate counting, *J. Symbolic Logic*, **79(2)** (2014), 496–525. pp. 160, 209, 293

[129] S. R. Buss, L. Kolodziejczyk and K. Zdanowski, Collapsing modular counting in bounded arithmetic and constant depth propositional proofs, *Trans. AMS*, **367** (2015), 7517–7563. pp. 334, 479

[130] S. R. Buss and J. Krajíček, An application of boolean complexity to separation problems in bounded arithmetic, *Proc. London Mathematical Society*, **69(3)** (1994), 1–21. pp. 304, 305, 435

[131] S. R. Buss, J. Krajíček, and G. Takeuti, On provably total functions in bounded arithmetic theories R^i_3, U^i_2 and V^i_2, in: *Arithmetic Proof Theory and Computational Complexity*, eds. P. Clote and J. Krajíček (1993), 116–161, Oxford Press.

[132] S. R. Buss and P. Pudlák, How to lie without being (easily) convicted and the lengths of proofs in propositional calculus, in: *Computer Science Logic 94*, eds. Pacholski and Tiuryn, Springer Lecture Notes in Computer Science, **933** (1995), 151–162. pp. 62, 335

[133] S. R. Buss and G. Turan, Resolution proofs of generalized pigeonhole principles, *Theoretical Computer Science*, **62(3)** (1988), 311–317. p. 292

[134] S. Cavagnetto, Propositional proof complexity and rewriting, Ph.D. thesis, Charles University in Prague (2008). p. 160

[135] Y. Chen and J. Flum, A logic for PTIME and a parameterized halting problem, in: *Fields of Logic and Computation*, eds. A. Blass, N. Dershowitz and W. Reisig, Lecture Notes in Computer Science, **6300** (2010), 251–276. p. 471

[136] Y. Chen and J. Flum, On p-optimal proof systems and logics for PTIME, in: *Proc. ICALP: Automata, Languages and Programming*, eds. S. Abramsky, C. Gavoille, C. Kirchner, F. Meyer auf der Heide and P. G. Spirakis, Lecture Notes in Computer Science, **6199** (2010), 321–332. p. 471

[137] C. L. Chang and R. C.-T. Lee, *Symbolic Logical and Mechanical Theorem Proving*, Academic Press (1973). p. 62

[138] M. Chiari and J. Krajíček, Witnessing functions in bounded arithmetic and search problems, *J. Symbolic Logic*, **63(3)** (1998), 1095–1115. pp. 293, 305, 350, 436, 437

[139] M. Chiari and J. Krajíček, Lifting independence results in bounded arithmetic, *Archive for Mathematical Logic*, **38(2)** (1999), 123–138. pp. 303, 305, 436

[140] V. Chvátal, Edmonds polytopes and a hierarchy of combinatorial problems, *Discrete Mathematics*, **4** (1973), 305–337. pp. 115, 121, 130

[141] V. Chvátal and E. Szemerédi, Many hard examples for resolution, *J. ACM*, **35(4)** (1988), 759–768. pp. 282, 434

[142] M. Clegg, J. Edmonds and R. Impagliazzo, Using the Groebner basis algorithm to find proofs of unsatisfiability, in: *Proc. 28th Annual ACM Symp. on Theory of Computing* (STOC) (1996), 174–183. pp. 112, 131, 350, 406

[143] A. Church, A note on the Entscheidungsproblem, *J. Symbolic Logic*, **1** (1936), 40–41. pp. 1, 32

[144] P. Clote and J. Krajíček, eds., *Arithmetic Proof Theory and Computational Complexity*, Oxford University Press (1993). p. 2

[145] P. Clote and E. Kranakis, *Boolean Functions and Models of Computation*, Springer (2002). pp. 6, 79, 112, 132, 157, 209, 231, 350, 406

[146] A. Cobham, The intrinsic computational difficulty of functions, in: *Proc. Conf. on Logic, Methodology and Philosophy of Science*, ed. Y. Bar-Hillel (1965), 24–30. pp. 233, 234, 235, 257

[147] S. A. Cook, The complexity of theorem proving procedures, in: *Proc. 3rd Annual ACM Symp. on Theory of Computing* (STOC) (1971), 151–158. pp. 26, 37

[148] S. A. Cook, A hierarchy for nondeterministic time complexity, *J. Computational Systems Science*, **7(4)** (1973), 343–353. p. 25

[149] S. A. Cook, Feasibly constructive proofs and the propositional calculus, in: *Proc. 7th Annual ACM Symp. on Theory of Computing* (STOC) (1975), 83–97. pp. 3, 163, 165, 233, 235, 244, 258, 435, 468

[150] S. A. Cook, Relativized propositional calculus, preprint at https://arxiv.org/abs/1203.2168 (2012). p. 91

[151] S. A. Cook and L. Fontes, Formal theories for linear algebra, *Logical Methods in Computer Science*, **8** (2012), 1–31. p. 232

[152] S. A. Cook and A. Haken, An exponential lower bound for the size of monotone real circuits, *J. Computer and System Science*, **58(2)** (1999), 326–335. p. 387

[153] S. A. Cook and J. Krajíček, Consequences of the provability of NP $\subseteq$ P/*poly*, *J. Symbolic Logic*, **72(4)** (2007), 1353–1371. pp. 157, 160, 471, 479

[154] S. A. Cook and T. Morioka, Quantified propositional calculus and a second-order theory for NC^1, *Archive for Mathematical Logic*, **44(6)** (2005), 711–749. p. 91

[155] S. A. Cook, and P. Nguyen, *Logical Foundations of Proof Complexity*, Cambridge University Press (2009). pp. 6, 79, 87, 183, 196, 209, 210, 230, 259, 404, 436, 478

[156] S. A. Cook and R. A. Reckhow, The relative efficiency of propositional proof systems, *J. Symbolic Logic*, **44(1)** (1979), 36–50. pp. 33, 34, 38, 39, 62, 79, 114, 183, 231

[157] S. A. Cook and M. Soltys, Boolean programs and quantified propositional proof systems, *Bull. Logic Section*, University of Lodz, **28(3)** (1999), 119–129. pp. 85, 86, 90, 92

[158] S. A. Cook and M. Soltys, The proof complexity of linear algebra, *Ann. Pure and Applied Logic*, **130** (2004), 277–323. pp. 232, 434

[159] S. A. Cook and N. Thapen, The strength of replacement in weak arithmetic, *ACM Trans. on Computational Logic*, **7(4)** (2006). p. 259

[160] S. A. Cook and I. Tzameret, Uniform, integral and efficient proofs for the determinant identities, in: *Logic in Computer Science (LICS), Proc. 32nd Annual ACM/IEEE Symp.*, **32** (2017), 1–12. p. 435

[161] W. Cook, C. R. Coullard and G. Turán, On the complexity of cutting plane proofs, *Discrete Applied Mathematics*, **18** (1987), 25–38. p. 132

[162] D. Cox, J. Little and D. O'Shea, *Ideals, Varieties, and Algorithms*, Springer (2007). p. 131

[163] W. Craig, Three uses of the Herbrand–Gentzen theorem in relating model theory and proof theory, *J. Symbolic Logic*, **22(3)** (1957), 269–285. pp. 14, 37

[164] S. Dantchev and S. Riis, On relativisation and complexity gap for resolution-based proof systems, in: *Proc. 12th Annual Conf. of the EACSL Computer Science Logic*, Springer (2003). pp. 270, 291

[165] S. Dash, Exponential lower bounds on the lengths of some classes of branch-and-cut proofs, *Mathematics of Operations Research*, **30(3)** (2005), 678–700. p. 407

[166] M. Davis, G. Logemann and D. Loveland, A machine program for theorem proving, *Commun. ACM*, **5(7)** (1962), 394–397. pp. 96, 112

[167] M. Davis, Y. Matiyasevich and J. Robinson, Hilbert's tenth problem: diophantine equations: positive aspects of a negative solution, in: *Mathematical Developments Arising from Hilbert Problems, Proc. Symp. in Pure Mathematics*, ed. F. E. Browder (1976), 323–378. p. 38

[168] M. Davis and H. Putnam, A computing procedure for quantification theory, *J. ACM*, **7(3)** (1960), 210–215. pp. 96, 112

[169] W. Diffie and M. Hellman, New directions in cryptography, *IEEE Trans. Information Theory*, **22** (1976), 423–439. p. 404

[170] C. Dimitracopoulos and J. Paris, Truth definitions for Δ_0 formulae, in: *Logic and Algorithmic*, Monographie No. **30** de L'Enseignement Mathematique, (1982), 318–329. p. 183

[171] C. Dimitracopoulos and J. Paris, The pigeonhole principle and fragments of arithmetic, *Zeitschrift f. Mathematikal Logik u. Grundlagen d. Mathematik*, **32** (1986), 73–80. p. 182

[172] M. Dowd, Propositional representations of arithmetic proofs, Ph.D. thesis, University of Toronto (1979). pp. 62, 231, 258

[173] J. Edmonds, Paths, trees, and flowers, *Canad. J. Math.*, **17** (1965), 449–467. p. 37

[174] H. B. Enderton, *A Mathematical Introduction to Logic*, Academic Press (2001). p. 37

[175] P. Erdös, Some remarks on the theory of graphs, *Bull. AMS*, **53** (1947), 292–294. pp. 275, 320

[176] J. L. Esteban and J. Toran, Space bounds for resolution, in: *Proc. 16th Symp. on Theoretical Aspects of Computer Science* (STACS) (1999), 551–561. pp. 113, 157

[177] J. L. Esteban and J. Toran, A combinatorial characterization of treelike resolution space, *Information Processes Letters*, **87(6)** (2003), 295–300. p. 113

[178] R. Fagin, Generalized first-order spectra and polynomial-time recognizable sets, in: *Complexity of Computation*, ed. R. Karp, SIAM–AMS Proc., **7** (1974), 27–41. p. 21

[179] U. Feige, J. H. Kim and E. Ofek, Witnesses for nonsatisfiability of dense random 3CNF formulas, in: *Proc. 47th IEEE Annual Symp. on Foundations of Computer Science* (FOCS) (2006), 497–508. pp. 160, 471

[180] A. Fernandes and F. Ferreira, Groundwork for weak analysis, *J. Symbolic Logic*, **67** (2002), 557–578. p. 232

[181] A. Fernandes, F. Ferreira and G. Ferreira, Analysis in weak systems, in: *Logic and Computation: Essays in Honour of Amilcar Sernadas*, College Publications (2017), 231–261. p. 232

[182] F. Ferreira and G. Ferreira, The Riemann integral in weak systems of analysis, *J. Universal Computer Science*, **14** (2008), 908–937. p. 232

[183] Y. Filmus, P. Hrubeš and M. Lauria, Semantic versus syntactic cutting planes, in: *Proc. 33rd Symp. on Theoretical Aspects of Computer Science* (STACS) (2016). p. 405

[184] Y. Filmus, M. Lauria, M. Mikša, J. Nordström and M. Vinyals, From small space to small width in resolution, *ACM Trans. Computational Logic*, **16(4)** article 28. p. 113

[185] Y. Filmus, M. Lauria, J. Nordström, N. Thapen and N. Ron-Zevi, Space complexity in polynomial calculus, *SIAM J. Computing*, **44(4)** (2015), 1119–1153. p. 293

[186] Y. Filmus, T. Pitassi and R. Santhanam, Exponential lower bounds for AC^0-Frege imply super-polynomial Frege lower bounds, in: *Proc. 38th International Colloq. on Automata, Languages and Programming* (ICALP), eds. L. Aceto, M. Henzinger and J. Sgall, Lecture Notes in Computer Science, **6755** (2011), 618–629. pp. 61, 62

[187] N. Fleming, D. Pankratov, T. Pitassi and R. Robere, Random $\Theta(\log n)$-CNFs are hard for cutting planes, in: *Proc. 58th Annual Symp. on Foundations of Computer Science* (FOCS) (2017), 109–120. p. 434

[188] G. Frege, *Begriffsschrift: Eine der arithmetischen nachgebildete Formelsprache des reinen Denkens*, Halle (1879). pp. 1, 37, 41, 52, 62

[189] H. Friedman, The complexity of explicit definitions, *Adv. Mathematics*, **20** (1976), 18–29. p. 381

[190] H. Friedman, On the consistency, completeness, and correctness problems, unpublished preprint (1979). p. 461

[191] E. Friedgut, Sharp thresholds of graph properties and the k-SAT problem, *J. American Matheamtical Society*, **12(4)** (1999), 1017–1054. p. 282

[192] M. Furst, J. B. Saxe and M. Sipser, Parity, circuits and the polynomial-time hierarchy, *Math. Systems Theory*, **17** (1984), 13–27. pp. 29, 183

[193] N. Galesi and N. Thapen, Resolution and pebbling games, in: *Proc. 8th International Conf. on Theory and Applications of Satisfiability Testing* (SAT), Lecture Notes in Computer Science, **3569** (2005), 76–90. pp. 113, 293

[194] M. R. Garey, D. S. Johnson and L. Stockmeyer, Some simplifed NP-complete graphs problems, *Theoretical Computer Science*, **1** (1976), 237–267. p. 160

[195] M. Garlík, A new proof of Ajtai's completeness theorem for nonstandard finite structures, *Archive for Mathematical Logic*, **54(3–4)** (2015), 413–424. p. 454

[196] M. Garlík, Construction of models of bounded arithmetic by restricted reduced powers, *Archive for Mathematical Logic*, **55(5)** (2016), 625–648.

[197] M. Garlík and L. Kolodziejczyk, Some subsystems of constant-depth Frege with parity, preprint (2016). pp. 157, 336

[198] G. Gentzen, Die Widerspruchsfreiheit der reinen Zahlentheorie, *Mathematische Annalen*, **112** (1936), 493–565. p. 64

[199] K. Ghasemloo, Uniformity and nonuniformity in proof complexity, Ph.D. thesis, University of Toronto (2016). pp. 63, 184

[200] K. Gödel, Über formal unentscheidbare Sätze der Principia Mathematica und verwandter Systeme I, *Monatshefte fur Mathematik und Physik*, **38** (1931), 173–198. pp. 1, 469

[201] K. Gödel, a letter to John von Neumann from 1956, reprinted in: *Arithmetic, Proof Theory and Computational Complexity*, eds. P. Clote and J. Krajíček, Oxford University Press (1993). p. 2

[202] A. Goerdt, Cutting plane versus Frege proof systems, in: *Proc. Conf. Computer Science Logic* (CSL 1990), eds. E. Börger, H. Kleine Büning, M. M. Richter , and W. Schonfeld, Lecture Notes in Computer Science, **533**, Springer (1991), 174–194. p. 231

[203] A. Goerdt, Davis–Putnam resolution versus unrestricted resolution, *Ann. Mathematics and Artificial Intelligence*, **6** (1992), 169–184. p. 295

[204] O. Goldreich and L. Levin, Hard-core predicates for any one-way function, in: *Proc. 21st ACM Symp. on Theory of Computing* (STOC) (1989), 25–32. p. 363

[205] R. E. Gomory, An algorithm for integer solutions of linear programs, in: *Recent Advances in Mathematical Programming*, eds. R. L. Graves and P. Wolfe (1963), 269–302. pp. 115, 121

[206] D. Grigoriev, Tseitin's tautologies and lower bounds for Nullstellensatz proofs, in: *Proc. IEEE 39th Annual Symp. on Foundations of Computer Science* (FOCS) (1998), 648–652. pp. 350, 351

[207] D. Grigoriev, Linear lower bound on degrees of Positivstellensatz calculus proofs for the parity, *Theoretial Computer Science*, **259(1-2)** (2001), 613–622. pp. 132, 347

[208] D. Grigoriev, Complexity of Positivstellensatz proof for the knapsack, *Computational Complexity*, **10(2)** (2001), 139–154. p. 351

[209] D. Grigoriev, E. A. Hirsch and D. V. Pasechnik, Complexity of semi-algebraic proofs, *Moscow Mathematical J.*, **2(4)** (2002), 647–679. pp. 131, 132, 351, 352, 408

[210] D. Grigoriev and N. Vorobjov, Complexity of Null and Positivstellensatz proofs, *Ann. Pure and Applied Logic*, **113(1)** (2001), 153–160. pp. 132, 351

[211] J. Grochow and T. Pitassi, Circuit complexity, proof complexity and polynomial identity testing, in: *Conf. on Proc. Foundations of Computer Science* (FOCS) (2014). pp. 152, 159, 160

[212] Y. Gurevich, Towards logic tailored for computational complexity, in: *Proc. Logic Colloq. 1983*, Lecture Notes in Mathematics, **1104** (1984), 175–216. p. 381

[213] D. Gutfreund, R. Shaltiel and A. Ta-Shma, If NP languages are hard on the worst-case, then it is easy to find their hard instances, *Computational Complexity*, **16(4)** (2007), 412–441. p. 479

[214] P. Hájek and P. Pudlák, Metamathematics of first-order arithmetic, in: *Perspectives in Mathematical Logic*, Springer (1993). pp. 159, 434, 478

[215] G. Hajós, Über eine Konstruktion nicht n-färbbarer Graphen, *Wiss. Zeitschr. Martin Luther Univ. Halle-Wittenberg*, **A 10** (1961). p. 154

[216] A. Haken, The intractability of resolution, *Theoretical Computer Science*, **39** (1985), 297–308. pp. 3, 292, 294

[217] A. Haken, Counting bottlenecks to show monotone P $\neq$ NP, in: *Proc. 36th IEEE Symp. on Foundations of Computer Science* (FOCS) (1995), 36–40. pp. 359, 382

[218] J. Hanika, Search problems and bounded arithmetic, Ph.D. thesis, Charles University, Prague (2004). p. 435

[219] J. Hanika, Herbrandizing search problems in bounded arithmetic, *Mathematical Logic Quarterly*, **50(6)** (2004), 577–586. p. 435

[220] J. Hartmanis, P. M. Lewis and R. E. Stearns, Hierarchies of memory limited computations, in: *Proc. IEEE Conf. on Switching Circuit Theory and Logic Design*, Ann Arbor (1965), 179–190. pp. 26, 38

[221] J. Hartmanis and R. E. Stearns, On the computational complexity of algorithms, *Trans. AMS*, **117** (1965), 285–306. pp. 25, 38

[222] J. Hastad, Almost optimal lower bounds for small depth circuits. in: *Randomness and Computation*, ed. S. Micali, (1989), 143–170. pp. 38, 299, 300, 304

[223] J. Hastad, On small-depth Frege proofs for Tseitin for grids, in: *Proc. the IEEE 58th Annual Symp. on Foundations of Computer Science* (2017), 97–108. pp. 335, 437

[224] J. Van Heijenoort, ed., *From Frege to Gödel (A Source Book in Mathematical Logic, 1879–1931)*, Harvard University Press (1977). p. 1

[225] J. Henzl, Weak formal systems, M.Sc. thesis, Charles University, Prague (2003). p. 159

[226] J. Herbrand, Recherches sur la théorie de la démonstration, *Travaux Soc. Sciences et Lettres de Varsovie, Class III, Sciences Mathematiques et Physiques*, **33** (1930). p. 258

[227] G. Higman, Subgroups of finitely presented groups, *Proc. Royal Soc. London*, **A 262** (1961), 455–475. p. 153

[228] D. Hilbert and W. Ackermann, *Principles of Mathematical Logic*, Chelsea (1950). Translation of 1938 German edition. p. 39

[229] D. Hilbert and P. Bernays, Grundlagen der Mathematik. I, in: *Die Grundlehren der mathematischen Wissenschaften*, **40**, Springer (1934). pp. 37, 39

[230] D. Hilbert and P. Bernays, Grundlagen der Mathematik. II, in: *Die Grundlehren der mathematischen Wissenschaften*, **50**, Springer (1939). pp. 37, 39

[231] E. A. Hirsch and A. Kojevnikov, Several notes on the power of Gomory–Chvátal cuts, *Ann. Pure and Applied Logic*, **141(3)** (2006), 429–436. p. 406

[232] W. Hodges, *Building Models by Games*, Dover (2006). pp. 289, 454

[233] P. Hrubeš, Lower bounds for modal logics, *J. Symbolic Logic*, **72(3)** (2007), 941–958. p. 410

[234] P. Hrubeš, A lower bound for intuitionistic logic, *Ann. Pure and Applied Logic*, **146** (2007), 72–90. p. 410

[235] P. Hrubeš, On lengths of proofs in non-classical logics, *Ann. Pure and Applied Logic*, **157(2-3)** (2009), 194–205. p. 410

[236] P. Hrubeš and P. Pudlák, A note on monotone real ciruits, *Information Processing Letters*, **131** (2017), 15–19. p. 406

[237] P. Hrubeš and P. Pudlák, Random formulas, monotone circuits, and interpolation, in: *Proc. 58th IEEE Symp. on Foundations of Computer Science* (FOCS) (2017), 121–131. pp. 380, 382, 434

[238] P. Hrubeš and I. Tzameret, Short proofs for the determinant identities, *SIAM J. Computing*, **44(2)** (2015), 340–383. p. 435

[239] L. Huang and T. Pitassi, Automatizability and simple stochastic games, In: *Proc. 38th International Colloq. on Automata, Languages and Programming* (ICALP), L. Aceto, M. Henzinger and J. Sgall Zurich, Lecture Notes in Computer Science, **6755** (2011), 605–617. p. 470

[240] N. Immerman, Nondeterministic space is closed under complementation, *SIAM J. Computing*, **17(5)** (1988), 935–938. p. 26

[241] R. Impagliazzo, V. Kabanets, and A. Wigderson, In search of an easy witness: exponential time vs. probabilistic polynomial time, *J.Computer Systems Science*, **65(4)** (2002), 672–694. p. 433

[242] R. Impagliazzo and J. Krajíček, A note on conservativity relations among bounded arithmetic theories, *Mathematical Logic Quarterly*, **48(3)** (2002), 375–377. pp. 231, 304, 305

[243] R. Impagliazzo and M. Naor, Efficient cryptographic schemes provably as secure as subset sum, *J. Cryptology*, **9(4)** (1996), 199–216.

[244] R. Impagliazzo, T. Pitassi and A. Urquhart, Upper and lower bounds for tree-like cutting planes proofs, in: *Proc. Conf. on Logic in Computer Science* (LICS) (1994), 220–228. pp. 382, 389, 406

[245] R. Impagliazzo and P. Pudlák, A lower bound for DLL algorithms for SAT, in: *Proc. 11th Symp. on Discrete Algorithms* (2000), 128–136. pp. 113, 293

[246] R. Impagliazzo, P. Pudlák, and J. Sgall, Lower bounds for the polynomial calculus and the Groebner basis algorithm, *Computational Complexity*, **8(2)** (1999), 127–144. pp. 346, 350, 352

[247] R. Impagliazzo and N. Segerlind, Counting axioms do not polynomially simulate counting gates, in: *Proc. IEEE 42nd Annual Symp. on Foundation of Computer Science* (FOCS) (2001), 200–209. pp. 335, 336

[248] R. Impagliazzo and N. Segerlind, Constant-depth Frege systems with counting axioms polynomially simulate Nullstellensatz refutations, *ACM Trans. Computational Logic*, **7(2)** (2006), 199–218. p. 231

[249] R. Impagliazzo and A. Wigderson, P = BPP unless E has sub-exponential circuits: derandomizing the XOR lemma, in: *Proc. 29th Annual ACM Symp. on Theory of Computing* (STOC) (1997), 220–229. pp. 432, 433

[250] D. Itsykson, A. Knop, A. Romashchenko and D. Sokolov, On OBDD based algorithms and proof systems that dynamically change order of variables, in: *Leibniz International Proc. in Informatics*, **66** (2017), 43:1–43:14. pp. 159, 408

[251] D. Itsykson and A. Kojevnikov, Lower bounds of static Lovász–Schrijver calculus proofs for Tseitin tautologies, *J. Math. Sci.*, **145** (2007), 4942–4952. p. 351

[252] D. Itsykson and D. Sokolov, Lower bounds for splittings by linear combinations, in: *Proc. Conf. on Mathematical Foundations of Computer Science* (2014), 372–383. pp. 157, 402, 409

[253] G. D. James, The module orthogonal to the specht module, *J. Algebra*, **46(2)** (1977), 451–456. p. 349

[254] G. D. James, *The Representation Theory of the Symmetric Groups*, Lecture Notes in Mathematics, **682** (1978). p. 349

[255] E. Jeřábek, Dual weak pigeonhole principle, Boolean complexity, and derandomization, *Ann. Pure and Applied Logic*, **129** (2004), 1–37. p. 478

[256] E. Jeřábek, Weak pigeonhole principle, and randomized computation, Ph.D. thesis, Charles University, Prague (2005). pp. 157, 159, 258, 336, 478

[257] E. Jeřábek, Approximate counting in bounded arithmetic, *J. Symbolic Logic*, **72(3)** (2007), 959–993. pp. 250, 258, 478

[258] E. Jeřábek, On independence of variants of the weak pigeonhole principle, *J. Logic and Computation*, **17(3)** (2007), 587–604. p. 258

[259] E. Jeřábek, Approximate counting by hashing in bounded arithmetic, *J. Symbolic Logic*, **7493** (2009), 829–860. pp. 250, 258, 478

[260] E. Jeřábek, A sorting network in bounded arithmetic, *Annals of Pure and Applied Logic*, **162(4)** (2011), 341–355. pp. 158, 478

[261] E. Jeřábek and P. Nguyen, Simulating non-prenex cuts in quantified propositional calculus, *Mathematical Logic Quarterly*, **57(5)** (2011), 524–532. p. 91

[262] J. P. Jones and Y. Matiyasevich, Basis for the polynomial time computable functions, in: *Proc. Conf. on Number Theory*, Banff, Alberta (1988), 255–270. p. 257

[263] S. Jukna, *Boolean Function Complexity*, Springer (2012). p. 38

[264] M. Karchmer and A. Wigderson, Monotone circuits for connectivity require super-logarithmic depth, in: *Proc. 20th Annual ACM Symp. on Theory of Computing* (STOC) (1988), 539–550. p. 367

[265] M. Karchmer and A. Wigderson, On span programs, in: *Proc. Conf. on 8th Structure in Complexity Theory*, IEEE (1993), 102–111. p. 407

[266] R. M. Karp, Reducibility among combinatorial problems, in: *Complexity of Computer Computations*, eds. R. E. Miller and J. W. Thatcher, Plenum (1972), 85–103. pp. 37, 121

[267] R. M. Karp and R. J. Lipton, Some connections between nonuniform and uniform complexity classes, in: *Proc. 12th Annual ACM Symp. on Theory of Computing* (STOC) (1980), 302–309. p. 38

[268] S. Kleene, *Introduction to Metamathematics*, North Holland (1952). p. 62

[269] J. Köbler, J. Messner, Complete problems for promise classes by optimal proof systems for test sets, in: *Proc. 13th Annual IEEE Conf. on Computational Complexity* (CCC 98) (1998), 132–140. p. 468

[270] J. Köbler, J. Messner and J. Torán, Optimal proof systems imply complete sets for promise classes, *Infinite Computation*, **184(1)** (2003), 71–92. p. 469

[271] L. Kolodziejczyk, P. Nguyen and N. Thapen, The provably total NP search problems of weak second-order bounded arithmetic, *Ann. Pure and Applied Logic*, **162** (2011), 419–446. p. 436

[272] J. Krajíček, On the number of steps in proofs, *Ann. Pure and Applied Logic*, **41** (1989), 153–178. pp. 60, 62

[273] J. Krajíček, Speed-up for propositional Frege systems via generalizations of proofs, *Comments. Mathematicae Universitas Carolinae*, **30(1)** (1989), 137–140. p. 62

[274] J. Krajíček, Exponentiation and second-order bounded arithmetic, *Ann. Pure and Applied Logic*, **48(3)** (1990), 261–276. pp. 196, 258

[275] J. Krajíček, Fragments of bounded arithmetic and bounded query classes, *Trans. AMS*, **338(2)** (1993), 587–598. pp. 304, 305

[276] J. Krajíček, Lower bounds to the size of constant-depth propositional proofs, *J. Symbolic Logic*, **59(1)** (1994), 73–86. pp. 3, 43, 80, 209, 210, 292, 300, 301, 304, 334, 381

[277] J. Krajíček, On Frege and extended Frege proof Systems. in: *Feasible Mathematics II*. eds. P. Clote and J. Remmel, Birkhauser (1995), 284–319. pp. 209, 335, 454

[278] J. Krajíček, *Bounded Arithmetic, Propositional Logic, and Complexity Theory*, Encyclopedia of Mathematics and Its Applications, **60**, Cambridge University Press (1995). pp. 6, 60, 61, 62, 80, 87, 112, 114, 170, 179, 183, 196, 203, 205, 209, 210, 230, 231, 236, 257, 258, 291, 292, 301, 304, 335, 336, 441, 454, 478

[279] J. Krajíček, A fundamental problem of mathematical logic, *Ann. Kurt Gödel Society, Collegium Logicum*, **2** (1996), 56–64. p. 6

[280] J. Krajíček, Interpolation theorems, lower bounds for proof systems, and independence results for bounded arithmetic, *J. Symbolic Logic*, **62(2)** (1997), 457–486. pp. 80, 107, 113, 157, 293, 371, 381, 382, 409

[281] J. Krajíček, On methods for proving lower bounds in propositional logic, in: *Logic and Scientific Methods*, eds. M. L. Dalla Chiara *et al.*, Kluwer Academic (1997), 69–83. p. 6

[282] J. Krajíček, Lower bounds for a proof system with an exponential speed-up over constant-depth Frege systems and over polynomial calculus, in: *Proc. 22nd International Symp. on Mathematical Foundations of Computer Science*, eds. I. Prívara and P. Růžička, Lecture Notes in Computer Science, **1295**, Springer (1997), 85–90. pp. 157, 335, 336

[283] J. Krajíček, Interpolation by a game, *Mathematical Logic Quarterly*, **44(4)** (1998), 450–458. pp. 405, 406

[284] J. Krajíček, Discretely ordered modules as a first-order extension of the cutting planes proof system, *J. Symbolic Logic*, **63(4)** (1998), 1582–1596. pp. 157, 158, 406, 408, 478

[285] J. Krajíček, Extensions of models of PV, in: *Proc: Logic Colloq.'95*, eds. J. A. Makowsky and E. V. Ravve, ASL/Springer Series Lecture Notes in Logic, **11** (1998), 104–114. p. 454

[286] J. Krajíček, On the degree of ideal membership proofs from uniform families of polynomials over a finite field, *Illinois J. Mathematics*, **45(1)** (2001), 41–73. p. 349

[287] J. Krajíček, Uniform families of polynomial equations over a finite field and structures admitting an Euler characteristic of definable sets, *Proc. London Mathematical Society*, **3(81)** (2000), 257–284. pp. 349, 455

[288] J. Krajíček, On the weak pigeonhole principle, *Fundamenta Mathematicae*, **170(1–3)** (2001), 123–140. pp. 108, 113, 209, 291, 335, 437

[289] J. Krajíček, Tautologies from pseudo-random generators, *Bull. Symbolic Logic*, **7(2)** (2001), 197–212. pp. 437, 439, 480
[290] J. Krajíček, Dehn function and length of proofs, *International J. Algebra and Computation*, **13(5)** (2003), 527–542. p. 160
[291] J. Krajíček, Dual weak pigeonhole principle, pseudo-surjective functions, and provability of circuit lower bounds, *J. Symbolic Logic*, **69(1)** (2004), 265–286. pp. 157, 258, 292, 293, 433, 437, 439, 440
[292] J. Krajíček, Diagonalization in proof complexity, *Fundamenta Mathematicae*, **182** (2004), 181–192. pp. 158, 439, 464, 470
[293] J. Krajíček, Implicit proofs, *J. Symbolic Logic*, **69(2)** (2004), 387–397. pp. 92, 158, 258, 474
[294] J. Krajíček, Combinatorics of first-order structures and propositional proof systems, *Archive for Mathematical Logic*, **43(4)** (2004), 427–441. pp. 159, 291, 455
[295] J. Krajíček, Hardness assumptions in the foundations of theoretical computer science, *Archive for Mathematical Logic*, **44(6)** (2005), 667–675. p. 480
[296] J. Krajíček, Structured pigeonhole principle, search problems and hard tautologies, *J. Symbolic Logic*, **70(2)** (2005), 619–630. pp. 291, 335, 436, 439
[297] J. Krajíček, Proof complexity, in: *Proc. European Congress of Mathematics* (ECM), ed. A. Laptev, European Mathematical Society (2005), 221–231. p. 6
[298] J. Krajíček, Substitutions into propositional tautologies, *Information Processing Letters*, **101(4)** (2007), 163–167. pp. 439, 441
[299] J. Krajíček, An exponential lower bound for a constraint propagation proof system based on ordered binary decision diagrams, *J. Symbolic Logic*, **73(1)** (2008), 227–237. pp. 159, 408
[300] J. Krajíček, A proof complexity generator, in: *Proc. 13th International Congress of Logic, Methodology and Philosophy of Science (Beijing, 2007)*, eds. C. Glymour, W. Wang, and D. Westerstahl (2009), 185–190. pp. 430, 437, 439
[301] J. Krajíček, A form of feasible interpolation for constant depth Frege systems, *J. Symbolic Logic*, **75(2)** (2010), 774–784. p. 410
[302] J. Krajíček, On the proof complexity of the Nisan–Wigderson generator based on a hard NP ∩ coNP function, *J. Mathematical Logic*, **11(1)** (2011), 11–27. pp. 383, 439
[303] J. Krajíček, A note on propositional proof complexity of some Ramsey-type statements, *Archive for Mathematical Logic*, **50(1–2)** (2011), 245–255. pp. 291, 292, 335
[304] J. Krajíček, *Forcing with Random Variables and proof Complexity*, London Mathematical Society Lecture Note Series, **382**, Cambridge University Press (2011). pp. 6, 161, 210, 336, 437, 439, 452, 453, 454, 480
[305] J. Krajíček, A note on SAT algorithms and proof complexity, *Information Processing Letters*, **112** (2012), 490–493. p. 478
[306] J. Krajíček, Pseudo-finite hard instances for a student–teacher game with a Nisan–Wigderson generator, *Logical Methods in Computer Science*, **8(3)** (2012), 1–8. pp. 439, 454
[307] J. Krajíček, A saturation property of structures obtained by forcing with a compact family of random variables, *Archive for Mathematical Logic*, **52(1)** (2013), 19–28.
[308] J. Krajíček, On the computational complexity of finding hard tautologies, *Bulletin of the London Mathematical Society*, **46(1)** (2014), 111–125. pp. 468, 479
[309] J. Krajíček, A reduction of proof complexity to computational complexity for $AC^0[p]$ Frege systems, *Proceedings of the AMS*, **143(11)** (2015), 4951–4965. p. 336
[310] J. Krajíček, Consistency of circuit evaluation, extended resolution and total NP search problems, *Forum of Mathematics, Sigma*, **4** (2016), e15. DOI: 10.1017/fms.2016.13. pp. 158, 436, 437

[311] J. Krajíček, Expansions of pseudofinite structures and circuit and proof complexity, in: *Liber Amicorum Alberti*, eds. J. van Eijck, R. Iemhoff and J. J. Joosten, Tributes Series **30**, College Publications, London (2016), 195–203. p. 454

[312] J. Krajíček, Randomized feasible interpolation and monotone circuits with a local oracle, preprint (2016). pp. 157, 402, 408, 409

[313] J. Krajíček and I. C. Oliveira, Unprovability of circuit upper bounds in Cook's theory PV, *Logical Methods in Computer Science*, **13(1)** (2017). p. 479

[314] J. Krajíček and I. C. Oliveira, On monotone circuits with local oracles and clique lower bounds, *Chicago J. Theoretical Computer Science*, to appear. p. 409

[315] J. Krajíček and P. Pudlák, The number of proof lines and the size of proofs in first order logic, *Archive for Mathematical Logic*, 27 (1988), 69–84. p. 62

[316] J. Krajíček and P. Pudlák, On the structure of initial segments of models of arithmetic, *Archive for Mathematical Logic*, **28(2)** (1989), 91–98. pp. 163, 259

[317] J. Krajíček and P. Pudlák, Propositional proof systems, the consistency of first-order theories and the complexity of computations, *J. Symbolic Logic*, **54(3)** (1989), 1063–1079. pp. 34, 35, 38, 62, 435, 462, 468

[318] J. Krajíček and P. Pudlák, Quantified propositional calculi and fragments of bounded arithmetic, *Zeitschr. f. Mathematikal Logik u. Grundlagen d. Mathematik*, **36(1)** (1990), 29–46. pp. 248, 258

[319] J. Krajíček and P. Pudlák, Propositional provability in models of weak arithmetic, in: *Proc. Conf. on Computer Science Logic (Kaiserlautern '89)*, eds. E. Boerger, H. Kleine-Bunning and M. M. Richter, Lecture Notes in Computer Science, **440**, Springer (1990), 193–210. p. 454

[320] J. Krajíček and P. Pudlák, Some consequences of cryptographical conjectures for S^1_2 and *EF*, in: *Proc. Meeting on Logic and Computational Complexity* (Indianapolis, 1994), ed. D. Leivant, Lecture Notes in Computer Science, **960**, Springer (1995), 210–220. p. 382

[321] J. Krajíček and P. Pudlák, Some consequences of cryptographical conjectures for S^1_2 and *EF*, *Information and Computation*, **140 (1)** (1998), 82–94. pp. 382, 404, 409

[322] J. Krajíček, P. Pudlák, and J. Sgall, Interactive computations of optimal solutions, in: *Mathematical Foundations of Computer Science*, ed. B. Rovan, Lecture Notes in Computer Science, **452**, Springer (1990), 48–60. p. 258

[323] J. Krajíček, P. Pudlák and G. Takeuti, Bounded arithmetic and the polynomial hierarchy, *Ann. Pure and Applied Logic*, **52** (1991), 143–153. pp. 235, 239, 257, 258, 479

[324] J. Krajíček, P. Pudlák, and A. Woods, An exponential lower bound to the size of bounded depth Frege proofs of the pigeonhole principle, *Random Structures and Algorithms*, **7(1)** (1995), 15–39. pp. 318, 334, 335

[325] J. Krajíček, A. Skelley and N. Thapen, NP search problems in low fragments of bounded arithmetic, *J. Symbolic Logic*, **72(2)** (2007), 649–672. pp. 292, 436, 437

[326] J. Krajíček and G. Takeuti, On bounded $\sum^1_1$-polynomial induction, in: *Feasible Mathematics*, eds. S. R. Buss and P. J. Scott, Birkhauser (1990), 259–280. pp. 248, 258

[327] J. Krajíček and G. Takeuti, On induction-free provability, *Ann. Mathematics and Artificial Intelligence*, **6** (1992), 107–126. p. 435

[328] G. Kreisel, Technical report NO. 3, Applied Mathematics and Statistics Laboratories, Stanford University, unpublished (1961). p. 381

[329] B. Krishnamurthy and R. N. Moll, Examples of hard tautologies in the propositional calculus, in: *Proc. 13th Annual ACM Symp. on Theory of Computing* (STOC) (1981), 28–37. pp. 112, 291

[330] J. L. Krivine, Anneaux préordonnés, *J. d'Analyse Mathématique*, **12** (1964), 307–326. pp. 115, 132

[331] O. Kullmann, On a generalization of extended resolution, *Discrete Applied Mathematics*, **96-97** (1999), 149–176. p. 295

[332] O. Kullmann, Upper and lower bounds on the complexity of generalized resolution and generalized constraint satisfaction problems, *Ann. Mathematics and Artificial Intelligence*, **40(3–4)** (2004), 303–352. p. 295

[333] J. C. Lagarias, An elementary problem equivalent to the Riemann hypothesis, *American Mathematical Monthly*, **109(6)** (2002), 534–543. p. 38

[334] J. B. Lasserre, Global optimization with polynomials and the problem of moments, *SIAM J. Optimization*, **11(3)** (2001), 796–817. p. 131

[335] M. Laurent, A comparison of the Sherali–Adams, Lovász–Schrijver and Lasserre relaxations for 0-1 programming, *Mathematics of Operations Research*, **28** (2001), 470–496. p. 132

[336] M. Lauria, Short *Res** (*polylog*) refutations if and only if narrow *Res* refutations, unpublished notes available at `www.dsi.uniroma1.it/~lauria`, (2011). pp. 114, 209

[337] M. Lauria and J. Nordström, Tight size–degree bounds for sums-of-squares proofs, in: *Proc. 30th Conf. on Computational Complexity* (CCC'15), ed. D. Zuckerman (2015), 448–466. p. 352

[338] Dai Tri Man Le, *Bounded Arithmetic and Formalizing Probabilistic Proofs*, Ph.D. thesis, University of Toronto (2014). p. 479

[339] T. Lee and A. Shraibman, Disjointness is hard in the multi-party number-on-the-forehead model, *Computational Complexity*, **18(2)** (2009), 309–336. p. 407

[340] L. A. Levin, Universal sequential search problems, *Problems of Information Transmission* (translated from Problemy Peredachi Informatsii (Russian)), **9(3)** (1973), 115–116. p. 38

[341] L. Lovász, M. Naor, I. Newman and A. Wigderson, Search problems in the decision tree model, in: *Proc. 32nd IEEE Symp. on Foundations of Computer Science* (FOCS), (1991), 576–585. p. 112

[342] L. Lovász and A. Schrijver, Cones of matrices and set-functions and 0-1 optimization, *SIAM J. Optimization*, **1** (1991), 166–190. pp. 115, 130, 131

[343] J. Lukasiewicz, *Elements of Mathematical Logic*, Pergamon Press (1963). p. 41

[344] O. B. Lupanov, The synthesis of contact circuits, *Dokl. Akad. Nauk SSSR (NS)*, **119** (1958), 23–26. p. 38

[345] R. Lyndon, An interpolation theorem in the predicate calculus, *Pacific J. Mathematics*, **9(1)** (1959), 129–142. pp. 14, 37

[346] R. Lyndon and P. Schupp, *Combinatorial Group Theory*, Springer (1997). p. 160

[347] A. Maciel and T. Pitassi, Towards lower bounds for bounded-depth Frege proofs with modular connectives, in: *Proof Complexity and Feasible Arithmetics*, eds. P. Beame and S. Buss, DIMACS Series, **39** (1998), 195–227. p. 335

[348] A. Maciel, T. Pitassi and A. Woods, A new proof of the weak pigeonhole principle, *J. Computer Systems Sciences*, **64** (2002), 843–872. pp. 220, 231

[349] J. Maly, Jan Krajíček's Forcing Construction and Pseudo Proof Systems, M.Sc. Thesis, University of Vienna (2016). p. 161

[350] J. Maly and M. Müller, Pseudo proof systems and hard instances of SAT, preprint 2017. p. 161

[351] A. Maté, Nondeterministic polynomial-time computations and models of arithmetic, *J. ACM*, **37(1)** (1990), 175–193. p. 454

[352] J. Messner, On optimal algorithms and optimal proof systems, in: *Proc. Symp. on Theoretical Aspects of Computer Science* (STACS 1999), eds. C. Meinel and S. Tison, Lecture Notes in Computer Science, **1563** (1999), 541–550. p. 468

[353] J. Messner and J. Torán, Optimal proof systems for propositional logic and complete sets, in: *Proc. Symp. on Theoretical Aspects of Computer Science* (STACS 1998), eds. M. Morvan, C. Meinel, and D. Krob , Lecture Notes in Computer Science, **1373** (1998), 477–487. p. 468

[354] O. Mikle-Barát, Strong proof systems, M.Sc. thesis, Charles University, Prague (2007). pp. 159, 408

[355] M. Mikša and J. Nordström, A generalized method for proving polynomial calculus degree lower bounds, in: *Proc. 30th Annual Computational Complexity Conference* (CCC 2015), *Leibniz International Proceedings in Informatics*, **33** (2015), 467–487. pp. 350, 351

[356] T. Morioka, Logical approaches to the complexity of search problems: proof complexity, quantified propositional calculus, and bounded arithmetic, Ph.D. thesis, University of Toronto (2005). p. 91

[357] A. A. Muchnik, On two approaches to the classification of recursive functions (in Russian), in: *Problems of Mathematical Logic, Complexity of Algorithms and Classes of Computable Functions*, eds. V. A. Kozmidiadi and A. A. Muchnik (1970), 123–138. p. 257

[358] M. Müller and J. Pich, Feasibly constructive proofs of succinct weak circuit lower bounds, in: *Proc. Electronic Colloq. on Computational Complexity* (ECCC), TR17-144 (2017). pp. 441, 474, 478

[359] S. Müller, Polylogarithmic cuts in models of V^0, *Logical Methods in Computer Science*, **9(1:16)** (2013). p. 231

[360] S. Müller and I. Tzameret, Short propositional refutations for dense random 3CNF formulas, *Ann. Pure and Applied Logic*, **165(12)** (2014), 1864–1918. pp. 160, 471

[361] D. Mundici, Complexity of Craig's interpolation, *Fundamenta Informaticae*, **5** (1982), 261–278. p. 380

[362] D. Mundici, A lower bound for the complexity of Craig's interpolants in sentential logic, *Archiv fur Math. Logik*, **23** (1983), 27–36. p. 380

[363] D. Mundici, Tautologies with a unique Craig interpolant, uniform vs. non-uniform complexity, *Ann. Pure and Applied Logic*, **27** (1984), 265–273. p. 380

[364] D. Mundici, NP and Craig's interpolation theorem, in: *Proc. Logic Colloq. 1982*, North-Holland (1984), 345–358. p. 380

[365] V. Nepomnjascij, Rudimentary predicates and Turing calculations, *Soviet Math. Dokl.*, **6** (1970), 1462–1465. p. 182

[366] N. Nisan, The communication complexity of the threshold gates, in: *Combinatorics, P. Erdös is Eighty*, **1**, eds. Miklós *et. al.*, Bolyai Mathematical Society (1993), 301–315. p. 388

[367] N. Nisan and A. Wigderson, Hardness vs. randomness, *J. Computer System Sciences*, **49** (1994), 149–167. pp. 432, 433, 438, 439

[368] J. Nordström, Short proofs may be spacious: understanding space in resolution, Ph.D. thesis, The Royal Institute of Technology, Stockholm (2008). p. 294

[369] J. Nordström, Narrow proofs may be spacious: separating space and width in resolution, *SIAM J. Computing*, **39(1)** (2009), 59–121. pp. 113, 294

[370] J. Nordström, On the relative strength of pebbling and resolution, *ACM Trans. Computational Logic*, **13(2)**, article 16 (2012). p. 294

[371] J. Nordström, Pebble games, proof complexity and time-space trade-offs, *Logical Methods in Computer Science*, **9(15)** (2013), 1–63. p. 294

[372] J. Nordström, On the interplay between proof complexity and SAT solving, *ACM SIGLOG News*, **2(3)** (2015), 19–44. p. 294

[373] J. Nordström and J. Hastad, Towards an optimal separation of space and length in resolution, *Theory of Computing*, **9**, article 14 (2013), 471–557. p. 294

[374] P. S. Novikov, On the algorithmic unsolvability of the word problem in group theory, *Trudy Mat. Inst. Steklova*, **44** (1955), 143. p. 153

[375] R. O'Donnell, SOS is not obviously automatizable, even approximately, in: *Proc. Conf. on Innovations in Theoretical Computer Science* (ITCS) (2017), 59:1–59:10. p. 471

[376] M. de Oliveira and P. Pudlák, Representations of monotone Boolean functions by linear programs, in: *Proc. 32nd Computational Complexity Conf.* (CCC 2017), Leibniz International Proceedings in Informatics, **79** (2017), 3:1–3:14. pp. 395, 407

[377] C. Papadimitriou, *Computational Complexity*, Addison Wesley (1994). p. 37

[378] C. Papadimitriou, The complexity of the parity argument and other inefficient proofs of existence, *J. Computer and System Sciences*, **48(3)** (1994), 498–532. p. 436

[379] R. Parikh, Existence and feasibility in arithmetic, *J. Symbolic Logic*, **36** (1971), 494–508. pp. 163, 165, 166, 182

[380] J. B. Paris, O struktuře modelu omezené E_1 indukce (in Czech), *Časopis pěstování matematiky*, **109** (1984), 372–379. p. 182

[381] J. Paris and C. Dimitracopoulos, Truth definitions for Δ_0 formulas, in: *Logic and Algorithmic, l'Enseignement Mathematique*, **30** (1982), 318–329. p. 182

[382] J. Paris and C. Dimitracopoulos, A note on undefinability of cuts, *J. Symbolic Logic*, **48** (1983), 564–569. p. 182

[383] J. B. Paris, W. G. Handley and A. J. Wilkie, Characterizing some low arithmetic classes, *Colloquia Mathematica Soc. J. Bolyai*, **44** (1984), 353–364. p. 182

[384] J. B. Paris and L. Harrington, A mathematical incompleteness in Peano Arithmetic, in: *Handbook of Mathematical Logic*, ed. J. Barwise, North-Holland (1977). p. 414

[385] J. B. Paris and L. Kirby, Σ_n-collection schemes in arithmetic, in: *Proc. Logic Colloq. '77*, North-Holland (1978), 199–209. p. 182

[386] J. B. Paris and A. J. Wilkie, Models of arithmetic and rudimentary sets, *Bull. Soc. Mathem. Belg.,* **B33** (1981), 157–169. pp. 182, 478

[387] J. B. Paris and A. J. Wilkie, Δ_0 sets and induction, in: *Proc. Jadwisin Logic Conf.*, (1983), 237–248. pp. 182, 183, 478

[388] J. B. Paris and A. J. Wilkie, Some results on bounded induction, in: *Proc. 2nd Easter Conf. on Model Theory* (1984), 223–228. pp. 182, 478

[389] J. Paris and A. J. Wilkie, Counting problems in bounded arithmetic, in: *Methods in Mathematical Logic*, Lecture Notes in Mathematics, **1130** (1985), 317–340. pp. 3, 61, 163, 165, 167, 169, 170, 182, 183, 209, 210, 221, 231, 258, 307, 336, 454, 478

[390] J. B. Paris and A. J. Wilkie, Counting Δ_0 sets, *Fundamenta Mathematica*, **127** (1987), 67–76. pp. 183, 259, 478

[391] J. B. Paris and A. J. Wilkie, On the scheme of induction for bounded arithmetic formulas, *Ann. Pure and Applied Logic*, **35** (1987), 261–302. pp. 166, 183, 435, 478, 480

[392] J. Paris, A. J. Wilkie and A. Woods, Provability of the pigeonhole principle and the existence of infinitely many primes, *J. Symbolic Logic*, **53(4)** (1988), 1235–1244. pp. 38, 183, 218, 220, 231, 259, 336

[393] S. Perron, Examining fragments of the quantified propositional calculus *J. Symbolic Logic*, **73(3)** (2008), 1051–1080. p. 91

[394] S. Perron, Power of non-uniformity in proof complexity, Ph.D. thesis, University of Toronto (2009). p. 91

[395] J. Pich, Hard tautologies, M.Sc. thesis, Charles University, Prague (2011). p. 439

[396] J. Pich, Nisan–Wigderson generators in proof systems with forms of interpolation, *Mathematical Logic Quarterly*, **57(4)** (2011), 379–383. p. 439
[397] J. Pich, *Complexity Theory in Feasible Mathematics*, Ph.D. thesis, Charles University, Prague (2014). p. 478
[398] J. Pich, Circuit lower bounds in bounded arithmetic, *Ann. Pure and Applied Logic*, **166(1)** (2015), 29–45. p. 479
[399] J. Pich, Logical strength of complexity theory and a formalization of the PCP theorem in bounded arithmetic, *Logical Methods in Computer Science*, **11(2)** (2015). p. 478
[400] T. Pitassi, P. Beame, and R. Impagliazzo, Exponential lower bounds for the pigeonhole principle, *Computational Complexity*, **3** (1993), 97–308. pp. 318, 334, 335
[401] T. Pitassi and R. Raz, Regular resolution lower bounds for the weak pigeonhole principle, *Combinatorica*, **24(3)** (2004), 503–524. p. 293
[402] T. Pitassi and R. Santhanam, Effective polynomial simulations, in: *Proc. 1st Symp. on Innovations in CS* (2010), 370–382. pp. 38, 469
[403] T. Pitassi and N. Segerlind, Exponential lower bounds and integrality gaps for tree-like Lovász–Schrijver procedures, *SIAM J. Computing*, **41(1)** (2012), 128–159. p. 349
[404] T. Pitassi and A. Urquhart, The complexity of the Hajos calculus, *SIAM J. Discrete Mathematics*, **8(3)** (1995), 464–483. pp. 155, 160
[405] V. R. Pratt, Every prime has a succinct certificate, *SIAM J. Computing*, **4** (1975), 214–220. p. 409
[406] D. Prawitz, *Natural Deduction. A Proof-Theoretic Study*, Stockholm (1965). p. 79
[407] P. Pudlák, A definition of exponentiation by bounded arithematic formula, *Comment. Mathematicae Universitas Carolinae*, **24(4)** (1983), 667–671. p. 183
[408] P. Pudlák, Cuts, consistency statements and interpretations, *J. Symbolic Logic*, **50** (1985), 423–441. pp. 259, 470
[409] P. Pudlák, On the length of proofs of finitistic consistency statements in first-order theories, in: *Proc. Logic Colloquium 84*, North Holland (1986), 165–196. pp. 259, 461, 462
[410] P. Pudlák, Improved bounds to the length of proofs of finitistic consistency statements, *Contemporary Mathematics*, **65** (1987), 309–331. pp. 259, 469, 470
[411] P. Pudlák, Ramsey's theorem in bounded arithmetic, in: *Proc. Computer Science Logic '90*, eds. E. Borger, H. Kleine Buning, M. M. Richter and W. Schonfeld, Lecture Notes in Computer Science, **533** (1991), 308–317. p. 292
[412] P. Pudlák, Lower bounds for resolution and cutting planes proofs and monotone computations, *J. Symbolic Logic*, **62(3)** (1997), 981–998. pp. 378, 382, 386, 387, 405
[413] P. Pudlák, The lengths of proofs, in: *Handbook of Proof Theory*, ed. S. R. Buss, Elsevier (1998), 547–637. pp. 6, 382
[414] P. Pudlák, On the complexity of propositional calculus, in: *Sets and Proofs, Proc. Logic Colloq. '97*, Cambridge University Press (1999), 197–218. pp. 132, 157, 158, 395, 407
[415] P. Pudlák, Proofs as games, *American Math. Monthly* (June–July 2000), 541–550. p. 293
[416] P. Pudlák, On reducibility and symmetry of disjoint NP-pairs, *Theoretical Computer Science*, **295** (2003), 323–339. pp. 383, 468, 471
[417] P. Pudlák, Consistency and games – in search of new combinatorial principles, in: *Proc. Logic Colloq. '03*, eds. V. Stoltenberg-Hansen and J. Vaananen, Association for Symbolic Logic (2006), 244–281. p. 436
[418] P. Pudlák, Twelve problems in proof complexity, in: *Proc. 3rd International Computer Science Symp. in Russia* (CSR) (2008), 13–27. p. 6

[419] P. Pudlák, Quantum deduction rules, *Ann. Pure and Applied Logic*, **157** (2009), 16–29. p. 160

[420] P. Pudlák, A lower bound on the size of resolution proofs of the Ramsey theorem, *Information Processing Letters*, **112(14–15)** (2012), 610–611. p. 292

[421] P. Pudlák, *Logical Foundations of Mathematics and Computational Complexity, A Gentle Introduction*, Springer (2013). pp. 6, 480

[422] P. Pudlák, Incompleteness in the finite domain, *Bulletin of Symbolic Logic*, **23(4)** (2017), 405–441. p. 469

[423] P. Pudlák and J. Sgall, Algebraic models of computation, and interpolation for algebraic proof systems, in: *Proof Complexity and Feasible Arithmetic*, ed. S. Buss, DIMACS Series **39** (1998), 279–295. pp. 392, 407

[424] P. Pudlák and N. Thapen, Random resolution refutations, preprint, in: *Proc. 32nd Computational Complexity Conf.* (CCC 2017), Leibniz International Proceedings in Informatics, **79** (2017), 1:1–1:10. pp. 156, 160, 293

[425] P. Raghavendra and B. Weitz, On the bit-complexity of sum-of-squares proofs, in: *Proc. 44th International Colloq. on Automata, Languages, and Programming* (ICALP) (2017), 80:1–80:13. p. 471

[426] R. Raz, Resolution lower bounds for the weak pigeonhole principle, *J. Association for Computing Machinery*, **51(2)** (2004), 115–138. p. 292

[427] R. Raz and I. Tzameret, Resolution over linear equations and multilinear proofs, *Ann. Pure and Applied Logic*, **155(3)** (2008), 194–224. p. 157

[428] R. Raz and A. Wigderson, Probabilistic communication complexity of Boolean relations, in: *Proc. IEEE 30th Annual Symp. on Foundations of Computer Science* (FOCS) (1989), 562–567. pp. 389, 408

[429] R. Raz and A. Wigderson, Monotone circuits for matching require linear depth, *J. ACM*, **39(3)** (1992), 736–744. pp. 360, 382, 389

[430] A. A. Razborov, Lower bounds for the monotone complexity of some Boolean functions, *Doklady Akademii Nauk SSSR*, **281(4)** (1985), 798–801. English translation in *Soviet Math. Doklady*, **31** (1985), 354–357. pp. 30, 288, 382

[431] A. A. Razborov, Lower bounds on the size of bounded depth networks over a complete basis with logical addition, *Matem. Zametki*, **41(4)** (1987), 598–607. pp. 30, 329

[432] A. A. Razborov, An equivalence between second-order bounded domain bounded arithmetic and first order bounded arithmetic, in: *Arithmetic, Proof Theory and Computational Complexity*, eds. P. Clote and J. Krajíček, Oxford University Press (1993), 247–277. p. 196

[433] A. A. Razborov, On provably disjoint NP-pairs, Basic Research in Computer Science Center, Aarhus, RS-94-36 (1994), unpublished report. p. 382

[434] A. A. Razborov, Bounded arithmetic and lower bounds in Boolean complexity, in: *Feasible Mathematics II*, eds. P. Clote and J. Remmel, Birkhauser (1995), 344–386. p. 478

[435] A. A. Razborov, Unprovability of lower bounds on the circuit size in certain fragments of bounded arithmetic, *Izvestiya RAN.*, **59(1)** (1995), 201–224. pp. 368, 381, 382, 440, 478

[436] A. A. Razborov, Lower bounds for the polynomial calculus, *Computational Complexity*, **7(4)** (1998), 291–324. pp. 343, 350

[437] A. A. Razborov, Improved resolution lower bounds for the weak pigeonhole principle, in: *Proc. Electronic Colloq. on Computational Complexity*, TR01-055 (2001). p. 293

[438] A. A. Razborov, Resolution lower bounds for the weak functional pigeonhole principle, *Theoretical Computer Science*, **303(1)** (2003), 233–243. p. 293

[439] A. A. Razborov, Resolution lower bounds for perfect matching principles, *J. Computer and System Sciences*, **69(1)** (2004), 3–27. pp. 293, 438, 439

[440] A. A. Razborov, Pseudorandom generators hard for *k*-DNF resolution polynomial calculus resolution, *Ann. Mathematics*, **181(2)** (2015), 415–472. pp. 438, 439, 440

[441] A. A. Razborov, On space and depth in resolution, preprint (2016). p. 113

[442] A. A. Razborov, Proof Complexity and Beyond, *SIGACT News*, **47(2)**, (2016), 66–86. p. 6

[443] A. A. Razborov, On the width of semi-algebraic proofs and algorithms, *Mathematics of Operations Research*, **42(4)** (2017), 1106–1134. p. 352

[444] A. A. Razborov and S. Rudich, Natural proofs, *J. Computer System Sciences*, **55(1)** (1997), 24–35. pp. 381, 433, 478

[445] R. A. Reckhow, On the lengths of proofs in the propositional calculus, Ph.D. thesis, University of Toronto (1976). pp. 52, 62, 79

[446] S. Riis, Independence in bounded arithmetic, D.Phil. thesis, Oxford University (1993). pp. 291, 454

[447] S. Riis, *Count(q)* does not imply *Count(p)*, *Ann. Pure and Applied Logic*, **90** (1997), 1–56. p. 230

[448] S. Riis, A complexity gap for tree-resolution, *Computational Complexity*, **10(3)** (2001), 179–209. p. 291

[449] R. Ritchie, Classes of predictably computable functions. *Trans. AMS*, **106** (1963), 139–173. p. 182

[450] M. Rivest, A. Shamir and L. Adleman, A method of obtaining digital signatures and public-key cryptosystems, *ACM Communications*, **21** (1978), 120–126. p. 364

[451] R. Robere, T. Pitassi, B. Rossman and S. A. Cook, Exponential lower bounds for monotone span programs, in: *Proc. 57th IEEE Symp. on Foundations of Computer Science* (FOCS) (2016), 406–415. pp. 393, 407

[452] A. Robinson, On ordered fields and definite functions, *Math. Ann.*, **130** (1955), 257–271. p. 132

[453] J. A. Robinson, A machine-oriented logic based on the resolution principle, *J. ACM* (1965), **12(1)**, 23–41. p. 112

[454] S. Rudich, Super-bits, demi-bits, and NP/*qpoly*-natural proofs, in: *Proc. 1st Int. Symp. on Randomization and Approximation Techniques in Computer Science*, Lecture Notes in Computer Science, **1269** (1997), 85–93. p. 437

[455] Z. Sadowski, On an optimal deterministic algorithm for SAT, in: *Computer Science Logic*, G. Gottlob E. Grandjean and K. Seyr, Lecture Notes in Computer Science, **1584** (1998), 179–187. p. 468

[456] Z. Sadowski, Optimal proof systems, optimal acceptors and recursive presentability, *Fundam. Inform.*, **79(1–2)** (2007), 169–185. p. 469

[457] P. Šanda, Implicit propositional proofs, M.Sc. thesis, Charles University, Prague (2006). p. 158

[458] J. E. Savage, Computational work and time on finite machines, *J. ACM*, **19(4)** (1972), 660–674. p. 28

[459] W. J. Savitch, Relationships between nondeterministic and deterministic tape complexities, *J. Computer and System Sciences*, **4** (1970), 177–192. p. 26

[460] C. Scheiderer, Sums of squares of polynomials with rational coefficients, *J. EMS*, **18(7)** (2016), 1495–1513. p. 231

[461] G. Schoenebeck, Linear level Lasserre lower bounds for certain *k*-CSPs, in: *Proc. 49th IEEE Symp. on Foundations of Computer Science* (FOCS) (2008), 593–602. pp. 347, 351

[462] H. Scholz, Ein ungel ostes Problem in der symbolischen Logik, *J. Symbolic Logic*, **17** (1952), 160. p. 2
[463] D. Scott, A proof of the independence of the continuum hypothesis, *Mathematical Systems Theory*, **1** (1967), 89–111. p. 196
[464] N. Segerlind, New separations in propositional proof complexity, Ph.D. thesis, University of California, San Diego (2003). pp. 336, 408
[465] N. Segerlind, Nearly exponential size lower bounds for symbolic quantifier elimination algorithms and OBDD-based proofs of unsatisfiability, in: *Proc. Electronic Colloq. on Computational Complexity*, TR07-009 (2007). p. 408
[466] C. E. Shannon, The synthesis of two-terminal switching circuits, *Bell System Technology J.*, **28** (1949), 59–98. p. 28
[467] H. D. Sherali and W. P. Adams, A hierarchy of relaxations and convex hull characterizations for mixed-integer 0–1 programming problems, *Discrete Applied Mathematics*, **52(1)** (1994), 83–106. p. 131
[468] M. Sipser, The history and status of the P versus NP question, in: *Proc. 24th Annual ACM Symp. on Theory of Computing* (STOC) (1992), 603–618. pp. 2, 38
[469] M. Sipser, *Introduction to the Theory of Computation*, Cengage Learning, 3rd ed. (2005). p. 37
[470] J. R. Shoenfield, *Mathematical Logic*, Association for Symbolic Logic (1967). pp. 37, 159, 258
[471] A. Skelley, Propositional PSPACE reasoning with Boolean programs versus quantified Boolean formulas, in: *Proc. 31st International Colloq. on Automata, Languages and Programming* (ICALP), Springer Lecture Notes in Computer Science, **3142** (2004), 1163–1175. p. 92
[472] A. Skelley, Theories and proof systems for PSPACE and the EXP-time hierarchy, Ph.D. thesis, University of Toronto (2005). p. 92
[473] A. Skelley and N. Thapen, The provably total search problems of bounded arithmetic, *Proc. London Mathematical Society*, **103(1)** (2011), 106–138. p. 305
[474] R. Smolensky, Algebraic methods in the theory of lower bounds for Boolean circuit complexity, in: *Proc. 19th Annnal ACM Symp. on Theory of Computing* (STOC) (1987), 77–82. pp. 30, 329
[475] C. Smorynski, The incompleteness theorem, in: *Handbook of Mathematical Logic*, ed. J. Barwise, Studies in Logic and the Foundations of Mathematics, North Holland (1989), 821–866. pp. 259, 469
[476] R. M. Smullyan, *Theory of Formal Systems*, Annals of Mathematical Studies, **47**, Princeton University Press (1961). pp. 2, 182, 196
[477] R. M. Smullyan, *First-Order Logic*, Springer (1968). p. 113
[478] D. Sokolov, Dag-like communication and its applications, in: *Proc. Electronic Colloquium in Computational Complexity*, TR16-202 (2016). p. 409
[479] M. Soltys, The complexity of derivations of matrix identities, Ph.D. thesis, University of Toronto (2001). p. 232
[480] M. Soltys and N. Thapen, Weak theories of linear algebra, *Archive for Mathematical Logic*, **44(2)** (2005), 195–208. p. 232
[481] P. M. Spira, On time–hardware complexity of tradeoffs for Boolean functions, in: *Proc. 4th Hawaii Symp. on System Sciences* (1971), 525–527. pp. 17, 37
[482] R. Statman, Complexity of derivations from quantifier-free Horn formulae, mechanical introduction of explicit definitions, and refinement of completeness theorems, in: *Proc. Logic Colloquium '76*, North-Holland (1977), 505–517. p. 79
[483] R. Statman, Bounds for proof-search and speed-up in the predicate calculus, *Ann. Mathematical Logic*, **15** (1978), 225–287. p. 79

[484] G. Stengle, A Nullstellensatz and a Positivstellensatz in Semialgebraic Geometry, *Mathematische Annalen*, **207(2)** (1974), 87–97. pp. 115, 132

[485] R. Szelepcsényi, The method of forcing for nondeterministic automata, *Bull. European Association for Theoretical Computer Science*, **33** (1987), 96–100. p. 26

[486] G. Takeuti, *Proof Theory*, Dover (1975); 2nd edn. (2003). pp. 79, 158, 159

[487] G. Takeuti, RSUV isomorphism, in: *Arithmetic, Proof Theory and Computational Complexity*, eds. P. Clote and J. Krajíček, Oxford University Press (1993), 364–386. p. 196

[488] N. Thapen, Higher complexity search problems for bounded arithmetic and a formalized no-gap theorem, *Archive for Mathematical Logic*, **50(7–8)** (2011), 665–680. pp. 305, 436

[489] N. Thapen, A tradeoff between length and width in resolution, *Theory of Computing*, **12(5)** (2016), 1–14. p. 294

[490] B. A. Trakhtenbrot, The impossibility of an algorithm for the decidability problem on finite classes, *Proc. USSR Academy of Sciences* (in Russian), **70(4)** (1950), 569–572. pp. 233, 447

[491] B. A. Trakhtenbrot, A survey of Russian approaches to Perebor (brute-force searches) algorithms, *J. IEEE Ann. History of Computing*, **6(4)** (1984), 384–400. p. 38

[492] G. C. Tseitin, On the complexity of derivations in propositional calculus, in: *Studies in Mathematics and Mathematical Logic, Part II*, ed. A. O. Slisenko (1968), 115–125. pp. 3, 62, 94, 97, 111, 112, 280

[493] G. C. Tseitin and A. A. Choubarian, On some bounds to the lengths of logical proofs in classical propositional calculus (in Russian), *Trudy Vyčisl Centra AN Arm SSR i Erevanskovo Univ.*, **8** (1975), 57–64. p. 62

[494] A. Turing, On computable numbers, with an application to the Entscheidungsproblem, *Proc. London Mathematical Society*, Series 2, **42** (1936–1937), 230–265. pp. 1, 32, 37

[495] I. Tzameret, Algebraic proofs over noncommutative formulas, *Information and Computation*, **209(10)** (2011), 1269–1292. p. 160

[496] A. Urquhart, Hard examples for resolution, *J. ACM*, **34(1)** (1987), 209–219. pp. 281, 293

[497] A. Urquhart, The complexity of propositional proofs, *Bull. Symbolic Logic*, **194** (1995), 425–467. pp. 6, 79, 112

[498] A. Urquhart, The depth of resolution proofs, *Studia Logica*, **99** (2011), 349–364. p. 113

[499] J. Vaananen, Pseudo-finite model theory, *Matematica Contemporanea*, **24** (2003), 169–183. p. 454

[500] L. van den Dries, *Tame Topology and o-Minimal Structures*, Cambridge University Press (1998). p. 132

[501] J. von Neumannn, *Collected Works*, Pergamon Press (1963). p. 230

[502] K. W. Wagner, Bounded query classes, *SIAM J. Computing*, **19(5)** (1990), 833–846.

[503] Z. Wang, Implicit resolution, *Logical Methods in Computer Science*, **9(4–7)** (2013), 1–10. p. 158

[504] I. Wegener, *The Complexity of Boolean Functions*, Wiley-Teubner Series in Computer Science (1987). p. 38

[505] I. Wegener, *Branching Programs and Binary Decision Diagrams – Theory and Applications*, SIAM Monographs in Discrete Mathematics and Its Applications (2000). p. 146

[506] R. Williams, Improving exhaustive search implies superpolynomial lower bounds, *SIAM J. Computing*, **42(3)** (2013), 1218–1244. p. 480

[507] A. Woods, Some problems in logic and number theory, and their connections, Ph.D. Thesis, University of Manchester (1981). pp. 38, 259

[508] C. Wrathall, Rudimentary predicates and relative computation, *SIAM J. Computing*, **7** (1978), 194–209. p. 182

[509] A. C.-C. Yao, Theory and applications of trapdoor functions, in: *Proc. 23rd Annual IEEE Symp. on Foundations of Computational Science* (FOCS) (1982), 80–91. p. 429

[510] A. C.-C. Yao, Separating the polynomial-time hierarchy by oracles, in: *Proc. 26th Annual IEEE Symp. on Foundations of Computational Science* (FOCS) (1985), 1–10. pp. 38, 304

[511] S. Žák, A Turing machine hierarchy, *Theoretical Computer Science*, **26** (1983), 327–333. p. 38

[512] D. Zambella, Notes on polynomially bounded arithmetic, *J. Symbolic Logic*, **61(3)** (1996), 942–966. p. 183

Special Symbols

Each special symbol is listed by the section where it was defined.

Sec. 1.1 At, $\top$, $\bot$, ℓ^1, ℓ^0, TAUT, SAT, UNSAT, $\mathbf{tt}_\alpha$, DNF, CNF, $\overline{a} \leq \overline{b}$, kCNF, kDNF, kSAT, $T \models \alpha$, NAND, $\oplus$, $\text{MOD}_{m,j}$, $\text{TH}_{n,k}$, ℓdp.

Sec. 1.2 C_L, F_L, R_L, $\mathbf{A} \models B$, Σ_1^1, $s\Sigma_1^1$, $\langle B(i_1, \ldots, i_k)\rangle_n$, $\langle B(i_1, \ldots, i_k, j)\rangle_{n,\mathbf{A}}$.

Sec. 1.3 $\text{time}_M(w)$, $t_M(n)$, $\mathbf{N}_L$, Time(f), NTime(f), P, NP, E, NE, EXP, NEXP, $\text{space}_M(w)$, Space(f), NSpace(f), s_M, L, NL, PSPACE, NPSPACE, $co\mathcal{C}$, $\leq_p$ (reduction), BPP, Λ.

Sec. 1.4 Def_C, $\oplus_n$, P/$poly$, AC^0, $\text{AC}^0[m]$, TC^0, NC^1, mP/$poly$, $\text{Clique}_{n,k}$, P_T.

Sec. 1.5 $\mathbf{s}_P(\alpha)$, $\equiv_p$, $P \geq_p Q$, $P \geq Q$, PHP_n.

Ch. 2 Introduction $\mathbf{k}(\pi)$, $\mathbf{k}_P(\alpha)$.

Sec. 2.1 $\vdash_F$, $\mathbf{w}(\pi)$.

Sec. 2.2 F^*, $\mathbf{h}(\pi)$, $\mathbf{h}_F(\alpha)$.

Sec. 2.4 SF, EF, $Ext(A)$, F_d, $\ell(\pi)$, $\ell_F(A)$.

Sec. 2.5 dp(A).

Sec. 3.1 LK, PK, LK^-, LK^*.

Sec. 3.4 LK_d, LK_d^*, $\text{LK}_{d+1/2}$.

Sec. 4.1 Σ_∞^q, Σ_i^q, Π_i^q, G, G_i, G_i^*, TAUT_i, $\leq_p^i$, $\equiv_p^i$, $s\Sigma_1^q$.

Sec. 4.2 $F(P)$, BPF.

Sec. 4.3 $Sk[A]$, A_{Sk}, S^2, S^2F.

Sec. 5.1 R, R^*.

Sec. 5.2 Search($\mathcal{C}$).

Sec. 5.3 $\mathbf{w}(\mathcal{C})$, $\mathbf{w}(D)$, $\mathbf{w}(\pi)$.

Sec. 5.4 R_w, R^*_w, $\mathbf{w}_{R_w}(\mathcal{C} \vdash A)$, $\mathcal{C} \vdash_k A$, $C \upharpoonright \ell = \epsilon$, $\mathcal{C} \upharpoonright \ell = \epsilon$.

Sec. 5.5 $\mathbf{h}(\pi)$, $\mathbf{h}_R(\mathcal{C})$, $\mathbf{Tsp}(\pi)$, $\mathbf{Csp}(\pi)$, $\mathbf{Tsp}_R(\mathcal{C})$, $\mathbf{Csp}_R(\mathcal{C})$.

Sec. 5.6 $\bigvee D$, $\neg D$, $\bigwedge \neg D$.

Sec. 5.7 DNF-R, R(f), R(k), R(log).

Sec. 5.8 *ER*.

Sec. 6.1 EC, EC/**F**.

Sec. 6.2 PC, PC/**F**, NS, NS/**F**, $\mathbf{deg}(\pi)$, PC*.

Sec. 6.3 CP, CP_c.

Sec. 6.4 SAC, SOS.

Sec. 6.5 LS, LS_+.

Sec. 6.6 $\mathbf{rk}(\pi)$.

Sec. 6.7 $\mathrm{PC}_<$, LS^d, LS^d_+, LS_*, $\mathrm{LS}_{*,split}$, $\mathrm{LS}^d_{+,\times}$.

Sec. 7.1 $P \vee Q$, R(CP), LK(CP), R(LIN), R(PC), R(PC_d), PCR.

Sec. 7.2 *CF*, *WF*, WPHP, dWPHP.

Sec. 7.3 $[P, Q]$, *iP*, *iER*.

Sec. 7.4 MLK, OBDD, PA, L_{PA}, s_n, $\underline{n}$, Δ_0, Σ^0_1, $\ulcorner w \urcorner$, $Taut(x)$, P_{PA}.

Sec. 7.5 ENS, UENS, IPS.

Sec. 7.6 R, NP/$k(n)$.

Sec. 7.7 Res-lin, GC, LS + CP, $F^c_d(\mathrm{MOD}_p)$, LJ, R(OBDD).

Sec. 8.1 $I\Delta_0$, Ω_1, $\omega_1(x)$, $|x|$, Δ_0-PHP.

Sec. 8.2 $L_{PA}(\alpha)$, $\langle \dots \rangle_n$, LKB.

Sec. 8.3 $Th_L(\mathbf{N})$.

Sec. 8.4 Sat_2, Ref_R, Fla_d, Prf_d, Sat_d, Ref_d, Prf_Q, Prf^0_Q, Sat^0, Sat, Ref_Q, $Ref_{Q,d}$.

Sec. 8.6 DNF-TAUT.

Sec. 9.1 L_{BA}, $len(x)$, $(x)_i$, $bit(i,x)$, BASIC, $\langle x, y \rangle$, $L(f)$, BASIC(α), $L(\alpha)$, BASIC(f), $L(\alpha, f)$, BASIC(α, f).

Sec. 9.2 $\mathcal{P}_b(\mathbf{N})$, $\mathbf{N}^2$, $\mathbf{M}^2$, $L_{BA}(\#)$, $x\#y$, BASIC(#), $\mathbf{M}(\mathbf{K})$, $nb(A)$, $\mathbf{K}(\mathbf{M})$.

Sec. 9.3 $\Sigma^{1,b}_0$, $I\Sigma^{1,b}_0$, V^0_1, $\Sigma^{1,b}_0$-CA, $X \leq x$, $\Sigma^{1,b}_1$, $\Pi^{1,b}_1$, $s\Sigma^{1,b}_1$, $s\Pi^{1,b}_1$, V^1_1, Σ^b_∞, T_2, LIND, PIND, S_2, S^i_2, $T \cap i_2$, Σ^b_i, Π^b_i, $s\Sigma^b_i$, $s\Pi^b_i$, T^i_2, S^i_2, $S^1_1(\alpha)$, V^1_2, U^1_1, U^1_2.

Sec. 9.4 $\Delta^{1,b}_1$, LNP.

Sec. 10.1 $F(\mathrm{MOD}_m)$, $F_d(\mathrm{MOD}_m)$, $F(\oplus)$, Q_m, $Q_{m,i}$, $Q_2\Sigma_0^{1,b}$, $Sat_d(\oplus)$, $IQ_2\Sigma_0^{1,b}$.

Sec. 10.2 $C_{n,k}$, FC, FC_d, $\exists^=$, $\exists^=\Sigma_0^{1,b}$, $I\exists^=\Sigma_0^{1,b}$, VTC^0.

Sec. 10.3 $Enum(X,x,Y)$.

Sec. 10.5 $\forall_1^{\leq}\bigwedge(\vee)$, $\forall_1^{\leq}\bigwedge(\vee)$-LNP.

Sec. 11.1 $Count^m(x,R)$, $e\perp f$, $rem(R)$, $Count_n^m$.

Sec. 11.2 $\mathrm{PHP}(R)$.

Sec. 11.4 $\mathrm{WPHP}(s,t,R)$, WPHP_n^{2n}, $\mathrm{WPHP}_n^{n^2}$, $onto\mathrm{WPHP}(s,t,R)$, $onto\mathrm{WPHP}_n^{n^2}$.

Sec. 11.6 $Eval$.

Sec. 11.7 $\neg\mathrm{PHP}_n$, $Eval_d$.

Sec. 12.1 PV, $s_0(x)$, $s_1(x)$, LRN, $Tr(x)$, $x\frown y$, $Less(x,y)$, L_{PV}, PV1, PV_1, $S_2^1(\mathrm{PV})$, $\mathrm{BASIC}(L_{\mathrm{PV}})$.

Sec. 12.3 $\|\ldots\|^n$.

Sec. 12.4 RFN_P, $Sat(x,z)$, RFN_{ER}.

Sec. 12.5 $i\text{-}RFN_P$, Sat_i, $AxSk[A]$, A_{Sk}, $Sk[T_2]$.

Sec. 12.6 $\mathrm{dWPHP}(g)$, APC_1, APC_2, RFN_{WF}.

Sec. 12.7 RFN_{iER}.

Sec. 13.1 $\mathrm{RAM}(R,k)$, RAM_n, $Tour(R,n)$, $TOUR_n$.

Sec. 13.2 Φ_{Sk}, L_{Sk}, Φ^{rel}, C^ρ, $\mathrm{RAM}(n,k)$, $\mathrm{RAM}^U(n,k)$, $Cli(H)$, $Ind(H)$, $\mathrm{RAM}^f(n,k)$.

Sec. 13.3 deg_E, E-PHP, $\partial_A(I)$, E-WPHP_n^m, bPHP, $TSE_{G,f}$, $e(G)$, $b\mathrm{PHP}_n$.

Sec. 13.4 $\mathcal{D}_{n,m}^k$, $\tau_{\overline{b}}(A)$.

Sec. 13.5 $\mathrm{Clique}_{n,\omega}$, $\mathrm{Color}_{n,\xi}$.

Sec. 13.6 Peb_G.

Sec. 13.7 GER, FC_{reason}^w.

Sec. 14.1 $S^{d,n}$, $S_i^{d,n}$, $Q+d$, $\neg(E\text{-}onto\mathrm{WPHP}_n^{n^2})$.

Sec. 14.2 $Var(S_{\overline{i}}^{d,n})$, $\mathbf{R}_{k,d,n}^+(q)$, $\mathbf{R}_{k,d,n}^-(q)$, $T_{\overline{i}}^{d,n,\ell}$.

Sec. 15.1 $\mathrm{dp}(A)$, $\neg onto\mathrm{PHP}_n$, $h(T)$, $Maps$, $\alpha||\beta$, $\alpha\perp\beta$, $H\triangleleft T$, $S\times T$, $S(H)$, H_φ, S_φ.

Sec. 15.2 α^ρ, H^ρ, D^ρ, R^ρ, n_ρ, $h(H)$, φ^ρ.

Sec. 15.4 r_k, $Field$, $Field_n$.

Sec. 15.5 $Partitions$, $\neg Count_n^3$, x_α.

Sec. 15.6 $\mathrm{AC}^0[2]$-R(PC), $\mathrm{AC}^0[2]$-R(LIN), $Ax[f_1,\ldots,f_m;z]$, $\mathsf{disj}_{m,a}$, $E_{m,i,a}$, ENS, $\mathcal{E}_t$.

Sec. 15.7 $\mathrm{PK}_d^c(\oplus)$, UENS.

Sec. 16.1 *Orbit(X)*, *Next(v)*, $Partitions_n$, $Partitions_{5n}$.

Sec. 16.2 $\neg\mathrm{WPHP}_n^m$, $\hat{S}$, *Maps*, $Maps_t$, $Maps^*$, $\hat{S}_t$, V_t, Δ_t, C_t, B_t.

Sec. 17.1 FI.

Sec. 17.2 BMS_n^+, BMS_n^-, BMP_n^+, BMP_n^-, FALSI, FALSI*, PROV(Q), OWF, OWP, RSA_i, $Hall_n$.

Sec. 17.4 $CC(R)$, $KW[U,V]$, $KW^m[U,V]$, (G, lab, S, F).

Sec. 17.5 $CC(A)$, $MCC_U(A)$, $\tilde{A}$.

Sec. 17.7 $sel(x,y,z)$.

Sec. 17.9 $\dot{\bigvee}$.

Sec. 18.2 $CC^{\mathbf{R}}(R)$, $MCC_U^{\mathbf{R}}(A)$, $C_\epsilon^{pub}(R)$, $\mathrm{R}_m^>$.

Sec. 18.6 $\mathbf{P_r}$.

Sec. 18.8 R(OBDD), CLO.

Sec. 19.2 Con_Q, $P + Ref_Q$.

Sec. 19.3 TFNP, $\mathrm{Def}_C^{n,s}$, $\mathrm{Def}_C^{0,s}$, $\neg Con_{ER}(s,k)$.

Sec. 19.4 $\tau(C)_{\overline{b}}$, $\tau(g)_{\overline{b}}$, $H(g)$, PRNG, SPRNG, Gad_f, $CV_{\ell,k}$, $nw_{k,c}$.

Sec. 19.5 $\mathbf{tt}_{s,k}$.

Sec. 19.6 PLS, GLS, RAM, BT.

Sec. 20.1 Sat_0, *Taut*, Prf_P, Pr_P, $\subseteq_{cf}$, $\forall_1 Th_{\mathrm{PV}}(\mathbf{N})$, RFN_P.

Sec. 20.2 L_{ps}.

Sec. 20.4 $\mathbf{M}_n$, L_{all}, L_n, Ω, $K(F)$, $\langle A\rangle$, $\mathcal{B}$, $[\![A]\!]$.

Sec. 21.2 $\leq_p$ (for disjoint pairs).

Sec. 21.3 $\ulcorner\varphi\urcorner$, Prf_T, Pr_T, Con_T, $Pr_T^y(x)$, $Con_T(y)$, $\vdash_m$.

Sec. 21.4 $\dot{x}$, *BigTaut*.

Sec. 21.5 (A,P), $(A,P) \succeq (B,Q)$.

Sec. 22.3 $i_\infty EF$.

Sec. 22.6 $Alg(P)$.

Index